10th International Conference on Wind Turbine Noise (WTN 2023)

Dublin, Ireland
21 - 23 June 2023

ISBN: 978-1-7138-8291-6

Printed from e-media with permission by:

Curran Associates, Inc.
57 Morehouse Lane
Red Hook, NY 12571

Some format issues inherent in the e-media version may also appear in this print version.

Institute of Noise Control Engineering/Europe (INCE/Europe)
Riverside House
4 Oakland Vale
New Brighton, Merseyside CH45 1LQ
United Kingdom

Phone: +44 (0)151 638 0181
Fax: +44 (0)151 639 5212

Mail@inceeurope.org

Additional copies of this publication are available from:

Curran Associates, Inc.
57 Morehouse Lane
Red Hook, NY 12571 USA
Phone: 845-758-0400
Fax: 845-758-2633
Email: curran@proceedings.com
Web: www.proceedings.com

TABLE OF CONTENTS

INCE EUROPE

10th International Conference

Wind Turbine Noise 2023

Trinity College Dublin, Ireland

21 - 23 June 2023

PROCEEDINGS

Welcome to Wind Turbine Noise 2023

Welcome to Dublin and the Tenth International Conference on Wind Turbine Noise organized by INCE/Europe.

After our enforced excursion into the world of remote conferences during the covid pandemic we are back on site though we will be livestreaming the conference for those who cannot join us.

We have a wide range of papers again this year and we have three substantial discussion forums of the type we had in Lisbon in 2019. The conclusions from some of those forums will be published later on the website.

You can find the conference programme on the website.

www.windturbinenoise.eu

As a delegate you will have access to all the technical sessions including lunch and refreshment breaks, you can join us for a drink at O'Neill's pub on Tuesday evening prior to the conference and you can attend the conference dinner on Wednesday evening at The Harbour Master.

We hope you enjoy the conference, meet colleagues and learn more about wind turbine noise.

Dick Bowdler

Chair of the Organising Committee

Organised by INCE-Europe

mail@inceeurope.org

www.inceeurope.org

Organisation

Chair
Dick Bowdler, UK

INCE/Europe
Jean Tourret, France, President
Geoff Leventhall, UK
Andy McKenzie, UK

Administrator
Cathy Mackenzie, UK

Co-Chairs
Bo Søndergaard, Denmark
Sabine von Hunerbein, UK
Mark Bastasch, USA

Committee Members
Mark Bastasch, USA
Franck Bertagnolio, Denmark
Norm Broner, Australia
Sylvia Broneske, UK
Matthew Cand, UK
Malcolm Hayes, UK
Michaela Herr, Germany
Cordula Hornung, Germany
Brian Howe, Canada
David Michaud, Canada
Bo Søndergaard, Denmark
Kathrin Stahl, Germany
Frits van den Berg, Netherlands
Sabine von Hunerbein, UK

Irish Committee Representatives
Simon Jennings
Damian Kelly
John Kennedy

Proceedings

These Proceedings are the collected papers that will be the subject of presentations at the conference. They are arranged in alphabetical order of the lead author.

The proceedings also include an abstract from Julia Kirch Kirkegaard who was principle speaker at the forum on Impact on People.

There is no paper for Ashutosh Sharma

10th International Conference
on
Wind Turbine Noise
Dublin – 21st to 23rd June 2023

Noise prediction for wind energy turbines based on CAA methods

Christina Appel, Institute of Aerodynamics and Flow Technology, Wind energy, German Aerospace Center (DLR), Lilienthalplatz 7, 38108 Braunschweig, Germany. Christina.Appel@DLR.de

Summary

The here presented work is part of the HGF (Helmholtz Association of German Research Centres) -funded DLR projects ViSion (Validation of Simulation Tools for the Description of Wind Turbines) and LAiSA (**L**astadaptive & **A**eroakust**i**sche **A**nalyse). In both projects we take part at the IEA Wind Task 39 – Quiet Wind Turbine Technology, code benchmark (1). The initial goal is to predict the turbulent boundary layer trailing edge noise (TBL-TEN) and the turbulent inflow noise (TIN) dominating the overall sound radiation for a 2.3 MW NM80 wind turbine. A comparison with other codes within the framework of IEA Wind Task 39 benchmark is shown in Bertagnolio et al. (2).

A process chain, originally developed by Rautmann (3), is used that breaks down the 3D rotor into individual segments, for each of which 2D CAA (computational aeroacoustics) simulations are performed to calculate the TBL-TEN contribution of each slice. Here this tool chain was picked up and supplemented: The rotation of the rotor, the associated amplitude modulation and the variation of the distance to arbitrary evaluation points for the individual rotor blade sections are considered. With the spatial information of each 2D slice related to the chosen observer point and data about the atmospheric turbulence a turbulent inflow noise model is applied. The individual contributions of TBL-TEN and TIN of the blade segments of the entire rotor are finally summarized and averaged over the rotor revolution. In addition to, e.g. the certification position according to IEC-64100-11 standard, arbitrary positions distributed in space can be specified. Third octave spectra can be exported at single observer points varying over the rotor revolution and OASPL values at spatial distributed positions in order to evaluate noise signatures.

1. Introduction

Sound radiated from a wind turbine originates from numerous noise sources. On the one hand there are mechanical sources, like the gear box and the generator. On the other hand, various aero-acoustic sources of sound are responsible for the sound that is perceived at an observer.

Comparing these two dominant types of noise, the aero-acoustic noise outweighs the overall sound emission of a wind turbine (4). The aero-acoustic noise can be split up into airfoil self-noise and turbulence inflow noise. The self-noise sources, in turn, can be split up into a wide range of sound mechanisms, which are differentiated in the literature (5). These are vortex shedding noise, separation noise, tip vortex noise, and turbulent boundary-layer trailing-edge noise. For multi megawatt turbines, the TBL-TEN turned out to be the prominent contributor to aero-acoustically generated noise (6). Hence, this paper will concentrate on trailing edge noise, but will show up an extension with commonly used semi-empirical turbulence inflow noise models, like it is done in e.g. the software NAFNoise (7) from the National Renewable Energy Laboratory (NREL).

The objective in the project LAiSA is to build a modular, fast prediction tool for the aero-acoustic footprint on the environment of a wind turbine, which can be coupled easily to the blade element method VAST (Versatile Aeromechanic Simulation Tool) developed by the DLR Institute of Simulation and Software Technology and to tools of the DLR Institute of Atmospheric Physics to cover the realistic inflow conditions and noise propagation in complex terrain. The presented method starts with a computational process chain (3) for predicting TBL-TEN by means of modelled stochastically turbulence (8).

To ensure the speed of the procedure, the rotor blades are abstracted as 2D airfoil sections, where a small number, in the order of 8-10, of sections in the outer third of the blade are already sufficient for a convergent solution. The position of the cuts should be chosen with respect to the spanwise change of the blade airfoil geometry. Finally, the result of this step is the directivity of TBL-TEN for each section, which is then the input for the Turbine acoustic prediction (TAP) tool.

In the next step the radiation angle and distance to an observer is calculated over the rotor revolution. The related factors from the precalculated directivity functions for each section are taken and the OASPL and third octave band values are summed up. The contributions of distance correction, 2D/ 3D correction (9) and convective amplification as a result of rotor blade rotation (10) are added. After all, the results are presented as spectra at single observer points or OASPL signatures of observer groups. The output is available time resolved over the rotor revolutions or averaged.

2. Trailing edge noise prediction via a 2D process chain

The 2D process chain for TBL-TEN was originally established by Rautmann (11) to provide an automated acoustic prediction for trailing edge noise on 2D profiles with the aim to obtain acoustically optimized airfoils. Note, the process chain does not contain an optimizer, but it supports the user to identify the acoustic drivers. The process chain operates via bash scripting the input parameters (e.g airfoil geometry, Reynolds number, angle of attack, chord length) and process parameters (e.g. number of iterations, simulated real time and post processing options). Two different mesh generators for RANS and CAA meshing are involved. For simulations the DLR CFD Code TAU and the DLR CAA code PIANO with the stochastic sound source model FRPM (8) (Fast random particle mesh method) are applied. Finally, a collection of python post processing tools is controlled by the process chain, to generate spectra and directivity outputs. The process chain has been also applied to operate an emulation tool set to provide acoustic predictions by means of a data base (12).

2.1 RANS Simulations

In order to generate a TBL-TEN prediction for a turbine, several characteristic profile sections (slice c1 – c16) are extracted from the rotor blade in the area of the outer third (Figure 1). Referred to the IEA Task 39 benchmark this is part of the benchmark description (1). The operating conditions of the present case are listed in Table 1.

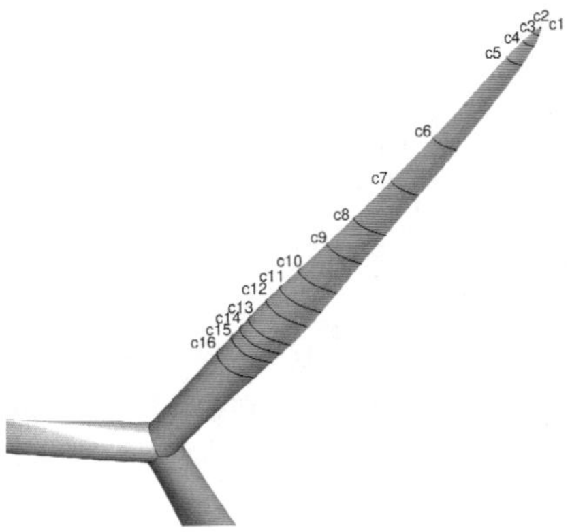

Figure 1 Radial slices

The effective angles of attack and the local incident flow velocities for the blade sections were determined in a first attempt by matching the flow conditions and pressure distribution from a 3D CFD simulation. Later this was done by a much more efficient blade element method (BEM), here the tool Qblade (13). For the flow conditions determined in this way, 2D RANS simulations are performed using the DLR TAU code. A direct use of the 3D RANS calculations is not possible due to the strongly different grid requirements for the use as CAA input. The 2D CFD meshes have been generated with IcemCFD and combine a structured near field and an unstructured farfield, Figure 2. The boundary layer resolution is done carefully with a $y+\leq 1$ to yield a sufficient resolution of the viscous sublayer. The structured part contains 118 points normal to the surface and 558 points around the airfoil. In total a CFD mesh contains 520.000 grid points. The applied meshing in done with respect to the best practice suggestions taken from (3). Though the 2D process chain works with parametric inputs, the CFD meshes for each slice are quite similar resolved. For turbulence modelling the Menter SST model (14) has been used. Exemplary the turbulent kinetic energy for the slices c1 -- c8 is shown in Figure 3. All plots in Figure 3 are scaled equally. The RANS simulations are performed using dimensioned quantities, while the CAA calculations are dimensionless. Therefore, the profile length in Figure 3 is already normalized to 1.0 with the respective chord length. One can see the increase of the profile thickness in the inner region. At the same time, the relative values of the TKE decrease due to the lower flow velocities. For section c8, the maximum TKE value is already four times lower than in the outer region.

Air density	*1.231 kg/m2*
Air temperature	*19 °C*
Wind speed	*6.1 m/s*
No wind shear	*-*
Turbulence intensity of the atmosphere	*8.96%*
Turbulence length scale of the atmosphere	*39 m*
Rotor speed	*12.3 rpm*
Pitch angle	*+0.15°*
Yaw angle	*0°*
Transition position suction side	*0.065 chord*
Transition position pressure side	*0.2 chord*

Table 1 Environment conditions

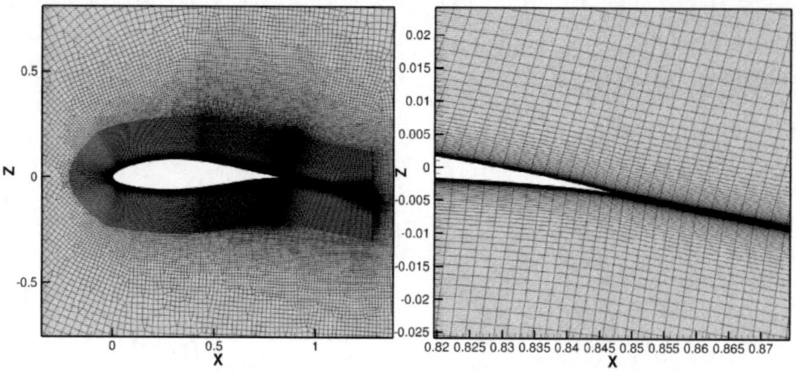

Figure 2 CFD mesh (left) and details near trailing edge (right)

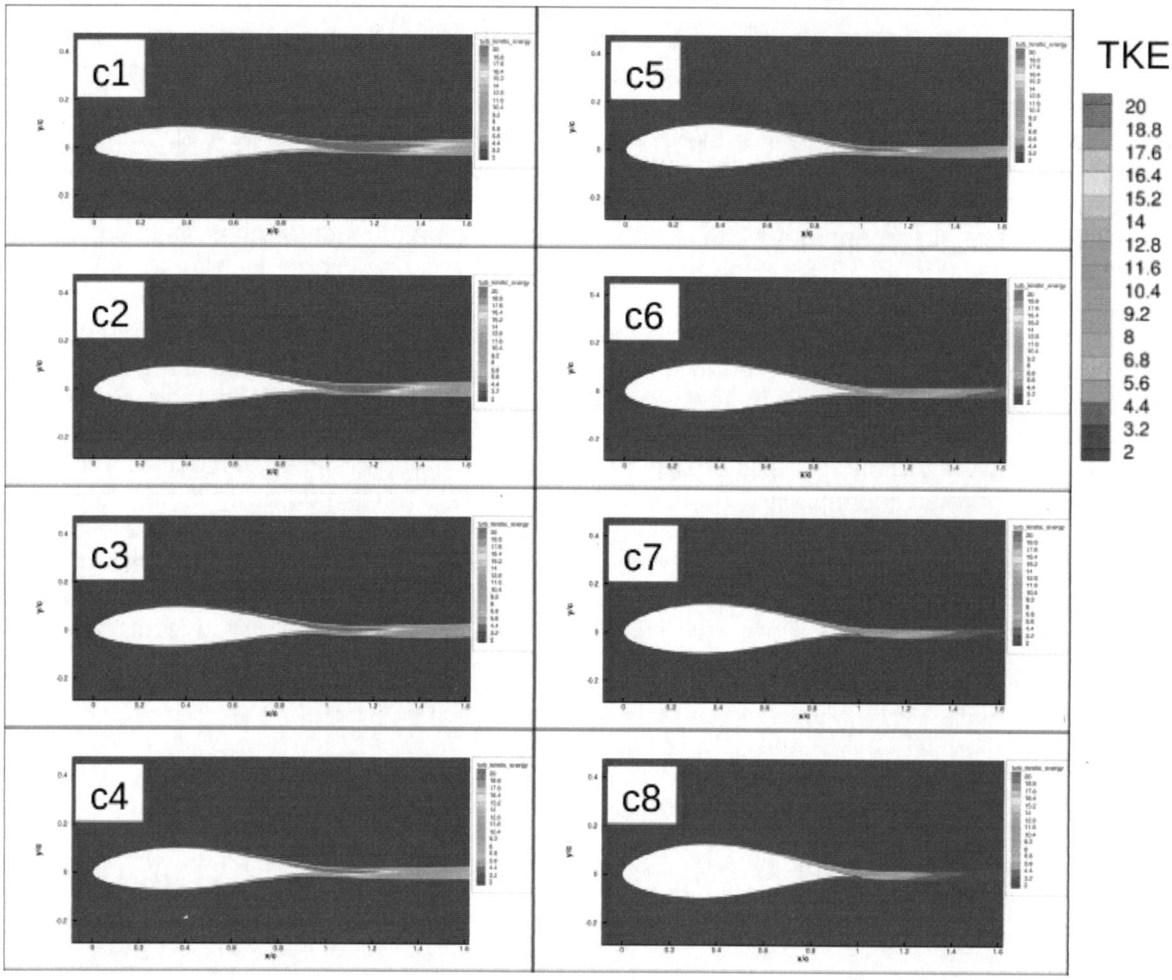

Figure 3 Distribution of turbulent kinetic energy (TKE) in m^2/s^2 for the slices c1 – c8.

2.2 CAA Simulations

The acoustic simulations are based on the APE-4 (acoustic perturbation equation) system with dominating vortex sound source. Acoustical source generation is obtained using the FRPM method (8). For the solution of the perturbation equation system the DLR CAA code PIANO is employed. The spatial discretization is based on the DRP scheme of Tam & Webb (15) and the explicit time integration is done with an alternating LDDRK scheme (16). For the acoustic vortex source, the fluctuating linearized Lamb vector L_i' is modelled. Here the index 0 denotes the magnitudes of the base flow, the index t those of the turbulence statistics and the prime indicates fluctuating quantities. ϵ_{ijk} is the Levi-Civita symbol.

$$L_i' = -\epsilon_{ijk}\omega_j^0 v_k^t - \epsilon_{ijk}\omega_j^t v_k^0 \quad [1]$$

In the sound propagation calculation with the DLR code PIANO, the sound field is simulated based on the modelled stochastic sound sources. The base flow is also considered in the propagation. For the modelling of the vortex sound sources, the rotation of the fundamental flow is determined from the RANS solution, the turbulent velocities v_k^t are modelled with FRPM and from this the fluctuating vorticity ω_j^t is calculated too. The simulations are evaluated at virtual microphones distributed in a circle with a radius of 2.5 chord lengths around the trailing edge. This allows not only the evaluation of single spectra but also the determination of directivity characteristics. The used CAA grids, generated with the DLR MegaCads grid generator, have been set up parametric, with 1100000 mesh points, and a resolution of 0.0005 chord at the trailing edge and 0.01 chord in the farfield region. This equals a maximal resolved frequency of about 30 kHz in the outer rotor section and 1.7 kHz for the most inner considered blade section.

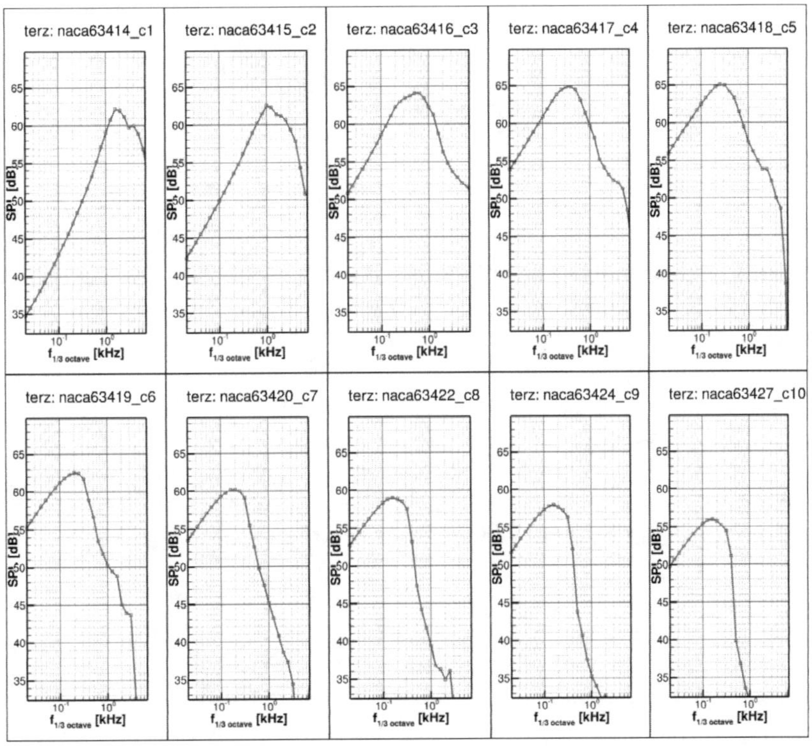

Figure 4 Third-octave spectra of TBL-TEN from 2D CAA calculations. Contributions of each slice, 90° below the trailing edge, perpendicular to the flow vector, normalized to a distance of 1m and a span width of 1m.

In *Figure 4* the third-octave spectra for the slices c1 -- c10, from 2D simulations by means of PIANO/FRPM, are shown. The spectra are normalized to a distance of 1m and a span width of 1m. From the rotor blade tip (c1) to the inner area (c10) one can see that the maximum of the third-octave spectra shifts from about 1.1 kHz to 150 Hz to lower frequencies. The maximum SPL values are between 56 and 65 dB. The highest values are reached at the outer slices c3, c4 and c5. Even further outboard, the flow velocities are higher, but the chord lengths are significantly lower and thus the contribution to trailing edge noise is lower too. Even further inboard the maximum SPL decreases noticeably. These slices will be skipped for the extrapolation of the whole turbine. Additional, flow separations at the inner slices with high thickness may occur, which have to treated with unsteady RANS calculations and averaging, to ensure proper CAA input data.

3. Summation and Extrapolation – TAP Tool

After the 2D CAA simulations have been carried out for all required sections, the extrapolation for the entire turbine can be performed. In addition to the geometric boundary conditions such as tower height, radial position of the selected 2D sections, number of blades, precone and tilt angle, as well as rotation and wind speed, the directivities from the CAA simulation are read in. Furthermore, observer positions or groups of evaluation points for noise maps are defined.

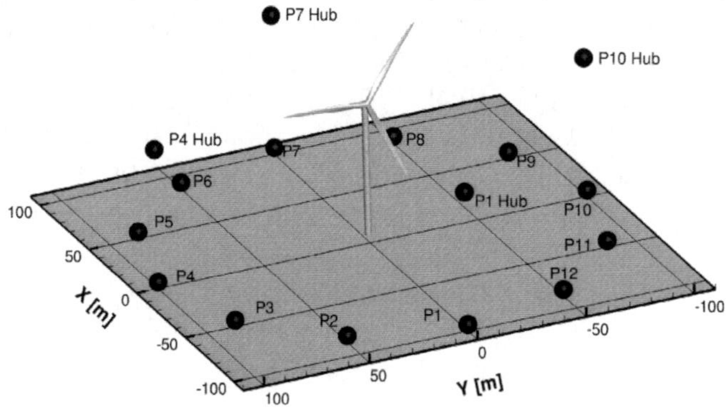

Figure 5 Observer Positions and notation

The observer positions are defined arbitrarily in space in a stationary coordinate system originating in the rotor hub. For the IEA Task 39 benchmark the chosen observers and notation are shown in Figure 5. Further, a rotating coordinate system whose origin is located at the respective trailing edge point of the blade element is defined. The task consists now of carrying out the transformation from one to the other coordinate system and to indicate the observer points with respect to the rotating system. Finally convert it to polar coordinates.
This complex transformation is decomposed into several simple rigid body movements in the form of rotations and displacements. To accomplish this, an affine transformation matrix is used in which all transformations can be mapped with one matrix multiplication each. In detail, the following motion steps are considered:

- Rotation around the tilt angle
- Rotation of the rotor
- Rotation around precone angle
- Rotation around pitch angle
- Shift along radial direction to the position of the slice
- Rotation around twist angle
- Displacement from the threading point (0.75 c) to the trailing edge

This result can be expressed in terms of the polar angles ϕ and θ and the distance r to the observer. Figure 6 from (11) after Ffowcs Williams and Hall (17) illustrates the relationship.

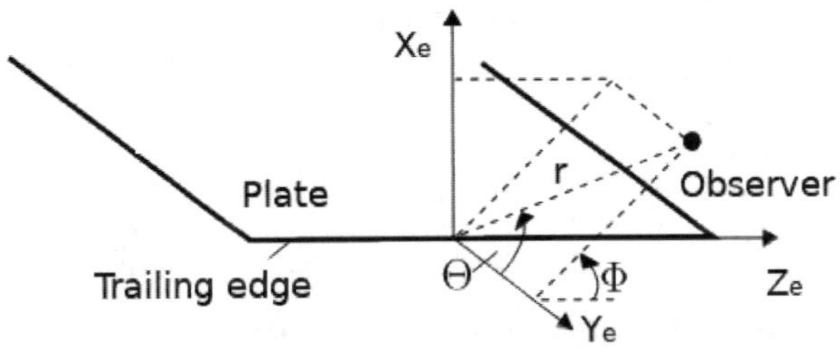

Figure 6 Flow over a semi-infinite plate (11). Polar angles ϕ and $\theta$with respect to an observer.

With this information, the acoustic contribution of the respective segment can be calculated. The beam angle is given by $\theta + \alpha_{eff}$. This allows to choose the related values from the underlying CAA directivity. The reduction due to distance is calculated using $20\,log(r/r_{Mic})$, Eq.[2], where r_{Mic} is the radius of the microphone circle in the CAA simulation and r is the distance trailing edge to observer point. The CAA directivity is 2D but the blade segment should be 3D. This is corrected accordingly to (17): the sound intensity of a semi-infinite plate scales with $sin(\phi)$, Eq.[3]. Furthermore, a 2D/3D correction, Eq.[4] of the far-field sound pressure levels is performed for the individual blade elements, considering the span of the blade segments. This is explained in more detail in Ewert et al. (9). The constant C in Eq.[4] is set to 1.4 according to Fassmann et al.(12). The chord length is c. The influence of the segment width is considered with Eq.[5], the reference width $b_{ref} = 1.0$m.

$$L_{p,distance} = 20\,log\left(\frac{r_{mic}}{r}\right) \quad [2]$$

$$L_{p,sin\,\phi} = 10\,log(sin\,\phi) \quad [3]$$

$$L_{p,2D/3D} = 10\,log\left(\frac{C}{2\pi}\cdot\frac{c}{r_{mic}}Ma_{piano}\right) \quad [4]$$

$$L_{p,width} = 10\,log\left(\frac{b}{b_{ref}}\right) \quad [5]$$

The convective amplification due to rotor motion is considered as a scaling of the directivity D according to Brooks and Burley (10). Here ξ is the angle between the vector of blade leading edge flow and the line connecting the trailing edge to the observer. Ma_{piano} is the Mach number of the flow at slice position.

$$D \propto \frac{2\,sin(\theta/2)^2\,sin\,\phi^2}{(1-Ma_{piano}\,cos(\xi))^4} \quad [6]$$

The numerator in the Eq.[6] describes the directivity of high-frequency trailing edge noise. It was analytically derived for trailing edge noise from a semi-infinite flat plate (5) but was also found to be valid for finite airfoils (18) assumed that the angle θ is not too close to 180° and the acoustic wavelength is smaller than the airfoil chord. For low frequency noise, where the acoustic wavelength is much larger than the airfoil chord, the $sin\,\theta/2^2$ term changes into $sin\,\theta^2$. As suggested in (19) respective (10) an exponent of 4.0 is used for the power law.

$$L_{p,Amp} = 40\,log\left(\frac{2\,sin(\theta/2)^2\,sin\,\phi^2}{(1-Ma_{tot}\,cos(\xi))^4}\right) \quad [7]$$

Finally, all terms of L_p can be added, while $L_{p,piano}$ equals to the value from the computed directivities from PIANO.

$$L_p = L_{p,piano} + L_{p,width} + L_{p,distance} + L_{p,2D/3D} + L_{p,sin\,\phi} + L_{p,Amp} \quad [8]$$

This has to be done for all blades and all blade sections. The results can now be presented as OASPL values or 1/3 octave spectra at single or observer positions or grouped positions to yield a noise map. In *Figure 7* this is done for a 200m times 200m horizontal map on the ground and for a vertical map in a downwind distance of 100m. The OASPL values are averaged over the rotor revolution.

According to the observer positions depicted in *Figure 5* the resulting TBL-TEN spectra are given in *Figure 8*. Each plot shows a set of curves corresponding to the azimuth angle of the rotor. Note there is a small variance with the azimuth angle for positions directly downwind the wind turbine (P1, P7), no variance for downwind hub position (P1hub, P7hub), and a large variance for the most lateral positions (P4, P10, P4 hub, P10 hub).

Figure 7 OASPL noise map for the Task 39 benchmark case at different azimuth angles.

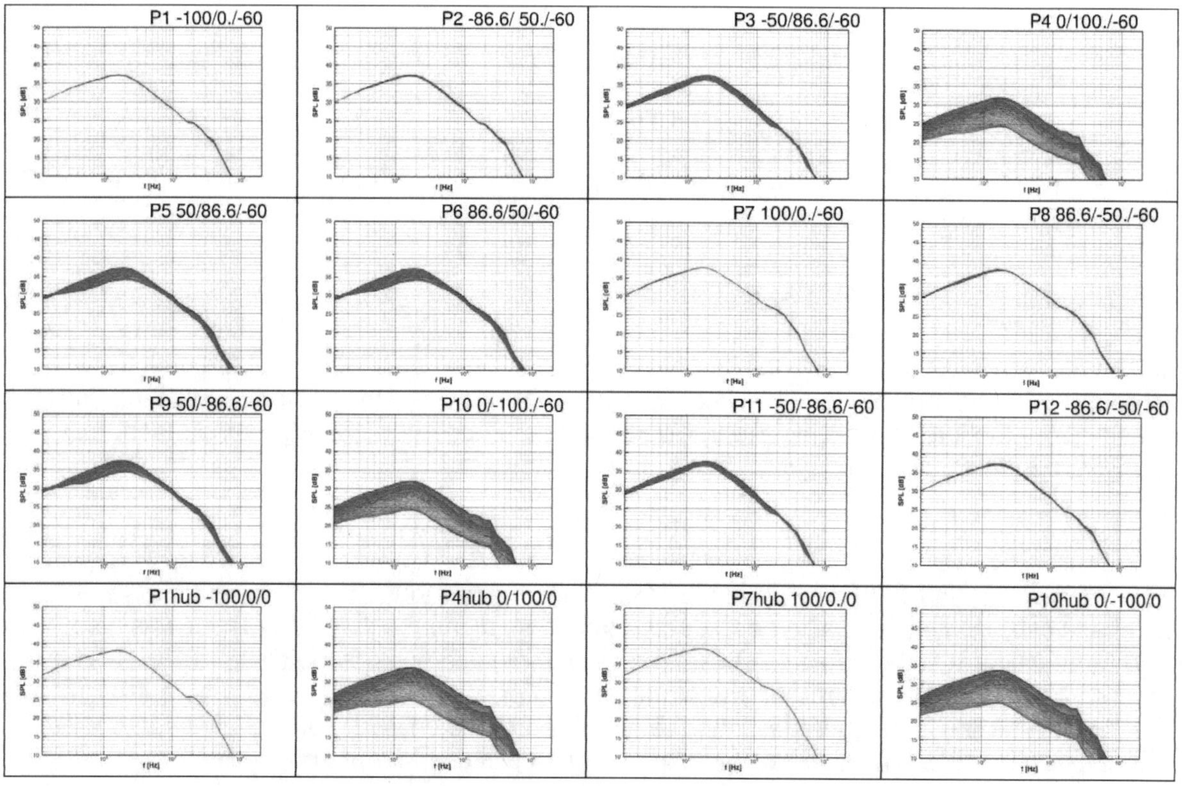

Figure 8 TBL-TEN spectra for benchmark observer positions.

4. Turbulent inflow noise via semi-empiric modelling

There are several methods to model turbulent inflow noise. On the one hand, there are CAA applications, which, however, require a high numerical effort. The high effort, e.g. for the very efficient CFD/CAA method used here, already becomes clear during the generation of the base flow including atmospheric turbulence. Not only the complete wind turbine including the inflow has to be recorded transiently, but also the near-wall regions on the rotor blades have to be resolved carefully, so that reasonable CAA results can be expected. In addition, the rotor circulation has also to be considered. The actual task of CAA simulation starts only after that. Following from that, for fast predictions on complete wind turbines, only TIN modelling can be considered.

One the other hand there is the analytical model by Amiet and Patterson (20) for TIN. Here the different characteristics of high and low frequency part of the TIN are considered and a smooth transition between both parts is achieved by applying a low frequency correction to the high frequency part. Furthermore, a bunch of simplified semi-empirical models based on the complex model of Amiet and Patterson can be found in literature (21) (22) (23)(25). These models employ basically an equation to model the third octave spectra for the high frequency part (index h) dependent on the wavenumber based on the chord length, as necessary normalized with the wavenumber of energy containing eddies from the inflow, and a low frequency correction (LFC, index L). Due to good results compared with the complex Amiet model, low computational effort and easy implementation, among others the semi-empirical models of Lowson(24) and Moriarty et. al.(25) have been chosen to be implemented in the TAP tool to provide the TIN contribution. In detail the following model equations are applied:

$$SPL_{1/3}^{L} = SPL_{1/3}^{H} + 10\log_{10}(LFC) \quad [9]$$

For the Lowson model:

$$SPL_{1/3}^{H} = 10\,log_{10}\left[\frac{\rho_0^2 c_0^2 \Lambda d}{2r^2} Ma^3 u'^2 K_1^3 (1+K_1^2)^{-7/3}\right] + 58.4 \quad [10]$$

with Λ the turbulent length scale of the atmosphere, $\rho_0$the density, c_0 the speed of sound, d the span, I the turbulence Intensity at the blade, based on the local flow velocity, u'^{2} the mean square turbulence level, $D_L$the low frequency directivity and K_1 the wave number with chord length and the local flow velocity U at the blade.

$$K_1 = \frac{\pi f c}{U} \quad [11]$$

The low frequency correction is defined as:

$$LFC = 10S^2 Ma K_1^2 \beta^{-2} \quad [12]$$

with

$$S^2 = \left(\frac{2\pi K_2}{\beta^2} + \left(1 + 2.4\frac{K_1}{\beta^2}\right)^{-1}\right)^{-1} \quad [13]$$

$$\beta^2 = 1 - Ma^2 \quad [14]$$

Several differences are present in the implemented model of Moriarty:

$$SPL_{1/3}^{H} = 10\log_{10}\left[\frac{\rho_0^2 c_0^4 \Lambda d}{2r^2} Ma^5 K_2^3 I^2 (1+K_2^2)^{-7/3} D_L\right] + 78.4 \quad [15]$$

Here the LFC contains a term to take the effective angle of attack into account:

$$LFC = 10S^2 Ma\left(1 + 9\alpha_{eff}^2\right) Ma K_2^2 \beta^2 \quad [16]$$

$$S^2 = \left(\frac{2\pi K_3}{\beta^2} + \left(1 + 2.4\frac{K_2}{\beta^2}\right)^{-1}\right)^{-1} \quad [17]$$

And two different wavenumbers considering the span and the turbulent length scale are employed:

$$K_2 = \frac{8\pi f \Lambda}{3U} \quad [18]$$

$$K_3 = \frac{\pi f b}{U} \quad [19]$$

In both models finally the $SPL_{1/3}$ is given by:

$$SPL_{1/3} = SPL_{1/3}^{H} + 10\log_{10}\left(\frac{LFC}{(1+LFC)}\right) \quad [20]$$

The low frequency directivity function is in both models the same:

$$D_L = \frac{\sin\theta^2 \sin\phi^2}{(1+Ma\cos\theta)^4} \quad [21]$$

To give an impression of the shape, D_L is plotted in Figure 9 in polar coordinates. With the information about the angles θ and ϕ according to a certain observer position, the value of D_L can be examined at every radial and azimuthal position of the blades.

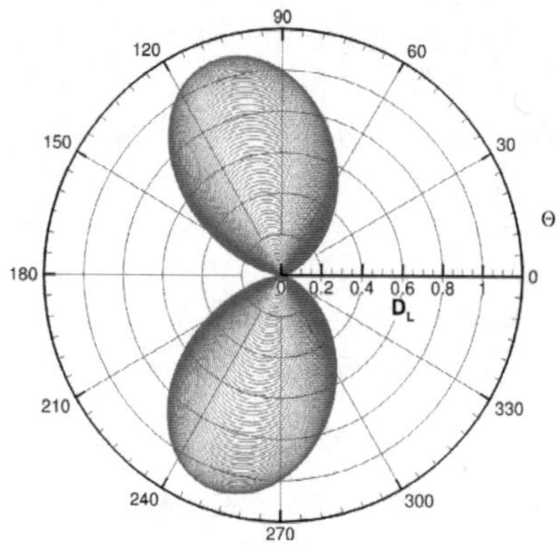

Figure 9 Exemplary directivity function for TIN

The turbulent length scale Λ is direct proportional to the SPL value for TIN, but is not trivial measure on site and shows a broad value range. The model by Moriarty assumes this value as the 2.45 times the hub height, up to a maximum of 73.5m. For the model of Lowson the length scale is set to a free parameter to be chosen by atmosphere or experimental data. Especially in wind tunnel validation experiments, the length scale can be defined exactly by e.g. the size of wind tunnel flow straightener meshes.

Wetz et al. (26) show up the complexity of atmospheric boundary layer measurements and the estimation of parameters like Λ or turbulence intensity. Within the observation period the length scale varies between 29 and 329m, which equals terms of decibels following Eq.[15] a range of more than 10dB. In Figure 10 the resulting TIN spectra for the Lowson and Moriarty model are shown. The turbulence length scale varies between 39m (value from IEA Task 39 Benchmark) and 100m (parameter from (23)).

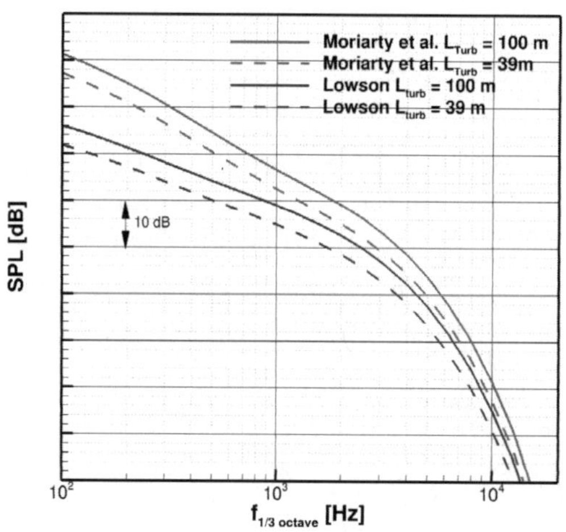

Figure 10 Dependency of turbulence length scale on TIN spectra for models by Lowson and Moriarty

5. Combination of TBL-TEN and TIN sources

The last step is the superposition of both sources. Since both parts are resolved over the rotor revolution, the final result can also be displayed accordingly or output in averaged form. The resulting 1/3 octave spectra are shown in Figure 11. Here the TIN model by Lowson is applied. The red curve set is again the TBL-TEN, the green curves are the TIN spectra. As the levels are much higher than the TEN noise, the superposition curves in black hide the green lines nearly completely.

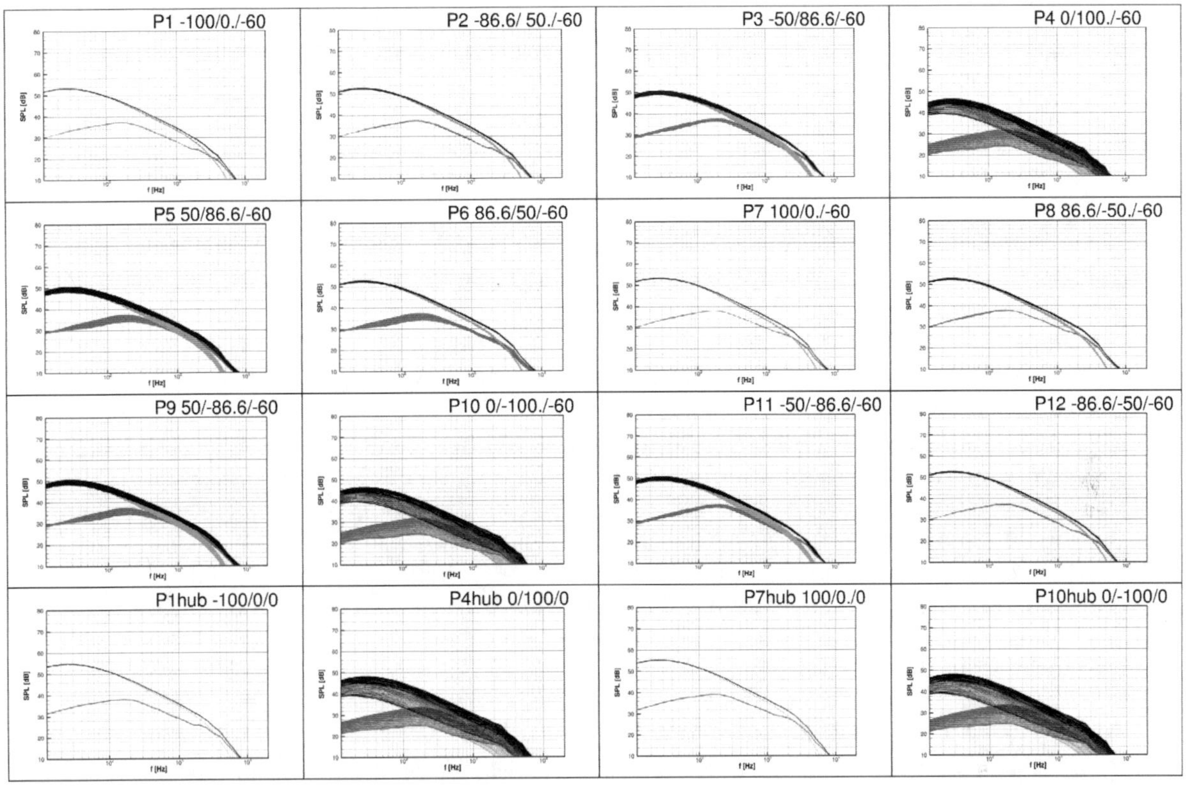

Figure 11 Superposition (black) of TBL-TEN (red) and TIN (green) spectra. TIN model by Lowson.

The TIN dominates the superposed spectra up to a frequency of 300Hz. Due to the impact of the turbulence length scale and the complexity to measure these data from experiments it is difficult to assess the results. On the other hand, there are experimental data sets published by Moriarty et al. and Buck et al. (21) (27), which shows similar behaviour.

6. Conclusions and Outlook

In the current work a high-fidelity CAA method for TBL-TEN noise is combined with extrapolation tool and semi-empirical models for TIN. The efficiency of the CAA method could be kept high and the numerical effort low by using 2D slices of a rotor blade. A complete run for one operational point of a wind turbine takes about 8h on 64 cpus, but could be also done on a desktop machine spending some more time. The extrapolation for several 100 observer points takes only a couple minutes.

Next action items are the implementation of OASPL output for TIN part, to add this to the noise maps too. Furthermore, additional semi-empirical and improved TIN models should be implemented, taking the real atmospheric conditions into account. Moreover, a separation noise model module is planned to be implemented. Also, the option to have more than one wind turbine at the same run in the extrapolation tool TAP is interesting to cover the acoustic superposition with the rotor azimuth for a group of turbines. Last but not least the validation with experimental field data is an important issue, hence we are looking forward to the first acoustic measurements from the DLR research wind farm WiValdi at Krummendeich.

References

(1) Bertagnolio, F., Madsen, H., Hornung, C., Schepers, G.: IEA Wind Task 39 – Quiet Wind Turbine Technology, Wind Turbine Noise Code Benchmark (Phase I), 2019.

(2) Bertagnolio, F., Fischer, A., Appel, C., Herr, M., Seel, F., Lutz, T., Boorsma, K., Schepers, G., Restrepo Botero, M., Casalino, D., Sucameli, C., Bortolotti, P., Wind turbine noise code benchmark: A comparison and verification exercise. 10th International Conference on Wind Turbine Noise, Dublin – 21st to 23th June 2023

(3) Rautmann, C., Dierke, J., Ewert, R., Hu, N.: Generic Airfoil Trailing-Edge Noise Prediction using Stochastic Sound Sources from Synthetic Turbulence, AIAA 2014-3298

(4) Wagner, S., Bareiss, R., Guidati, G.: Wind turbine noise, Vol. 16823 of Research and Development, Springer, Berlin, 1996.

(5) Brooks, T. F., Pope, D. S., Marcolini, M. A., "Airfoil Self-Noise and Prediction," Hampton and VA. Langley Research Center,1989.

(6) Oerlemans, S., Fisher, M., Maeder, T., Kögler, K., "Reduction of Wind Turbine Noise Using Optimized Airfoils and Trailing-Edge Serrations," AIAA Journal, Vol. 47, No. 6, 2009, pp. 1470–1481.

(7) Moriarty, P., NAFNoise User's Guide, Documentation, https://nwtc.nrel.gov/system/files/nafnoise.pdf, 2005.

(8) Ewert, R.: RPM - the fast Random Particle-Mesh method to realize unsteady turbulent sound sources and velocity fields for CAA applications, AIAA 2007-3506.

(9) Ewert, R., Appel, C., Dierke, J., Herr, M. RANS/CAA Based Prediction of NACA 0012 Broadband Trailing Edge Noise and Experimental Validation, AIAA 2009-3269.

(10) Brooks, T. F., Burley, C. L.: Rotor Broadband Noise Prediction with Comparison to Model Data, AIAA 2001-2210.

(11) Rautmann, C.: Numerical Simulation Concept for Low-Noise Wind Turbine Rotors, DLR Forschungsbericht 2017-35.

(12) Faßmann, B., Reiche, N., Ewert, R., Herr, M., Delfs, J.: Evaluation of Wind Turbine Noise based on Numerical Simulation Methods, AIAA 2018-3924.

(13) Marten, D.: QBlade Guidelines v 0.6, (2012), TU Berlin

(14) Menter, F. R.: Two-equation eddy-viscosity turbulence models for engineering AIAA Journal, Vol. 32, No. 8, August 1994.

(15) Tam, C., Webb, J.: Dispersion-relation-preserving finite difference schemes for computational acoustics. J Comp Phys 107, 262–281, 1993.

(16) Hu, F. Q., Hussaini, M. Y., Manthey, J. L.: Low-dissipation and low-dispersion Runge–Kutta schemes for computational acoustics. J Comp Phys 124, 177–191, 1996.

(17) Ffowcs Williams, J. E., Hall, L.H.: Aerodynmaic sound generation by turbulent flow in the vicinity of a scattering plane. JFM, 40, No. 4, 1970. J Comp Phys 1996;124:177–91.applications, AIAA Journal, Vol. 32, No. 8 1994.

(18) Hutcheson, F.V., Brooks, T.F.: Effects of angle of attack and velocity on trailing edge noise, AIAA 2004-1031, 2004.

(19) Oerlemans, S., Sijtsma, P., Mendez Lopez, B.: Location and quantification of noise sources on a wind turbine, JSV 299 (2007) 869–883.

(20) Paterson, R. W., Amiet, R. K., Noise and Surface Pressure Response of an Airfoil to incident Turbulence, Journal of Aircraft, vol. 8, no. 14, 1977.

(21) Moriarty, P., Migliore, P., Semi-Empirical Aeroacoustics Noise Prediction Code for Wind Turbines, 2003.

(22) Buck, S., Oerlemans, S., Palo, S., Experimental Validation of a Wind Turbine Turbulent Inflow Noise Prediction Code," AIAA Journal, pp. 1495-1506, 2018.

(23) Lowson, M., Assessment and Prediction Of Wind Turbine Noise, Flow Solutions Report, vol. 92, no. 19, 1992.
(24) Lowson, M., Theory and experiment for wind turbine noise. AIAA 94-0119, 1994.
(25) Moriarty, P., Guidati, G., Migliore, P., Recent Improvement of a Semi-Empirical Aeroacoustic Prediction Code for Wind Turbines, AIAA 2004-3041, 2004
(26) Wetz, T., Zink, J., Bange, J., Wildmann, N., Analyses of Spatial Correlation and Coherence in ABL flow with a fleet of UAS, https://doi.org/10.21203/rs.3.rs-2033943/v1
(27) Buck, S., Oerlemans, S., Palo, S., Experimental validation of a wind turbine turbulent inflow noise prediction code. AIAA Journal, Vol. 56, No. 4, April 2018

INCE EUROPE

10th International Conference
on
Wind Turbine Noise
Dublin – 21st to 23rd June 2023

A comparison of wind data sources for wind farm noise compliance assessments

Jason Baldwin, TNEI: jason.baldwin@tneigroup.com
James Mackay, TNEI: james.mackay@tneigroup.com

Summary

This paper considers a number of sources of wind data that can be used for wind farm noise compliance assessments with reference to the relevant guidance used in the UK and the Republic of Ireland.

A brief summary of existing guidance and previous papers of note is presented, and information detailing the impact of under estimating wind speeds with relevance to several different types of noise level limits is provided.

Analysis is presented which considers noise data collected during a noise compliance survey undertaken at four locations surrounding a wind farm (>10 turbines in size) situated in the UK. The noise data are correlated with various sources of wind data and the resulting average measured levels are then compared with predicted noise levels. Variation from the predicted levels is calculated and a commentary is provided on which sources correlate most closely, where the correlation is poor, consideration is given to why this may be.

1. Introduction

Wind farm noise limits in the UK and Republic of Ireland are typically set in accordance with ETSU-R-97 *'The assessment and rating of noise from wind farms'*[1] and the *'Wind Energy Development Guidelines'* 2006[1] (WEDG2006)[2], respectively. The UK Institute of Acoustics document *'A good practice guide to the application of ETSU-R-97 for the assessment and rating of wind turbine noise'* (IOA GPG)[3] is used to supplement the guidance in ETSU-R-97 and WEDG2006.

The limits proposed in ETSU-R-97 and WEDG2006 are set relative to wind speeds measured at the wind farm site. Where limits are set relative to wind speeds, it is clearly important that the methods used to measure wind speeds are accurate and appropriate.

This paper:

- outlines why accurate measurement of wind speed during compliance monitoring assessments is increasingly important in the UK and the Republic of Ireland (by reviewing the way in which limits are being set to account for cumulative considerations);
- considers some of the approaches that can be used to measure or derive wind speeds (with reference to previous papers and good practice guidance);
- considers how data from various sources (e.g. multiple wind turbines on the same site) can be considered;
- presents the results of a compliance monitoring assessment, undertaken at four noise monitoring locations, using several different options; and
- compares the results of each method to highlight factors that may require consideration when designing noise compliance surveys.

2. Potential approaches which can be used to set noise limits

When using ETSU-R-97 and WEDG2006 noise limits are usually[2] set 5 dB(A) above background noise levels subject to fixed minimum limits which apply when background noise levels are low. The limits established in accordance with ETSU-R-97 and WEDG2006 are sometimes referred to as 'Total Noise Limits' (TNL) which should not be exceeded by the combined operation of all wind farms in an area. The TNL can be considered to represent a 'noise budget'. To account for the presence of proposed, consented or operational wind farms in an area it is sometimes necessary to set limits for an individual wind farm at levels below the TNL (to account for the situation where a proportion of the TNL has been allocated to, and can realistically be used by, other wind farm developments). A noise limit allocated to a specific wind farm is sometimes referred to as a 'Site Specific Noise Limit' (SSNL). The need to consider cumulative noise continues to increase as more wind farm are proposed / consented and the use of SSNLs is therefore likely increase in the future.

For wind farm sites where noise has been a key design consideration, the margin between the consented noise limit and predicted noise levels can be small (in extreme cases it can be zero). In some cases, particularly in Northern Ireland, noise limits are also set relative to predicted noise levels. A buffer is sometimes added, for example limits may be set 2 dB above the predicted levels, but in some cases, limits are set equal to predicted levels. Where limits are set relative to predicted levels noise limits tend to reduce, sometimes quite significantly, at lower wind speeds.

---

[1] The WEDG are subject to on ongoing review. A draft update was issued for consultation in 2019 but at time of writing the WEDG2006 continue to be used.

[2] ETSU-R-97 also includes the option of a 'simplified' limit which is 35 dB at all wind speeds up to 10 m/s.

Figure 1 illustrates three different noise limits which could be applied to a single proposed wind farm and compares the limits to predicted levels;

1. the development could be allocated the entire TNL;
2. it could be allocated an SSNL (which may be appropriate if another wind farm is also proposed in the area); or
3. it could be allocated a limit which is based upon the predicted noise levels plus a margin, in this case 2 dB has been added.

Figure 1 – Predicted noise levels compared to example noise limits

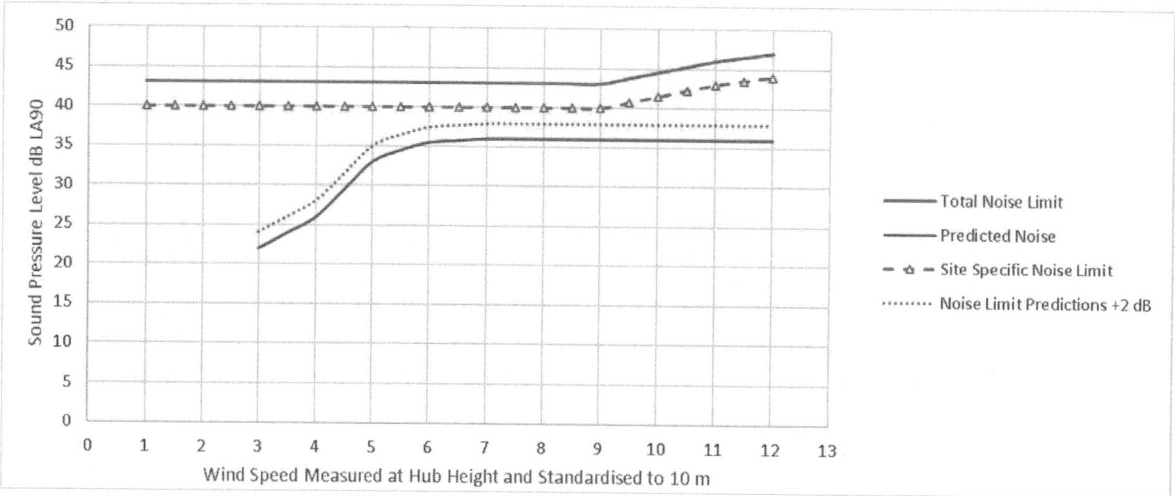

It can be seen that the margin between the predicted noise levels and the three noise limits varies considerably. The margin can have a significant influence on the complexity of any noise compliance monitoring, should it be required.

In the UK, compliance monitoring is only usually required in the event of a complaint. In the Republic of Ireland, a noise compliance assessment is increasingly common requirement of new wind farm consents. When undertaking measurements to determine compliance with the consented noise limits, it is important to ensure that the source of wind speed is both accurate and appropriate. A number of options are available (these are discussed further below) and an error in the measured wind speed of just 0.5 m/s can have a very significant impact on the results of the compliance assessment, particularly where limits have been set relative to predicted levels.

Figure 2 illustrates how the noise level measured during a compliance assessment might vary from predicted level if the approach used to measure wind speed underestimated actual speeds by 0.5 m/s. The resulting curve (shown in light blue) is shifted to the left on the graph and whilst compliance with the TNL and the SSNL would be achieved, it can be seen that the measured levels would exceed the Noise Limit based on Predictions + 2 dB between 3.5 and 4.5 m/s.

Figure 2 – Impact of 0.5 m/s error in wind speed measurement

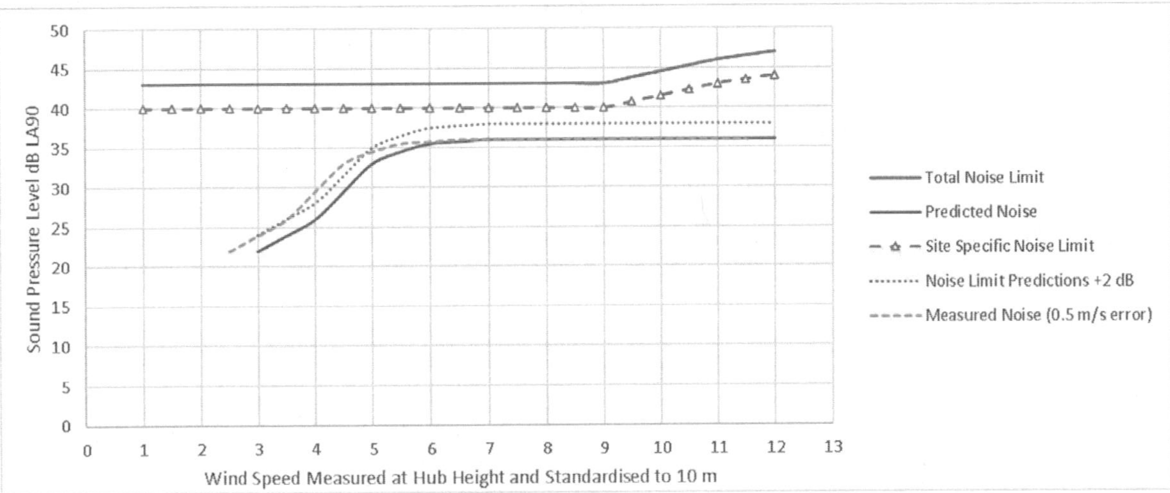

As noted above, noise limits are sometimes set *equal* to predicted levels. If this approach was taken for this example site, the measured levels would exceed the noise limits at all wind speeds below 7 m/s.

3. Relevant Guidance / Previous Papers)

The importance of utilising accurate and appropriate wind speed measurements is well established and widely understood. For example ETSU-R-97, which was published in 1996, provided guidance on wind speed measurement and analysis, stating on page 88 that:

'if the measurement of wind speed is from an anemometer which may be in the wake of a turbine in certain wind directions these data should also be removed'

The potential impacts associated with under (or over) estimating wind speed was considered by Bullmore *et al* (2007)[4]. The paper highlighted the fact that wind speeds can vary across large wind farm sites and concluded that:

'The findings have shown that the assumption of a single wind speed reference for all turbines that form a large wind farm site may overestimate the actual wind speed seen by each individual turbine. This is particularly the case for the turbines nearest to a location of interest which may be partly shielded by the further upwind turbines which experience uninterrupted (by the wind farm) and higher wind speed conditions. This means that a single wind speed reference will likely overestimate the sound emissions of the turbine nearest to a location of interest. This effect appears to be most significant at higher wind speeds for the sites studied.'

In 2014, Broneske[5] undertook a detailed comparison of the differences in the wind speed recorded by the met mast anemometer, turbine anemometer, SoDAR, and the wind speed derived from power at three wind farm sites. Broneske found that:

'the results of the average measured wind farm noise correlated with wind speed differ for different wind measurement methods. Not much difference has been found at wind farm site 1 between using the average nacelle anemometer wind speed and using the nacelle anemometer wind speed of just the wind turbine closest to the noise measurement location. However, it is expected that for more complex and larger wind farm sites the difference would be more noticeable.'

In July 2014, to UK Institute of Acoustics published Supplementary Guidance Note 5 (SG5)[6] to supplement the good practice guidance contained within the IOA GPG. SGN5 covered post completion measurements and provided a useful summary of the options available to measure wind speed at operational wind farm sites.

The IOA GPG made is clear that:

'it is crucial that the wind speed reference (hub height, standardised 10 metre, or measured 10 metre) for noise levels and noise limits is clearly and consistently defined, particularly when drafting conditions or assessing compliance'

SGN5 detailed that data will normally[3] be measured or calculated[4] at hub height and noted that the data will preferably be taken from a permanent on-site anemometry mast. The guidance goes on to note that, alternatively, data may be obtained from the power output of turbines where wind speed is derived using the turbines electrical power curve[5]. The use of LiDAR or SoDAR is also discussed, and also notes that the least preferable method, and only if all other options have been explored, is the use of data from the nacelle mounted anemometers.

SGN5 highlights that the use of data from turbine anemometers can be subject to error, and notes that it is essential that the measurements are corrected to account for the presence of the rotating blades in front of the anemometer. This finding is consistent with the conclusions made by Broneske who noted that:

'A maximum increase of 4 dB of the prevailing wind farm noise at standardised 10 m height wind speeds of 6 and 7 m/s has been found when comparing results with nacelle anemometer derived wind speeds against SoDAR derived wind speeds.'

It should be noted that regardless of which option is used in any compliance survey there are a number of additional points which need to be considered, including the siting of a mast / LiDAR / SoDAR and which turbine(s) will be considered if Option B is used.

The location of the wind monitoring device is an important consideration during any noise compliance campaign. Because a wind turbines noise output is primarily determined by the wind speed prevalent at the turbine location(s), should the wind monitoring method not capture the same conditions being experienced by the turbine(s), then shifts in the wind speed when correlating with the noise being experienced at a receptor will occur.

When using data derived from the power generated by the turbines (or when using data collected from the turbine anemometers, where that is considered to be a robust approach) it is necessary to consider which turbine(s) should be considered. In relation to this, SGN5 states that;

'In utilising turbine power output data as a proxy for wind speed, consideration should be given to the wind direction of interest and hence which turbine to reference i.e. within the diagram below, if the wind direction of interest was South Westerly then turbine 1,2 or 3 would be the reference turbine, whilst if the wind direction of interest was Northerly then turbine 1, 4 or 7 would be the reference turbine.'

---

[3] Unless noise limits are set relative to wind speeds measured directly at 10 m height (which is a method that the IOA GPG notes *'should only be adopted for smaller-scale developments'*.

[4] The IOA GPG provide a method for determining hub height wind spees using two measurements at lower heights.

[5] Using the guidance contained in BS EN 614000-11:2013.

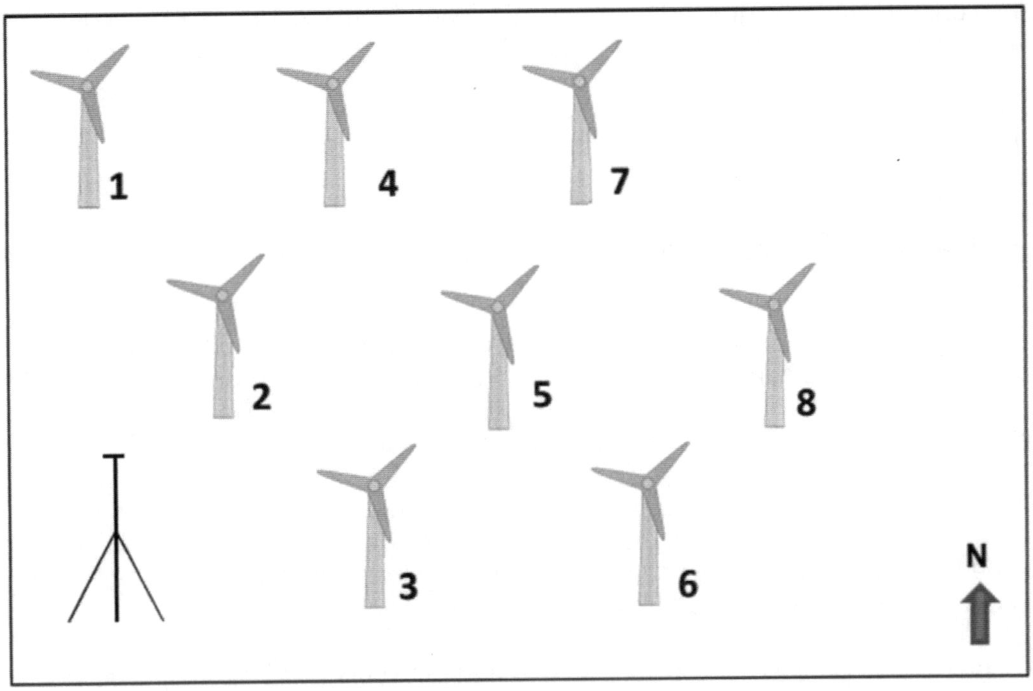

Where larger wind farms are being considered there is the potential for increased separation distances between the upwind (turbines furthest away), and downwind (closest turbines) to the noise monitoring location, which could result in differences in measured wind speed.

A working group was set up to establish a set of standardised techniques for the measurement of noise immissions from wind farms (informing document IEC61400-11-2). It is understood that the latest draft of this document states that for larger wind farms (> 5 turbines), the mean wind speed from the most sound relevant turbines at a particular noise monitoring location should be used (given that the wind speed experienced by these turbines will inform the sound levels being experienced at the closest noise monitoring location). The most sound relevant turbines are determined by comparing the overall predicted noise from the wind farm to the predicted noise with the least sound relevant turbines removed. When the difference in the two levels becomes >1dB, then the remaining turbines (the closest turbines) are used for deriving wind speed.

Additionally, for wind farms operating with 5 turbines or less, it is understood that the mean wind speed (derived from power) from all of the turbines should be used.

4. Comparison of wind speed data collected by different sources

This paper presents the results of a noise compliance survey undertaken at an operational wind farm in located in the UK. The data presented in this paper was collected in 2022 and covers a period of approximately 5 months. The site comprises a wind farm of >10 turbines and has a maximum output of approximately 30 MW. The wind farm is in a rural location where background noise levels are low, the turbines are located on high ground, which falls away in all directions towards the nearest noise sensitive receptors which are located to the North, East, South and West. The wind farm is notably oblong in shape, and stretches a larger area in length (North to South) than it does in width (East to West).

All of the turbines operate in unconstrained mode and the Wind Farm Operator supplied data from a SoDAR unit along with detailed data from the SCADA system which included, for each turbine:

- Nacelle orientation (degrees)
- Power output (kW)
- Rotor speed (r.p.m.)

All data were supplied in 10-minute periods.

Four Noise Monitoring Locations (NMLs) were included in the survey; one to the north, two to the east, and one to the south of the wind farm. Monitoring locations were sited between the wind farm and the closest noise sensitive receptors (to increase the contribution of turbine noise relative to background noise) and were carefully selected to minimise the potential for noise from watercourses and vegetation to influence the measurements.

Data were filtered to exclude invalid data (e.g. data collected where the wind turbines were not operating) and periods where rain was recorded. The data were also filtered to only consider periods where the NMLs were downwind of the wind turbines. Measured noise levels have also been normalised to further protect the identity of the site.

The measured noise levels included the contribution from turbine noise and noise from other sources (background noise). The data was not corrected for background noise (measurements taken where the wind farm was off), and so noise specific to the wind farm could not be derived. However, the signal to noise ratio was maximised through kit placement between the wind farm and NSRs. Observations on site suggest that the measured noise levels were dominated by turbine noise (except at very low wind speeds). It should however be noted that the specific noise levels from the wind farm will be lower than the levels presented.

Several of the various methods of deriving / measuring wind speed as highlighted in the 'Relevant Guidance / Previous Papers' section above have been considered when correlating the filtered datasets. The wind speed[6] sources used for comparison are detailed below:

1) Data measured by the LiDAR
2) Data measured by the anemometer (of the closest turbine to the NML)
3) Data derived by the power output (of the closest turbine)
4) Data derived by the power output (most distant turbine – which receives 'clean air' with no wake effects)
5) Data derived by the power output (primary turbines*)
6) Data derived by the power output (all turbines)

*Whilst the method outlined in the draft IEC61400-11-2 recommends calculating the most sound relevant turbines based on comparing overall predicted levels to those excluding the least dominant turbines; because this guidance has not published yet the approach has not been used. Instead turbines have been selected based on the guidance in the IOA GPG SGN5 which states:

'It is suggested that these [the primary turbines] *are determined such that the predicted difference between noise immission levels with all turbines operating and the primary turbines operating be less than 0.5 dB and that this is determined separately for each survey location.'*

---

[6] Data from a meteorological mast was not available.

5. Results

One of the four Noise Monitoring Locations (NML1) has been used to provide detailed comparison of the analysis results, whilst a comparison of polynomial trends will be provided for the remaining three NMLs.

A comparison of the filtered measured noise data correlated with each of the measured/ derived wind speed sources above (Source 1-6) and the predicted noise levels[7] at each of the Noise Monitoring Locations is presented in the following pages;

[7] Predictions have been undertaken in accordance with the guidance in the IOA GPG.

Figure 3 - Source 1 NML2 - LiDAR Measured Wind Speed

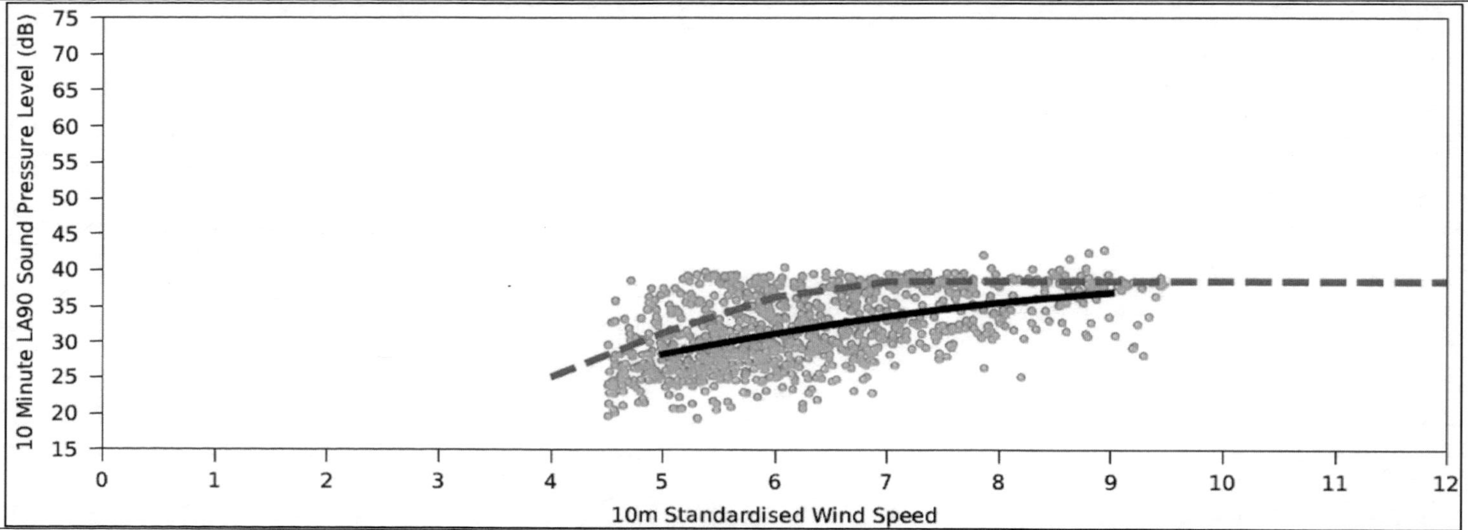

Figure 4 - Source 2 NML2 - Anemometer Measured Wind Speed (Closest Turbine)

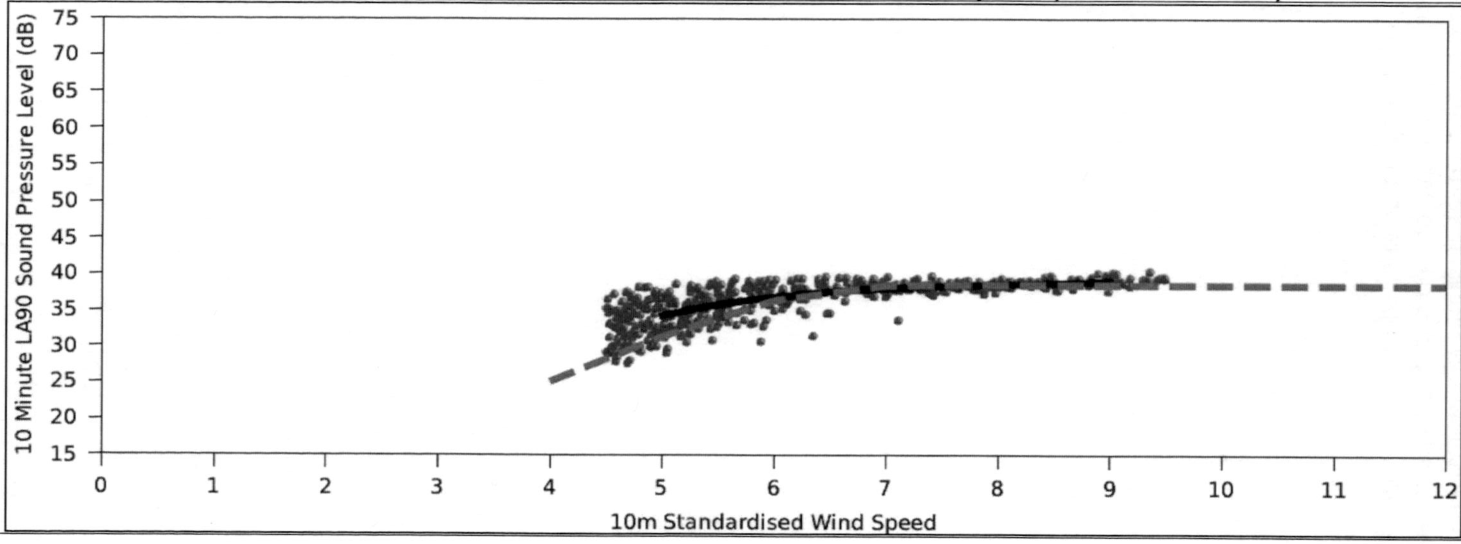

Figure 5 - Source 3 NML2 - Wind Speed Derived from Power Output (Closest Turbine)

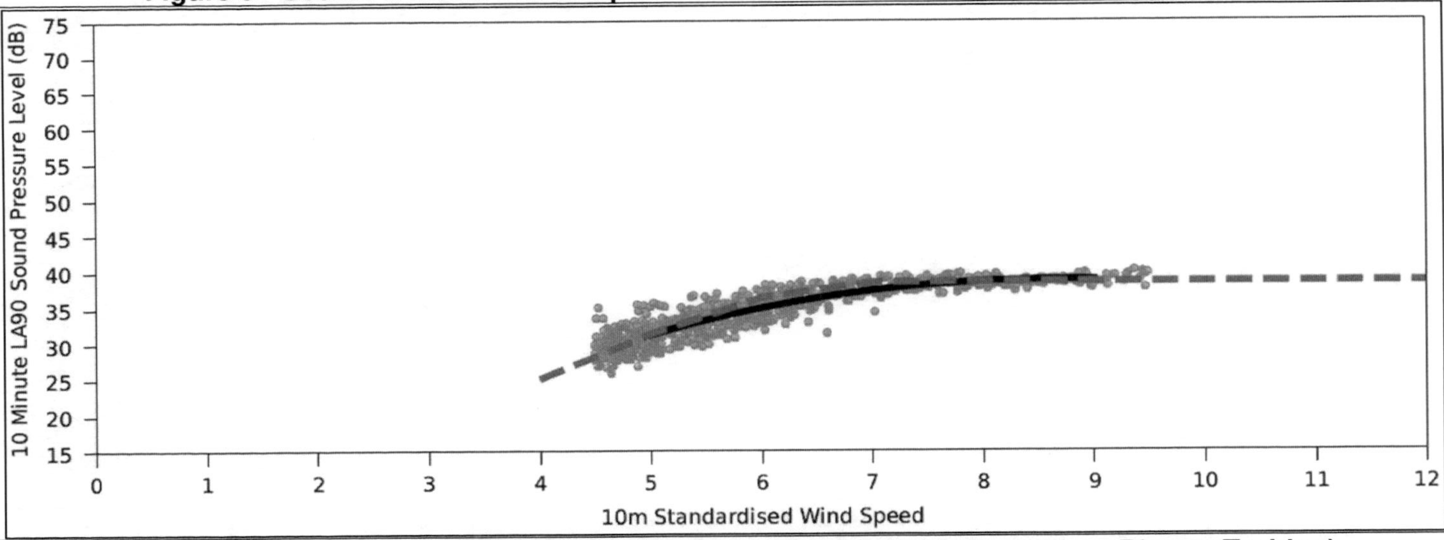

Figure 6 - Source 4 NML2 - Wind Speed Derived from Power Output (Most Distant Turbine)

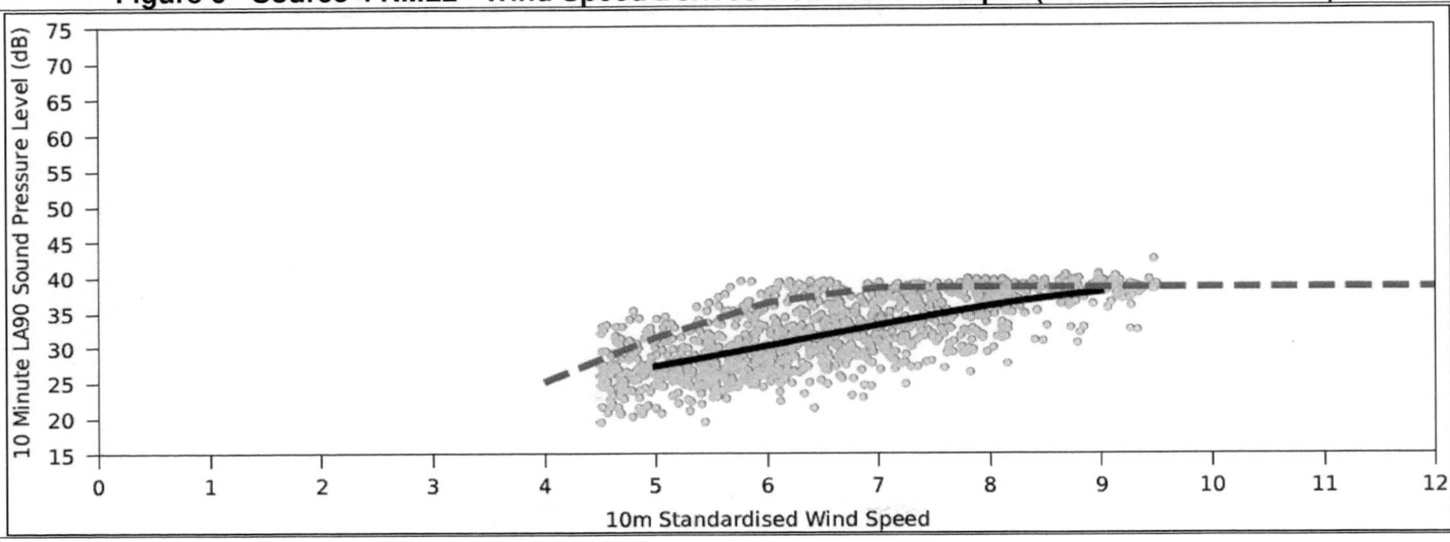

Figure 7 - Source 5 NML2 - Wind Speed Derived from Power Output (Acoustically Dominant Turbines)

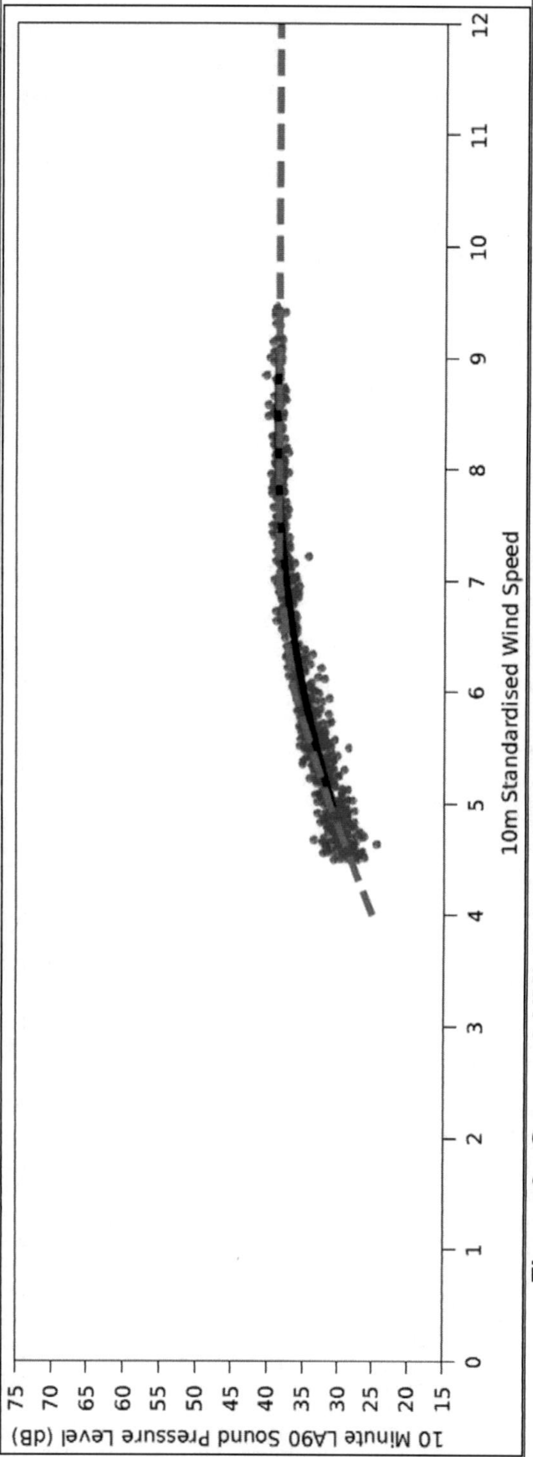

Figure 8 - Source 6 NML2 - Wind Speed Derived from Power Output (All Turbines)

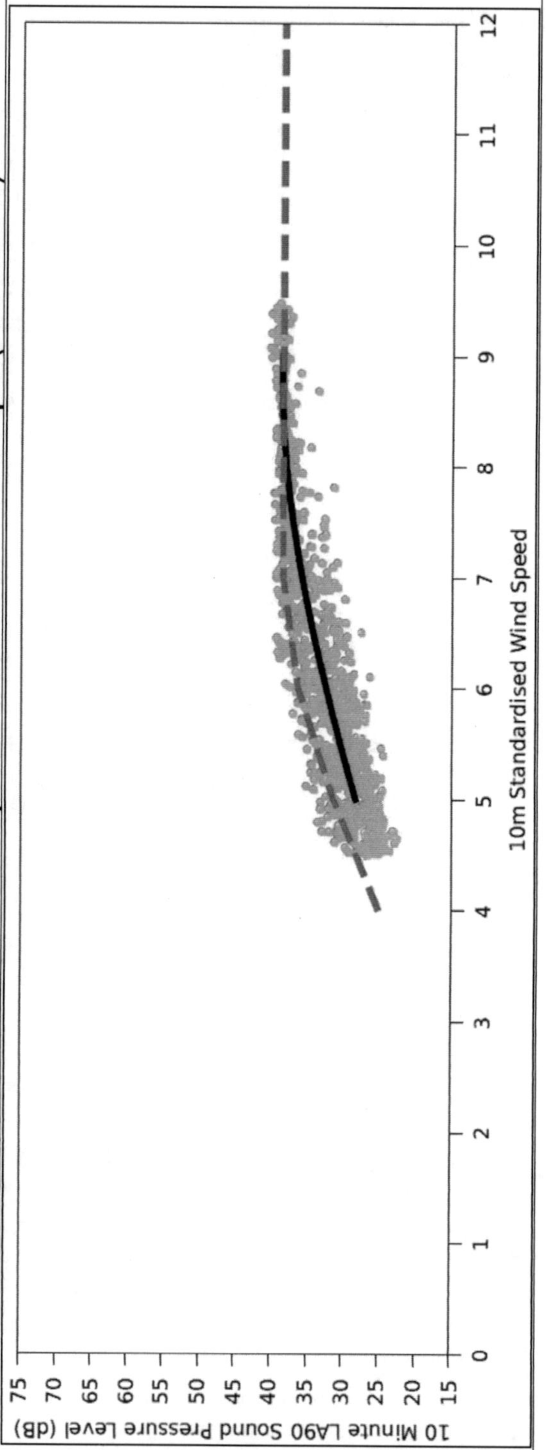

The results show that for this particular location;

1. There is a notably high degree of scatter where the sources of wind speed measurement / derivation are located further away. This is true for the LiDAR, the most distant 'clean air' turbine, and, to a lesser extent, the 'All' Turbines sources. There is a poor correlation with relation to the turbine noise predictions, with the exception that measured data points do not increase above the predicted noise levels.
2. There is good correlation of data at the higher windspeeds (7m/s and above) when considering the closest turbine anemometer. Data are clustering at the lower wind speeds (<7m/s), which may be the influence of the moving blades on the turbine anemometer. The wind speeds are shifting notably to the left on the wind speed axis suggesting that the anemometer may be underestimating the wind speed.
3. There is a high degree of good correlation when using both the closest and the primary turbine datasets. The measured data points are streamlined, with little to no scatter, and follow the trend of the noise predictions very closely.

The trends (or lines of best fit), have been overlaid in Figures 9-12 below to show a comparison of the results at each of the remaining locations;

Figure 9 – Comparison of datasets NML1

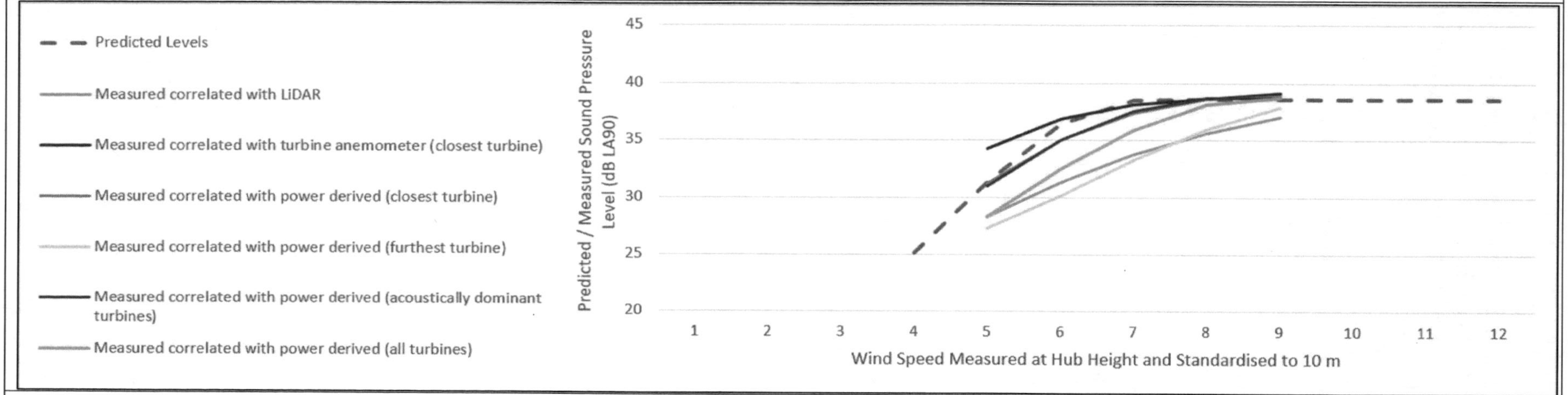

Figure 10 – Comparison of datasets NML2

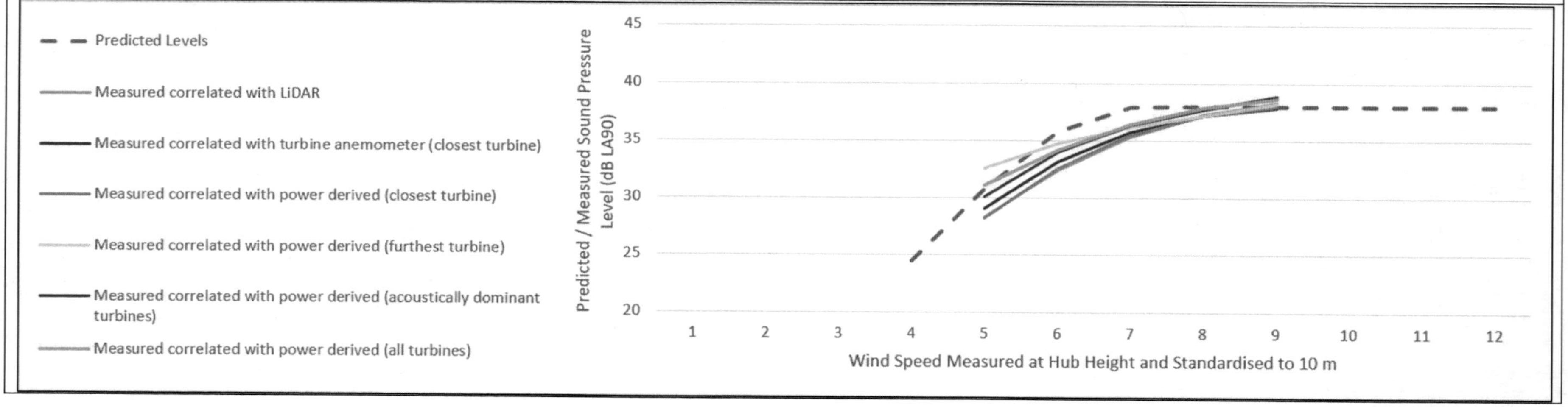

Figure 11 – Comparison of datasets NML3

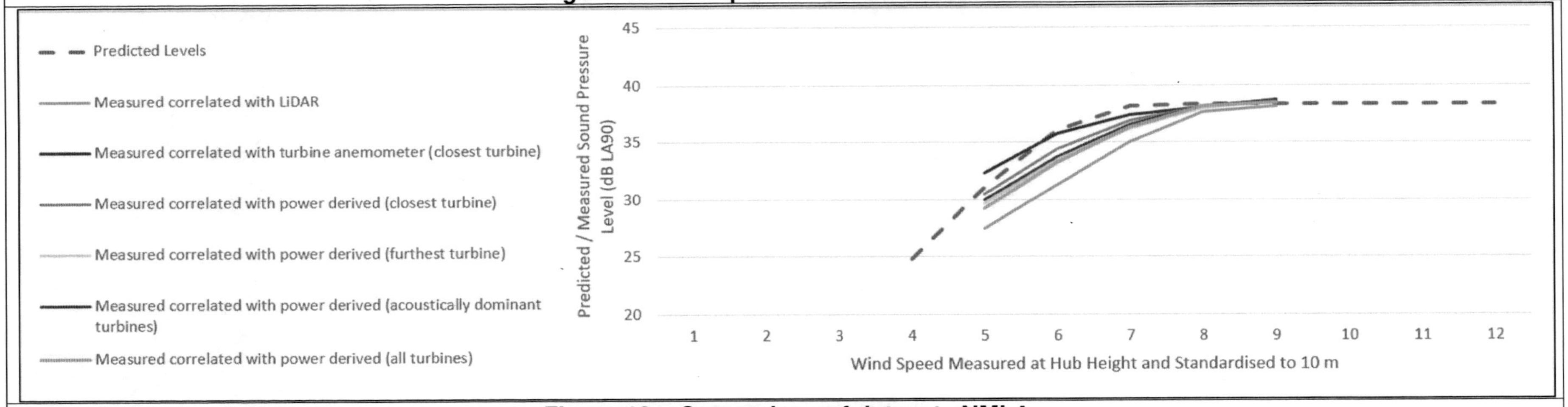

Figure 12 – Comparison of datasets NML4

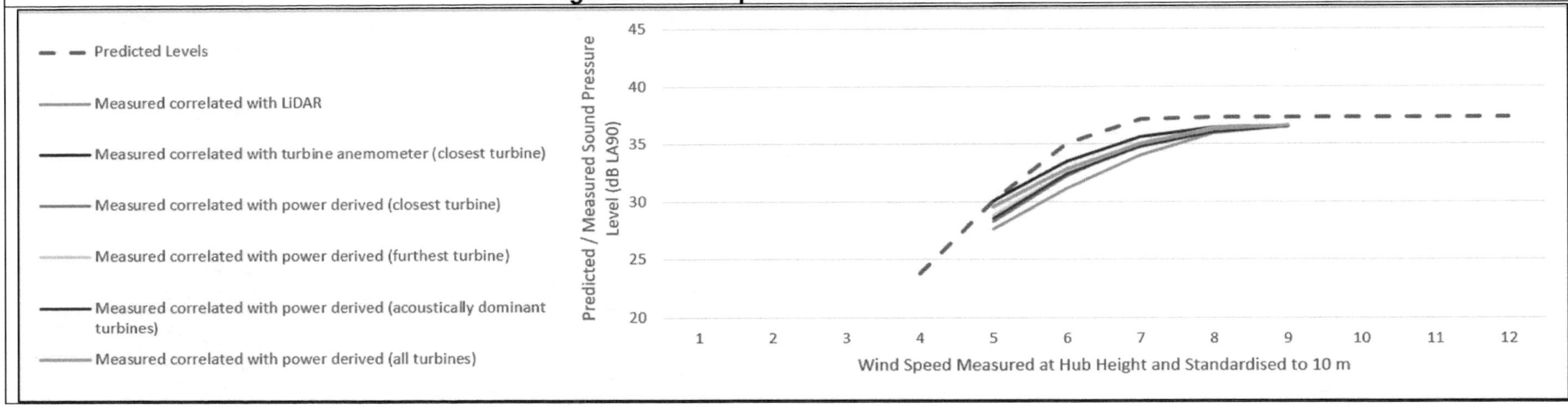

A comparison of how closely each of the trends match the predicted noise has been provided in Table 1 below. The average of the difference for each wind speed bin has been taken, and assigned a rank from 1-6 for comparison purposes;

Table 1 - Comparison of Wind Speed Source with Predictions

Variation	NML1	NML2	NML3	NML4
Measured correlated with LiDAR	0.73	1.46	1.04	0.97
Measured correlated with turbine anemometer (closest turbine)	0.62	0.30	0.02	0.40
Measured correlated with power derived (closest turbine)	0.74	0.20	0.28	0.71
Measured correlated with power derived (furthest turbine)	0.13	1.58	0.51	0.65
Measured correlated with power derived (acoustically dominant turbines)	0.29	0.21	0.42	0.72
Measured correlated with power derived (all turbines)	0.19	0.83	0.55	0.56

Rank	NML1	NML2	NML3	NML4	Sum
Measured correlated with LiDAR	5	5	6	6	22
Measured correlated with turbine anemometer (closest turbine)	4	3	1	1	9
Measured correlated with power derived (closest turbine)	6	1	2	4	13
Measured correlated with power derived (furthest turbine)	1	6	4	3	14
Measured correlated with power derived (acoustically dominant turbines)	3	2	3	5	13
Measured correlated with power derived (all turbines)	2	4	5	2	13

The results for the other 3 NMLs indicate that;

1. The LiDAR shows the least correlation with noise predictions, which is indicated by it's generally high variation / low ranking on comparison to the other wind speed source methods;
2. The closest turbine anemometer ranks best overall for low variation/ high ranking, with comparison to predictions;
3. Generally, the wind speed derived from the various power methods show similar variation. Notably for NML1 where the clean air turbine is located furthest away[8] the level of variation is high, but this is not the case for NML2 where its variation is low (both are located on adjacent ends of the wind farm), with both the clean air turbines chosen to be at similar distances.
4. The closest power derived turbine shows highest variation at NML2, but generally low variation at the other locations.

6. Discussion and Recommendations

The results indicate poor correlation with the Lidar at all locations. There is a high degree of scatter at all locations, it could be in this particular instance that both LiDAR location (relative to the turbines) and wake effects are contributing, however the relative distances between LiDAR location and NMLs does indicate different wind speeds are being experienced at different parts of the site, potentially leading to increased scatter.

The clean air turbines used at NML1 and NML2 were quite some distance from the NMLs (the wind farm is oblong in shape). At NML1 the variation and scatter was high, whilst at NML2 it was low. A potential reason for this is that more data was filtered out to consider downwind directions at NML2 than at NML1 (NML2 is located the south of the turbines, whilst NML1 is located to the north). Scatter is still being seen in the dataset and should more data be measured then there is potential for the variation to be higher, and likely more similar to NML1.

The clean air turbines showed good correlation at NML3 and 4, which will likely be attributed to the fact that the clean air turbines are much closer to the locations than NML1 and 2 are (due to them being located in the centre of the site). There is still a degree of scatter present in the plots, but the data do correlate well with predicted noise levels.

With the exception of NML2, the closest turbine, and the most acoustically dominant turbines showed high degrees of correlation with the predicted noise levels. The datasets showed a lack of scatter, and were consistently streamlined.

For each of the NMLs the turbine anemometer wind speed showed least variation compared to turbine noise predictions. Each of the locations exhibited large amounts of datapoints that were clustering at the lower wind speeds (around 5-6 m/s). This may be a result of the moving blades on the turbines, as has been identified as a potential concern in previous literature. This can have the effect of exaggerating noise levels at the lower wind speeds in comparison to the other (power derived) wind speed methods (as was also found by Broneske). This is particularly important where limits are set relative to predicted levels, as shown in Figure 2. Regardless of this, the line of best fit still averaged out to closely match predicted levels at these wind speeds; this of course could vary significantly on a site by site basis.

---

[8]As mentioned previously the wind farm is oblong in shape, and NML2 sits on the tip end of the wind farm. More wind turbines sit between the clean air turbine and closest turbines as a result at this location.

7. Conclusions

This paper has indicated that the source of wind speed measurements has the potential to have an impact on the results of compliance surveys, particularly where noise limits are set relative to wind farm noise levels.

The results also suggest, for the site considered, that no one method provided consistently better levels of correlation with predicted levels. The analysis did find that data collected by a LiDAR unit resulted in the weakest correlation with predicted levels whilst the best correlation occurred when data from the closest turbine / the primary turbines were used. This finding suggests that the guidance in the IOA GPG (which suggests that the most distant turbine should be used) may not always be appropriate.

The assessment suggests that when using a LiDAR / SoDAR unit, siting needs to be carefully considered to take account of the location relative to the turbines and the potential for the measurements to be influenced by wake effects.

Consideration of the results obtained using the wind speeds obtained by the wind turbine anemometer suggest that this approach may underestimate the wind speeds received by the turbine. Such trends may however be location (and turbine model) specific so this may not be case for other wind farms.

It should be noted that, overall, the variation between the results obtained using the various sources was generally quite small and further work could consider how such variation compares with wider levels of uncertainty associated with the noise assessment. A case could be made that an assessment using any of the methods presented in this paper is valid.

Whilst not the focus on this study it should be noted that the analysis indicates that predictions undertaken using the guidance included in the IOA GPG provide appropriate predictions of wind turbine noise (noting as outlined above that the measured levels presented in this paper include the influence of background noise meaning that the specific noise levels from the turbines will be lower than those presented).

References

[1] ETSU for the DTI (Department of Trade and Industry, "The Working Group on Noise from Wind Turbines ETSU-R-97 The Assessment and Rating of Noise from Wind Farms'," 1996.

[2] H. a. L. G. (. Department of Environment, "https://www.gov.ie/en/publication/f449e-wind-energy-development-guidelines-2006/," 2006. [Online]. [Accessed 19 01 2022].

[3] UK Institute of Acoustics, "Good Practice Guidance on the application of ETSU-R-97 for wind turbine noise assessment," 2013.

[4] Bullmore et al, Second International Meeting on Wind Turbine Noise, 20-21 September 2007, Lyon, France, "Wind Farm Noise Predictions: The Risks of Conservatism".

[5] Broneske, S, Internoise 2014, 16-19 November 2014, Melbourne, Australia, "Wind turbine noise measurements - How are results influenced by different methods of deriving wind speed".

[6] UK Institute of Acoustics, "A Good Practice Guide to the Application of ETSU R 97 for the Assessment and Rating of Wind Turbine Noise SGN5: Post Completion Measurements," 2014.

10th International Conference
on
Wind Turbine Noise
Dublin – 21st to 23rd June 2023

Evaluation of real noise effect inside the dwelling with open windows, a methodology

Andrea Bartolazzi, a.bartolazzi@studiorinnovabili.it

Summary

In several countries, noise limits for wind energy depend on the comparison between background noise and total noise. Inside a dwelling with open windows a fundamental difference between background noise and turbine noise comes from the fact that background noise is omnidirectional while turbine noise has a definite direction. With background noise the difference between noise outside the dwelling and noise inside the dwelling with open windows is significant and literature has analyzed it. A different situation happens with turbine noise because the correction factor depends on the relative geometrical position between the window and the single turbine. The sum of the effects of all the different turbines on the single window creates a specific effect that is highly dependent on the direction of the window. To analyze the noise inside the dwelling with open windows we propose both an acoustic model that represents the situation and a measurement. We compare the results of three contemporary noise measurements of a real case. The three sound level meters are positioned inside the dwelling without direct visibility towards the turbines, on the window plane with direct visibility towards the turbines, and outside the dwelling. This allows the verification of the parameters of the acoustic model. The result is an assessment of the influx of the relative geometrical position between the window and the turbine. This result is later used to propose a methodology of calculation of the noise inside the dwelling that considers separately every couple turbine-window. A further result is a proposal of calibration of the parameters of this methodology.

1. Introduction

The aim of this article is to analyse noise inside dwellings with open windows and the effect of surrounding turbines. The reason lies in the fact that inside a house to protect oneself from the noise of wind turbines there are relatively inexpensive methods if one can keep the windows closed. However, in many cases one cannot close windows or does not want to close them.
In cases where the user wants to keep the windows open, the calculation of noise transmission from the wind turbine to the reception should be treated considering background noise in a different way than turbine noise.

1.1 Problem definition.

The problem is to calculate how much is the noise contribution from a group of turbines placed around a receiver if we measure inside with open windows. It is also essential to be able to calculate the background noise level inside the receiver with turbines off. This is needed to calculate the difference between the total noise, the sum of the background noise and the noise of the turbines, and the background noise.
The shape of the receiver is considered rectangular and with one window per wall.
The goal is to figure out both how much is the total noise inside the house and what the difference of noise between the background noise with the turbines off and the background noise with turbines on.

1.2 Why it is complex

The calculation is quite complex because the background noise is omnidirectional. The background noise is produced mainly by vegetation moved by wind. Therefore, noise around the dwelling comes from all directions. For this reason, the walls of the house even with the windows open help to reduce the inside background noise consistently. The literature speaks of values in the range of 6 to 10 decibels [2]. The quantity of background noise reduction inside the house clearly depends on where the sources are located. For country houses usually vegetation is all around the house. In general, we do not know exactly where the sources are located. Doing a background noise source analysis is a complex task. The background noise of interest is mainly the wind-proportional background noise, while the occasional background noise from non-wind-related events is less interesting and less relevant to the evaluation of wind farms. That part of background noise is very well correlated with [4] wind intensity at the receiver.
Turbine noise has a different situation because we know exactly where turbines are located. We also know quite well how much noise they produce as a function of the wind at the hub. We know less precisely the correlation between the wind at the receiver and the wind at the hub. The aerodynamic models that are used to perform this correlation are quite complex. They depend on the stability of the air, the set of wind obstacles at the receiver, and the macro-turbulence in the area. In particular, the exponential coefficient describing the wind shear typically goes from values around 0.10-0.15 in the daytime to 0.3-0.5 at night when the air is more stable. Regarding obstacles, it is evident that if you measure the wind from the protected side of the house the intensity will be lower. Regarding macro-turbulence we are referring to the case where the house is in an area with complex aerodynamic characteristics, as for example a narrow valley. In this case the wind flow could have a vein detachment, and a macro-turbulence of the area is generated. The correlation between the wind on the ground at the receiver and the wind at altitude or hub height becomes much more complicated. Methods that consider the nonlinear elements of the Navier-Stokes equation are typically used to predict it, but uncertainty is still high.

1.3 Why it is interesting

The reasons for interest related to this calculation are mainly because wind turbine noise is most difficult to mask. At the same time in summer seasons people enjoy keeping open windows. When windows can be closed, there are relatively inexpensive methods to lower the noise inside the house. On the other hand, in cases where you want to keep the windows open during a beautiful summer day or night, the noise is very annoying.
For homeowners in areas occupied by wind turbines the change is strong. They are perhaps used to summer nights when noise level can also be very low around 25 dB or less. After the plant is built, they may end up with noise levels 15 dB higher.
Clearly it would be good to have the opportunity to verify this situation before the plant is built. But this is very complex for well-known reasons. Field measurements inside homes before construction are practically unfeasible. Measurements would be extremely complex given the

number of windows in each house. In addition, an extensive measurement campaign at a receiver would inevitably trigger a contrary reaction from the homeowner regardless of whether the homeowner is heavily damaged. The results of this measurement campaign moreover would not be very reliable because the variability of measurements, weather conditions, and measurement points largely undermines the meaning of the analysis.

2. Method

First, we propose a new calculation method based on the separate treatment of every couple turbine window.

2.1 Traditional calculation

Traditionally, the calculation to check noise at open windows is based on calculating the contribution of all turbines outside the receiver and then subtract a correction factor. This correction factor considers the effect of the walls of the dwelling that reduce the sound pressure level inside the house.
This analysis on field is typically done by measuring background noise outside dwellings. The calculation of turbine noise contribution is typically done following ISO 9613 including ground and air absorption but finally referring to free-field values i.e., outside. The two values both refer to the outside situation. It is mathematically correct to add them logarithmically to get the total value you would have once the park is built.
The more uncertain step is to carry the calculated outdoor value back to the interior of the dwelling. To do this one could use the noise correction factor values from literature. As explained above, however, a larger uncertainty is inherent in this step.
In mathematical terms:

$$L_T = 10log_{10}\left(\sum_{i}^{N_T} 10^{\frac{L_i}{10}}\right)$$

$$L_E = 10log_{10}\left(10^{\frac{L_B}{10}} + 10^{\frac{L_T}{10}}\right)$$

$$L_{E\ OpenWindows} = L_E - \Delta L_{out\_in}$$

L_T is the noise level of turbines at the receiver
L_B is the background noise level at the receiver
L_i is the contribution of every turbine at the receiver
N_T is the number of turbines
L_E is the environmental noise which is the addition of turbine and background noise at the receiver
$L_{E\ Openwindows}$ is the environmental noise level inside the receiver with open windows
ΔL_{out_in} is the correction factor used to consider noise reduction due to receiver walls

2.2 Proposed methodology

This is why a methodology is proposed that considers each window-turbine pair separately. The dwelling is supposed rectangular with one window on every external wall. If for example, the dwelling has an external wall and a window on the north side, a turbine on the north of the dwelling

would be just in front of that window. The angle α is zero. The noise coming form that turbine is fully received inside the dwelling. The correction factor can be set at 0 dB.
Inversely if the turbine is on the south of the dwelling, the angle α and the correction factor will be set to $c(\alpha)$. The parameter c, the correction factor can be depending on the angle α or be a fix level as for example a value up to 10dB.
For the turbine noise the correction factor depends on the angle between the direction windows-turbine and the direction of the windows

For $\alpha_{ij} \leq \alpha_L$

$$\Delta L_{ij} = 0$$

For $\alpha_{ij} > \alpha_L$

$$\Delta L_{ij} = c(\alpha)$$

Given that in this model the correction factor due to obstacles depends on the relative position between the turbine and the window/wall for each turbine-window pair there will therefore be a different correction factor. The calculation therefore will be done differently considering each turbine-wall pair separately and giving a different value to the correction factor coefficient. In mathematical terms:

$$L_{T\ OpenWindows} = 10log_{10}\left(\sum_{i}^{N_T} 10^{\frac{L_i - \Delta L_{ij}}{10}}\right)$$

L_i is the contribution of every turbine outside the receiver
N_T is the number of turbines
$L_{T\ Openwindows}$ is the turbine noise level inside the receiver *j* with open windows
ΔL_{ij} is the correction factor used to consider noise reduction of the turbine *i* towards window/wall *j*

The angle within which a turbine is noisy inside the window is one of the main parameters of this calculation. By choosing rectangular dwellings and a maximum angle of 45°, one is representing the situation in which each turbine projects its noise inside a window or the window of the following wall. A limit angle α_L of 45° is also coherent with past analysis [5] that show how the visibility of the turbine from the position inside the receiver is limited due to the internal geometry of most receivers.
Therefore, the two basic parameters of this calculation are the correction factor value for turbines that are not direct visibility from inside the window, and the angle of view to be considered.
One of the objectives of this analysis is to identify correct values for these two parameters.
For the background noise we consider valid the same model with fixed correction factor.

$$L_{B\ OpenWindows} = L_B - \Delta L_{out\_in}$$

Then all the contributions from the different turbines and the background noise are added together.

$$L^*_{E\ OpenWindows} = 10log_{10}\left(10^{\frac{L_{B\ OpenWindows}}{10}} + 10^{\frac{L_{T\ OpenWindows}}{10}}\right)$$

$L^*_{E\ Openwindows}$ is the environmental noise inside the dwelling at open windows

3. Experiment

Objective of experimental results is to calibrate the given model by calculating a reasonable correction factor for turbines out of the visibility of the SLM.

3.1 Test setup

To test the effect of the visibility of a turbine from inside a receiver with the windows open, we placed three Sound Level Meters at a window during power-off tests on a wind farm. The three Sound Level Meters were located outside the house, on the window front, and inside the house in an area not directly visible from any of the turbines.

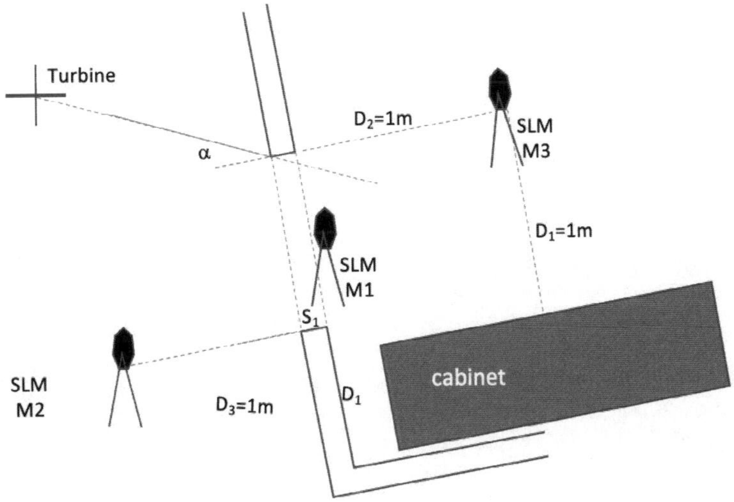

Figure 1 – Measurement Set-up

3.2 Results obtained

The result shows that the indoor Sound Level Meter (M3) receives less noise both during operation of wind park and during park shutdown.
According to the conceptual scheme shown above, the internal Sound Level Meter records a lower noise than the external Sound Level Meter (M2) because it enjoys greater noise protection due to the walls. At the time of turbine shutdown, however, there will be a decrease in noise on both the external Sound Level Meter and the enternal Sound Level Meter. The noise difference of 2-3 dB is one of the parameters we were looking for. It can be argued that the large difference in value from the literature may be due to reflection on the walls particularly on the cabinet wall. Therefore, it is true that the internal sound level meter records only slightly less noise than the external one, but it is also true that probably some of the noise is reflected from the cabinet.

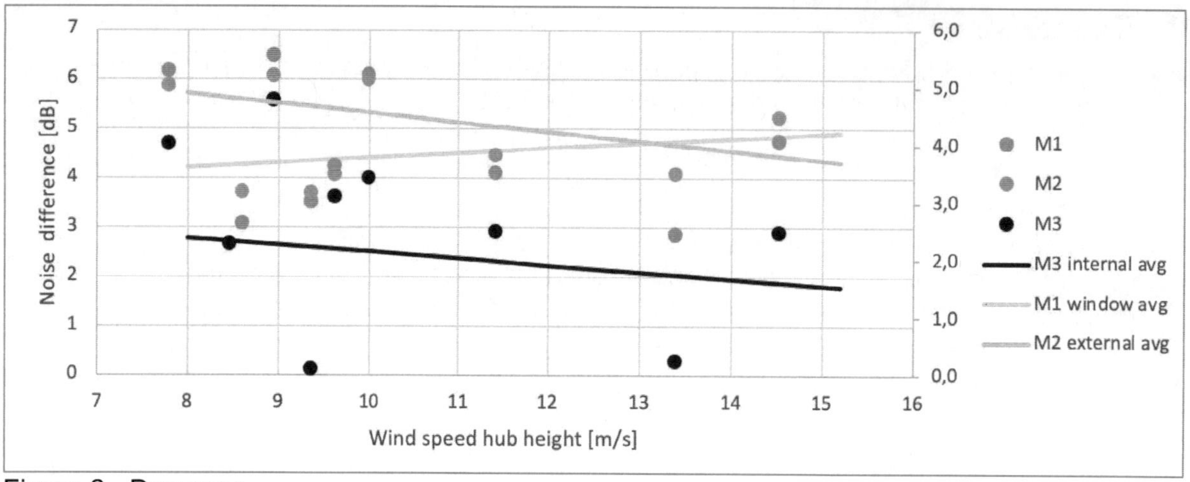

Figure 2 - Day case

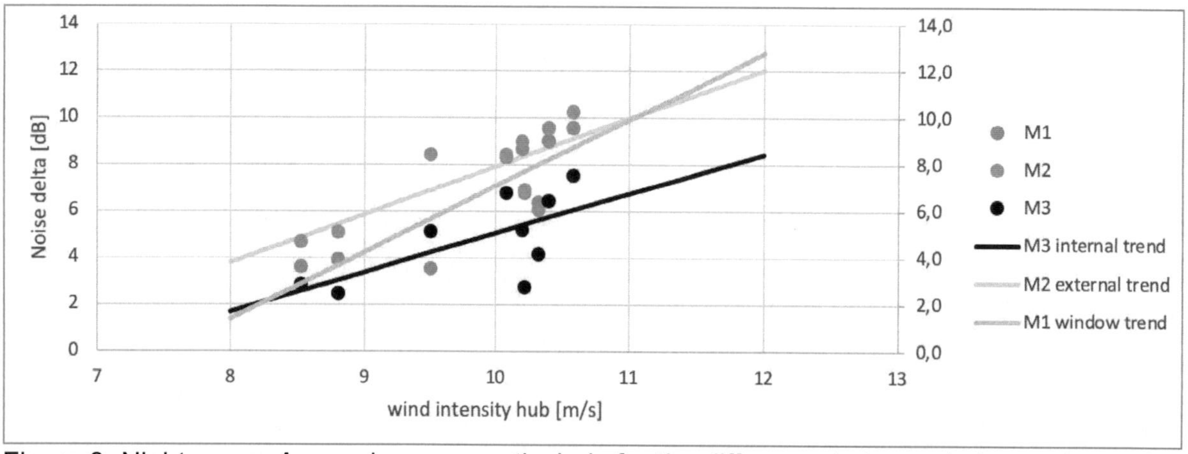

Figure 3- Night case - As can be seen particularly for the difference between indoor noise and outdoor noise at night there is always a difference of at least 2 dB up to almost 4 dB for higher wind speeds. The main reason for this difference is that while the M1 and M2 meters directly receive the noise of at least one turbine in the case of the M3 meter this does not directly receive the noise.

4. Conclusions

We can conclude that a reasonable level for the correction factor is around 3dB. The measured correction factor is much lower than what could be expected from literature.
An angle α_L of 45° is also coherent with past analysis [5] that show how the visibility of the turbine from the position inside the receiver is limited due to the internal geometry of most receivers.

The calculation of noise at open windows can be seen as a sub-case of the calculation at the receiver i.e., that situation where the building obstructs the transmission of part of the noise from the wind power plant. A further area of interest is the estimate of noise in case the SLM is placed outside the dwelling but without direct visibility from part of the turbines.

References

[1] Intern. Standard IEC 61400-11 (2012) Wind turbine systems. Part 11: acoustic noise measurement techniques

[2] UNI / TS 11143 (2013) Metodo per la stima dell'impatto e del clima acustico per tipologia di sorgenti

[3] Dick Bowdler, What is the Real Background Noise?, Proceedings of WTN Conference, Lyon (F), 2007

[4] O. Fegeant. Wind-induced vegetation noise. part 1: A prediction model. Acta Acustica, 85:228–240, 1999; O. Fegeant. Wind-induced vegetation noise. part 2: Field measurements. Acta Acustica, 85:241–249, 1999;

[5] A. Bartolazzi, C. Pompili, A model to calculate the delta between internal noise with open windows vs external noise, Proceedings of WTN Conference Europe, 2021

INCE EUROPE

10th International Conference on Wind Turbine Noise Dublin – 21st to 23rd June 2023

A year in review: ANSI/ACP's new standard for wind turbine sound predictions

Mark Bastasch, Jacobs, Mark.Bastasch@Jacobs.com

Summary

Common methods for modeling wind turbine sound rely on sound power levels from International Electrotechnical Commission (IEC) 61400-11 and propagation algorithms from International Organization for Standardization (ISO) 9613-2. Typical approaches in the United States yield relatively similar results; nonetheless, standardization of prediction methods was pursued to facilitate a robust and repeatable process that increases regulatory confidence and comprehension. American Clean Power Association (ACP) is recognized by the American National Standards Institute (ANSI) as an Accredited Standards Developer. ACP established a wind turbine sound modeling standard working group and reached consensus on a preferred method for predicting sound levels during the siting and permitting process. ANSI approved this standard in April of 2022 and made it available on the ANSI website in the fall of 2022. A primary goal of this new standard was to establish a uniform method of predicting future project sound levels such that predevelopment sound assessment results and predictions used in research can be readily compared. Initial indications are that this goal and the corresponding goals presented in this paper are being achieved.

1. Introduction

In 2021, ACP Standards established a working group to develop a standard for modeling sound levels from wind turbines with the goal of ensuring predictions were repeatable, uniform, and more easily understood by all interested parties. Over an approximate 10-month period, the working group held regular 1- to 2-hour virtual meetings and worked collaboratively to develop the standard. As acoustical modeling of wind energy projects in the United States historically relied on sound power levels based on IEC 61400-11 and the ISO 9613-2 propagation methodology, the working group focused on standardizing implementation of these methods. The group reached consensus on a methodology for predicting sound levels, which in turn served as the basis for the new ANSI/ACP standard. ANSI found the standard complied with their essential requirements and approved the formal standard in the April 2022. The standard was subsequently published on ANSI's website in the fall of 2022. Additional outreach efforts to ensure practitioners were aware of the standard included webinars and participation in various technical conferences. This paper summarizes the standard and initial feedback.

2. Summary of Working Groups Findings

The following discussion summarizes the standardized modeling parameters established in the standard. The apparent wind turbine model-specific sound power level should be determined in accordance with IEC 61400-11 for the downwind location. Using this sound power level as the basis, propagation modeling should be conducted in accordance with the ISO 9613-2 standard (Acoustics—Sound Attenuation During Propagation Outdoors Part 2: General Method of Calculation) using either of two sets of parameters, identified as Option 1 and Option 2:

Option 1

A. Ground factor of G=0.
B. While modeled receptor height does not influence model results with G=0, it can be stated as 1.5 meters for consistency with existing standards with respect to the microphone position for measurements or as 4 meters for consistency with Option 2.
C. Turbine modeled at hub height using vendor's apparent downwind sound power level specified consistent with IEC 61400-11.
D. No other model adjustments.

Option 2

A. Ground factor of G=0.5.
B. Receptor height of 4 meters.
C. Turbine modeled at hub height using vendor's apparent downwind sound power level specified consistent with IEC 61400-11.
D. A model adjustment of + 2 decibels.

The difference between Options 1 and 2 is small (tenths of a decibel), a difference that is neither reliably measured nor discernible. Option 2 is provided as some jurisdictions have established a precedent of adjusting model inputs or results and requiring such adjustments to Option 1 is not supported. Using a receptor height of 4 meters in Option 2 should not influence the typical microphone measurement height.

Atmospheric absorption per ISO 9613-1 for conditions of 10 degrees Celsius (50 degrees Fahrenheit) and 70 percent humidity is recommended for both options.

2.1 Informative Annex

As regulatory approaches in the United States vary from state to state and in certain states from county to county, the standard includes a brief appendix of additional commentary on a range of topics that the working group expected would assist those conducting the analysis.

For jurisdictions that require the evaluation of L_{eq} durations outside of the 10-minute to 1-hour timeframe, additional adjustments to the predicted value are likely to be warranted. Averaging times longer than 1 hour are noted to yield sound levels less than those predicted given that the standard relies on the full-acoustic output and conservative meteorological conditions.

Reference is made to the forthcoming IEC 61400-11-2 (Wind energy generation systems – Part 11-2: Measurement of wind turbine noise characteristics in receptor position), which is expected to provide more thorough guidance on field measurements at far field locations when released. As such, measurements were considered outside the scope of this standard during development.

Apparent sound power levels based on IEC 61400-11 are typically readily available from the turbine vendors for wind speeds up to 15 meters per second (m/s). These data are stated to be appropriate as the highest sound power levels are generally achieved between 8 and 10 m/s.

When a project utilizes more than one type of turbine whose highest sound power levels are achieved at different wind speeds, one may wish to consider what may occur simultaneously rather than utilize the highest sound power level of each turbine model simultaneously.

C-weighted and octave band sound levels are also briefly discussed. Several cited studies found that C- and A-weighted levels were highly correlated.

When considering the potential influence of topography, moving the source height from hub height to tip height as well as comparisons to flat ground predictions are noted to potentially assist in evaluating reasonableness of predicted shielding. The standard does not implement a concave ground adjustment.

3. Goals of the Standard

In the United States, this new standard does not supersede permit requirements, nor does it invalidate previous or potential future studies. Rather, the goals of this standardization effort are to:

- Serve as a guide for practitioners.
- Establish a uniform method of predicting future projects sound levels at typical setback distances for utility-scale on-shore projects.
- Reduce confusion and facilitate a uniform basis for comparing predicted results during permitting and in the peer-reviewed scientific literature.
- Develop a robust and repeatable process that bolsters regulatory confidence in modeling results.
- Provide reasonably conservative results for realistic worst-case conditions.
- Recognize that other calculation methods have been used and that this standard does not indicate flaws or errors in other methods.

3.1 Are the goals being achieved?

While the standard has been widely available for approximately 6 months as of this manuscript, initial indications are that it is achieving the stated goals. Consultants and acoustical practitioners are implementing the standard and referencing it in regulatory filings and permitting studies for new projects. Future research efforts in the United States are anticipated to at least provide standardized results should they choose to implement another prediction methodology. Debates over appropriate prediction methodologies appear to have subsided since this standard was published. Implementation has been reasonably well received and the overall goals are being achieved.

4. Conclusions

A new ANSI standard (ACP 111-1, Wind Turbine Sound Modeling) was established in April 2022 to standardize the prediction of wind turbine sound levels. This new standard gained recognition over the next several months and became more widely available in the fall of 2022. ANSI/ACP 111-1 draws on existing standard methods in ISO 9613-1, ISO 9613-2, and IEC 61400-11 to establish a uniform method of predicting future project sound levels such that predevelopment sound assessment results and predictions used in research can be readily compared. The standardization of predictions ensures a robust and repeatable process that bolsters regulatory confidence in the results and reduces potential confusion. Initial feedback from practitioners indicates that ANSI/ACP 111-1 is achieving its intended goals and has been well received.

5. Acknowledgements

This effort would not have been possible without the collaboration, dedication, and insight of the following working group members (listed alphabetically):

- Payam Ashtiani
- Mark Bastasch
- Kaveh Habibi
- Michael Hankard
- Kenneth Kaliski
- Richard Lampeter
- Krispian Lowe
- Marcel Mibus
- Chris Ollson
- Justin Puggioni
- Erik Sloth

Special thanks are owed to the American Clean Power Standards team for their assistance throughout the working group and standard development process.

References

American National Standards Institute (ANSI). 2022. *Standard ACP 111-1, Wind Turbine Sound Modeling*. April 1.

International Electrotechnical Commission (IEC). 2012. *IEC 614000 Wind turbines – Part 11: Acoustic noise measurement techniques*. Edition 3.0. 2012-11.

International Organization for Standardization (ISO). 1996. *ISO 9613-2:1996 Acoustics – Attenuation of sound during propagation outdoors – Part 2: General method of calculation*, Edition 1. 1996-12.

International Organization for Standardization (ISO). 1993. *ISO 9613-1:1993 Acoustics – Attenuation of sound during propagation outdoors – Part 1: Calculation of the absorption of sound by the atmosphere*, Edition 1. 1993-06.

International Electrotechnical Commission (IEC). *IEC 61400 Wind energy generation systems - Part 11-2: Measurement of wind turbine noise characteristics in receptor position*, https://www.iec.ch/, Unpublished as of preparation of this manuscript.

10th International Conference
on
Wind Turbine Noise
Dublin – 21st to 23rd June 2023

Guidelines for Environmental Assessment of Wind Farm Noise in Chile

Bastián-Monarca, N.A. Acústica Austral: nbastian@acusticaustral.cl
Arenas, J.P. Universidad Austral de Chile: jparenas@uach.cl

Summary

If investors plan to build a wind farm or other type of power plant in Chile, an Environmental Impact Assessment must be completed if the project produces more than 3 MW. This assessment must consider a noise and vibration impact study, among others. Related to the noise part, according to law 19.300, there are different environmental protection objects, such as population's health, natural renewable sources (native wildlife), the system of life and customs of human groups, tourist value of an area, and cultural heritage. Therefore, an environmental assessment of wind farm noise must consider all these environmental protection objects. This work presents guidelines for assessing environmental noise in these protection objects, including an application to a real project in Chile. In this sense, the noise regulation decree DS38 enacted by the Chilean Ministry of the Environment is recommended for assessing the effects on the population's health. For natural renewable sources (native wildlife), several international studies are used for different species. For the system of life and customs of human groups and cultural heritage, the Spanish Royal Decree 1367/2007 is recommended. Not exceeding background noise is advised to assess the tourist value. The results of several projects in Chile have shown that a wind farm project can have a potential noise impact area at a distance of 2 km around the wind farm, which is very similar to the potential impact area related to shadow flicker.

1. Introduction

Environmental assessment of wind turbine noise (WTN) is a considerable challenge worldwide. In Chile, the current noise regulation applicable to WTN is the DS38 enacted by the Ministry of the Environment of Chile (DS38) [1]. For new projects, the Chilean Service of Environmental Assessment (SEA) published a specific guide to evaluating the wind farm noise of new projects presented at Chile's System of Environmental Impact Assessment (SEIA) [2]. This guide is based on the noise regulation decree DS38 and gives several recommendations for background noise measurement techniques and prediction methods for wind farm noise [3]. In addition, according to law 19.300 of Chile [4], there are different environmental protection objects, such as population health, natural renewable sources (native wildlife), the system of life and customs of human groups, tourist value of an area, and cultural heritage. Therefore, a wind farm noise environmental assessment must evaluate all these environmental protection objects. This work shows guidelines for assessing environmental noise in these protection objects and presents some applications to real projects in Chile.

2. Environmental protection objects

2.1 Population health

2.1.1 Wind Turbine Noise and Construction

For assessing the effects on the population's health caused by WTN and the construction of a wind farm, it is recommended to use the noise regulation decree DS38 enacted by the Chilean Ministry of the Environment [1].

The regulation defines four types of land use for urban and rural areas. Noise limits are defined as maximum permissible A-weighted equivalent sound pressure levels, Leq in dBA for daytime (07:00 to 21:00 hrs) and nighttime (21:00 to 07:00 hrs). The limit values in urban sites vary between 45 dBA for the night in exclusive residential areas and 70 dBA for industrial-restricted areas. In rural areas, the noise limit is determined as the lowest value among the following two conditions:

i) The background noise plus 10 dBA or.

ii) Either 65 dBA for daytime or 50 dBA for nighttime.

For the background noise measurement, the DS38 established that "*The continuous equivalent sound pressure level (Leq) should be measured continuously, until the reading stabilized, recording the value of Leq every 5 minutes. Reading shall be understood as stabilized when the arithmetic difference between two consecutive registers is less than or equal to a value of 2 dBA. The considered level will be the last of the recorded levels. In any case, the measurement shall not last more than 30 minutes*".

A corrected sound pressure level is used for characterizing the source's emission. This value comes from the analysis of the descriptor LAeq 1 minute, and the maximum level LAmax recorded in a measurement period of 3 minutes. A comparison between both descriptors and the background noise level is necessary to define the final noise level that will characterize the noise from the source.

Also, it is essential to highlight the following criteria that appear in the DS38:

1) The measurement must be made in the most unfavorable condition for the receiver (worst-case condition).
2) The background noise must be measured under the same conditions as the noise source.

2.1.2 Transportation noise

Both the Transit Noise and Vibration Impact Assessment Manual of the Federal Transit Administration of the U.S. Department of Transportation (FTA) [10] and the Regulation of the Swiss Confederation on noise protection (OPB 814.41) [11] are recommended for the SEA [12] to assess the effects on the population's health caused by transportation noise.

The FTA establishes three land use categories. Category 1 (high sensitivity) and 3 (institutional) use the noise metric Leq 1hr in dBA, and category 2 (residential) uses the noise metric Ldn. This noise metric includes a 10 dB penalty for nighttime noise. According to Fig. 1, the noise limits depend upon the existing noise exposure.

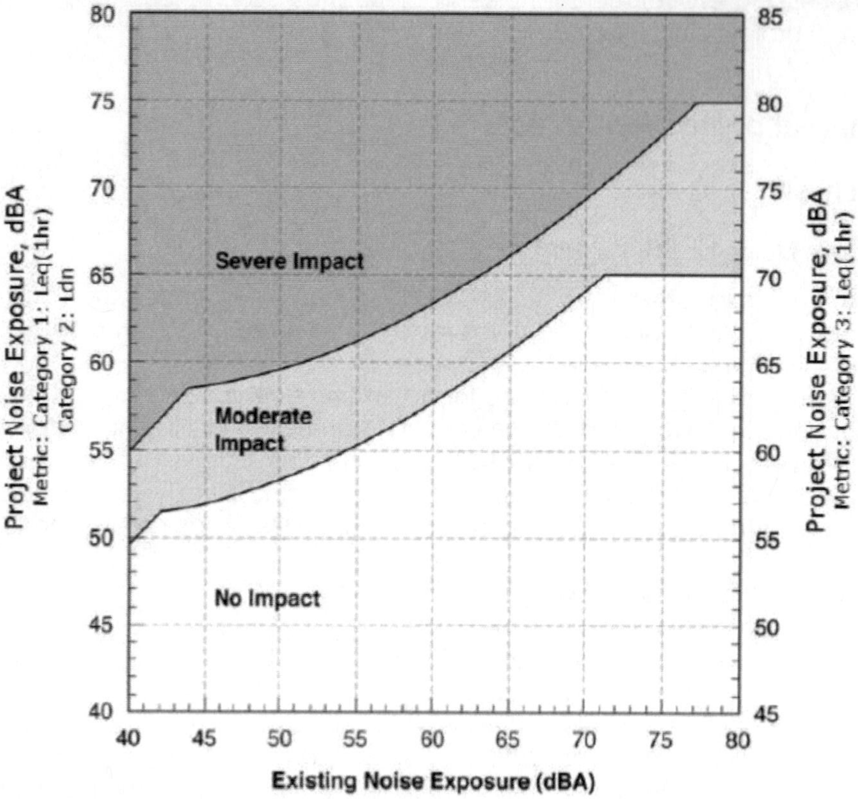

Figure 1: Noise Impact Criteria for Transit Projects. **Source:** [10].

The Swiss Confederation establishes four degrees of sensitivity related to the type of land use and defines noise limits for the day (between 06:00 and 22:00 hrs.) and night (between 22:00 and 06:00 hrs.). For projects that use an existing route, the noise limits vary between 45 and 60 dBA at night and between 55 and 70 dBA at day.

2.2 Natural renewable sources (native wildlife)

For natural renewable sources (native wildlife), several international studies are used for distinct species. In this sense, the SEA published a specific guide to evaluating noise on native wildlife for projects presented at the SEIA [5] (see Table 1).

Table 1: Recommended noise limits for wildlife. **Source:** [5].

Vertebrate taxonomic group	Effect description	Effect type	Type of source	Limit	Reference
Amphibians	Changes in frequency components of vocalizations	Behavioral	Environmental	62 dB(C) average	[6]
	Reduction in chorus tenure and duration by male anurans	Behavioral	Transportation	72 dB(A) average	[7]
Reptiles	Difficult for localization	Behavioral	Transportation	75 dB(C) average	[7]
Birds	Changes in frequency	Behavioral	Environmental	60 dB(A) average	[8]

Vertebrate taxonomic group	Effect description	Effect type	Type of source	Limit	Reference
	components of vocalizations				
	Decline in reproductive success	Behavioral	Environmental	58 dB(A) average	
			Industrial and Transportation	68 dB(A) average	
	Effects on physiology and development	Physiological	Transportation	60 dB(A) SPL max	[7]
	Increase in vigilance and alert behavior	Behavioral	Military	80 dB(A) SPL max 63 dB(A) average	
	Hearing damage	Physiological	Single Impulse (e.g., blast)	140 dB(A) SPL max	[8]
	Temporarily elevated threshold		Industrial and Transportation	93 dB(A) SPL max	
	Disruption of foraging in gleaning bats	Behavioral	Transportation	80 dB(A) average	
Mammals	Reduced reproductive efficiency	Behavioral	Industrial and Construction	68 dB(A) average	[7]
	Short-term increase in heart rates and shifts in resting and movement behaviors of ungulates	Physiological - Behavioral	Military	85 dB(Z) average	

SPL: Sound pressure level.

2.3 System of life and customs of human groups and Cultural heritage

2.3.1 Wind Turbine Noise and Construction

The Spanish Royal Decree 1367/2007 is recommended for both the system of life and customs of human groups and cultural heritage. This regulation defines six types of land use. Noise limits are defined as Day Sound Level (Ld), Evening Sound Level (Le), and Night Sound Level (Ln). The limit values vary between 45 and 55 dBA (Ln) for new infrastructure (such as wind farms).

2.3.2 Transportation noise

For assessing the effects of transportation noise on both the system of life and customs of human groups and cultural heritage, both the Transit Noise and Vibration Impact Assessment Manual of the Federal Transit Administration of the U.S. Department of Transportation (FTA) [10] and the Regulation of the Swiss Confederation on noise protection (OPB 814.41) [11] are recommended for the SEA [12].

2.4 Tourist value of an area

2.4.1 Wind Turbine Noise and Construction

There is no specific regulation for this kind of environmental protection object. Therefore, keeping noise levels below background noise is advised to assess the tourist value.

2.4.2 Transportation noise

For assessing the effects of transportation noise on the tourist value of an area, both the FTA [10] and the OPB 814.41 [11] are recommended for the SEA [12].

2.5 Summary

A summary of the different regulations recommended for each environmental protection object is presented in Table 2.

Table 2: Summary of recommended regulation/criteria for each environmental protection objects.

Environmental protection object	Type of source	Regulation/criteria
Population health	WTN and construction	D.S. N°38/11 [1]
	Transportation	FTA [10] or OPB 814.41 [11]
Natural renewable sources (native wildlife)	WTN and construction	International criteria (see Table 1)
	Transportation	International criteria (see Table 1)
The system of life and customs of human's groups	WTN and construction	Spanish Royal Decree 1367/2007 [9]
	Transportation	FTA [10] or OPB 814.41 [11]
Tourist value of an area	WTN and construction	Not exceeding background noise
	Transportation	FTA [10] or OPB 814.41 [11]
Cultural heritage	WTN and construction	Spanish Royal Decree 1367/2007 [9]
	Transportation	FTA [10] or OPB 814.41 [11]
WTN: Wind Turbine Noise		

3. Application to a real project in Chile

This section presents a case study of a wind farm project planned to be constructed and operated in the Biobío (Chile) region. The project will have 63 wind turbines and will produce 470 MW. The wind turbines have a hub height of 165 m and a rotor diameter of 190 m.
In total, the following numbers of receivers for each Environmental Protection Object were studied:

- **Population health:** 192 receivers.
- **Natural renewable sources (native wildlife):** 9 sites of interest.
- **The system of life and customs of human's groups:** 32 receivers.
- **Tourist value of an area:** 2 receivers.
- **Cultural heritage:** 3 receivers.

In addition, several recommendations given in the specific WTN guide of the SEA [2] were also followed. They are summarized next:

- The standard verification was performed for three different wind speed ranges (6-8, 8-10, and 10-12 m/s) at the hub height.
- Eight continuous background noise measurements were carried out over 14 days for the same wind speed ranges at the hub height.
- The method described in ISO 9613-2 [16] was used with specific parameters, Immission height of 4m, Humidity of 70%, Temperature of 10°C, Wind direction Downwind, and Ground factor of 0.5. These parameters have been adopted from international regulations [17, 18] and different previous studies conducted by the MMA [19, 20].
- A special windscreen that complies with IEC 61400-11 [21] and the recommendations given by the Institute of Acoustics [22] was employed for all noise measurements. In particular, windscreens ACO Pacific model WS7-80T were used.
- The noise measurement was conducted using sound level meters Norsonic models Nor1531 and Nor139. The sound level meters were calibrated with a sound level calibrator Norsonic model Nor1255.
- The wind speed and direction at the hub height data were provided in 10 minutes averages for the study period. The measurements were made at the hub height with the SODAR system.

3.1 Results

A summary of the impact area (distance at which the noise limit is accomplished) for each environmental protection object is presented in Table 3 for each phase of the project lifecycle (construction, operation, and closure).

Table 3: Summary of impact area for each environmental protection object.

Environmental protection object	Type of source	Period	Maximum radial distance per phase (m)		
			Construction	Operation	Closure
Population health	WTN and construction	Day	839	638	977
		Night	1385	1113	---*
	Transportation	Day	32	20	7
Natural renewable sources (native wildlife)	WTN and construction	Day	883	390	675
		Night	665	390	---*
	Transportation	Day	174	171	170
The system of life and customs of human's groups	WTN and construction	Day	493	288	394
		Night	1219	1165	---*
	Transportation	Day	32	20	7
Tourist value of an area	WTN and construction	Day	2137	1591	1.495
	Transportation	Day	136	100	12
Cultural heritage	WTN and construction	Day	493	288	394
		Night	1219	1165	---*
	Transportation	Day	136	100	12
---*: Does not apply					

The impact area can be vast for different environmental protection objects (see Table 3). In this case, the greater maximum radial distance is for the tourist value of an area. The criteria adopted did not exceed background noise, and the distance obtained was 2137 m for Construction, 1591 m for Operation, and 1495 m for Closure (all for daytime). The second maximum radial distance occurs for the population's health at night. In this case, the distances obtained were 1385 m for Construction and 1113 m for Operation (this distance was very similar for both the system of life and customs of human groups and cultural heritage). Thus, a wind farm can have a potential noise impact area of 2 km. This value is similar to the potential impact area of shadow flicker, which can be between 2-2.5 km according to the Guideline of Australia – New South Wales [13], Austria [14], and Germany (distance in which rotor blade covers at least 20% of the sun disk is approximately 2 km) [15].

Furthermore, noise maps for two environmental protection objects are shown in Figs. 2 and 3.

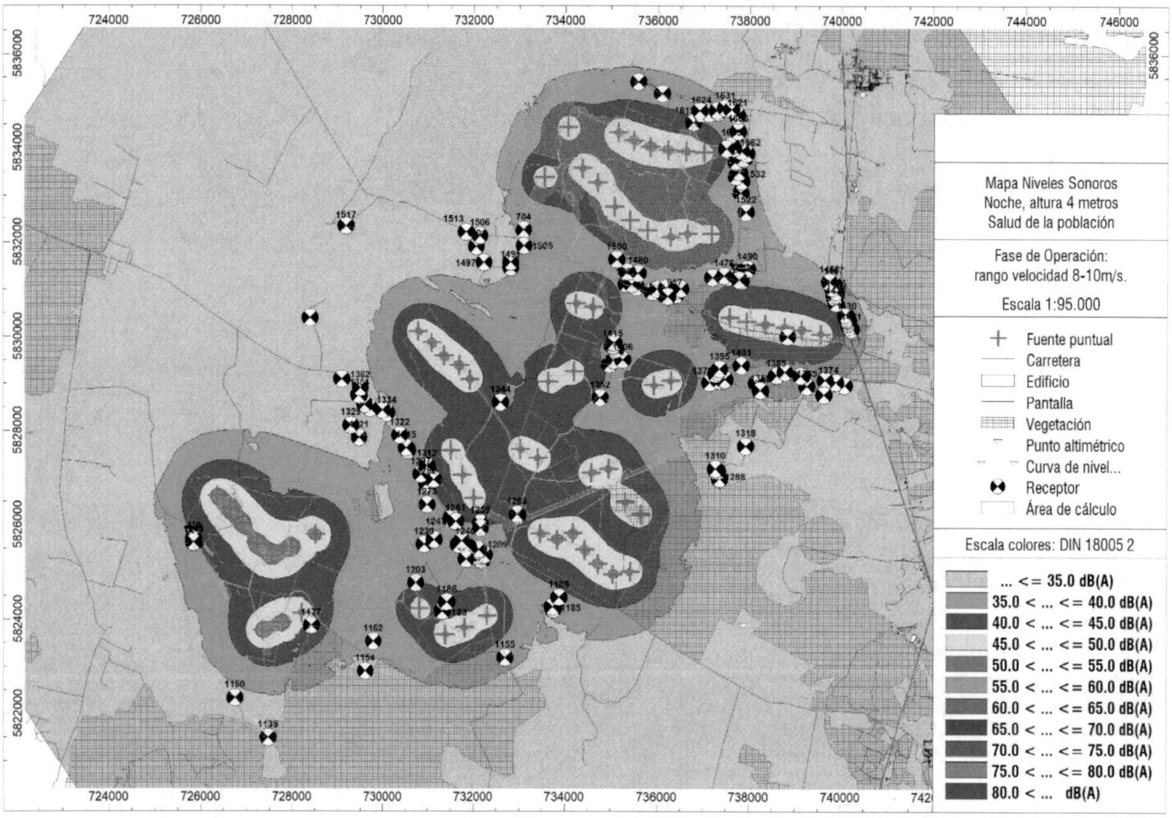

Figure 2: Noise map for the operation phase of the wind farm project (nighttime, people health).

4. Conclusions

This work described how different environmental protection objects are considered in Chile in a noise impact study which is part of an environmental impact assessment. The assessment of the effects on the population's health is made using the noise regulation decree DS38 enacted by the Chilean Ministry of the Environment. For natural renewable sources (native wildlife), several international studies are used for different species. For the system of life and customs of human groups and cultural heritage, the Spanish Royal Decree 1367/2007 is recommended. Not exceeding background noise is advised to assess the tourist value. The guide of the Federal Transit Administration or the OPB841.14 is recommended to assess the effect of transportation noise on all environmental protection objects. The results of an actual project in Chile show that a wind farm project can have a potential noise impact area at a distance of 2 km around the wind farm, which is very similar to the potential impact area related to shadow flicker.

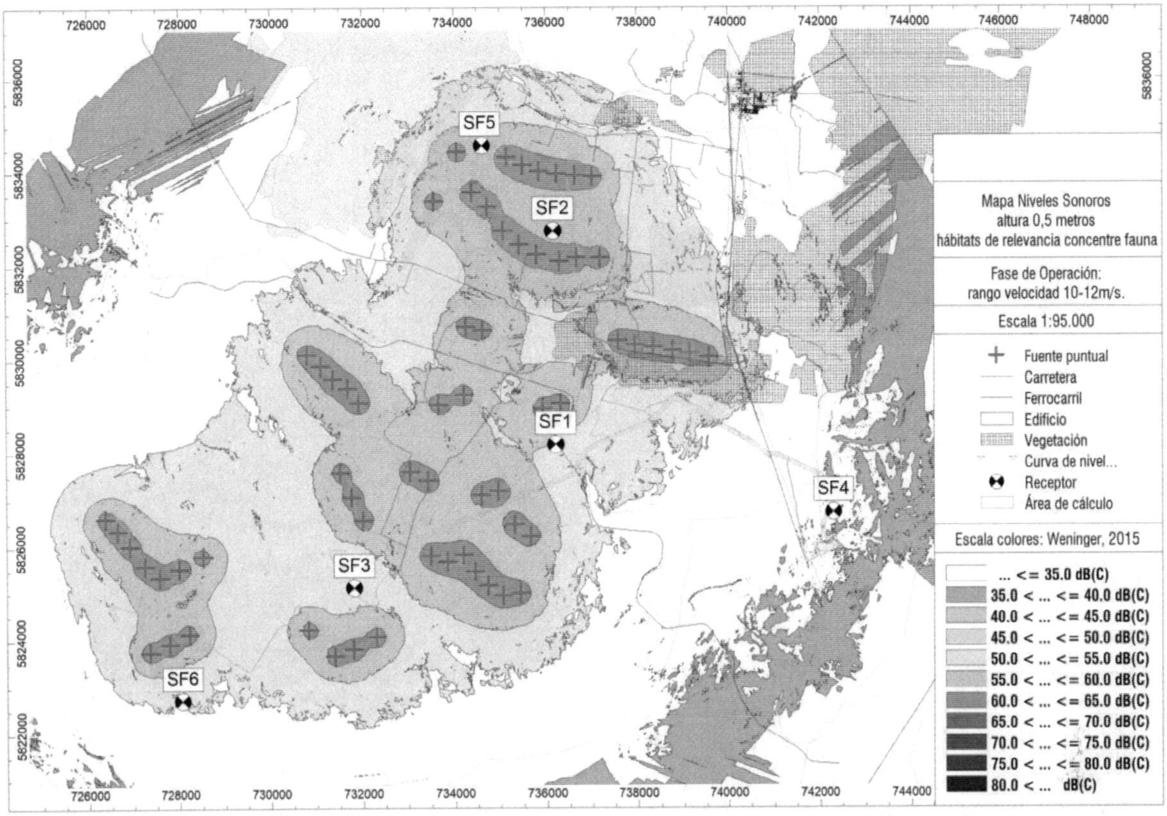

Figure 3: Noise map for the operation phase of the wind farm project (nighttime, natural renewable sources [native wildlife]).

References

[1] Decreto Supremo N°38 del Ministerio del Medio Ambiente (MMA) (2011). *Norma de emisión de ruidos generados por fuentes que indica*. Ministerio del Medio Ambiente, Gobierno de Chile.

[2] Servicio de Evaluación Ambiental (SEA) (2020). *Guía para la Aplicación del DS N°38 de 2011, del Ministerio del Medio Ambiente, que Establece Norma de Emisión de Ruidos Generados por Fuentes que Indica, para Proyectos de Parques Eólicos en el SEIA*.

[3] Bastián-Monarca, N.A. & Arenas, J.P. (2022). *Challenges of Environmental Assessment of Wind Farm Noise in Chile: Background Noise. 28th International Congress on Sound and Vibration ICSV28*. Singapore.

[4] Ley 19.300 sobre Bases Generales del Medio Ambiente, Gobierno de Chile (1994).

[5] Servicio de Evaluación Ambiental (SEA). *Criterio de evaluación en el SEIA: Evaluación de impactos por ruido sobre fauna nativa* (2022).

[6] Shieh, B.S., Liang, S.H., Chen, C.C., Loa, H.H. & Liao, C.Y. (2012). *Acoustic adaptations to anthropogenic Noise in the cicada Cryptotympana takasagona Kato (Hemiptera: Cicadidae).* Acta Ethologica 15, 33-38.

[7] Shannon, G., McKenna, M.F., Angeloni, L.M., Crooks, K.R., Fristrup, K.M., Brown, E., Warner, K.A., Nelson, M.D., White, C., Briggs, J., McFarland, S. and Wittemyer, G. (2016). *A synthesis of two decades of research documenting the effects of noise on wildlife*. Biol. Rev. (2016), 91, pp. 982-1005. doi: 10.1111/brv.12207.

[8] Dooling, R. & Popper, A. (2007). *The Effects of Highway Noise on Birds*. Environmental BioAcoustics LLC Rockville, MD 20853. Prepared for: The California Department of Transportation Division of Environmental Analysis. Prepared Under Contract 43A0139 Jones and Stokes Associates.

[9] Ministerio de la Presidencia (España) (2007). *Real Decreto 1367/2007, de 19 de octubre, por el que se desarrolla la Ley 37/2003, de 17 de noviembre, del Ruido, en lo referente a zonificación acústica, objetivos de calidad y emisiones acústicas.*

[10] U.S. Department of Transportation, Federal Transit Administration (FTA) (2018). *Transit Noise and Vibration Impact Assessment Manual*. FTA Report No. 0123.

[11] Norma Suiza 814.41 OPB (1986). *Reglamento de la Confederación Suiza Sobre la Protección Contra el Ruido*.

[12] Servicio de Evaluación Ambiental (SEA)(2019). *Guía Para la Predicción y Evaluación de Impactos por Ruido y Vibración en el SEIA*. Gobierno de Chile.

[13] State of New South Wales Department of Planning & Infrastructure (2011). *Draft NSW Planning Guidelines Wind Farms*.

[14] Gschiel, E, Amt der Burgenland Landesregierung (2014). *Umweltmedizinische Begutachtung am Beispiel Windkraft*.

[15] Länderausschuss für Immissionsschutz (2002). *Hinweise zur Ermittlung und Beurteilung der optischen Immissionen von Windenergieanlagen (WEA-Schattenwurf-Hinweise)*.

[16] International Organization of Standardization (ISO) (1987). *ISO 1996-2:1987* Acoustics – Description and measurement of environmental noise – Part 2: Acquisition of data pertinent to land use.

[17] Standards New Zealand (2010). *NZS 6808:2010 Acoustics – Wind Farm Noise*. Wellington, New Zealand.

[18] UK Department of Trade and Industry (1996). *ETSU-R-97: The assessment & rating of noise from wind farms*.

[19] Parra, J.D., Valdebenito, I. & Lobos, V.H. (2017). *Addressing a Management Strategy of Wind Farms Noise Control in Chile*, 7th International Conference on Wind Turbine Noise, Rotterdam, Netherlands.

[20] Parra, J.D., Valdebenito, I. & Oyarzún, C. (2019). *Applying a general noise regulation on Wind Turbines: an administrative interpretation to a specific noise source*, 8th International Conference on Wind Turbine Noise, Lisbon, Portugal.

[21] International Electrotechnical Commission (IEC) (2006). *IEC 61400-11: Wind Turbine Generator Systems – Part 11: Acoustics Noise Measurement Techniques,* International Standard.

[22] UK Institute of Acoustics (2013). *A good practice guide to the application of ETSU-R-97 for the assessment and rating of wind turbine noise*.

INCE EUROPE

**10th International Conference
on
Wind Turbine Noise
Dublin – 21st to 23rd June 2023**

Detection and assessment of amplitude modulated noise of wind turbines

Andrea Bauerdorff – German Environment Agency: andrea.bauerdorff@uba.de
Till Kühner – Dr. Kühner GmbH: kuehner@umweltakustik.de
Alexander Alaimo Di Loro – deBakom GmbH: info@debakom.de
Dirk Schreckenberg – ZEUS GmbH: schreckenberg@zeusgmbh.de
Christoph Pörschmann – Cologne University of Applied Sciences: christoph.poerschmann@th-koeln.de

Summary

Wind turbines are important for a sustainable energy supply and their expansion contributes significantly to achieving independence from fossil imports. The noise emitted by these turbines is often discussed in politics, the media and the public. In order to prevent harmful effects on human beings and the environment, various industrial and commercial installations are subject to licensing in Germany. This includes wind turbines with a height of more than 50 m (4. BImSchV). Principally, installations subject to licensing shall be established and operated in such a way that this does not involve harmful effects on the environment or other hazards, considerable disadvantages and considerable nuisance to the general public and neighbourhood (BImSchG). The sound from wind turbines is subject to both strong spatial and temporal fluctuations in its amplitude and frequency composition. It depends on meteorological conditions and the location of the wind turbine. In addition, it depends on the type of turbine, the rotational speed and the operating mode. In connection with wind turbines, it is often discussed that this kind of sound is perceived as whoosh noise. This is an amplitude-modulated sound generated by wind turbines. In this paper a research project commissioned by the German Environment Agency, the generation of amplitude-modulated sound and its influence on the noise perception of residents living near wind turbines is presented and discussed.

1. Introduction

When it comes to the expansion of wind energy the topic of noise often leads to discussions. The focus of these discussions is on issues that deal with both the mitigation as well as the effects of noise on the health and quality of life of the population.

The sound from wind turbines is subject to both strong spatial and temporal fluctuations in its amplitude and frequency composition. It depends on meteorological conditions and the location of the wind turbine. In addition, it depends on the type of turbine, the rotational speed and the operating mode. In context with wind turbines, it is often discussed that this kind of sound is perceived as whoosh noise. This is an amplitude-modulated sound generated by wind turbines. A frequently argued theory is that this noise characteristic, which is specific to wind turbines, leads to an increased perception and annoyance of the residents. The multi-year research project "*Noise effects of the use of land-based wind energy*" (Schmitter, S; Alaimo Di Loro, A; Hemmer, D et al. 2022), commissioned by the German Environment Agency, which is presented here, investigated the noise effects caused by wind turbines on residents. The main goal of this project was to find out how often, over what periods of time, and to what extent wind turbines cause amplitude modulations. Furthermore, it was of special interest within the scope of the project to find out whether these modulations are audible and measurable in the surrounding neighbourhood. For this purpose, long-term sound measurements were carried out in five different study areas in Germany. These measurements were conducted over at least six weeks in the immission area and two weeks in the emission area in accordance with the German Standard DIN EN 61400-11 (DIN 2006). Based on these measurements, the sound-specific influences of wind turbines were analyzed. An algorithm was developed to find and quantify amplitude modulations in the measurement signal. To assess the annoyance caused by amplitude-modulated noise, annoyance surveys were carried out and evaluated among residents in the study areas. In addition, listening tests were carried out under laboratory conditions at three locations.

2. Measurements

Continuous acoustic measurements were carried out for the recording of amplitude-modulated noise. For this purpose, permanent measuring stations were set up at five carefully selected sites with wind turbines in the vicinity of residential areas. At each site, the noise of the wind turbines was measured over a period of at least 6 weeks. At 3 locations, emission measurements were carried out simultaneously over a period of 2 weeks.

During one of the measurement campaigns, additional measurements were specifically carried out for infrasound contributions. In addition to the direct evaluation, these measurements were also used to determine results for infrasound from the measurements in the other study areas.

2.1 Windfarm constellations

An important criterion in the selection of the measurement sites was to ensure measurements at a wide variety of wind farm constellations. For this purpose, the measurements took place at five different locations all over Germany (Figure 1).

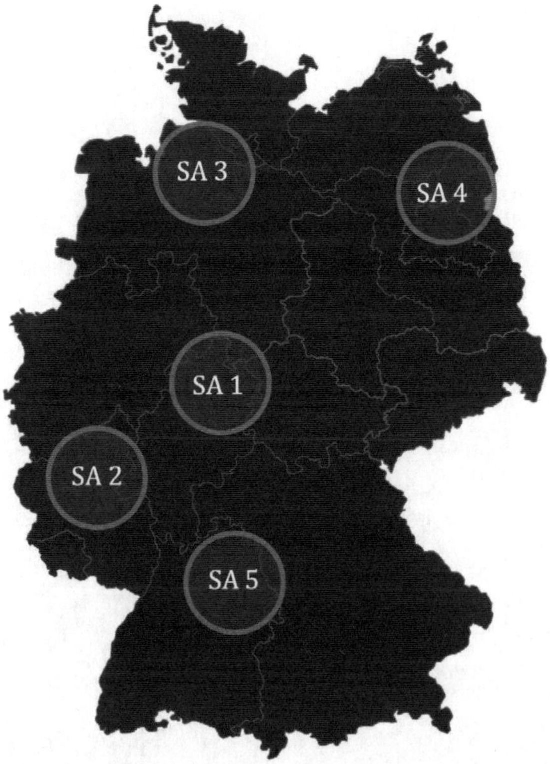

Source: own presentation, deBakom GmbH

Figure 1: Locations of the study areas (SA) in Germany

The individual sites differed in the following characteristics:

a) Number of wind turbines (1 to 21 wind turbines)
b) Wind turbine type (four different manufacturers with a total of six different models)
c) Hub height of the wind turbines (approx. 100 m to approx. 140 m)
d) Power range of the wind turbines (2 MW to 3 MW)
e) Rotor diameter (approx. 80 m to approx. 135 m)
f) Topographic position (flat to hilly terrain)
g) Distance of the measuring equipment to the wind turbine for immission measurements (approx. 800 m to 1500 m)
h) Season of the measurement (spring to winter)

2.2 Measurement set up

For the measurements, deBAKOM long-term measurement systems were used, with calibrated Class 1 microphones sensor for meteorological data at several heights (Figure 2). The microphones were placed on a mast, to get measurements in agreement with German environmental noise regulations (BImSchG, TA Lärm 1998, FGW 2021, DIN 2006). To reduce the wind noise at the external microphone, a secondary wind screen was used. The measurement data were continuously recorded and stored.

Source: own Presentation, deBakom GmbH

Figure 2: Measurement set-up in the immission range of the wind turbines

3. Measurement and evaluation of amplitude modulation

Noise caused by wind turbines, which is often perceived as a "whoosh," is commonly referred to amplitude modulation, which is a periodic rise and fall in sound pressure level. It is known, however, that noise generated by wind turbines is also subject to other temporally irregular fluctuations, which are produced among other things by propagation, wind or interference. These fluctuations are perceived by residents but are not usually described as "whooshing" and are usually not directly related to rotation frequency. Here, the term amplitude modulation is used for level fluctuations in connection with the rotational frequency. Figure 3 shows an example of a rapid rise and fall of the level in a 1.2 s cycle.

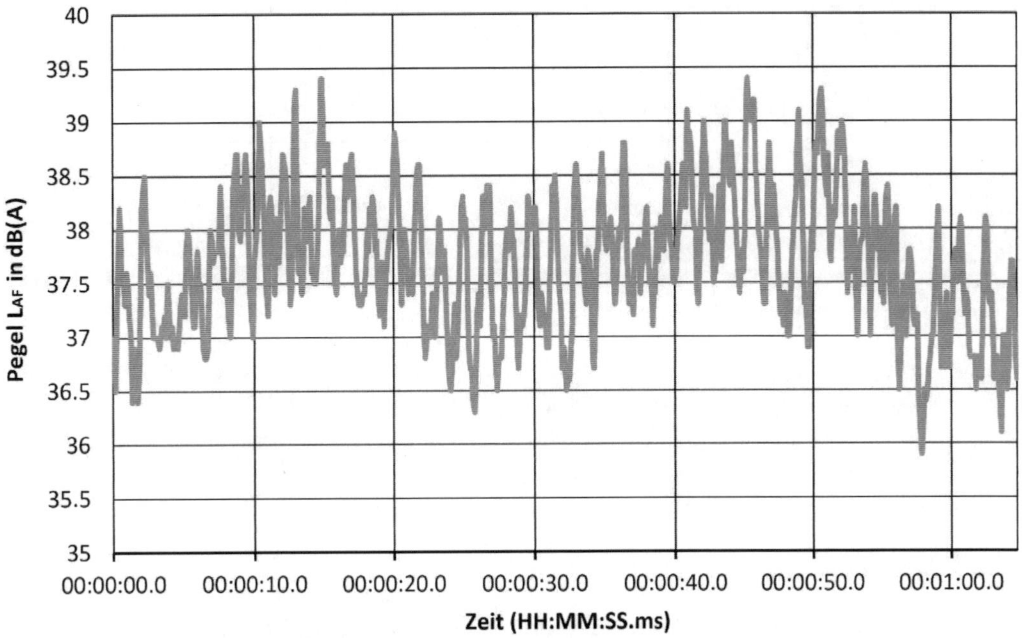

Source: own presentation, Dr. Kühner GmbH

Figure 3: Sample fluctuation in volume levels due to amplitude modulation

Pegel L_{AF} in dB(A) = Level L_{AF} in dB(A)
Zeit (HH:MM:SS.ms) = Time (HH:MM:SS.ms)

3.1 Quantisation of amplitude modulations

The measured data were examined for the occurrence of amplitude modulation. For this purpose, an algorithm was designed to determine the depth and frequency of the modulations. At 10 s intervals, for a window of the sound pressure level series, a frequency analysis is performed to find the frequency with the maximum modulation. To make sure that this modulation frequency matches modulations from a wind turbine, the fact that the rotational speed of wind turbines can only change slowly, is used to filter out modulations that cannot be due to a wind turbine. Figure 4 shows a time series comparing the frequency of modulations expected from the logged rotational speed from a wind turbine with the frequency modulations determined from the recorded sound pressure level modulations. Even if the rotational speed of the wind turbine is not known, the series of modulation frequencies can be used to consistently determine the rotational speed of the wind turbine causing the modulations and filter out modulations from other sources.

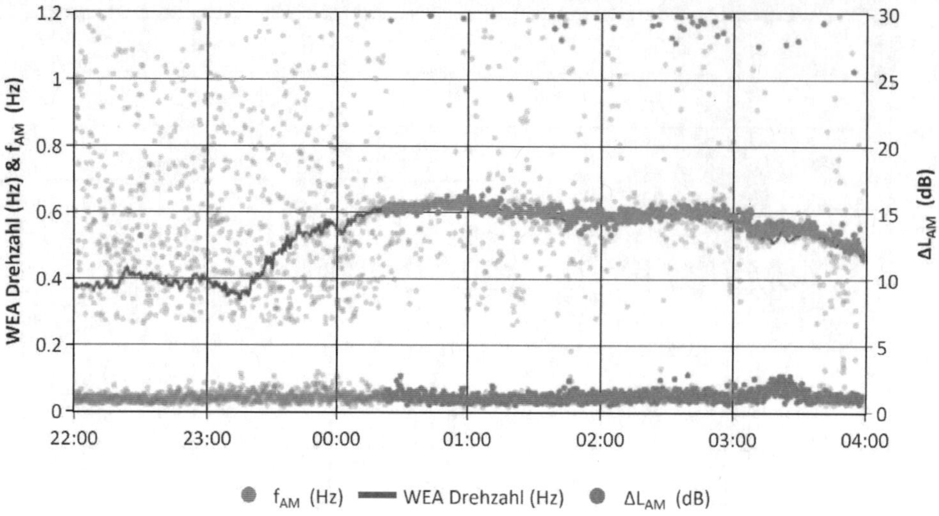

Source: own presentation, deBAKOM GmbH

Figure 4: Frequency of maximum modulation for 10s windows vs time

ΔL_{AM} = modulation depth; f_{AM} = modulation frequency

WEA Drehzahl (Hz) & f_{AM} (Hz) = Wind turbine rotational speed (Hz) & f_{AM} (Hz))
WEA Drehzahl (Hz) = Wind turbine rotational speed (Hz)

3.2 Extraneous noise

When evaluating the measurement data, care was taken to ensure that only the data in which no disturbing wind, rain or extraneous noise occurred was used.
Some examples of extraneous noises that had to be filtered out are:

- mooing cows, singing birds
- cars and aircraft

3.3 local neighbourhood soundscrickets (in one measurement campaign)Results for modulation depths

Excluding all times with extraneous noise, table 1 shows the frequency of amplitude modulations occurring in the measurement locations.

Table 1: Frequency and depth of amplitude modulations in residential areas at night

Measurement location	Frequency of occurrence of AM in %	ΔL_{AM95} in dB	ΔL_{AM50} in dB	ΔL_{AM05} in dB
Immission range SA 1	10.8	1.1	2.0	4.2
Immission range SA 2	47.4	1.3	2.4	4.7
Immission range SA 3	1.7	0.6	1.4	5.5
Immission range SA 4	42.0	0.9	1.5	3.3
Immission range SA 5	22.3	0.8	1.6	2.9

It was found that the modulation depth ΔL_{AM} for all locations and over all power ranges of the wind turbines was on average approx. 1.5 dB to 2.5 dB (Figure 5).

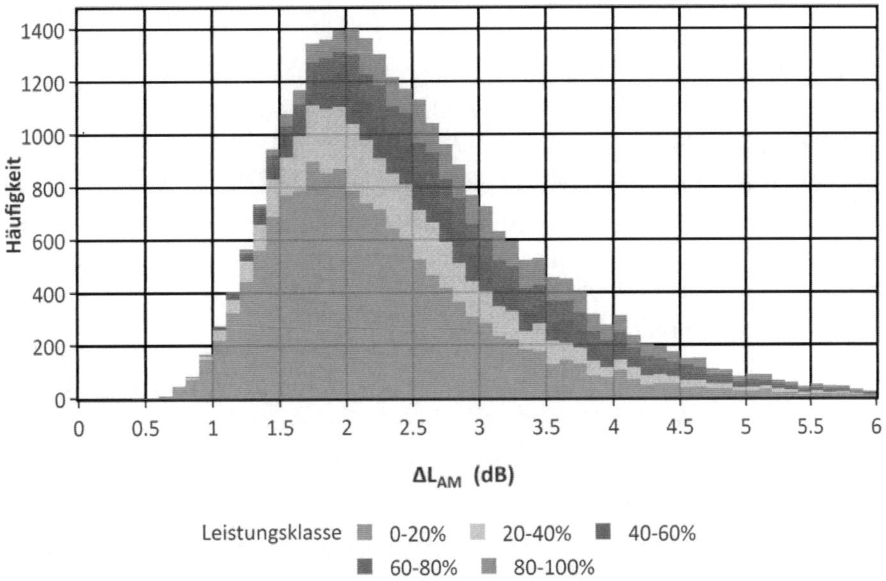

Source: own presentation, deBAKOM GmbH

Figure 5: Frequency distribution of modulation depth ΔL_{AM} in study area 2, classified by turbine power output

Häufigkeit = Frequency
Leistungsklasse = Power class

The evaluation of the measurement data showed that higher modulation depths occurred at sites with few wind turbines (1 WT, 3 WT).

3.4 Metrological dependences

For an investigation of the meteorological dependence of the occurring amplitude modulation, the measured data were grouped according to wind direction and electrical power of the wind turbines. When analyzing the data, a correlation was only recognizable for a single wind turbine. Here, the modulation depth ΔL_{AM} in the immission area increased by 1.2 dB with crosswind and increasing power. In the case of a downwind, the modulation depth increases by only 0.6 dB. In the study areas with several wind turbines this trend did not show up, so that no clear correlations between power and wind direction and modulation depth can be modulation depth could be identified.

4. Annoyance surveys

To better understand the degree to which noise from wind turbines contributes to annoyance, surveys were carried out with residents in the same areas as the measurement campaigns. The study consisted of a main survey by telephone or online, and additional in-depth interviews. The main survey was carried out with 468 persons from all five study areas, all living within 4 km of the wind farms in the study areas. Noise annoyance due to wind turbines was assessed by means of the verbal 5-point scale with 1 (not at all) to 5 (extremely annoyed or disturbed) as recommended by ISO/TS15666 (ISO 2021).The annoyance was rated as low on average, with a weak statistical correlation between rated noise level and annoyance. The relatively low level of noise annoyance due to wind turbines is apparently due to low noise exposure for wind turbines in relation to other noise sources in the study areas. However, a significant increase in the percentage of respondents indicating a high level of annoyance is observed as soon as a noise rating level L_r of approx. 35 dB is exceeded. If factors other than the noise rating level are considered, the direct effect of the noise rating level on the percentage of highly annoyed (% HA) is mitigated (Figure 6). Of these factors, the following have the greatest impact on annoyance of wind turbine noise:

- Attitude towards the local wind turbine
- Visual impact due to shadow casting and rotation of the rotors, blinking aviation-obstruction lighting, the sight of wind turbines in general and the perceived negative view of the impact wind turbines on the landscape.
- Perceived sound characteristics, such as 'whooshing'

In addition, respondents explicitly regarded the perception of the sound characteristics of wind turbines as whooshing as one of the most annoying characteristics of the wind turbine sound. As whooshing is often understood as the subjective perception of amplitude modulations (AM), this indicates a specific effect of the AM on annoyance due to wind turbine noise. This is confirmed by the result of the study that the number of periodic AM in the study areas corresponds with the average degree of annoyance due to wind turbine noise in these areas, i.e., in those areas with a higher number of periodic AM the average noise annoyance is higher.

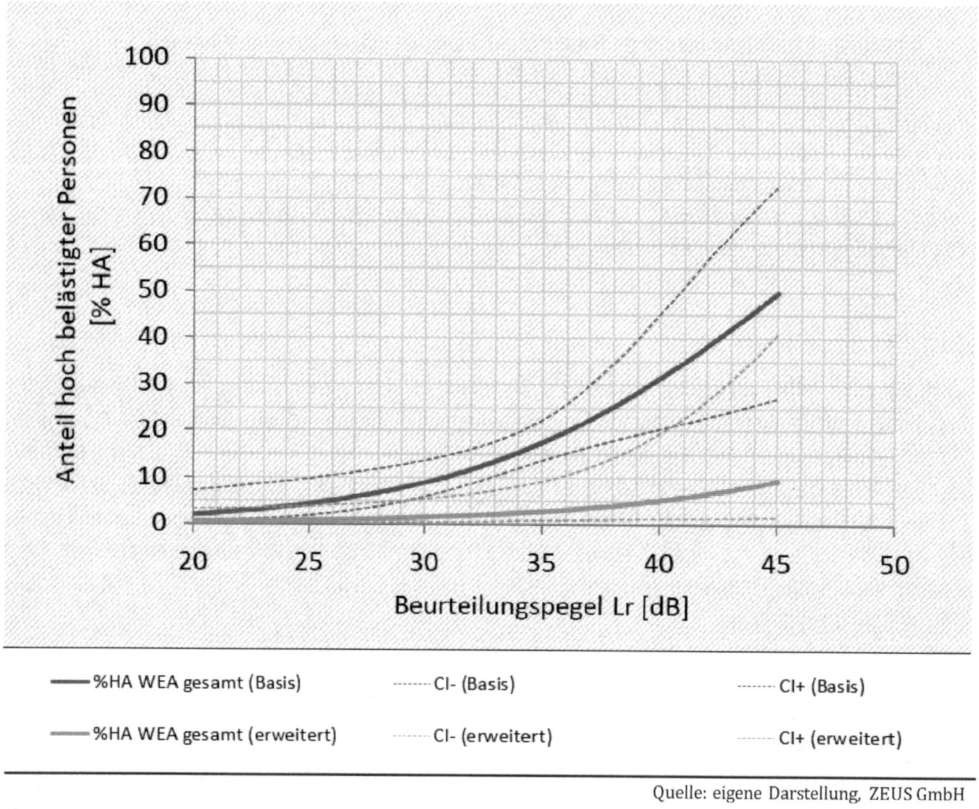

Source: own presentation, ZEUS GmbH

Figure 6: Percentage of persons who are highly annoyed (% HA) by wind turbine noise, total

% HA = % highly annoyed; WT = wind turbine; CI-/+ = lower/upper limit of the confidence interval of the exposure-response curve; Basis: Influencing factor noise rating level L_r unadjusted; Extended: Influencing factors noise rating level L_r, noise sensitivity, attitude towards wind turbines, perceived stress, visual impact of wind turbines, sound characteristics

Anteil hoch belästigter Personen [%HA]	= Percentage of highly annoyed persons [%HA]
Beurteilungspegel Lr [dB]	= Noise rating level Lr [dB]
%HA WEA gesamt (Basis)	= %HA WT total (basic)
%HA WEA gesamt (erweitert)	= %HA WT total (extended)
CI-/+ (basis)	= CI-/+ (basic)
CI-/+ (erweitert)	= CI-/+ (extended)

5. Performance and results of listening experiments

We conducted the listening experiments at three of the five study areas with subjects who had already participated in the surveys. In addition, we conducted the experiments at the TH Köln with a non-exposed group of participants, most of whom were students or research assistants. In total 79 persons participated in the experiments. We assessed annoyance according to ISO/TS 15666 (ISO 2021) using the ICBEN 11-point scale ranging from "not at all annoying" to "extremely annoying" which was tested and found to be suitable for investigating the annoyance of WTs in a laboratory study (Schäffer, B et al. 2016). During the listening experiments, we presented signals with different modulation depths of AM, which were extracted from the audio recordings of two measurement locations and presented with different immission levels. Furthermore, the listening experiments included recordings of time-constant AM and time-varying AM.

As the main result, we found a strong effect of the level on annoyance, which did not vary much between the different groups of subjects in the different study areas, and even the listening experiments with the control group showed only slightly different results (Figure 7). The listening experiments also revealed that annoyance caused by AMs increases rapidly as soon as the AMs become audible. The strength of the AM has a much weaker effect on annoyance than the mere audibility of the AM. In contrast, the relationship between annoyance and level is much easier to describe, as annoyance increases almost uniformly with level. Furthermore, time-varying stimuli showed no significant variation in perceived annoyance between stimuli with increasing or decreasing AM over time.

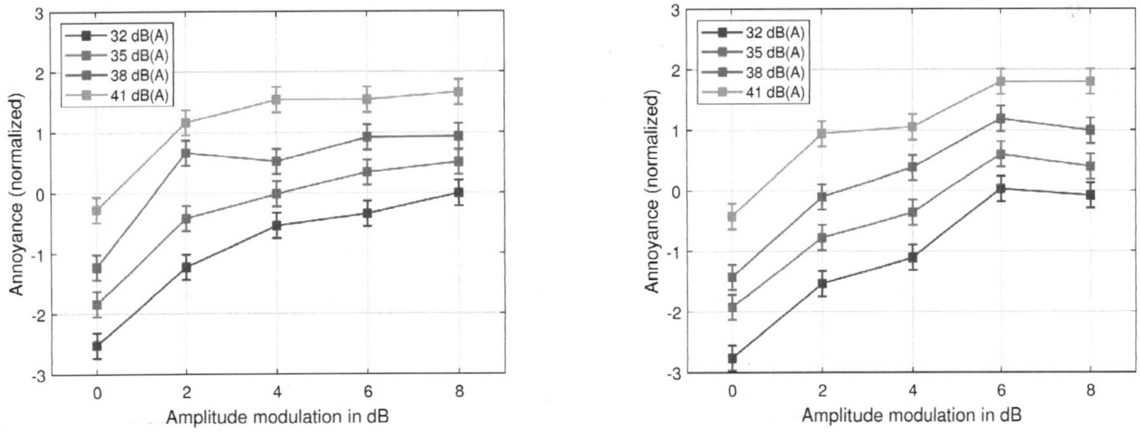

Source: own presentation, TH Köln

Figure 7: Normalized annoyance as a function of AM (x-axis), immission level (colour), and measurement location. Shown are the normalized annoyance values averaged over subjects and 95%

within-subject confidence intervals based on the error term of the AM main effect.

6. Conclusions

In summary, it can be stated that amplitude modulation occurred at all immission measuring points investigated. However, the frequency varied strongly between the locations (between 1.7 % and 50 % of the evaluable measuring times) and was considerably lower than at the emission measuring points. It could be determined that parameters such as wind farm constellation, turbine type, distance of the immission measuring point to the WT, topography or season influence the frequency. A clear correlation could not be determined.

The evaluation of the measurement data over all study areas showed that the modulation depth was in the median range of 1.5 dB to 2.5 dB at both the immission and emission measurement points. There was a tendency for a higher modulation depth to be present when the number of turbines was low and the distance between the wind turbine and the immission measurement point was relatively small.

Both the annoyance survey and the listening experiments revealed that amplitude modulation can increase noise annoyance to residents.

In conclusion, it can be stated that the findings so far show tendencies with regards to the occurrence of amplitude modulations, especially at greater distances, and their contribution to noise annoyance for residents, but statistically significant statements can only be made on the basis of further investigations.

References

BImSchG *Bundes-Immissionsschutzgesetz* (Federal immission Control Act) - Act on the Protection against Harmful Effects on the Environment Caused by Air Pollution, Noise, Vibrations and Similar in the version promulgated on 17 May 2013 (BGBl. I p. 1274), as last amended by Article 2 paragraph 3 of the Act of 19 October 2022 (BGBI I p.1792).

4. BImSchV *Vierte Verordnung zur Durchführung des Bundes-Immissionsschutzgesetzes* (Fourth Ordinance on the Implementation of the Federal Immission Control Act), Ordinance on installations requiring a permit in the version published on May 31, 2017 (BGBl. I p. 1440), as amended by Article 1 of the Ordinance of 12 October 2022 (BGBl. I p. 1799).

DIN – Deutsches Institut für Normung (2019) *DIN EN 61400-11*; VDE 0127-11:2019-05, Windenergieanlagen - Teil 11: Schallmessverfahren (IEC 61400-11:2012 + A1:2018) Deutsche Fassung EN 61400-11:2013 + A1:2018 (Wind turbines - Part 11: Acoustic noise measurement techniques (IEC 61400-11:2012 + A1:2018); German version EN 61400-11:2013 + A1:2018)

FGW (2021) *Richtlinien für Windenergieanlagen - Teil 1: Bestimmung der Schallemissionswerte* (Technical Guidelines for wind turbines, part 1: determination of noise emission), Revison 19, last amended on 1 March 2021, Fördergesellschaft Windenergie und andere Dezentrale Energien, Berlin

ISO (2021) *ISO/TS 15666:2021*. Acoustics - Assessment of noise annoyance by means of social and socio-acoustic surveys. Geneva; Switzerland

Schäffer, B et al.(2016) *Short-term annoyance reactions to stationary and time-varying wind turbine and road traffic noise: A laboratory study*, The Journal of the Acoustical Society of America, vol. 139, no. 5, pp. 2949-2963, 2016

Schmitter, S; Alaimo Di Loro, A; Hemmer, D et al. (2022) *Noise effects of the use of land-based wind energy*, Umweltbundesamt, Texte 69/2022, Dessau-Roßlau,
Noise effects of the use of land-based wind energy | Umweltbundesamt

TA Lärm (1998) *Sechste Allgemeine Verwaltungsvorschrift zum Bundes-Immissionsschutzgesetz, Technische Anleitung zum Schutz gegen Lärm – TA Lärm* (Sixth General Administrative Provision to the Federal Immission Control Act, Technical Instructions on Noise Abatement - TI Noise) of 26 August 1998 (Joint Ministerial Gazette (GMBl.) No. 26/1998 p. 503), last amended on 1 June 2017 (BAnz AT 8 June 2017 B5)

10th International Conference
on
Wind Turbine Noise
Dublin – 21st to 23rd June 2023

Wind turbine noise code benchmark: A comparison and verification exercise

Franck Bertagnolio¶, Andreas Fischer
DTU Wind and Energy Systems, Roskilde, Denmark

Christina Appel, Michaela Herr
DLR, Institut für Aerodynamik und Strömungstechnik, Braunschweig, Germany

Ferdinand Seel, Thorsten Lutz
Institut für Aerodynamik und Gasdynamik, Universität Stuttgart, Stuttgart, Germany

Koen Boorsma, Gerard Schepers, Miguel Restrepo Botero
Wind Energy, TNO Energy Transition, Petten, The Netherlands

Damiano Casalino, Wouter van der Velden
Dassault Systèmes Deutschland GmbH, Stuttgart, Germany

Carlo R. Sucameli
Lehrstuhl für Windenergie, Technische Universität München, München, Germany

Pietro Bortolotti
National Wind Technology Center, National Renewable Energy Laboratory, Golden, CO, USA

Summary

In a number of institutions and companies, researchers and engineers are developing numerical models and frameworks that are used to predict the aerodynamic noise emissions from wind turbine rotors. The simulation codes range from empirically tuned engineering models to high-fidelity computational ones. Their common feature is the fact that they all specifically model the main aerodynamic noise mechanisms occurring at the rotating blades (namely, the turbulent boundary layer): trailing-edge and turbulent inflow noise. Nevertheless, different modelling techniques and implementations may generate different results, even when assessed on the same rotor design and operating conditions, which raises the question of the actual fidelity and reliability of these

¶Corresponding author: frba@dtu.dk

models. Trailing-edge noise is put at the forefront of the present study, as it is recognized to be the main source of audible noise from modern wind turbines.

The present benchmark aims at comparing the results from different modelling approaches and drawing some conclusions from these comparisons. This effort, denoted as Wind Turbine Noise Code benchmark, was initiated in 2019 as a joint activity between the IEA Wind Task 39 (Quiet Wind Turbine Technology) and Task 29 (Detailed Aerodynamics of Wind Turbines, now Task 47).

In addition to the investigation of the noise emissions themselves, the rotor aerodynamic characteristics are investigated, as they are the source of the noise generation mechanisms discussed herein.

A number of test cases are defined, and the aerodynamic and aeroacoustic predictions from the various models are compared. A fair agreement between the aerodynamic predictions is observed. There exist some discrepancies between the different noise prediction methods, but it is difficult to conclude if one methodology is better than another in order to design a wind turbine with noise as a constraint.

1. Introduction

It is a well-accepted fact that trailing-edge (TE) noise is the prominent source of aerodynamic broadband noise from wind turbines in the audible range [35]. Therefore, it is important for the wind industry to assess and subsequently mitigate (e.g. using serration) this particular source of noise in order to reduce the environmental impact of wind turbines and wind farms. Aerodynamic noise sources also include turbulent inflow (TI) noise, which is normally more dominant at lower frequencies than TE noise (at least for modern multimegawatt wind turbines), but it can also be audible.

The present work aims at comparing various simulation methods for predicting and quantifying these two main aerodynamic noise sources from wind turbines. Note that other noise sources such as mechanical/tonal noise, low-frequency tower-blade interaction, tip noise, etc., are not considered in the present study, although these can have a significant impact on the acoustic footprint of a wind turbine. In addition, atmospheric propagation effects (such as reflection, refraction, diffraction, and air absorption) are also neglected, despite their potential impact on the perceived noise at dwellings.

This work is conducted as part of the IEA Wind Technology Collaboration Programme. Various institutions from participating countries have contributed to the present comparisons by using their own simulation framework that can model wind turbine aerodynamic noise emissions. The goal is to compare the different methodologies and analyze the consistency (or the lack thereof) of the results when simulating the same rotor in the same operating conditions.

In the following, the context and objectives of the present study are discussed. The various modelling strategies that are used for the comparisons are reviewed. The first part of the study concentrates on the comparisons of the aerodynamic quantities that are essential for the prediction of wind turbine aerodynamically generated noise. Then, the actual noise predictions are considered, focusing on the relationship between the aerodynamic and acoustic results. The study is concluded with comparisons of some of the model results with actual field noise measurement data from a wind turbine.

2. Context and objectives

TE and TI noise are the respective results of the interaction of the airfoil boundary layer (BL) and atmospheric turbulence with the blades. A variety of numerical methods have been derived to model these phenomena, ranging from relatively simple empirical formulae to high-end computationally expensive simulation codes. In a long-term effort, various models for TE noise were investigated at the airfoil level in a series of comparison rounds as part of the Benchmark Problems for Airframe Noise Computation (BANC) [20, 21]. The present study attempts to compare a number of these models when considering a full wind turbine rotor, identify some potential pitfalls in this context, and possibly improve the use and prediction results of these methods in the future. For example, higher-fidelity models could be used to tune or improve lower-fidelity ones, which are more suited to the constraint of a rapid turnaround time typical of industrial design.

The first objective is to make sure that the underlying aerodynamic simulations of the rotor flow are sufficiently close to each other, so that the impact on noise predictions related to possible discrepancies in the aerodynamic input data is minimized. Therefore, the first part of the study concentrates on rotor aerodynamic characteristics.

The second objective is to compare acoustic results. An analysis is conducted in an attempt to identify 1) the reasons for discrepancies between similar methodologies if/when such discrepancies are observed and 2) trends between different modelling approaches, e.g. empirical vs. high-fidelity models.

3. Computational methods

The various computational frameworks used in the present article are described in this section. A rough categorization of the different methodologies is introduced here.

The *first* step in the prediction of aerodynamic noise from a turbine usually consists of calculating the aerodynamic flow field around the turbine's rotor. Two main methodologies can be applied here:

- The most popular engineering method for predicting a wind turbine rotor flow aerodynamic is the blade element momentum (BEM) method, originally derived by Glauert [18], which is based on mass and momentum conservation principles.
- The second option is to numerically solve the associated conservation equations (here, Navier-Stokes or Euler) using computational fluid dynamics (CFD). This is usually much more computationally expensive.

The *second* step consists of defining the TE noise modelling approach. Note that the prediction of this noise source from wind turbines requires detailed boundary layer characteristics along the blades, which are normally not provided by BEM methods.

Three approaches are generally adopted:

- *Empirical modelling:* in all cases, this amounts to using the well-known Brooks Pope Marcolini (BPM) model [10]. Note that this model can include various aerodynamic noise sources (e.g. tip noise, blunt TE noise), but only TE noise is considered here. The model is based on theoretical work for the scaling of TE noise and empirical fitting using a series of experiments on the NACA0012, during which aerodynamic and acoustic properties were measured.

- *Semi-empirical modelling:* the models are extensions of the original model named TNO * developed by Parchen [36]. The TNO TE noise model and its revised versions are a combination of Kraichnan theory for BL turbulence, including various assumptions for characterizing the turbulence, and a scattering model for the TE noise prediction using either Howe or Amiet theory. A flow solver (CFD or Xfoil) is typically used for determining the aerodynamic and turbulent flow inputs to the overall model. These methodologies will be denoted as semi-empirical or TNO-type models in the following.
- *High-fidelity modelling:* the models are based on high-performance computing for solving the main rotor flow field and the acoustic field, either jointly or separately.

In addition, each of the above methods uses a flow solver to compute the aerodynamics around the blades, which are in most cases used as inputs for the above noise models (except when the aerodynamic and acoustic calculations are coupled, e.g. for the Lattice-Boltzmann Method).

Furthermore, as far as TI noise modelling is concerned, the frameworks used by the participants of the present comparison exercise are all implementations of the Amiet TI model [1, 37]. Two main versions can be distinguished here. The first one is the complete model implementation that involves the computation of the unsteady lift from a flat plate. The second is based on its asymptotic approximation for higher frequencies. Note that a simpler version using Lowson's method can also be used for TI noise modelling [30]. Nevertheless, more elaborate modelling methods are available for predicting TI noise from wind turbines [26].

The various numerical frameworks from the different participating institutions are summarized below. For further details about these frameworks, the reader is referred to the IEA Wind Technology Collaboration Programme website and the report specifically related to the present work [4].

3.1. TNO - SILANT

The aeroacoustic calculation of TNO is divided into three programs: Blademode [7], RFOIL [28], and SILANT[33].

BladeMode is an in-house aeroelastic blade stability software using the BEM theory. It is used in a quasi-steady configuration for the present application. The resulting sectional angle of attack and Reynolds number distribution along the blade span are then used as input to the SILANT model. This program includes the noise calculation from turbulent TE noise and tip noise based on the BPM model [10] and TI noise using the model of Amiet [1] and Lowson [29]. The RFOIL2D panel code with interacting BL is used to provide the boundary layer displacement thickness at the TE of the airfoil sections along the blade. The data are stored as a look-up table for the SILANT model.

The resulting sectional noise source strengths are acoustically summed over the blades and rotor. In addition to calculating noise sources, SILANT can also include Doppler effects (and additional effects related to atmospheric propagation which are ignored in the present work).

3.2. NREL - OpenFAST

OpenFAST is a popular multi-physics solver developed and released by the National Renewable Energy Laboratory. OpenFAST integrates an aeroacoustic model that is described in Bortolotti et al. [8]. The model implements a conventional turbulent inflow model from Amiet [1], with the optional

*Note that the designation of the so-called TNO TE noise model originates from the institute where its conceptor worked at the time. It is the same TNO institute at which two of the authors of the present article are working. In order to avoid confusion, it must be made clear that these two authors use a different TE noise model in their computational framework, but that both models will be referred to as TNO in the figure captions.

correction defined by Moriarty et al. [34]. The model also implements the noise sources defined by Brooks et al. [10]. The models implemented in OpenFAST were subjected to a validation study operating a GE 1.5 MW wind turbine. The results are discussed in Bortolotti et al. [9] and Hamilton et al. [19].

3.3. TUM - Cp-Max AAM

This framework is the one described in [41]. It is based on the in-house-developed aeroservoelastic wind turbine solver Cp-Lambda, which implements a BEM formulation and provides the aerodynamic inputs necessary for the aeroacoustic calculations.

Several aeroacoustic models are implemented within this framework. For the purpose of the present paper, TE noise results are provided for two different models (the BPM model [10] and a version of the TNO model described in [41]). For both models, 2D boundary layer characteristics are obtained through XFoil. TI noise spectra are provided for two different formulations of the Amiet model. The first formulation is the full implementation of [37], while the second one corresponds to the approximations of the Amiet model for high and low frequencies. An additional low-frequency correction is included, as shown in [30].

3.4. IAG Stuttgart - IAGNoise+

The Institute of Aerodynamics and Gas Dynamics (University of Stuttgart, Germany) uses the IAGNoise+ noise prediction code. This semi-empirical model computes the generated TE noise based on 3D flow solutions from CFD simulations. In this work, Reynolds-averaged Navier Stokes (RANS) simulations using a k-ω SST turbulence model were run with the flow solver FLOWer.

IAGNoise+ employs a TNO-Blake-type model for the computation of TE noise [5]. Compared to a classical TNO-type model, the current implementation [22] includes the part of the wall pressure fluctuation source term that is associated with turbulence-turbulence interaction and usually neglected in the basic model. This inclusion allows for more accurate predictions at higher angles of attack, where slight to moderate flow separation occurs. Additionally, the anisotropy factor was adjusted to also include adverse pressure gradient effects. The IAGNoise+ prediction tool also offers a way to calculate inflow noise, based on the model proposed by Paterson and Amiet [38] with Moriarty's thickness correction [34].

3.5. DTU - HAWC2-Noise

This framework uses the HAWC2 code [27] as a basis. It is a time-domain multibody aeroelastic code used for the study and design of wind turbines. The blade element momentum theory by Glauert [18] is applied in order to calculate the aerodynamic loading [32].

The aerodynamic data are used as inputs to an acoustic module that can account for TI noise using the Amiet model [1], and TE noise using a version of the TNO model [17] for which scattering is accounted for using the Amiet model [2]. Both noise model formulations are in the spectral domain. Therefore, it is assumed that that the acoustic emissions are quasi-stationary (at each time step of the aeroelastic solver), and spectrograms can be obtained for each of the noise sources. The detailed aerodynamic characteristics of the turbulent BL, which are used as input to the TE noise model, are computed as a preprocessing step with the 2D RANS solver EllipSys2D at each discrete section along the blades.

3.6. 3DS wind turbine multi-fidelity approach

Wind turbine aerodynamic and acoustic calculations have been performed using the multi-fidelity framework *Opty∂B*-WTNOISE® [12, 42]. Three approaches have been used: one is based on a blade element momentum theory (BEMT) rotor aerodynamic calculation, and the other two rely on lattice Boltzmann method very large eddy simulation (LBM-VLES) scale-resolved transient flow simulations.

3.6.1. BEM-based methodology

The BEMT tool uses BEM theory with uniform inflow and tip-loss correction [11], and a viscous panel method available in *Opty∂B*-BEMT is used for defining the boundary layer flow on the blades [13, 14]. Wall pressure spectra are computed with semi-empirical formulations. On the suction side, a model is used, obtained by blending Schlinker's [40] model at low frequency with Kamruzzaman's [25] model at high frequency, and by recalibrating the overall energy to the Schlinker model value. On the pressure side, the Schlinker model is used. The Schlinker and Amiet model is used for TI noise [1].

3.6.2. 2.5D LBM/FW-H-based methodology

PowerFLOW® 2.5D simulations are performed by means of a fully automatic workflow fed with sectional coordinate profiles generated by *Opty∂B*-PFROTOR, and values of Mach number and angle of attack computed by *Opty∂B*-BEMT [12].

Simulations are carried out on extruded blade sections of fixed span of 0.1 m. For every radial strip selected by the user from the available blade segmentation, the PowerFLOW simulation generates a transient wall pressure file which is used by the frequency-domain FW-H solver *Opty∂B*-FWHFREQ executed by *Opty∂B*-WTNOISE. Full-blade noise spectra are recovered by *Opty∂B*-WTNOISE via an incoherent summation of sectional noise spectra, scaled by the ratio of the physical spanwise extension of the blade strip and the 2.5D simulated span.

3.6.3. 3D LBM/FW-H-based methodology

PowerFLOW 3D simulations are performed by means of a fully automatic workflow used for multicopter eVTOL, rotorcraft, fan, and wind turbine applications [12]. A series of simulations are carried out with mesh refinement in different blade strips where the turbulent scales are trigged by a trip. Similar to the 2.5D approach, the full turbine noise levels are recovered by incoherent summation of the individual strip contributions.

3.7. DLR - hybrid RANS-based CAA method PIANO/FRPM

An automatized 2D process chain for turbulent boundary layer trailing edge noise (TBL-TEN) [16, 39] is used to provide an acoustic prediction for trailing-edge noise of 2D profiles. Originally developed to assist low-noise airfoil design optimization, this method has been validated in detail within the BANC framework [20, 21]. The process chain operates via bash scripting the input parameters (like airfoil geometry, Reynolds number, angle of attack, chord length and process parameters, e.g. number of iterations, simulated real time and post processing options). The CFD code TAU, which is developed at the German Aerospace Center (DLR), is applied for the RANS simulations, and the DLR computational aeroacousics (CAA) code PIANO with the stochastic sound source model FRPM [15] (**F**ast **R**andom **P**article **M**esh method) is applied for the acoustic prediction.

In a second step, the results from the process chain are combined with DLR's TAP (**T**urbine **A**coustic **P**rediction) tool to extrapolate and summarize the data for a complete rotor [3]. Ongoing work includes the successive extension of TAP by additional semi-empirical source models for flow separation and TI noise. TI noise predictions applied herein are based on Hornung et al. [23].

4. Test case definitions and physical inputs

All the calculations presented in this article are based on the 2.3 MW wind turbine NM80. The use of the NM80 turbine geometry has been granted to the participants of Task 39 for the present study. This turbine was initially investigated as part of the DANAERO project [31]. It was further used as a reference turbine for the aerodynamic benchmark that was conducted as part of IEA Wind Task 29 (now Task 47) [6]. Some details of the turbine geometry can be found in the latter publications. Four test cases were defined for the present study, although only one of them will be considered in the present article. The main operational conditions of interest are the following:

- Test Case 1.1: Axisymmetric configuration (i.e. no rotor tilt), rigid structure, a wind speed of 6.1 m/s, turbulence intensity of 8.96%, and a rotor speed of 12.3 rpm. Additional information such as atmospheric conditions, blade pitch, etc., are also specified.

For the rotor noise calculations, a number of observer positions are defined. Twelve positions are defined on the ground around the turbine, equally distributed on a circle with a radius equal to the maximum height of the turbine (i.e. tower height plus half-rotor diameter), as depicted in Fig. 1. In addition, a single point is located at the same distance but on the rotor axis in the downstream direction. Note that in all noise calculations, atmospheric propagation effects and ground reflections are discarded, but the geometrical spreading is accounted for.

In addition, results from a noise measurement campaign conducted on a megawatt-size turbine will be considered.

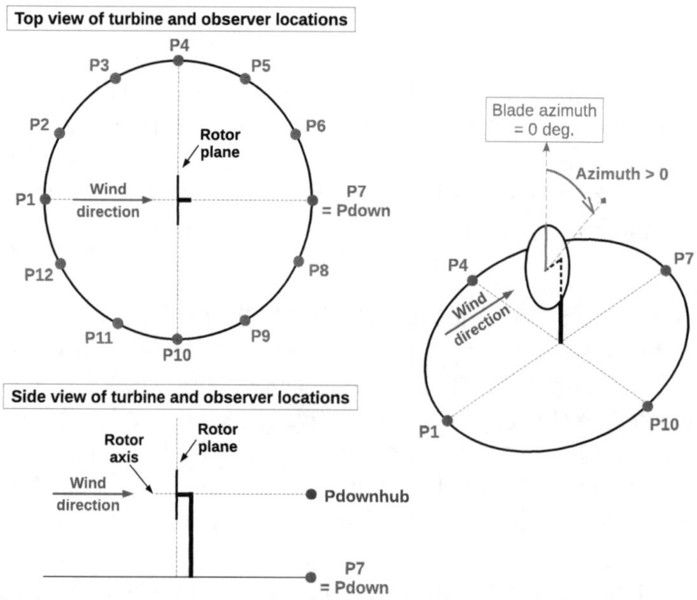

Fig. 1 Sketch of the observer locations around the turbine for the noise calculation results.

5. Comparison of aerodynamic results

As mentioned earlier, the aeroacoustic emissions of a wind turbine are highly dependent on the atmospheric inflow and resulting flow on the blades, which is computed using BEM theory or CFD in this work. Therefore, the first step for comparing numerical frameworks is comparing the aerodynamic data along the blades. Three spanwise locations along the blades were chosen for the comparisons: $r = 19\,\text{m}, 30\,\text{m}$ and $37\,\text{m}$ from the root of the blade.

Note that since both TI and TE noise are scaling with the Mach number (to a specific power depending on the mechanism), it is well-known that toward the blade tip, as the effective velocity becomes higher, the aerodynamic noise emissions increase. Consequently, this study focuses on BL characteristics on the outer part of the blades.

5.1. Incoming flow

The relative and effective (i.e. including rotor induction) inflow velocities, angles of attack, and lift and drag coefficients at the three spanwise locations are displayed in Fig. 2. The agreement between the inflow velocities is nearly perfect, which is consistent with the imposed rotor speed of the test case. Some discrepancies are observed between the calculated angles of attack, but these remain relatively small, within less than 1 deg, and these appear to become even smaller toward the tip of the blade. The lift coefficients present very small discrepancies as well, but the drag coefficients do depart more significantly.

Overall, all methods deliver similar results in terms of the aerodynamic loading on the turbine.

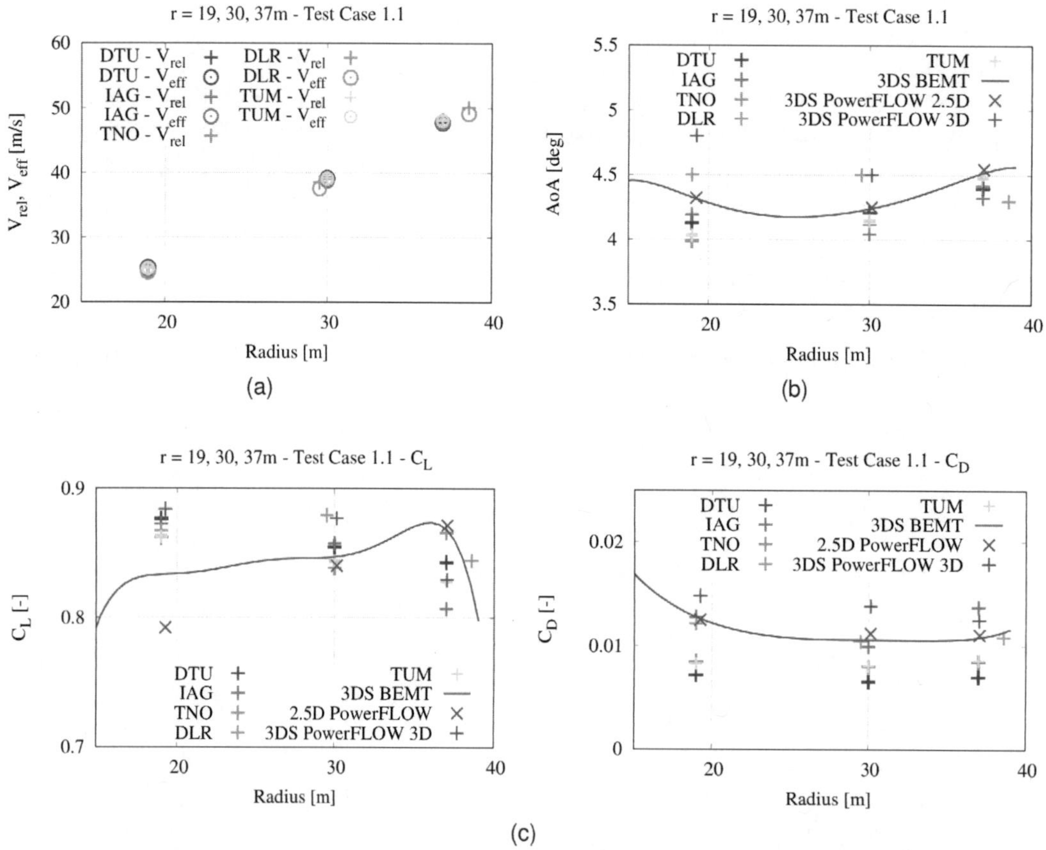

Fig. 2 Aerodynamic quantities of the incoming flow along the blade span: (a) relative and effective inflow velocity, (b) angle of attack, and (c) lift and drag coefficients.

5.2. Boundary layer thicknesses and profiles near the TE

It is well-known that the turbulent BL characteristics near the TE have a large impact on the TE noise emissions. These characteristics are investigated in the present section.

The BL thickness δ, BL displacement thickness δ^*, and BL momentum thickness θ are displayed in Fig. 3 for the suction and pressure sides. It can be observed that there is relatively good agreement between all methods, and that the discrepancies appear to be getting smaller toward the tip of the airfoil, which should contribute to a better convergence of the aerodynamic noise model results in the following sections.

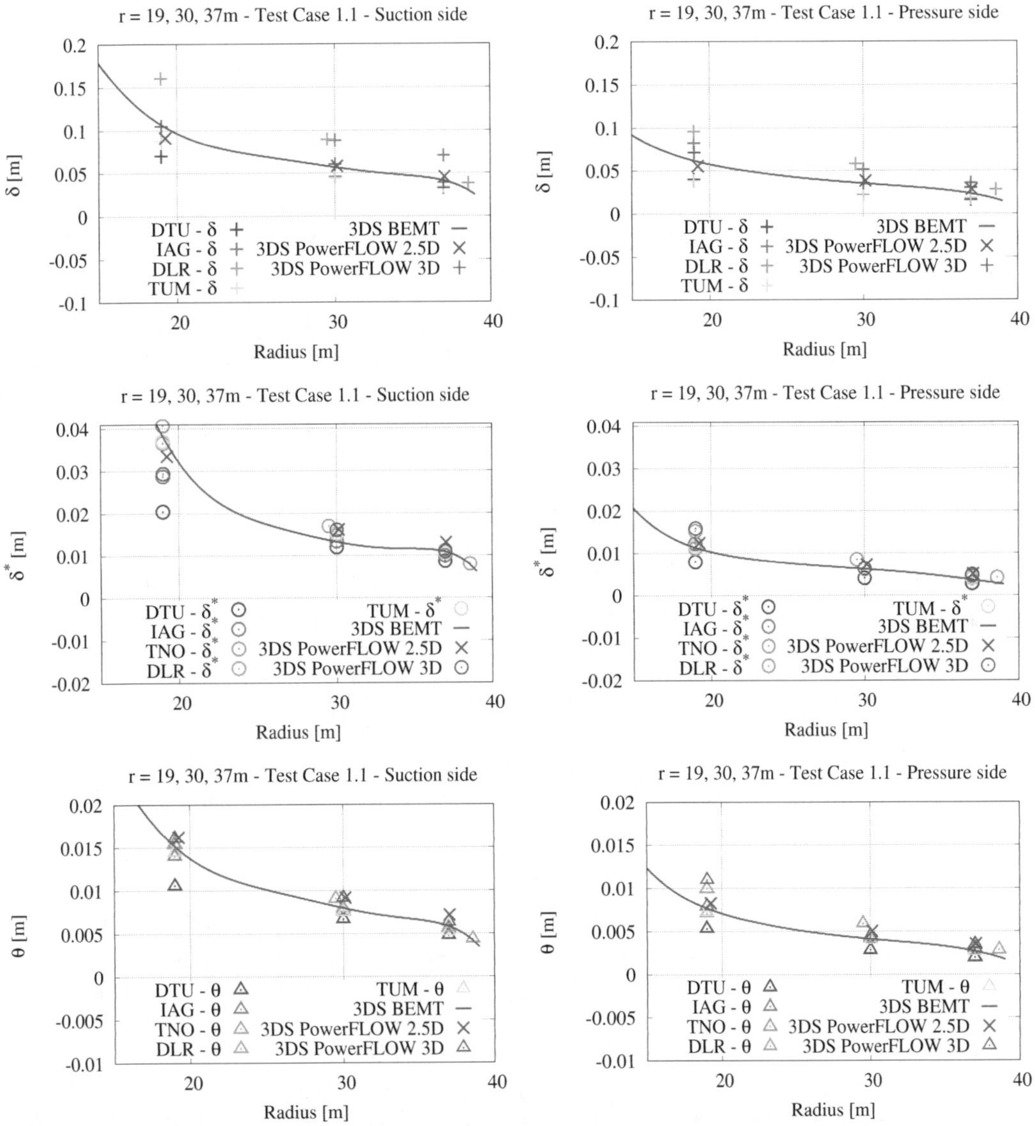

Fig. 3 Boundary layer thickness (top), displacement thickness (middle), and momentum thickness (bottom) along the blade span on the *suction* side of the airfoil at $x/C = 93\%$ (left) and *pressure* side at $x/C = 91\%$ (right).

The boundary layer profiles for BL velocity, turbulent kinetic energy, turbulence dissipation rate, and integral length scales, which are again important parameters influencing TE noise, are displayed in Fig. 4 for the suction and pressure sides at the outer spanwise section $r = 37\,\mathrm{m}$. Note that results from only a few methods are displayed here, as these quantities do not need to be explicitly calculated in some of the present numerical frameworks in order to compute TE noise.

There are noticeable discrepancies in the turbulent quantities. Note here that DLR and the Technical University of Denmark (DTU) use 2D CFD calculations to obtain the BL profiles at various sections along the span, whereas the University of Stuttgart Institute for Aerodynamic and Gas Dynamics (IAG) conducts a full 3D CFD simulation of the entire blade. This may affect the resulting computed BL profiles.

The impact of these turbulent BL quantities on the surface pressure spectra at the same location (see Section 5.3), and TE noise at the rotor level (see Section 6), is investigated in the following.

5.3. Surface pressure spectra near TE

The surface pressure spectra on the suction side at $x/C = 93\%$ and pressure side at $x/C = 91\%$ (i.e. relatively close to the TE) are displayed in Fig. 5. Since these spectra are characteristics of the turbulent flow in the vicinity of the TE, it is expected that they will have a large impact on the TE noise emission.

There is relatively good agreement between DTU, IAG, and the 3DS BEMT results above the peak frequency around 400–500 Hz on the suction side. The 3DS PowerFLOW results show higher spectral levels across the whole frequency range with slightly smaller slope above the peak frequency. It is noteworthy that all methods exhibit peak frequencies close to each other.

However, the discrepancies are larger on the pressure side. Nevertheless, all methods exhibit higher peak frequencies, which could be expected from the smaller BL thicknesses (see Fig. 3) and lower integral length scales (see Fig. 4). The spectra appear flatter above peak frequency for most methods. In addition, the spectral levels, e.g. at peak frequencies, are also lower in agreement with the observed lower turbulent kinetic energy levels on the pressure side (see Fig. 4).

When comparing high-fidelity model results to those that use the semi-empirical TNO model (or its variants) in Fig. 6, it is observed that the high-fidelity results (here, only PowerFLOW 2.5D and 3D) indicate a larger energy content in the low-frequency range, but also at high frequencies for the suction side. The semi-empirical methods also appear to converge on the suction side at higher frequencies.

The main takeaway from the present section is a lack of variety of methodologies for evaluating the surface pressure, which prevents the drawing of firmer conclusions. It is restricted here to three semi-empirical modelling approaches, with the LBM approach being the only one characterized by high modeling fidelity. Since surface pressure is the direct link between the boundary layer turbulent quantities and the noise emission, this is probably key to a better understanding of the discrepancies between the different noise models at the rotor level.

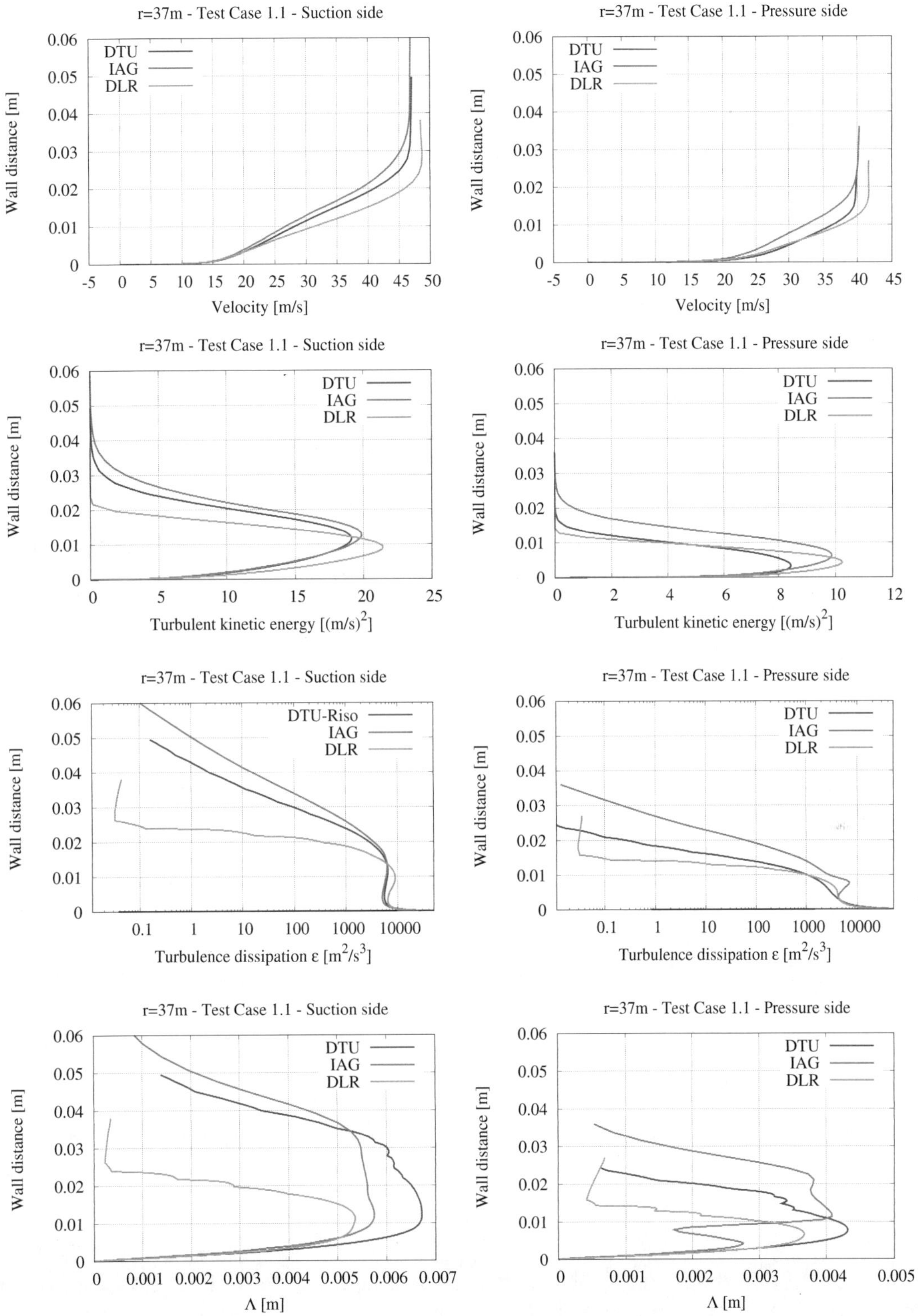

Fig. 4 Boundary layer profiles of: velocity (top row), turbulent kinetic energy (second row), turbulence dissipation rate (third row), and integral length scale (bottom row) at the blade span $r = 37$ m on the suction side of the airfoil at $x/C = 93\%$ (left) and pressure side at $x/C = 91\%$ (right).

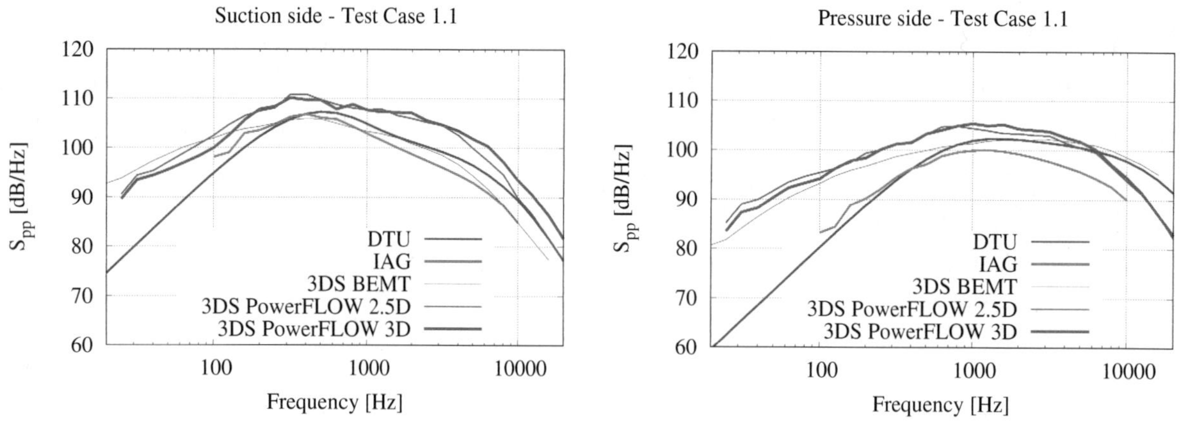

Fig. 5 Surface pressure spectra on the suction side at $x/C = 93\%$ **(left) and on the pressure side at** $x/C = 91\%$ **(right) at the blade span** $r = 37\,\mathrm{m}$**.**

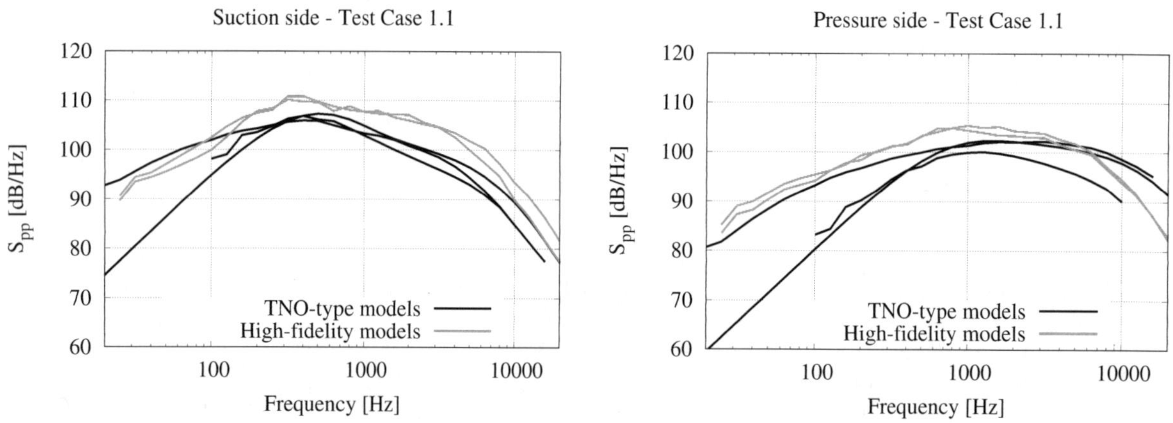

Fig. 6 Surface pressure spectra on the suction side at $x/C = 93\%$ **(left) and on the pressure side at** $x/C = 91\%$ **(right) at the blade span** $r = 37\,\mathrm{m}$**.**

6. Comparison of acoustic results

In this section, the aerodynamic noise emission of the full rotor is discussed.

6.1. Test case 1.1

The individual contributions from the TI and TE noise at the ground location downstream of the turbine are displayed in Fig. 7.

There is good agreement in the TI noise predictions in the high-frequency range, for $f \gtrsim 200$ Hz. This is to be expected for two separate reasons: first, all implementations are essentially similar, since they are based on the same Amiet model. Second, the model is built for a flat plate at no incidence. It follows that the actual blade shape or its angle of attack distribution does not influence the output of TI models and is mainly influenced by the velocity distribution along the blades, which is essentially identical for all the frameworks considered.

Below 200 Hz, two groups of prediction methods emerge: one predicts a continuous spectral slope toward lower frequencies while the other exhibits a higher energy bump in this frequency range. From Fig. 8, it is clear that the difference lies in the implementation of the full Amiet TI model, or its high-frequency asymptotic approximation, as discussed in Section 3. In addition, there is a larger spread of the results for the full Amiet models, which is attributed to the various implementations by the different participants. This highlights the dependency of rotor noise on the specific airfoil noise models, and this would probably require further investigations at the airfoil level.

Regarding TE noise, the spread of the model results is larger than for TI noise, as expected. The spectral slopes of the different models in the high-frequency range appear in good agreement, although an energetic spread with an amplitude slightly lower than 10 dB exists at any given high frequency. Looking toward the spectral peak, the peak frequency is in relative good agreement for all methods, with a spread amplitude of approximately 200 to 300 Hz. However, there is an even larger spread in the peak spectrum values. It is noteworthy that the high-fidelity methods, which exhibited larger energy levels for the wall-pressure spectra (see Section 5.3), now predict lower noise levels.

Looking at TE noise in Fig. 8, a number of features emerge that distinguish between high-fidelity, semi-empirical (TNO-type) and empirical (BPM) models, as discussed in Section 3. High-fidelity models predict lower spectral energy in the high-frequency range (beyond peak frequency), although a more noticeable energy bump at higher frequencies (above 1000 Hz) emerges. The latter is probably caused by the pressure-side TE noise contribution, since bluntness noise is not included in these models. This spectral bump is not clearly visible in the other modelling approaches, if it is indeed caused by the pressure-side TE noise contribution, even though it is a part of them. The trends for the different high-fidelity models are different for the low-frequency range. The empirical models consistently predict higher energy levels in the high-frequency range. The semi-empirical models lie somewhere in between for most of the spectral range, with some of them predicting lower energy for the pressure-side spectral bump at very high frequency.

The above comparisons, in particular for TE noise, indicate that there is a need to simplify the comparisons in order to trace back the origin of the observed discrepancies at the rotor level. It could be implemented by coming back to a simpler configuration with a rotating airfoil section of limited span [4], or even to a static 2D airfoil [20, 21] for which comparisons with wind tunnel data are possible.

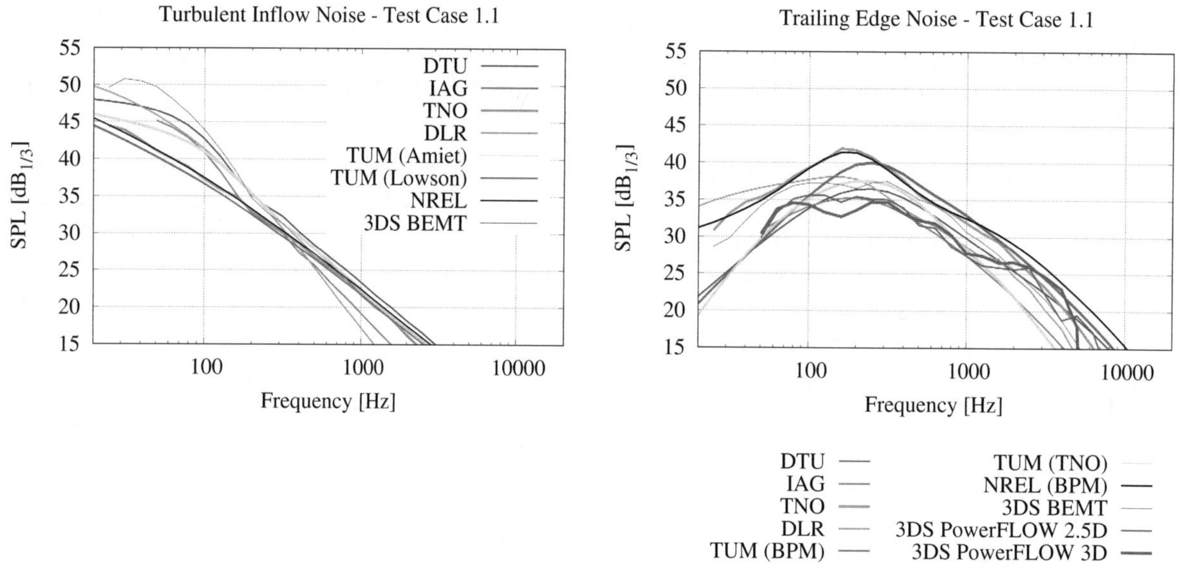

Fig. 7 TI noise (left) and TE noise (right) spectra at a location downstream of the rotor on the ground.

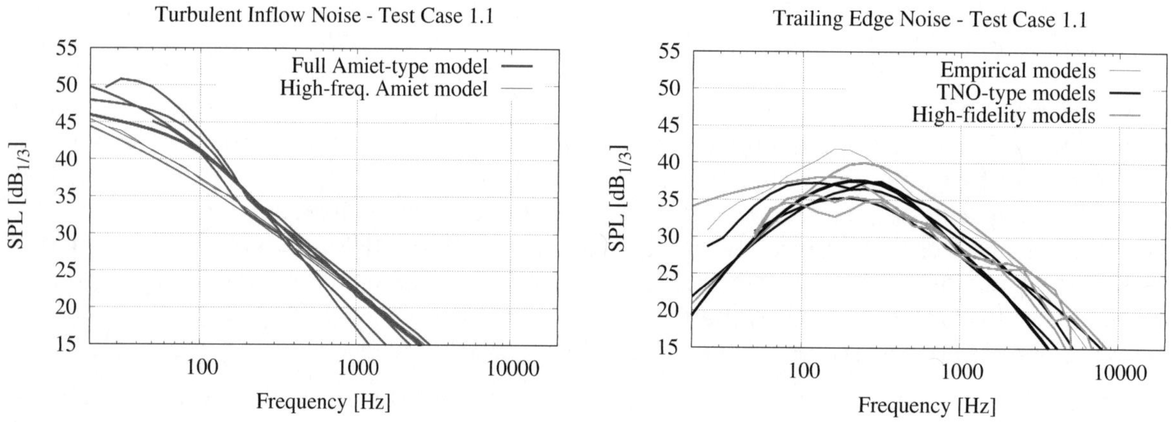

Fig. 8 TI noise (left) and TE noise (right) spectra at a location downstream of the rotor on the ground.

6.2. Wind turbine noise model directivity pattern

Test-case 1.1 is also used to investigate the directivity pattern around the wind turbine on the ground. As mentioned earlier, the noise spectra are predicted at various locations distributed around the turbine. These spectra are A-weighted and integrated across frequency and displayed in Fig. 9.

All models predict a reduction of noise in the plane of the rotor, although with various amplitudes relative to the upstream and downstream directions. This is an expected result given the more dipole-like behavior of TI noise, explaining the sharper deficit observed in the figure for this specific noise mechanism. The cardioid directivity pattern for TE noise, at airfoil level, similarly leads to a rotor plane noise deficit, which appears less pronounced.

Furthermore, the directivity pattern appears symmetric with respect to the rotor plane. Common sense would suggest that the perceived noise is higher downstream of the rotor, but it must be reminded that atmospheric effects are not included here.

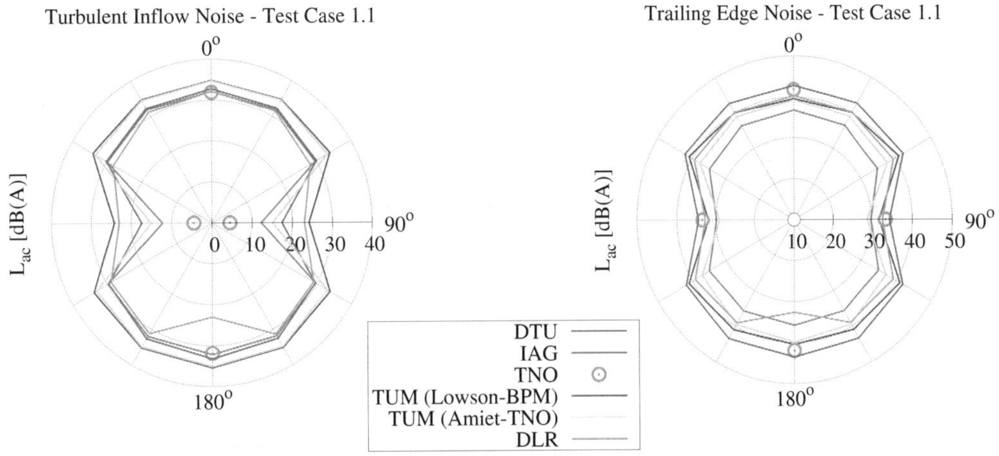

Fig. 9 A-weighted integrated spectra of TI noise (left) and TE noise (right) around the turbine on the ground.

6.3. Comparison with noise field measurements

The measurement of wind turbine noise for site assessment is often conducted according to the IEC 61400-11 standard [24]. The NM80 turbine that has been considered in the present work has been acoustically assessed using that standard. In the present section, the model results are compared to these field noise measurements.

The measured noise spectrum at a wind speed of 8 m/s is compared with six different models in Fig. 10. Note that the TI contribution for the DLR results is an extrapolation of lower wind speed data, and that it might slightly underestimate the actual noise level in the frequency range 100–400 Hz, but the results are unaffected above peak frequency at 500 Hz.

A higher energy spectral bump is observed in the measurements in the frequencies ranging from 100 to 300 Hz. It is attributed to mechanical noise, as a spectral tone at the center frequency of 137 Hz has been clearly identified during the same measurement campaign. This is compatible with the observed local energy peaks maxima located at the 1/3 octave band center frequencies: 125 Hz for the fundamental tone and 250 Hz and 500 Hz for the harmonics. Therefore, the models that do not account for mechanical noise are underpredicting the measurement data in this frequency range. Elsewhere there is a good agreement between models and measurements, which stay nearly within the ±2 dB uncertainty margin of the measurements, except at very low and very high frequencies. As expected from the results observed in the previous section, the Lowson-TNO approach underpredicts the noise levels in the low-frequency range corresponding to the TI noise contribution, and overpredicts in the high-frequency range where TE noise dominates.

The acoustic power curves of the A-weighted integrated spectra as a function of wind speed are displayed in Fig. 11. The model results remain again within ±2 dB of the measured noise levels, except at the wind speed of 6 m/s. Unfortunately, the measured spectrum is not available at this wind speed. The reason for this discrepancy remains unknown. It was checked that the design rotational speed and blade pitch at this wind speed did match the design electrical power output in the HAWC2 model. Therefore, it can only be surmised that the turbine controller is more aggressive/optimal than the design parameters available for the present comparisons, and that rotational speed is increased at this particular wind speed in reality to maximize the power output.

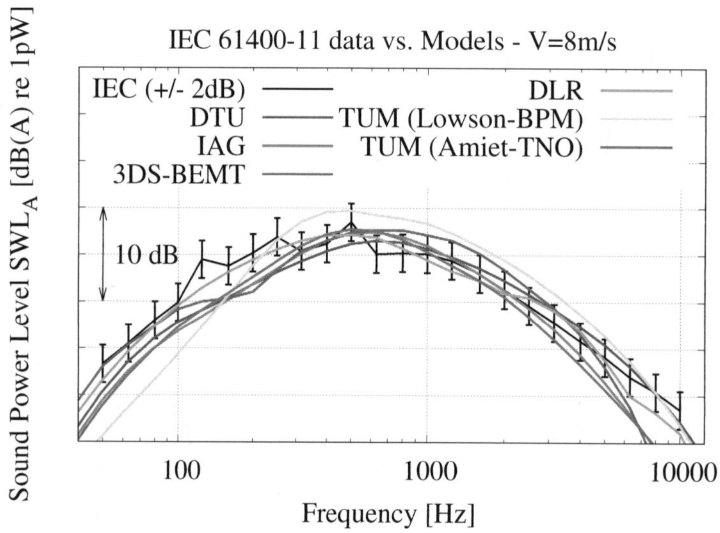

Fig. 10 A-weighted sound power spectrum of the measured noise using IEC 61400-11 standard measurement procedure versus model results for a wind speed of 8 m/s.

7. Conclusions

The comparisons presented in this paper highlight a number of discrepancies when evaluating wind turbine rotor noise with various methodologies based on airfoil TI and TE noise modelling.

It is observed that the use of a single identical model for TI noise (Amiet model) and its different implementations may yield significantly different noise predictions at the rotor level. Nevertheless, this model appears to be the main engineering approach for modelling this phenomenon, although higher-fidelity models exist.

Regarding the prediction of TE noise and the analysis conducted in the present work, a bottleneck is identified. It resides in connecting the aerodynamic quantities, in particular the turbulent

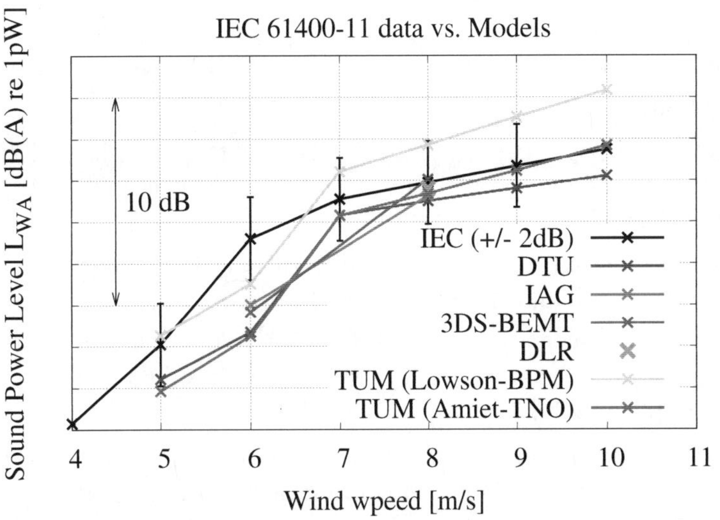

Fig. 11 Integrated A-weighted sound power as a function of wind speed using IEC 61300-11 standard measurement procedure versus model results.

boundary layer, to the rotor noise emission through the surface pressure. Indeed, the surface pressure models have not been thoroughly investigated, mainly because of a lack of available results. The use of simpler configurations is therefore suggested, e.g. limited spanwise section, or even 2D section in standstill, in order to identify the origin of these discrepancies.

A largely expected result from the present study is the presence of a noise level deficit in the rotor plane. This feature has been observed earlier in the field as well as in numerical predictions. The present contribution tends to confirm that there is a sharper deficit originating from TI noise, which would suggest that it is even more pronounced at lower frequencies.

Note that wind turbine designers in the industry, while mostly resorting to empirical models in the design loops for reducing turnover time, have access to a considerable amount of experimental data which can be used to tune and improve their modelling frameworks. The present study indicates that introducing semi-empirical models, which aim at accounting for more physical processes in the prediction tools, still suffer from relatively large discrepancies between each other when predicting rotor noise emissions. Therefore, it can be surmised that tuning or improving these models is still required. High-fidelity model results appear to somehow converge for TE noise within the high-frequency range, still with some discrepancies. Note, however, that only two high-fidelity approaches (for TE noise) were considered in the present work. Nevertheless, when comparing different models of varying fidelity with actual wind turbine noise measurements, it appears that all model results stay for the most part within the ± 2 dB uncertainty margin associated with the field measurement.

To conclude, real field conditions are difficult to reproduce within rotor noise models (e.g. blade leading edge erosion or fouling, atmospheric turbulence influencing the BL turbulence and subsequently TE noise, etc.). These are also difficult to identify (when comparing with measurements) and quantify. Therefore, many aspects remain to be considered for developing accurate prediction models for wind turbine rotor noise.

Acknowledgments

The present benchmark is conducted as a sub-task of the IEA Wind TCP research Task 39 (Quiet Wind Turbine Technology), as well as a part of IEA Wind TCP research Task 29 Phase IV (Analysis of Aerodynamic Measurements).

DTU participation was partly supported by the EUDP projects Jr.nr. 64016-0056 and Jr.nr. 134-21022 funded by the Danish Energy Agency (Energistyrelsen), as well as by countries participating to the Task for supporting the Operating Agent duties.

The DanAero projects were funded partly by the Danish Energy Authorities (EFP2007. Journal nr.: 33033-0074 and EUDP 2009-II. Journal nr.: 64009-0258), and partly by eigenfunding from the project partners Vestas, Siemens, LM, Dong Energy and DTU.

Financial support of the German Federal Ministry for Economic Affairs and Climate Action (BMWK) to the operation of Task 39 is also highly acknowledged.

The authors are thankful to Vestas Wind Systems A/S for providing the noise measurements of the NM80 turbine, as well as allowing their anonymized display in the present article.

This work was authored in part by the National Renewable Energy Laboratory, operated by Alliance for Sustainable Energy, LLC, for the U.S. Department of Energy (DOE) under Contract No. DE-AC36-08GO28308. Funding provided by the U.S. Department of Energy Office of Energy Efficiency and Renewable Energy Wind Energy Technologies Office. The views expressed in the article do not necessarily represent the views of the DOE or the U.S. Government. The U.S. Government retains and the publisher, by accepting the article for publication, acknowledges that the U.S. Government retains a nonexclusive, paid-up, irrevocable, worldwide license to publish or reproduce the published form of this work, or allow others to do so, for U.S. Government purposes.

References

[1] Amiet, R. K. (1975). Acoustic Radiation from an Airfoil in a Turbulent Stream. *J. Sound Vib.*, 41(4):407–420.

[2] Amiet, R. K. (1976). Noise due to Turbulent Flow Past a Trailing Edge. *J. Sound Vib.*, 47(3):387–393.

[3] Appel, C., Faßmann, B., and Lin, S. (2020). CAA-basierte Schallvorhersage für Windkraftanlagen. In *DAGA 2020 (Conference)*, Hannover, Germany. Deutsche Gesellschaft für Akustik e.V.

[4] Bertagnolio, F., Fischer, A., Seel, F., Lutz, T., Hornung, C., Boorsma, K., Schepers, G., Botero, M. R., Appel, C., Herr, M., Sucameli, C., Bortolotti, P., v. d. Velden, W., and Casalino, D. (2023). Task 29 T3.7 & Task 39 Wind Turbine Noise Code Benchmark - Preliminary Results. Retrieved from: https://usercontent.one/wp/iea-wind.org/wp-content/uploads/2021/05/report_WTNCBenchmark.pdf, Accessed: 2023-04-29, IEA Wind TCP.

[5] Blake, W. K. (1986). *Mechanics of Flow-Induced Sound and Vibration, Vol.I and II*, volume in Applied Mathematics and Mechanics. Frenkiel, F.N. and Temple, G. (eds.), Academic Press.

[6] Boorsma, K., Schepers, G., Aagard Madsen, H., Pirrung, G., Sørensen, N., Bangga, G., Imiela, M., Grinderslev, C., Meyer Forsting, A., Shen, W. Z., Croce, A., Cacciola, S., Schaffarczyk, A. P., Lobo, B., Blondel, F., Gilbert, P., Boisard, R., Höning, L., Greco, L., Testa, C., Branlard, E., Jonkman, J., and Vijayakumar, G. (2023). Progress in the validation of rotor aerodynamic codes using field data. *Wind Energy Science*, 8(2):211–230.

[7] Boorsma, K. and Schepers, J. (2011). Enhanced wind turbine noise prediction tool SILANT. In *Presented at the Fourth International Meeting on Wind Turbine Noise*, AIAA Paper 2010-645, Rome (Italy).

[8] Bortolotti, P., Branlard, E., Platt, A., Moriarty, P., Sucameli, C., and Bottasso, C. L. (2020). Aeroacoustics Noise Model of OpenFAST. Technical report, NREL, Boulder, USA. Available at: https://www.osti.gov/servlets/purl/1660130.

[9] Bortolotti, P., Guo, Y., Simley, E., Roadman, J., Hamilton, N., Moriartya, P. J., Sucameli, C. R., and Bertagnolio, F. (2021). Validation Efforts of an Open-Source Aeroacoustics Model for Wind Turbines: Preprint. Technical report, NREL, Boulder, USA. Available at: https://www.osti.gov/servlets/purl/1808274.

[10] Brooks, T. F., Pope, S. D., and Marcolini, M. A. (1989). Airfoil Self-Noise and Prediction. NASA Reference Publication 1218, Retrieved from: https://ntrs.nasa.gov/archive/nasa/casi.ntrs.nasa.gov/19890016302.pdf, NASA Langley Research Center, Hampton (VA). (Accessed: 2020-04-08).

[11] Casalino, D., Grande, E., Romani, G., Ragni, D., and Avallone, F. (2021). Definition of a benchmark for low reynolds number propeller aeroacoustics. *Aerospace Science and Technology*, 113.

[12] Casalino, D., Van der Velden, W., and Romani, G. (2022). A framework for multi-fidelity wind-turbine aeroacoustic simulations. In *28th AIAA/CEAS Aeroacoustics Conference, June 2022, Southampton, UK.*

[13] Drela, M. (1989). *XFOIL: An Analysis and Design System for Low Reynolds Number Airfoils. In: Mueller T.J. (eds) Low Reynolds Number Aerodynamics. Lecture Notes in Engineering, vol 54.* Springer, Berlin, Heidelberg.

[14] Drela, M. and Giles, M. B. (1987). Viscous-inviscid analysis of transonic and low reynolds number airfoils. *AIAA Journal*, 25(10):1347–1355.

[15] Ewert, R. (2018). RPM - the fast Random Particle-Mesh method to realize unsteady turbulent sound sources and velocity fields for CAA applications. In 13th *AIAA/CEAS Aeroacoustics Conference*, Conf. Proceedings, Rome, Italy. AIAA 2007-3506.

[16] Faßmann, B., Reiche, N., Ewert, R., Herr, M., and Delfs, J. (2018). Evaluation of Wind Turbine Noise based on Numerical Simulation Methods. In 18th *AIAA/CEAS Aeroacoustics Conf. (Proc.)*, Atlanta, Georgia. AIAA 2018-3924.

[17] Fischer, A., Bertagnolio, F., and Madsen, H. A. (2017). Improvement of TNO type trailing edge noise models. *European Journal of Mechanics B - Fluids*, 61:255–262.

[18] Glauert, H. (1935). *Airplane Propellers*, volume In: Aerodynamic Theory Volume IV (W. F. Durand, Ed.). Springer, Berlin, Heidelberg.

[19] Hamilton, N., Bortolotti, P. E., Jager, D., Guo, Y., Roadman, J. M., and Simley, E. (2021). Aeroacoustic Assessment of Wind Plant Controls. Technical report, NREL, Boulder, USA. Available at: https://www.osti.gov/servlets/purl/1785330.

[20] Herr, M., Ewert, R., Rautmann, C., Kamruzzaman, M., Bekiropoulos, D., Iob, A., Arina, R., Batten, P., Chakravarthy, S., and Bertagnolio, F. (2015). Broadband Trailing-Edge Noise Predictions - Overview of BANC-III Results. In 21th *AIAA/CEAS Aeroacoustics Conf. (Proc.)*, AIAA Paper 2015-2847, Dallas (TX).

[21] Herr, M., Kamruzzaman, M., and Bahr, C. (2016). BANC-IV-1: TBL-Trailing-edge noise. In *Fourth Workshop on Benchmark Problems for Airframe Noise Computations (BANC-IV)*, 22nd AIAA-CEAS Aeroacoustics Conference (Workshop), Lyon, France.

[22] Hornung, C., Lutz, T., and Krämer, E. (2019). A model to include turbulence-turbulence interaction in the prediction of trailing edge far field noise for high angles of attack or slightly separated flow. *Renewable Energy*, 136:945–954.

[23] Hornung, C., Scheit, C., Noffke, N., and Kamruzzamann, M. (2021). Turbulence Inflow Noise Prediction of Wind Turbine Rotors: The physically correct Representations of the Simplified Amiet and Lowson Model. In *WTN 2021*, Remote from Europe.

[24] IEC (2012). International Standard, Wind Turbines - Part 11: Acoustic Noise Measurement Techniques. IEC 61400-11, International Electrotechnical Commission, Geneva (CH). ISBN 978-2-83220-463-4.

[25] Kamruzzaman, M., Bekiropoulos, D., Lutz, T., Würz, W., and Krämer, E. (2015). A semi-empirical surface pressure spectrum model for airfoil trailing- edge noise prediction. *International Journal of Aeroacoustics*, 14(5-6):833–882.

[26] Klein, L., Gude, J., Wenz, F., Lutz, T., and Krämer, E. (2018). Advanced CFD-MBS coupling to Assess Low-Frequency Emissions from Wind Turbines. *Wind Energy Science*, 3:713–728.

[27] Larsen, T. J. and Hansen, A. M. (2007). How 2 HAWC2, The User's Manual. Tech. Rep. RISØ-R-1597(ver.3-1), Risø-DTU, Roskilde, Denmark. Available at `http://www.hawc2.dkdownload/hawc2-manual` (Accessed: 2021-03-11).

[28] Lindenburg, C. (2002). Bladmode, program for rotor blade mode analysis. Tech. Rep. ECN-C-02-050-r2, ECN, The Netherlands.

[29] Lowson, M. V. (1993). Assessment and prediction of wind turbine noise. ETSU W/13/00284/REP, UK.

[30] Lowson, M. V. and Ollerhead, J. B. (1969). A Theoretical Study of Helicopter Rotor Noise. *J. Sound Vib.*, 9(2):197–222.

[31] Madsen, H. A., Bak, C., Paulsen, U. S., Gaunaa, M., Fuglsang, P., Romblad, J., Olesen, N. A., Enevoldsen, P., Laursen, J., and Jensen, L. (2010). The DAN-AERO MW Experiments. In 48th *AIAA Aerospace Sciences Meeting Including The New Horizons Forum and Aerospace Exposition (Proceedings)*, AIAA Paper 2010-645, Orlando (FL).

[32] Madsen, H. A., Larsen, T. J., Pirrung, G. R., Li, A., and Zahle, F. (2020). Implementation of the blade element momentum model on a polar grid and its aeroelastic load impact. *Wind Energy Science*, 5(1):1–27. doi:10.5194/wes-5-1-2020.

[33] Montgomerie, B., Brand, A., Bosschers, J., and van Rooij, R. (1997). Three-dimensional effects in stall. Tech. Rep. ECN-C-96-079, ECN, The Netherlands.

[34] Moriarty, P., Guidati, G., and Migliore, P. (2005). Prediction of Turbulent Inflow and Trailing-Edge Noise for Wind Turbines. In *Proc. of the* 11th *AIAA/CEAS Aeroacoustics Conf.*, AIAA Paper 2005-2881, Monterey, CA.

[35] Oerlemans, S., Sijtsma, P., and López, B. M. (2007). Location and Quantification of Noise on a Wind Turbine. *J. Sound Vib.*, 299(5-6):869–883.

[36] Parchen, R. (1998). Progress report DRAW: A Prediction Scheme for Trailing-Edge Noise Based on Detailed Boundary-Layer Characteristics. TNO Rept. HAG-RPT-980023, TNO Institute of Applied Physics, The Netherlands.

[37] Paterson, R. W. and Amiet, R. K. (1976). Acoustic Radiation and Surface Pressure Characteristics of an Airfoil Due to Incident Turbulence. In 3rd *AIAA Aero-Acoustics Conference*, Conf. Proceedings, Palo Alto, CA.

[38] Paterson, R. W. and Amiet, R. K. (1977). Noise and Surface Pressure Response of an Airfoil to Incident Turbulence. *Journal of Aircraft*, 14(8):729–736.

[39] Rautmann, C. (2017). Numerical Simulation Concept for Low-Noise Wind Turbine Rotors. DLR Forschungsbericht FB-2017-35, DLR (Institute for Aerodynamics and Flow Technology), Brauschweig, Germany. Dissertation.

[40] Schlinker, R. H. and Amiet, R. K. (1981). Helicopter rotor trailing edge noise.Technical Report 1. NASA Contractor Report 3470, Retrieved from: https://ntrs.nasa.gov/archive/nasa/casi.ntrs.nasa.gov/19820003986.pdf, NASA Langley Research Center. (Accessed: 2020-04-08).

[41] Sucameli, C. R., Bortolotti, P., Croce, A., and Bottasso, C. L. (2018). Comparison of some wind turbine noise emission models coupled to BEM aerodynamics. *Journal of Physics: Conference Series*, 1037:022038.

[42] van der Velden, W. C., Casalino, D., and Romani, G. (2023). Full-Scale Serrated Wind Turbine Trailing Edge Noise Certification Analysis Based on the Lattice-Boltzmann Method. In *AIAA SCITECH 2023 Forum (Proceedings)*, AIAA 2023-0970, National Harbor (MD).

INCEEUROPE

10th International Conference
on
Wind Turbine Noise
Dublin – 21st to 23rd June 2023

Optimizing the locations of turbines in a wind farm according to acoustic and spatial constraints

Alexis Bigot, Sixense Engineering: alexis.bigot@sixense-group.com
Marwen Bejaoui, Sixense Engineering: marwen.bejaoui@sixense-group.com

Summary

Within the framework of environmental impact studies of wind farms in France, acoustic propagation software are used to define the regulatory exceedances of the noise emergence criterion. Based on these results, the acoustician then proposes a noise curtailment plan which defines the operating mode of each machine at different wind speeds and wind directions. Experience shows that noise curtailment plans lead to losses in electricity production ranging from 4-5% on average to 10-12% - and even more for very sensitive projects. For this reason, we have developed an algorithm that optimises not only the operating modes of the machines, but also their position. This optimisation algorithm, called OPPIO, finds a layout that minimises production losses while considering the displacement constraints of each wind turbine. It is achieved by coupling the output of the acoustic model (CadnaA) to a Mixed-Integer Nonlinear Programming (MINLP) model.

1. Introduction

The acoustic impact of wind farms is calculated using an acoustic propagation software. Based on these results, the acoustician then proposes a noise curtailment plan which defines the operating mode of each machine at different wind speeds and wind directions.
The optimization work therefore focuses on the operation of the machines and not on their position. In this article we describe an innovative methodology, named OPPIO, which allows to perform both types of optimizations simultaneously.

2. Methodology

The OPPIO methodology is based on the coupling of an acoustic propagation software with a constrained optimization algorithm. The propagation software used in this article is CadnaA (Datakustik) with calculations performed according to the ISO9613-2 standard, but other software and calculation standards (such as Nord2000, Harmonoise...) could also be used.

This methodology works as follows:

1. Construction of an acoustic propagation model including topography, building, noise sources in their initial position, and receivers.
2. Calculation of source-receiver transfer functions per octave band.
3. Definition of the constants and variables.
4. Definition of the constraints.
5. Definition of the objective function.
6. Solving the problem with a constrained optimization algorithm.

In the following, the number of wind turbines is called $N_{Machine}$, the number of acoustic receivers is called $N_{Receiver}$, the number of octave bands is called N_{Octave}, the number of wind speed bins and wind direction sectors are called $N_{WindSpeed}$ and $N_{WindDir}$, and the number of operating modes of the wind turbines is called N_{Mode} (from the standard full-power mode to the most restricted modes, including also a stop mode).

2.1 Source – Receiver Transfer Function

This transfer function gives for each receiver the sound pressure level per octave band generated by each source, for a reference sound power level (e.g. 100 dB per octave band). These transfer functions are derived from the CadnaA acoustic propagation model.

The advantage of these transfer functions is that the sound level can be easily recalculated for new machine positions:
For each source-receiver pair, and each octave band, the sound level for a new position is approximated as:

$$Lp = TF - Lw_{Ref} + Lw_{WT} + \Delta_{geo}$$

Where:

TF: source – receiver transfer function, for the initial position of the wind turbine
Lw_{Ref}: reference sound power level used (ex 100 dB/octave)
Lw_{WT}: sound power level of the wind turbine
Δ_{geo}: correction of the geometric spreading for the new position of the wind turbine

2.2 Constants of the problem

For a given site, the constants of the problem are:

- The initial positions (X,Y) of the machines, generally expressed in Lambert 93 coordinates.
- The position of the acoustic receptors used in the impact study.
- The acoustic thresholds to be respected for each receptor, which are derived from the initial state measurements and French regulation. These are matrices of dimensions ($N_{Receiver}$ x $N_{WindSpeed}$ x $N_{WindDir}$).
- The electrical production in kW of the machines: matrix called *PowerCurve*, of dimension (N_{Mode} x $N_{WindSpeed}$).
- The sound power in dB of the machines: matrix called *NoiseCurve* of dimension (N_{Mode} x $N_{WindSpeed}$ x N_{Octave}).
- The source-receiver transfer functions: matrices of dimension ($N_{Machine}$ x $N_{Receiver}$ x N_{Octave} $N_{WindSpeed}$ x $N_{WindDir}$).
- The wind speed rose: matrix called *WSRose* of dimension ($N_{WindSpeed}$ x $N_{WindDir}$).

2.3 Variables of the problem

The variables are the data for which the algorithm will try to find the optimal values to maximize (or minimize) the objective function. In our case the variables are:

- The operating mode of each machine: This is a matrix called *ModeFonc*, of dimension ($N_{Machine}$ x $N_{WindSpeed}$ x N_{Mode} x $N_{WindDir}$) and filled with 0 and 1.
- The new position of the machines, which are 2 x $N_{Machine}$ variables (real values).

2.4 Constraints

The constraints may vary for each project. But generally, they are:

- Constraints on the possible displacement of each wind turbine in X and Y, resulting from the intersection of all the project constraints.
- Minimum distance between machines.
- Acoustic constraints: the cumulative sound level of all the wind turbines at a given point must not exceed the acoustic threshold.
- Constraints linked to the supplier of the machines for the programming of the operating modes (not used in this paper).

2.5 Objective function

The function to be maximized is the average electricity production of the wind farm, which is expressed as a product of the operating mode and electricity power matrices, weighted by the probability of occurrence of the wind speeds and the wake losses. The expression of the objective function is:

$$\sum_{i=1}^{N_{Machine}} \sum_{j=1}^{N_{WindSpeed}} \sum_{k=1}^{N_{Mode}} \sum_{d=1}^{N_{WindDir}} ModeFonc(i,j,k,d).PowerCurve(j,k).WSRose(j,d).WakeLoss(i,j,k,d)$$

Note 1: This calculation of electrical production takes into account a simplified wake loss estimation between machines. This wake loss is calculated for the initial position of the wind turbines only. It was not possible to calculate it for the new positions of the turbines, because it would have introduced nonlinearity in the objective function. However, for small displacement, the error induced by this approximation should be small.
Note 2: The wind speed rose (occurrence in % of the wind speed for each wind direction) could be replaced by Weibull laws commonly used to estimate the electricity production. But it should not change the calculation results in terms of position.

2.6 Optimisation algorithm

Since the variables are both integer (machine modes) and real (machine positions), the algorithm used is a Mixed-Integer Nonlinear Programming (MINLP) model. The non-linearity of the problem comes from the presence of squares and square roots in the evaluation of distances in the constraints.

3. Results

3.1 Example on project A

The project A is a repowering project of 5 wind turbines. The new wind turbines are Nordex N131 3.6MW at 114m hub height.
The background noise was measured at 6 locations for 5 weeks, and the noise impact is calculated at the closest 9 receivers for the following situations:

- 2 wind directions: South-West [135°; 315°] and North-East [315°; 135°]
- 3 periods: Day 7h-21h, Evening 21h-22h, and Night 22h-7h

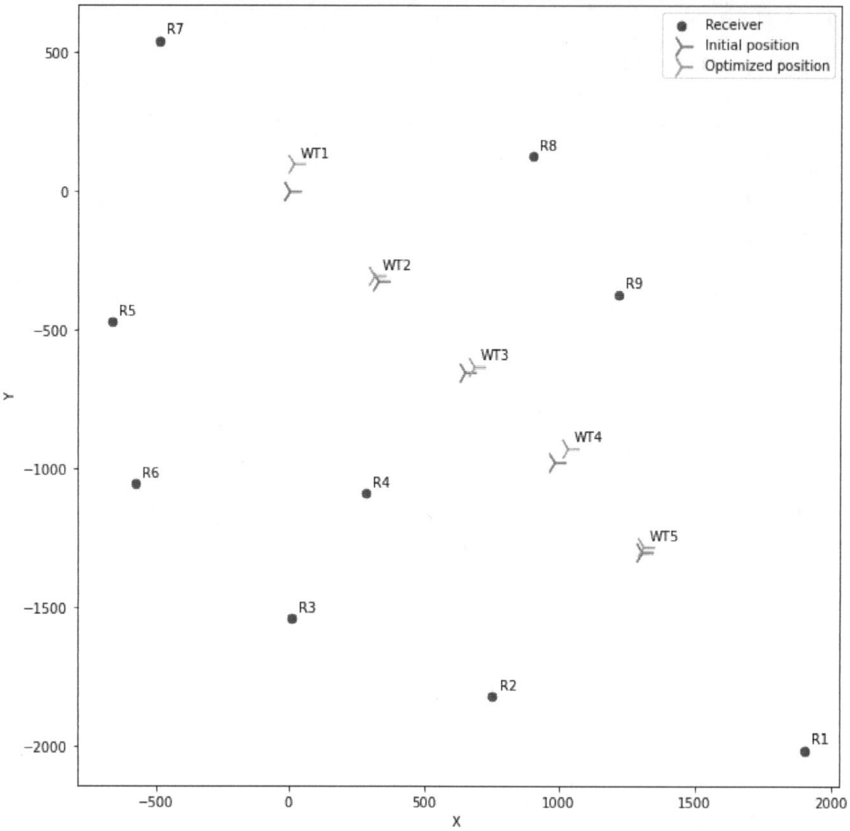

Figure 1: Initial position (in red) and optimized positions (in green) for project A

Turbine	Constraints in X	Constraints in Y	Displacements
WT1	[-20m ;+20m]	[0m ;+100m]	100.9m
WT2	[-20m ;+30m]	[0m ;+20m]	22.9m
WT3	[-15m ;+30m]	[0m ;+20m]	36.0m
WT4	[0m ;+50m]	[0m ;+50m]	70.6m
WT5	[0m ;+5m]	[0m ;+20m]	20.5m

Figure 2: Displacements calculated for project A

The optimization of the position of the wind turbines can be seen on figures 3 and 4. We can see more Full Power modes and less Stops of wind turbines, especially in South-West direction, and about 3% improvement in the production loss due to the acoustic curtailment plan in the night period.

Night Period (22h-7h) Wind direction South-West [135° ; 315°[											
Wind speed at hub height	5m/s	6m/s	7m/s	8m/s	9m/s	10m/s	11m/s	12m/s	13m/s	>13m/s	Production loss
WT1				Mode 5	Mode 10	Mode 5	Mode 5	Mode 7	Mode 5	Mode 5	
WT2			Mode 8	Mode 11	Mode 12	Mode 11	Mode 5	Mode 5	Mode 5	Mode 5	
WT3			Stop	Stop	Mode 12	Mode 12	Mode 11	Mode 8	Mode 6	Mode 5	16.0%
WT4			Mode 5	Mode 8	Mode 12	Mode 7	Mode 5	Mode 5			
WT5				Mode 5	Mode 7	Mode 5	Mode 5				

Night Period (22h-7h) Wind direction South-West [135° ; 315°[											
Wind speed at hub height	5m/s	6m/s	7m/s	8m/s	9m/s	10m/s	11m/s	12m/s	13m/s	>13m/s	Production loss
WT1			Mode 8	Mode 5	Mode 5	Mode 5	Mode 5	Mode 5	Mode 5	Mode 5	
WT2			Mode 12	Mode 8	Mode 12	Mode 7	Mode 5	Mode 5	Mode 5		
WT3			Mode 12	Stop	Mode 12	Mode 12	Mode 8	Mode 5	Mode 5	Mode 5	12.8%
WT4			Mode 12	Mode 8	Mode 9	Mode 5	Mode 5	Mode 5			
WT5			Mode 7	Mode 5	Mode 5	Mode 5	Mode 5				

Figure 3: Curtailment Plan for project A, South-West wind direction (night period), before optimization (table above) and after optimization (table below). In green: Full Power mode. In orange: Restricted mode. In Red: Stop.

Night Period (22h-7h) Wind direction North-East [315° ; 135°[											
Wind speed at hub height	5m/s	6m/s	7m/s	8m/s	9m/s	10m/s	11m/s	12m/s	13m/s	>13m/s	Production loss
WT1			Mode 5	Mode 7	Mode 5	Mode 5	Mode 5	Mode 5			
WT2			Mode 1	Mode 8	Mode 10	Mode 6	Mode 5				
WT3				Mode 6	Mode 7	Mode 5					5.8%
WT4			Mode 5	Mode 5	Mode 5	Mode 5					
WT5			Mode 10	Mode 8	Mode 5						

Night Period (22h-7h) Wind direction North-East [315° ; 135°[											
Wind speed at hub height	5m/s	6m/s	7m/s	8m/s	9m/s	10m/s	11m/s	12m/s	13m/s	>13m/s	Production loss
WT1				Mode 5	Mode 5	Mode 5	Mode 5	Mode 5			
WT2				Mode 6	Mode 7	Mode 5	Mode 5				
WT3			Mode 7	Mode 8	Mode 9	Mode 5					5.5%
WT4				Mode 5	Mode 5	Mode 5					
WT5			Mode 8	Mode 6	Mode 5						

Figure 4: Curtailment Plan for project A, North-East wind direction (night period), before optimization (table above) and after optimization (table below). In green: Full Power mode. In orange: Restricted mode. In Red: Stop.

3.2 Example on project B

The project B has 8 wind turbines Enercon E92 2.35MW at 78m hub height.
The background noise was measured at 8 locations for 4 weeks, and the noise impact is calculated at the closest 8 receivers for the following situations:

- 1 wind direction: North-West [270°; 360°]
- 2 periods: Day 7h-22h, and Night 22h-7h

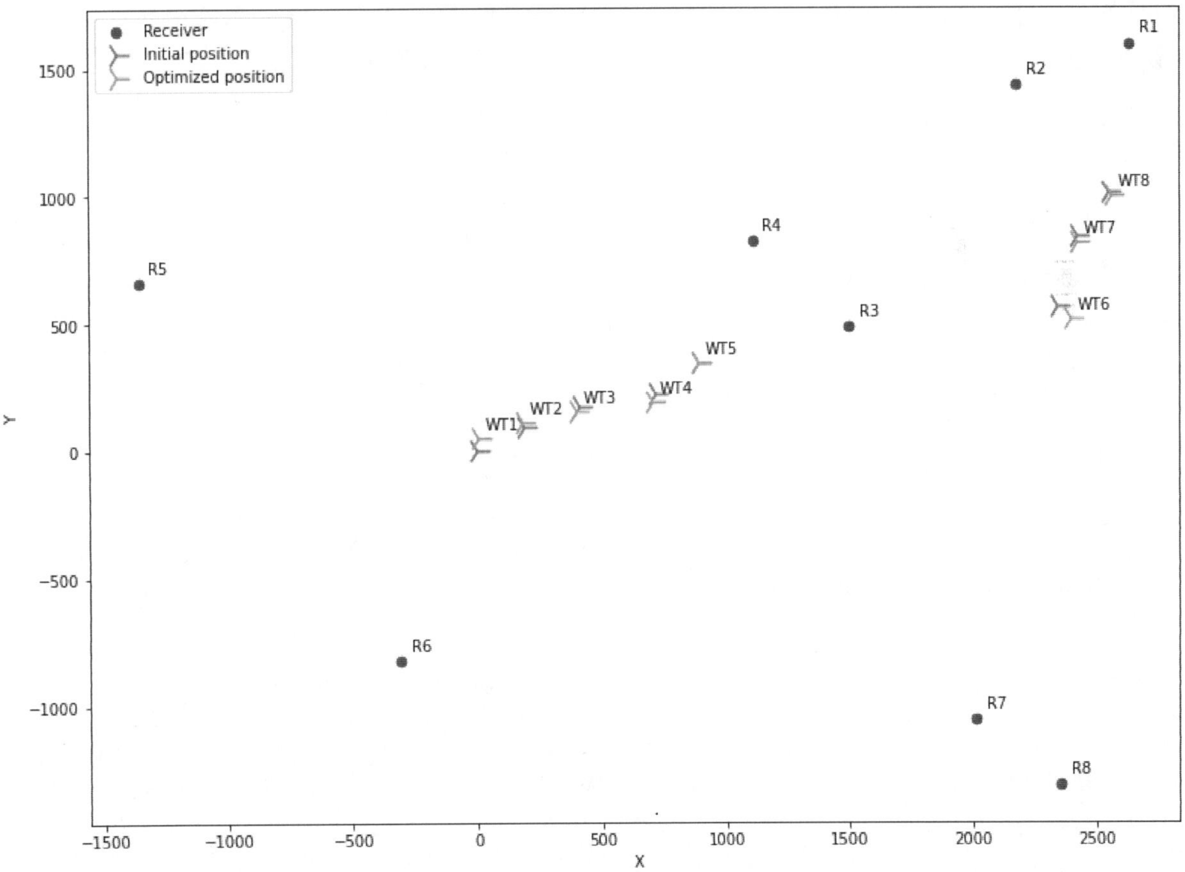

Figure 5: Initial position (in red) and optimized positions (in green) for project B

Turbine	Constraints in X	Constraints in Y	Displacements
WT1	[0m ;+50m]	[-140m ;+50m]	50.0m
WT2	[-60m ;+50m]	[-90m ;+22m]	23.6m
WT3	[-9m ;+30m]	[-15m ; 0m]	17.4m
WT4	[-13m ;+45m]	[-30m ; 0m]	32.7m
WT5	[0m ;+50m]	[0m ; 0m]	0m
WT6	[-10m ;+50m]	[-50m ;+50m]	70m
WT7	[0m ;0 m]	[-25m ;+60m]	21.6m
WT8	[-30m ;+15m]	[-15m ;+30m]	16.5m

Figure 6: Displacements calculated for project B

The optimization of the position of the wind turbines can be seen on figures 7. We can see more Full Power modes and less Stops of wind turbines (e.g. at 7m/s), and about 3% improvement in the production loss due to the acoustic curtailment plan in the night period.

Night Period (22h-7h) Wind direction North-West [270° ; 360°[									
Standardized wind speed at 10m	3m/s	4m/s	5m/s	6m/s	7m/s	8m/s	9m/s	>9m/s	Production loss
WT1							Mode 1	Mode 1	39.0%
WT2						Mode 2	Mode 1	Mode 1	
WT3			Mode 6					Mode 3	
WT4			Mode 6	Mode 5	Stop	Stop	Stop	Mode 6	
WT5			Mode 6	Stop	Stop	Stop	Stop	Stop	
WT6				Mode 5			Mode 2	Mode 5	
WT7						Mode 4	Mode 5	Mode 6	
WT8				Mode 6	Stop	Stop	Stop	Mode 6	

Night Period (22h-7h) Wind direction North-West [270° ; 360°[									
Standardized wind speed at 10m	3m/s	4m/s	5m/s	6m/s	7m/s	8m/s	9m/s	>9m/s	Production loss
WT1									36.0%
WT2						Mode 1	Mode 1	Mode 1	
WT3						Mode 1	Mode 1	Mode 4	
WT4			Mode 6	Mode 1	Mode 6	Stop	Stop	Mode 6	
WT5			Mode 6	Stop	Stop	Stop	Stop	Stop	
WT6					Mode 5			Mode 5	
WT7				Mode 1	Mode 5	Mode 3	Mode 5	Mode 6	
WT8				Mode 6	Mode 5	Stop	Stop	Mode 6	

Figure 7: Curtailment Plan for project B, North-West wind direction (night period), before optimization (table above) and after optimization (table below). In green: Full Power mode. In orange: Restricted mode. In Red: Stop.

4. Conclusion and perspectives

The methodology presented in this paper is based on simultaneous optimization of the operating modes and the positions of the wind turbines. It allows a significant gain on the average electricity production of the wind farm. For projects where acoustic losses are initially high, this gain can be up to 3% at night, and 1% for day/night combined.
In terms of input data, this requires knowledge of the latitudes of each machine in relation to their initial location, in addition to the data required for acoustic impact calculations. Simplifications had to be made in the mathematical formulation of the problem (e.g. correction of geometric decay, wake effects, calculation of the electricity production), but after verification by full recalculation, these errors appear to be small in terms of wind turbine location.

In the future, improvements can be made taking into account the programming constraints of wind turbine suppliers and solving some problems of non-convergence of the optimisation algorithm.

This approach modifies the usual process of wind farm development projects. Acoustic studies are very often carried out at the end of the project, i.e. once the site is virtually fixed. If the acoustic studies are to be given the opportunity to optimise and therefore slightly modify the position of the machines, this must be anticipated, as it will necessarily modify several parts of the Environmental Impact Assessment.

References

ISO9613-2: *Acoustics - Attenuation of sound during propagation outdoors - Part 2: General method of calculation.*

Ministère de la Transition Ecologique, *Protocole de mesure de l'impact acoustique d'un parc éolien terrestre* (2022) : https://www.ecologie.gouv.fr/sites/default/files/protocole-mesure-acoustique.pdf

B. Plovsing, J. Kragh, *Nord2000. Comprehensive Outdoor Sound Propagation Model. Part 1: Propagation in an Atmosphere without Significant Refraction. Hørsholm*: DELTA Acoustics & Vibration Report AV 1849/00 (2006).

B. Plovsing, J. Kragh, Nord2000. *Comprehensive Outdoor Sound Propagation Model. Part 2: Propagation in an Atmosphere with Refraction*. Hørsholm: DELTA Acoustics & Vibration Report AV 1851/00 (2006).

A. Bigot, P. Economou, C. Economou, *Wind turbine noise prediction using Olive Tree Lab Terrain*. 7th International Conference on Wind Turbine Noise (2017).

E. Klotz et al., *Practical guidelines for solving difficult mixed integer linear programs*. Surv. Oper. Res. Manag. Sci., 18 (2013).

10th International Conference
on
Wind Turbine Noise
Dublin – 21st to 23rd June 2023

Evaluation of acoustic measurements in relation to annoyance reports from residents in the vicinity of a wind farm

Esther Blumendeller, blumendeller@ifb.uni-stuttgart.de
Po Wen Cheng, cheng@ifb.uni-stuttgart.de
Stuttgart Wind Energy at University of Stuttgart, Germany

Summary
Within the framework of the interdisciplinary project Inter-Wind, the annoyance of residents of a wind farm is related to acoustic, ground motion and meteorological data, as well as wind turbine (WT) operational data. Measurements were carried out at a wind farm in southern Germany and in parallel at residential sites in the vicinity of the wind farm, while residents were able to report different levels of annoyance with a noise reporting app. This paper focuses on acoustic measurements in the low and infrasonic frequency range. The parallel evaluation allows filtering and assessment of the acoustic, but also meteorological and WT-operational data in relation to the annoyance periods. It has been shown, that ratings with at least somewhat annoyance are present at maximum rotational speeds of a WT, higher wind speeds at hub height and stable atmospheric conditions. Wind direction, air temperature and humidity cannot be related with annoyance reports in this study.

1 Introduction

The expansion of renewable energy is one of the major social tasks in Europe and therefore requires acceptance and support from the population. In the case of onshore WTs, the complaints of local residents are manifold and concern for example visual aspects, noise, vibrations and shadow flicker [1, 2]. In previous studies [3, 4, 5], residents of wind farms were provided with diaries or an app to document times of annoyance in relation to WTs. Here, different approaches were pursued for the combination with acoustic, meteorological or WT operational data. In order to understand the effects of WT noise on local residents, the Inter-Wind research project (grants 03EE2023A-D) analyses annoyance situations of local residents in an interdisciplinary approach. For this purpose, the interaction of sound (University of Stuttgart – Stuttgart Wind Energy (SWE)), ground motion (Karlsruhe Institute of Technology – Geophysical Institute (GPI)) and meteorology (Centre for Solar Energy – and Hydrogen Research Baden-Wuerttemberg (ZSW)) is being investigated in connection to psychological annoyance assessment (Medical

School Hamburg (MSH)). While measurements were carried out, the residents of the wind farm were able to document their annoyance via a noise reporting app and indicate e.g. the level of annoyance and noise description. This enables a systematic inspection of the measurement data at the times when the noise reporting app was used. In this contribution, the focus is on the acoustic, meteorological and WT operational data and the relation to reported annoyance.

2 Measurement campaigns and instrumentation

During three measurement campaigns in winter 2020, spring 2022 and winter 2022/2023, acoustic and meteorological data were collected in the vicinity of a wind farm. Wind farm Tegelberg with three GE 2.7 MW WTs is located on the Swabian Alb in southern Germany close to an escarpment. In total, acoustic measurement data of about 184 days are available from 2020-10-23 to 2020-12-15 (campaign Tegelberg 1), 2022-03-24 to 2022-05-11 (campaign Tegelberg 2) and 2022-11-10 to 2023-01-29 (campaign Tegelberg 3).

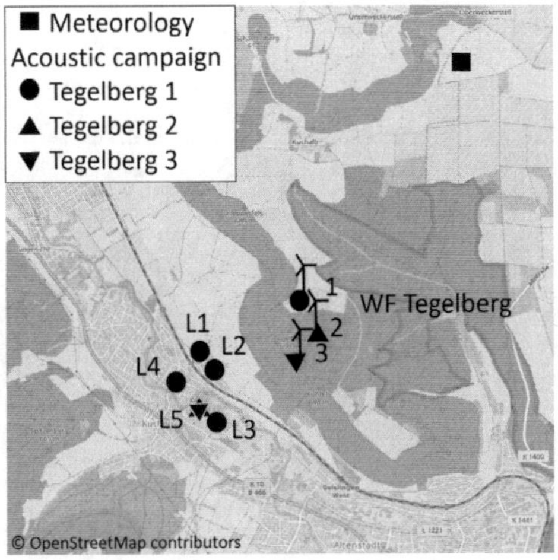

Figure 1. Map [6] with the locations of the wind turbines of wind farm Tegelberg, and measurement locations at the wind farm and in the municipality of Kuchen. Circles indicate positions during measurement campaign Tegelberg 1 in 2020, triangles indicate positions during measurement campaign Tegelberg 2 and 3 in 2022 and 2023. The square indicates the position of the meteorological measuring mast.

Figure 1 shows a map with the microphone, meteorological measuring mast and WT locations. Acoustic measurements were conducted with two G.R.A.S. 47AC 1/2 inch free-field condenser microphones at the outside locations (wind farm and outside buildings) and with a Brüel&Kjaer 4964 1/2 inch free-field infrasound microphone with G.R.A.S. 26CI preamplifier inside the buildings. During measurement campaign Tegelberg 1, one microphone was positioned 140 m south of WT 1, whereas the other two microphones were positioned outside and inside residential sites L1-L4 in the municipality of Kuchen in the direct vicinity of the wind farm at about 1 km distance. During campaign Tegelberg 2, the space around WT 2 was chosen as measurement location, where the microphone was positioned 50 m south of the WT 2. For this campaign, the microphone was chosen to be placed near the WT closest to the residents. Since WT 3 was not operating during the planned measurement time period, WT 2 was selected for the measurements at the wind farm. During the third campaign (Tegelberg 3), the microphone was positioned 50 m north-east of WT 3. Location L5, in a closed outdoor swimming pool, was chosen for the other two microphones, since sites L1-L4 were not available during the second and third campaign. Here measurements were conducted throughout the campaigns outside and inside a small clubhouse, where WT 2 and WT 3 were also visible.

Although the wooden building construction does not correspond to a typical residential building in Germany, it can still be used comparatively. Detailed information regarding wind farm and measurement instrumentation can be found in [7, 8].

During the measurement campaigns, the WTs were regularly switched off during pre-defined night time periods. This allows it to assign acoustic signals to the operation of the WTs. Furthermore 10 minute mean operational data was provided by the WT operator. The parameters of the WT operation and the meteorological data measured at hub height are shown for the measurement campaigns in Figure 2 a)–d). It can be seen, that all WTs mainly operated at rotational speeds above 5 rpm. While the wind was mainly coming from southern and south-western direction during the Tegelberg 1 and Tegelberg 3 campaigns, the main direction during the Tegelberg 2 campaign in spring time was south-east and west. The hub height wind speed varied around 5 m/s in all campaigns. It is also shown that rated power was reached only at about 4 % of the time.

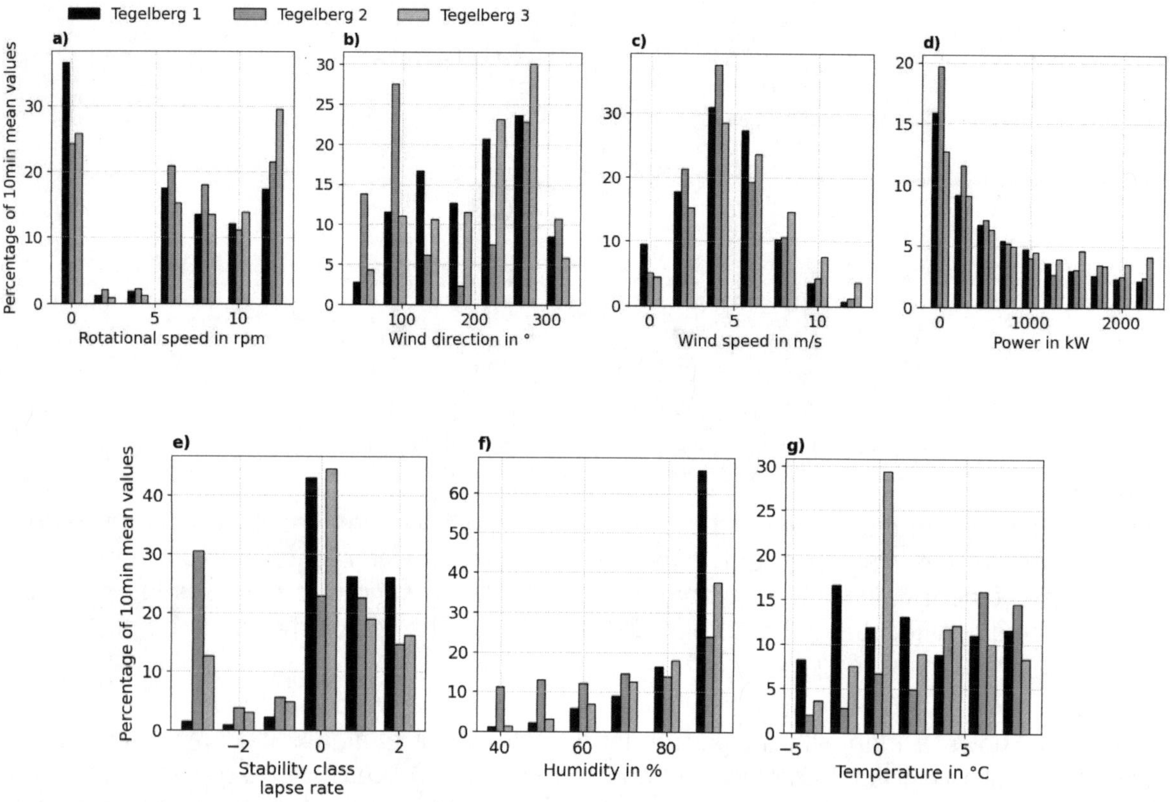

Figure 2. WT data of WT 2 and meteorological data from the measuring mast in 2.4 km distance to the wind farm. a) rpm, b) wind direction at hub height, c) wind speed at hub height, d) absolute power output, e) lapse rate indicating stable (1 – 3), neutral (0) and unstable (–1 – –3) atmospheric conditions, f) humidity in 3 m height, g) temperature in 3 m height.

Meteorological data is obtained from measurements at a measuring mast in 2.4 km distance to the wind farm at a test site (see Figure 1) at ZSW. The lapse rate γ based on the gradient of virtual potential temperature Θ_v at two different heights z (3 m and 96 m), corrected for air pressure, moisture and density effects is defined as

$$\gamma = \frac{\Delta\Theta_v}{\Delta z}. \tag{1}$$

This parameter is used to assess static atmospheric stability [9]. With a positive lapse rate, the temperatures near the ground are lower than at 96 m, which leads to less mixing of the air layers and thus stable conditions. If there are higher temperatures near the ground than at altitude and thus a negative lapse rate, this is an indication of unstable atmospheric conditions. An overview of the meteorological conditions during the three measurement campaigns can be seen in Figure 2 e)–g) for atmospheric stability, humidity and temperature. During the Tegelberg 1 campaign, atmospheric conditions were mainly neutral and stable. In the second and third campaign more time periods with unstable conditions were present. Figures 2 f) and g) show, that the winter months have higher humidity and lower temperatures compared to the spring months.

In order to assess the residential noise annoyance caused by WTs, resident surveys were conducted by the MSH. The aim was to understand how residents perceive WTs and WT immissions. In addition to the survey, an app was developed that allowed residents to report if and when they felt annoyed by WT noise. The use of the app was made possible during the same period as the acoustic and meteorological measurements, which enables an analysis at the times when residents documented annoyance. When they reported to hear WT noise, they were asked to rate the annoyance on a scale from 0 („not at all") to 4 („very"). During the second and third measurement campaign, residents were additionally asked to use the app regularly at a fixed time, even if there was no annoyance [10].

Data filtering and evaluation

The use of the app by residents during the same time as acoustic and meteorological measurements and together with available WT operating data, allows a precise analysis of these documented time points. The acoustic, meteorological and WT operating data of all measurement campaigns were filtered depending on the criteria listed in Table 1.

Table 1. Overview of the filter criteria and number of app reports for the three measurement campaigns. The level of annoyance is rated on a scale from 0 "not annoyed at all" to 4 "very annoyed".

Filter criteria	Level of annoyance	Reports Tegelberg 1	Reports Tegelberg 2	Reports Tegelberg 3
Sound heard + annoyed	2–4	79	25	22
Sound heard + not annoyed	0	–	17	1
No sound heard	–	1	113	36

In order to obtain detailed information about the spectral content of the WT sound, a narrowband analysis of the acoustic data is carried out. Narrowband spectra (spectrograms with scipy [11]) were calculated over periods of 10 minutes for a time length of T=10 s, Hanning window with window length N_{win}=T·fs for fs=20 kHz and 50 % overlap [8].

3 Results

The WT operational and meteorological parameters considered in Figure 2 are shown in Figure 3 for app ratings with somewhat to very annoyance levels of the residents. During the three measurement campaigns, more annoyance reports occurred at WT operation with maximum rotational speed and stable atmospheric conditions (Figure 3 a) and e)). In addition, higher wind speeds at hub height were present during reported annoyance (Figure 3 c)). The mean value of the wind speed of the measurement campaigns of 5.6 m/s is exceeded by 2 m/s for the annoyance reports during Tegelberg 1, by 3.5 m/s during Tegelberg 2 and by 2.7 m/s during

Tegelberg 3 campaign. For all other parameters such as wind direction, power output, temperature and humidity, no clear association with the annoyance reports can be identified.

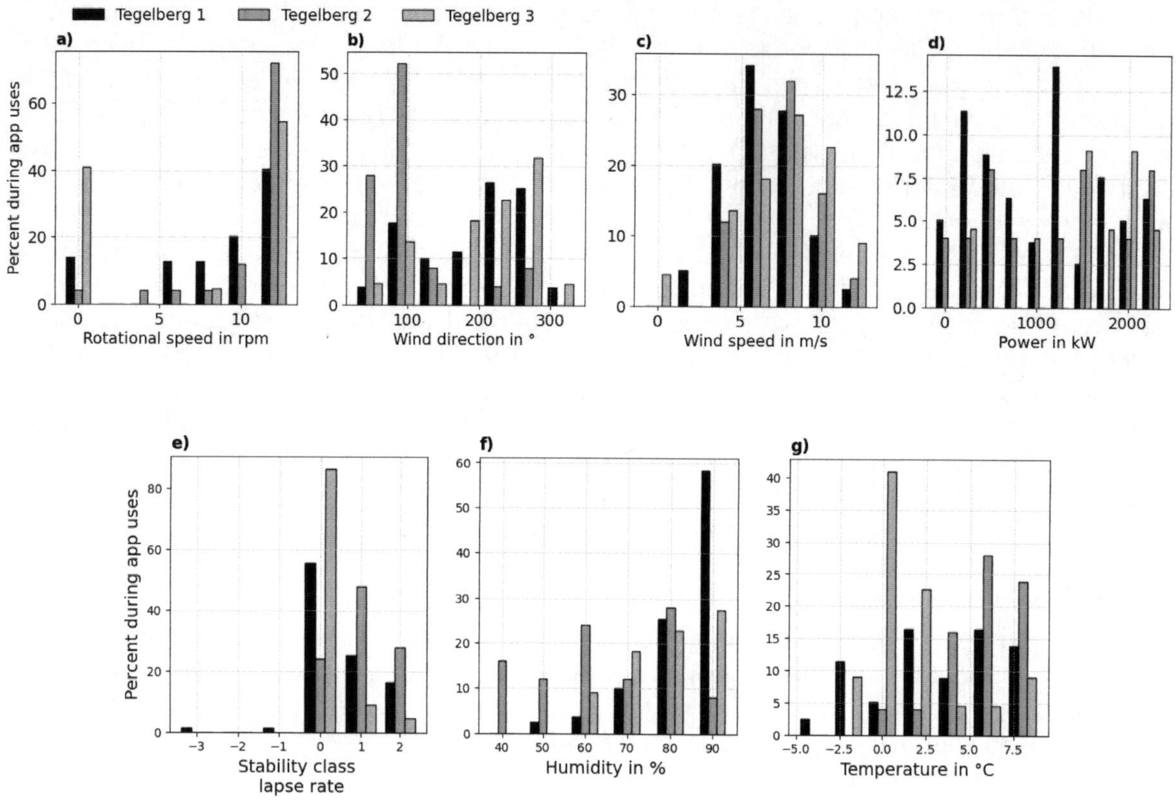

Figure 3. WT data of WT 2 and meteorological data during app reports with level of annoyance 2–4 during all measurement campaigns. a) rpm, b) wind direction, c) wind speed, d) absolute power output, e) lapse rate for stable (1 – 3), neutral (0) and unstable (–1 – –3) atmospheric conditions, f) humidity, g) temperature.

For a comparison of times with and without reported annoyance, the three filter criteria according to Table 1 are compared regarding WT operation and meteorology during the Tegelberg 2 measurement campaign. Figure 4 shows, that annoyance reports mainly occur during WT operation with maximum rotational speed. In contrast, it can be seen that in 50 % of the periods without perceived WT sound, the WT was out of operation. This indicates, that the app reports from residents can be attributed to WT operation. Furthermore, the wind speed for times with somewhat to very annoyance level is on average 9.1 m/s and around 5.4 m/s without present sound, from which a association of increased number of annoyance reports at higher wind speeds can be deduced. From Figure 4 e) it can be seen, that stable atmospheric conditions are associated with more annoyance reports. No association can be found between wind direction, humidity and temperature and annoyance reports.

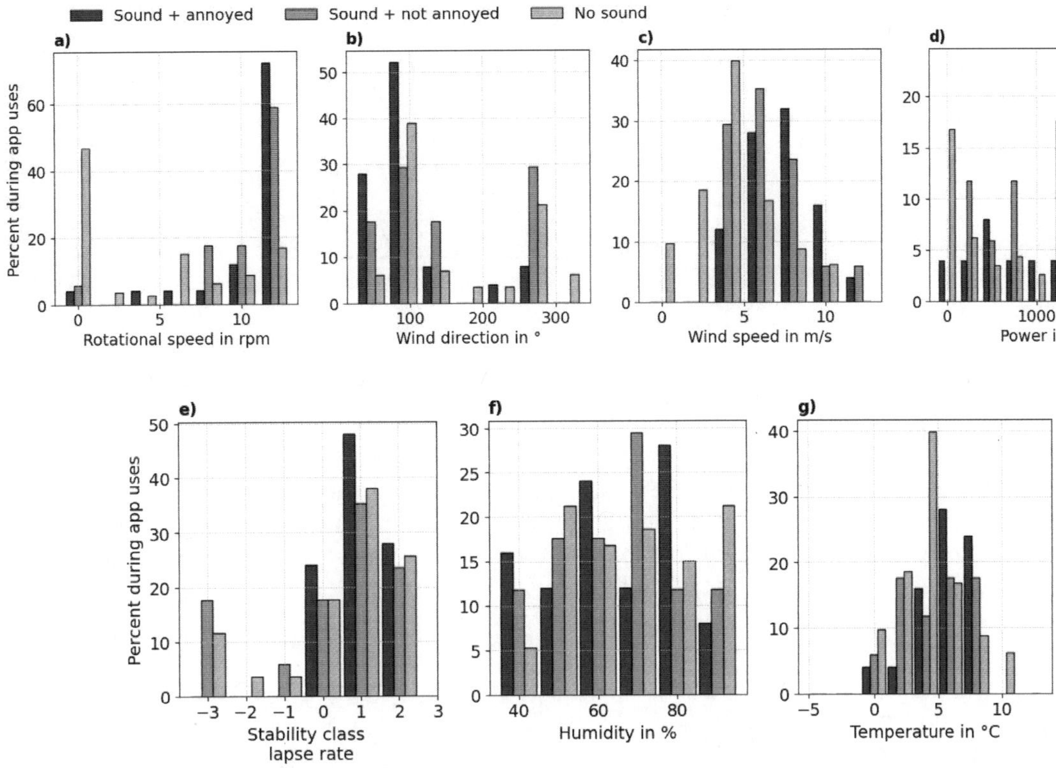

Figure 4. WT 2 operational and meteorological data during app reports of the Tegelberg 2 measurement campaign, separated by a) rpm, b) wind direction, c) wind speed, d) absolute power output, e) lapse rate for stable (1 – 3), neutral (0) and unstable (–1 – –3) atmospheric conditions, f) humidity, g) temperature.

Figure 5 a) shows the narrowband spectra for the microphone position at WT 2 for the three filter criteria according to Table 1 of the Tegelberg 2 measurement campaign. Different sound pressure levels (SPL) over the entire frequency range up to 200 Hz for the different degrees of annoyance rating can be identified. The background levels are obtained by filtering the acoustic data for similar hub height wind speeds corresponding to the wind speeds of the three annoyance criteria (as seen in Figure 4 b)) and for night time periods, to minimise extraneous noise.

A comparison of the frequency spectrum for the time points with annoyance reports (with annoyance level 2–4) with the spectra for varying WT rotational speeds in Figure 5 b) shows, that these level differences seem to depend on the range of rotational speed. The full-load operation at 12 rpm seems to be an important factor in annoyance ratings with somewhat to very levels.

Similar results are obtained for the measurement position in the building at a distance of 1 km from the wind farm (Figure 6 a) and b)). Especially in the frequency range below 20 Hz, different SPL for the different degrees of noise rating can be identified, which are also dependent on the rotational speed of the WT.

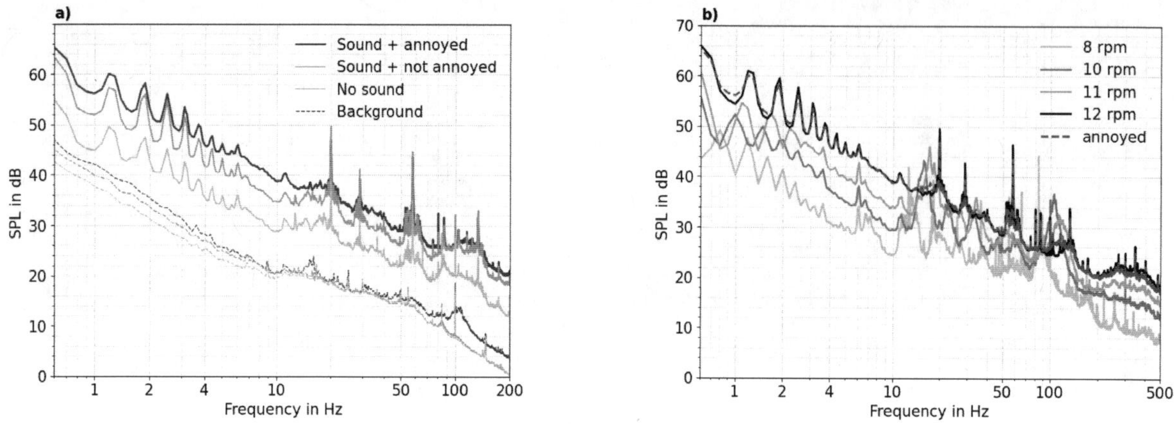

Figure 5. a) Averaged narrowband spectra of acoustic data measured at the wind farm during the Tegelberg 2 campaign and filtered for the three criteria according to Table 1. b) Narrowband spectra during nighttime from 20:00-23:50 UTC filtered for rotational speeds of WT 2 between 8 rpm and 12 rpm and spectrum during annoyance reports with somewhat to very annoyance level.

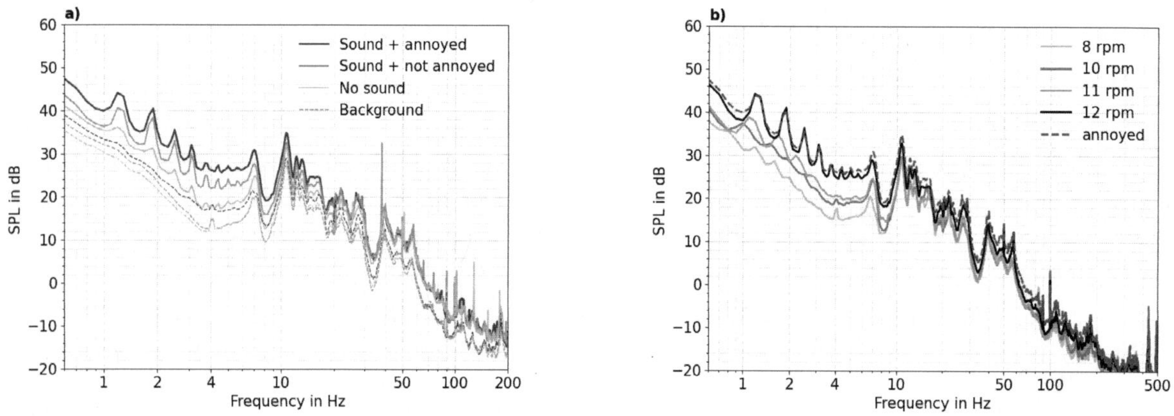

Figure 6. a) Averaged narrowband spectra of acoustic data measured inside the building during the Tegelberg 2 campaign and filtered for the three criteria according to Table 1. b) Narrowband spectra during nighttime from 20:00-23:50 UTC filtered for rotational speeds of WT 2 between 8 rpm and 12 rpm and spectrum during annoyance reports with somewhat to very annoyance level.

The 1/3-octave spectrum in Figure 7 a) enables a comparison with the human hearing threshold according to [12, 13]. SPL differences in the frequency range below 20 Hz are also recognisable here for the different degrees of noise rating. However, the SPL in this frequency range are far below the hearing threshold and exceed it only at about 50 Hz.

Figure 7 shows the averaged and A-weighted SPL for the frequency range 20 Hz–10 kHz for the three measurement positions and filter criteria. Slightly higher SPL are present at the wind farm during the times with annoyance rating 2–4. At a distance of 1 km outside and inside the building, the SPL are comparable for all filter criteria.

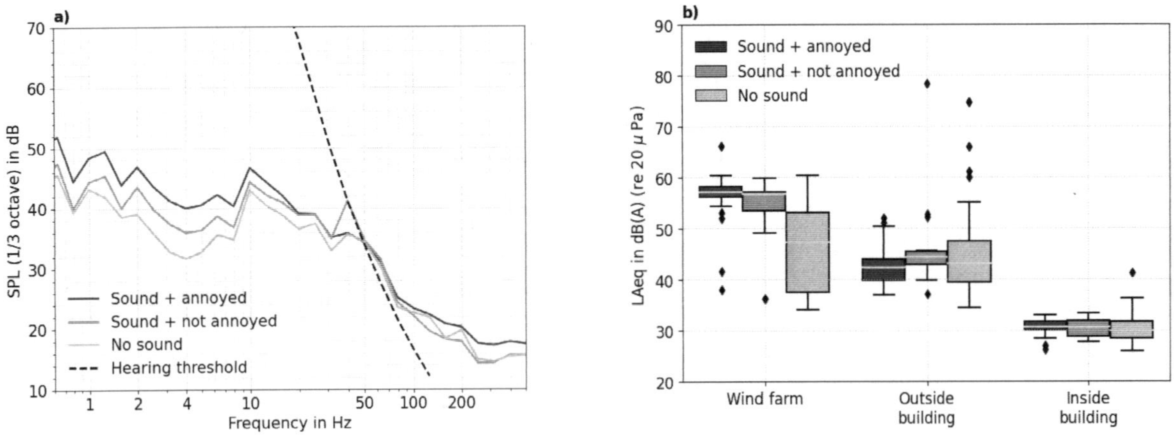

Figure 7. a) Averaged 1/3-octave spectra of acoustic data measured inside the building during the Tegelberg 2 campaign and filtered for the three criteria according to Table 1. b) Comparison of $L_{A,eq10min}$ (frequency range 20 Hz–10 kHz) for the three filter criteria at the three microphone positions. Light grey lines indicate the mean value.

4 Conclusion and outlook

In this paper, data filtering was demonstrated using annoyance times from a reporting app near a wind farm. The data filtering was performed for the acoustic, meteorological and WT operational data of three measurement campaigns in winter 2020, spring 2022 and winter 2022/2023. It was shown that ratings with at least somewhat annoyance are present at maximum rotational speeds of a WT, higher wind speeds at hub height and stable atmospheric conditions. Wind direction, air temperature and humidity cannot be related to annoyance reports in this study. During time periods of somewhat to very annoyed ratings, higher SPL are present in the frequency range below 20 Hz compared to time periods without annoyance, which seem to be dependent on the WT rotational speed. It has been shown, that the SPL in this frequency range are far below the human hearing threshold and only exceed the hearing threshold at 50 Hz. Therefore, this range is not considered relevant for annoyance. Finally, previous studies, e.g. [14, 15, 16] can be confirmed, that the equivalent and A-weighted SPL is not suitable for establishing a connection between WT noise and annoyance.

Acknowledgements

We thank the local authorities of the municipality Kuchen for their support, as well as the Stadtwerke Schwäbisch Hall and the KWA Contracting AG for providing operating data. Furthermore we thank Philipp Berlinger and Sebastian Putzke for supporting the acoustic measurements, Florian Müller (MSH) and Maayen Wigger (ZSW) for providing data. We acknowledge the support of the local residents, who enabled the installation of instruments on their property and within their houses. This study is supported by the Federal Ministry for Economic Affairs and Climate Action based on a resolution of the German Bundestag (grants 03EE2023B).

References

[1] van Kamp, I., & van den Berg, F. (2018). Health Effects Related to Wind Turbine Sound, Including Low-Frequency Sound and Infrasound. Acoustics Australia, 46(1), 31–57. https://doi.org/10.1007/s40857-017-0115-6.

[2] Pedersen, E., & Persson Waye, K. (2004). Perception and annoyance due to wind turbine noise—a dose–response relationship. The Journal of the Acoustical Society of America, 116(6), 3460–3470. https://doi.org/10.1121/1.1815091

[3] Cooper, S. (2014). The results of an acoustic testing program Cape Bridgewater wind farm.

[4] Hansen, K. L., Nguyen, P., Micic, G., Lechat, B., Catcheside, P., & Zajamšek, B. (2021). Amplitude modulated wind farm noise relationship with annoyance: A year-long field study. The Journal of the Acoustical Society of America, 150(2), 1198–1208. https://doi.org/10.1121/10.0005849

[5] Søndergaard, L. S., Volk, C. P., Hansen, T. R., Enggaard, L., Sørensen, T., & Pena, A. (2021). Wind farm neighbourship investigated by a daily app questionnaire combining weather, noise, and annoyance. 9th International Conference on Wind Turbine Noise.

[6] OpenStreetMap, https://www.openstreetmap.org, 2023-03-20.

[7] Gaßner, L., Blumendeller, E., Müller, F. J. Y., Wigger, M., Rettenmeier, A., Cheng, P. W., Hübner, G., Ritter, J., & Pohl, J. (2022). Joint analysis of resident complaints, meteorological, acoustic, and ground motion data to establish a robust annoyance evaluation of wind turbine emissions. Renewable Energy, 188, 1072–1093. https://doi.org/10.1016/j.renene.2022.02.081

[8] Blumendeller, E., Hofsäß, M., Goertlitz, A., & Cheng, P. W. (2022). Impact of wind turbine operation conditions on infrasonic and low frequency sound induced by on-shore wind turbines. Journal of Physics: Conference Series. https://doi.org/10.1088/1742-6596/2265/3/032048

[9] Platis, A., Hundhausen, M., Lampert, A., Emeis, S., & Bange, J. (2022). The Role of Atmospheric Stability and Turbulence in Offshore Wind-Farm Wakes in the German Bight. Boundary-Layer Meteorology, 182(3), 441–469. https://doi.org/10.1007/s10546-021-00668-4

[10] Müller, F. J. Y., Leschinger, V., Hübner, G., & Pohl, J. (2023). Understanding subjective and situational factors of wind turbine noise annoyance. Energy Policy.

[11] J. D. Hunter, "Matplotlib: A 2D Graphics Environment," in Computing in Science & Engineering, vol. 9, no. 3, pp. 90-95, May-June 2007. https://doi.org/10.1109/MCSE.2007.55.

[12] DIN 45680:1997-03: Measurement and assessment of low-frequency noise immissions in the neighbourhood.

[13] Møller, H., & Pedersen, C. S. (2004). Hearing at low and infrasonic frequencies. Noise and Health, 6(23), 37–57.

[14] Persson, K., Björkman, M., & Rylander, R. (1985). An Experimental Evaluation of Annoyance Due to Low Frequency Noise. Journal of Low Frequency Noise, Vibration and Active Control, 4(4), 145–153. https://doi.org/10.1177/026309238500400401.

[15] Persson, K., & Björkman, M. (1988) Annoyance due to low frequency noise and the use of the dB(A) scale. Journal of Sound and Vibration.

[16] Siponen, D. (2011). Noise annoyance of wind turbines. Technical Research Centre of Finland (VTT), 1–26.

10th International Conference
on
Wind Turbine Noise
Dublin – 21st to 23rd June 2023

Adaptive Tuned Mass Dampers for reduction of multiple resonances in variable speed wind turbine applications

Alexander Busch, ESM GmbH, Alexander.Busch@esm-gmbh.de
Sebastian Schmidt, ZF Friedrichshafen AG, Sebastian.Schmidt2@zf.com
Ben Marrant, ZF Wind Power Antwerpen NV, Ben.Marrant@zf.com
Tom van der Kamp, ZF Wind Power Antwerpen NV, Tom.vanderkamp@zf.com

Summary

In this paper, the application of an adaptive tuned mass damper on a wind turbine gearbox is presented. After the motivation and the actual problem definition, the basics of regular tuned mass dampers and the functionality of adaptive tuned mass dampers are introduced. Thereupon the impact of the device and achievable reductions of the vibration amplitude is shown in an experimental environment. In a first experiment at a shaker test set-up, the principal functionality is assessed. Furthermore, the adaptive tuned mass damper is applied on a 13 MW gearbox test rig to achieve environmental conditions with respect to vibrations as in the real wind turbine.

1. Introduction

Wind turbine OEMs are being faced with more stringent noise regulations for onshore wind turbine applications [Klu2002]. The overall sound pressure level of the wind turbine is determined by the aero-acoustic noise generated by the aerodynamic flow around rotor blades [Oer2007], [Jia2012]. The aero-acoustic noise is broadband in frequency [Oer2007]. This is opposed to mechanical noise, which is narrowband [Jia2012].

Typical sources of mechanical noise are gears, cooling fans, the generator, etc. [Jia2012]. In the case of geared drivetrain applications, the excitation occurs at discrete[1] mesh frequencies, inducing vibrations which are transmitted along the structure of a turbine and further amplified throughout the drivetrain and the remainder of the wind turbine at resonances and finally being radiated by the rotor, the tower or nacelle to the surroundings [Jia2012]. This noise mechanism is referred to as structure borne sound. As the excitation occurs only at one single frequency at a time, the noise produced is called a tone. When this narrowband tone protrudes from the

[1] In theory, the excitation is a perfect single frequency (discrete frequency), in practice the excitation frequencies will occur in a narrowband but can still be assumed a single frequency due to small bandwidth.

broadband aero-acoustic masking this is called a tonality. Although a tone is not determining the total sound pressure level of the wind turbine it can still be experienced as an undesirable sound [Jia2012], [Klu2002].

The gear mesh frequencies, at which the gears excite, are proportional to the rotational speed of the wind turbine drivetrain. These are referred to as gear mesh orders. When an excitation such as a gear mesh order crosses an eigenfrequency, related to an eigenmode, this results in a resonance [Gas2013].

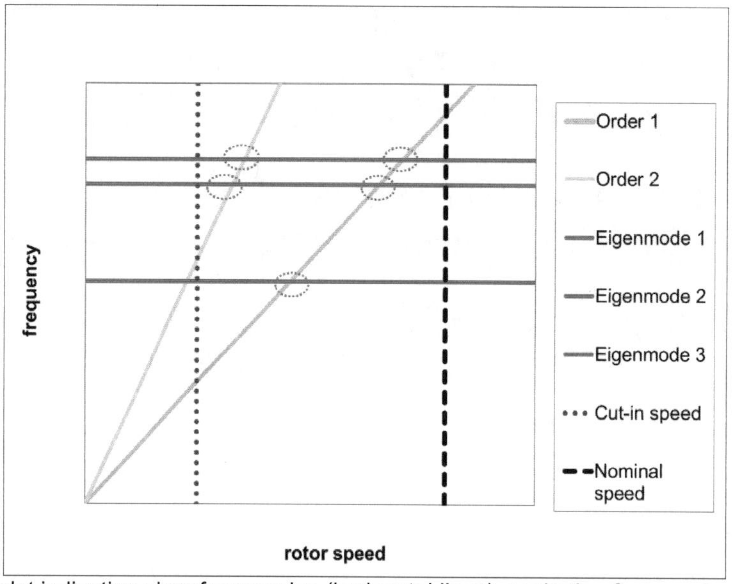

Figure 1: Campbell plot indicating eigenfrequencies (horizontal lines), excitation frequencies (skewed lines) and operating range (vertical lines).

One gear mesh order might excite multiple resonances. As a modern wind turbine makes use of variable speed operation [Hal2011], one should address all critical resonances in the operational speed range by:

- design optimizations or
- mitigation devices, e.g., dampers.

In the next sections theory and practice of an adaptive tuned mass damper are being explained as well as how the adaptive variant compares to a classic tuned mass damper. The adaptive tuned mass damper is being tested on a small-scale dedicated test bench as well as on a full-size gearbox.

2. Theory and simulation of adaptive tuned mass dampers

A common approach to address structure borne sound is the application of tuned mass dampers (TMD). This section introduces regular (passive) TMD and their working principle. Furthermore, it classifies adaptive tuned mass dampers (ATMD) as a subgroup of TMD and introduces into the basics of a TMD as well as the functionality of the ATMD.

2.1 Working principle of regular TMD

In applications with a high modal density, a variable speed, and multiple sources of excitation such as wind turbines, resonances cannot be avoided. In cases where it is not feasible to achieve acceptable amplitude reductions by means of design optimization tuned mass dampers are a means for amplitude reduction. This section will describe the basic working principle.

The target of a tuned mass damper is the compensation of the result of excitation forces or excitation torques with counter acting mass or inertia forces. In the example of a one mass oscillator with the mass *m,* the spring stiffness *n,* the damping *d* and an excitation force *F,* a tuned mass damper can be added as a second mass m_T with a spring stiffness n_T and a damping d_T creating a counter action force F_T (see *Figure 2*). The tuned mass damper is further characterized by a tuning frequency ω_T.

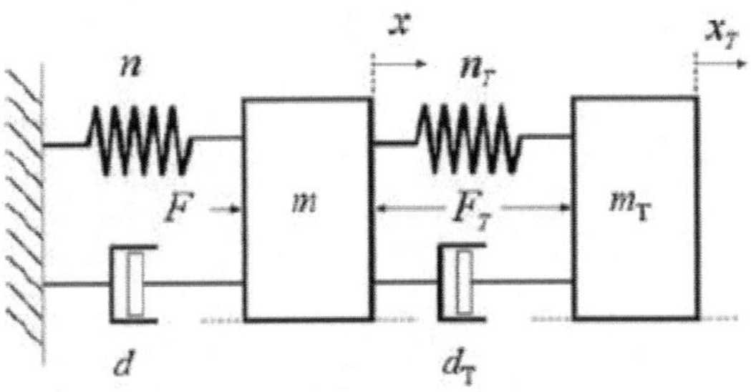

Figure 2: One-mass oscillator with tuned mass damper forming a two-mass oscillator [Zel2018].

Figure 3 shows the amplification functions of a system without tuned mass damper (grey) and two different designs of a tuned mass damper. The amplification function of the system with tuned mass dampers can be characterized by two peaks - one peak for both masses oscillating in phase, the other peak for both masses oscillating counter phase.

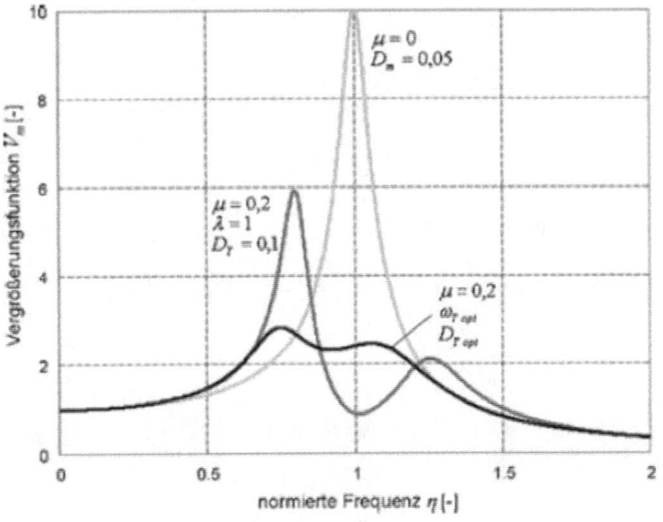

Figure 3: Effect of different tuned mass damper parameters [Zel2018].

In drive trains with gearboxes, a variable speed will cause a variable gear excitation. In the case of a variable gear excitation the aim is typically to find a tuned mass damper design such that the resulting amplifications of both peaks are similar in amplification. This can also be referenced as parameter set D_{Topt} and ω_{Topt} [Pet2001].

In case of lower damping (mid grey curve) amplification in the lower hand peak is increased whilst the amplification around η=1 is significantly lower (approximately 6 dB lower compared to the parameters D_{Topt} and ω_{Topt}). The following section will explain how this benefit is utilized in an adaptive tuned mass damper.

2.2 Adaptive tuned mass dampers as a special form of tuned mass dampers

Adaptive tuned mass dampers (ATMD) can be classified as a subset of regular, passive TMD since an ATMD consists of a mass, a spring element and damping. The term "adaptive" refers to the capability of adapting or tuning its resonance frequency. It is designed for variable single frequency excitations such as the gear mesh frequency and works with low inherent damping.

The frequency-dominating elements of a TMD are the mass and the spring. The adaptivity of an ATMD system can be therefore achieved either by a variable modal mass or a variable stiffness [Hill2002]. This paper focusses on a system designed with a constant mass to provide a fixed mass ratio and thus, a more balanced reduction of the main system throughout different frequencies [Pet2001].

Figure 4 shows the reduction of the amplitude of the main system or structure by a damping device. The figure apparently shows the typical behaviour of a TMD, tuned to the eigenfrequency of a main system. Even though this figure seems to show a TMD, we will consider it an ATMD for now. In this example, the best reduction is achieved in a small frequency range around 140 Hz.

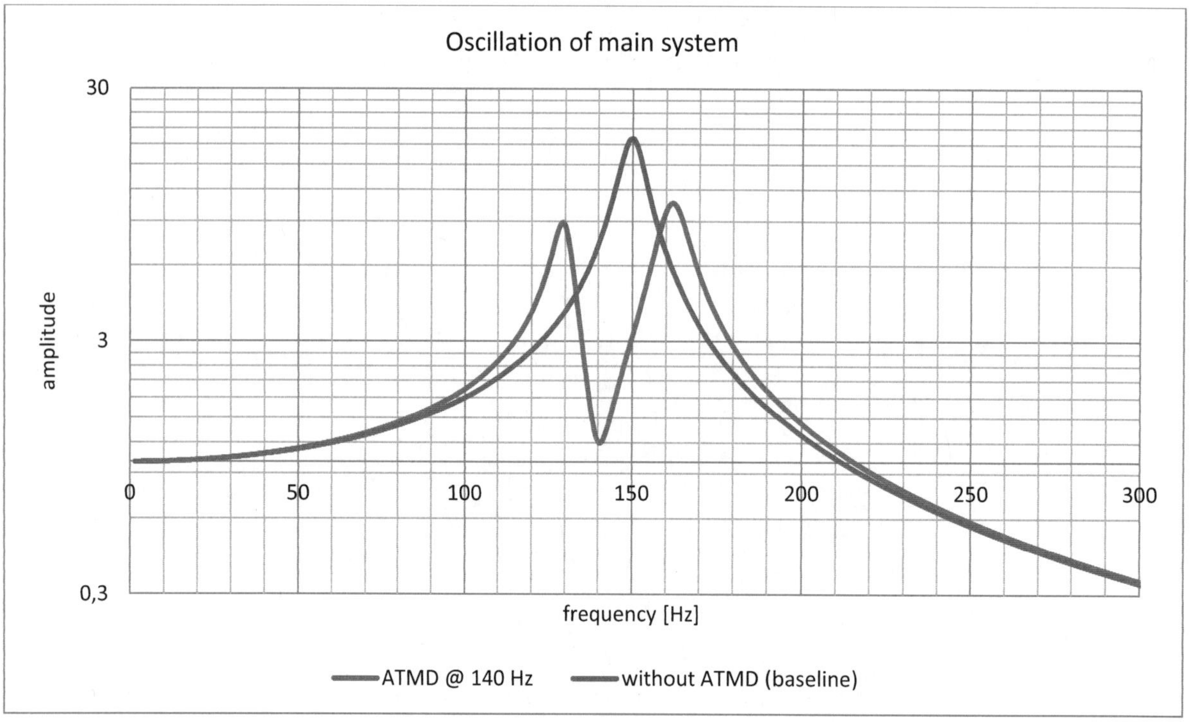

Figure 4: Simulated impact of a TMD / ATMD tuned to a fixed frequency.

By increasing the stiffness of the ATMD while operating, it is capable to apply the same behaviour as for 140 Hz for example to 160 Hz or 180 Hz as well. In fact, an ATMD can set its stiffness to any value between the minimum and maximum stiffness. Therefore, it is able to tune itself to any frequency between the highest and lowest frequency, which are defined by limitations due to geometry of the design. Or, simpler – an ATMD can be thought of as a continuum of independently working TMD. The overall reduction would then be the envelope of the minima (point of maximum reduction of the main system) caused by every single TMD.

The following figure shows the reduction when the ATMD is tuned to specific fixed frequencies between 140 Hz and 200 Hz (thin lines) and the reduction when the ATMD tracks the excitation from 140 Hz to 200 Hz producing the mentioned envelope.

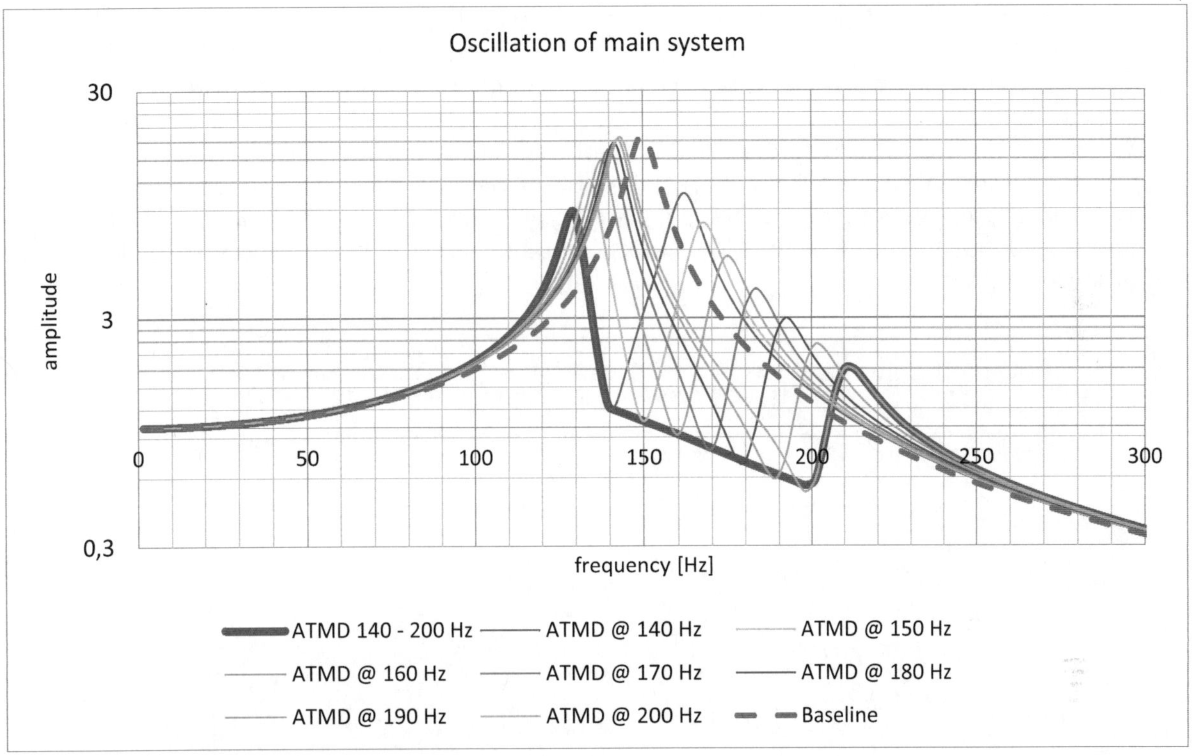

Figure 5: Impact of an ATMD tuned to various fixed frequencies and the reduction of the main system (red dashed curve) when the ATMD is operating and retuning itself to every frequency between 140 Hz and 200 Hz (blue curve). The mass ratio of main system and ATMD mass is approx. 4,7 %.

2.3 ATMD design

The ATMD (see *Figure 6*) consists of a steel mass, spring packages providing stiffness, support structure and an electrical unit (actuator and controller). The desired frequency is obviously determined by the mass - stiffness ratio. The actuator changes the ATMDs stiffness to tune it to a target frequency or to track a speed dependent excitation.

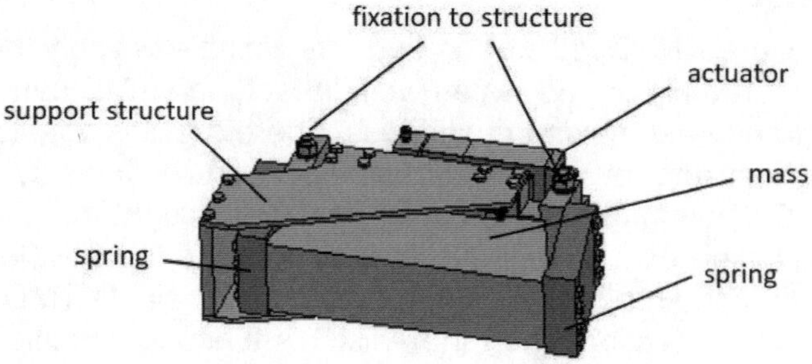

Figure 6: ATMD developed by ESM

A controller converts the input speed into a target frequency and controls the actuators movement accordingly. The operating frequency range can be adjusted depending on the application. Whenever an excitation (e.g., first order of gear mesh frequency) enters the operating frequency range, it can be tracked by the ATMD. Equivalently to the first order, the second and third order can be tracked as well, providing the possibility of a so called "split order tracking". In the split order tracking, the ATMD tracks a particular speed dependent excitation and tunes itself constantly to the corresponding frequency. As soon as this specific order is not close to an eigenfrequency anymore, the ATMD tracks another order, which is approaching an eigenfrequency.

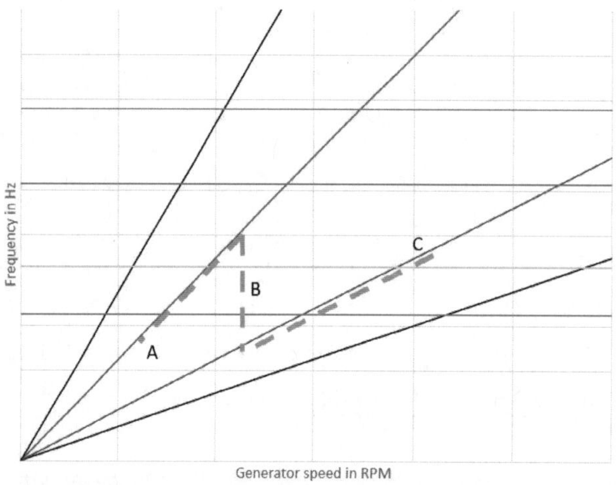

Figure 7: Simplified Campbell diagram; horizontal lines represent eigenfrequencies, diagonal lines indicate speed dependent excitations and dashed lines showing the split order tracking.

In Figure 7, the ATMD would track the red excitation (starting from speed A) while it is close to the yellow and red eigenfrequency (up to speed B).When the purple excitation gets closer to the red eigenfrequency, the ATMD switches to track the purple excitation up to speed C instead of the red excitation.

3. Measurements on a shaker test bench at ESM

In this test setup, the ATMD is mounted on a steel plate with a steel spring suspension representing the main system or structure to be reduced. Even though, the frequency range is different, the mass ratio is again 4,7 % as for the simulations. Two shakers are deployed to apply a unidirectional sine-sweep excitation to the system. Simultaneously, the excitation frequency is used as input (representing gearbox speed in the wind turbine application) to the ATMD controller determining the required movement of the actuator to bring the ATMD into tune and match the excitation frequency.

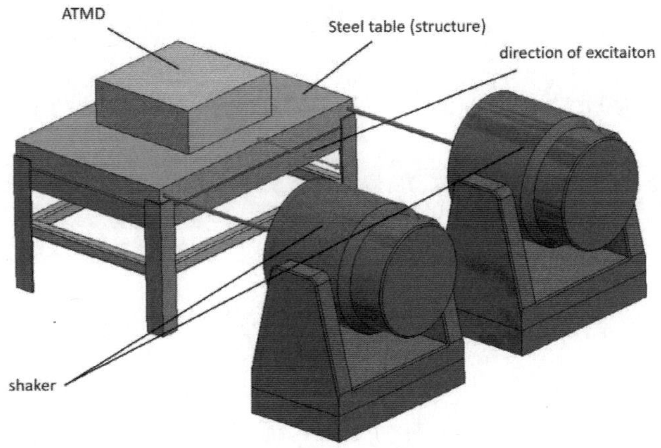

Figure 8: Test set-up for investigation of vibrational behaviour at ESM.

In contrast to the simulations for an exemplary frequency range shown above, the following measurements have been used as preparation for an application on a gearbox test bench. The amplitudes are normalized with respect to the maximum baseline amplitude and frequencies are normalized with respect to the maximum frequency.

To evaluate the performance of the ATMD, two experiments are done. In the first step, the ATMD is disabled and out of tune (means the operating frequency of the ATMD is far off the frequency range that is excited by the shakers). The structure is then excited at a desired frequency range by the shakers performing a sine sweep. This measurement without the impact of the ATMD is considered as the baseline measurement. In the second measurement, the same process is applied, but this time the ATMD is engaged and tracks the excitation. The results are shown in *Figure 9* below.

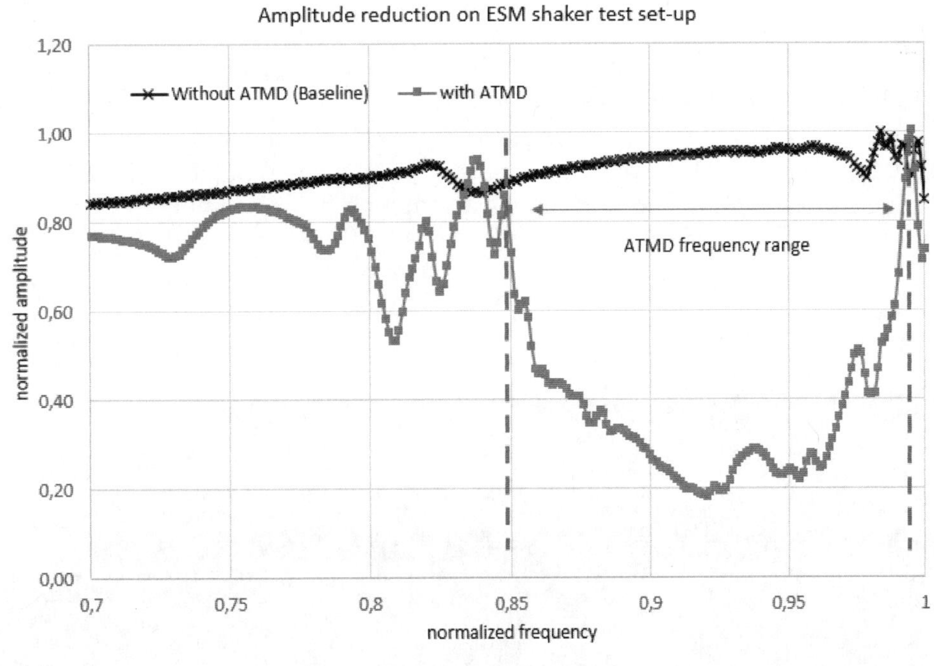

Figure 9: Comparison of vibration amplitudes on a shaker table with and without ATMD. The mass ratio is approx. 4.7 %. Axis are normalized to the maximum frequency and the maximum amplitude of the baseline measurement.

The vibration level of the structure is reduced to 0.2 which corresponds to -14 dB reduction, the average reduction in the ATMD frequency range is -8 dB.

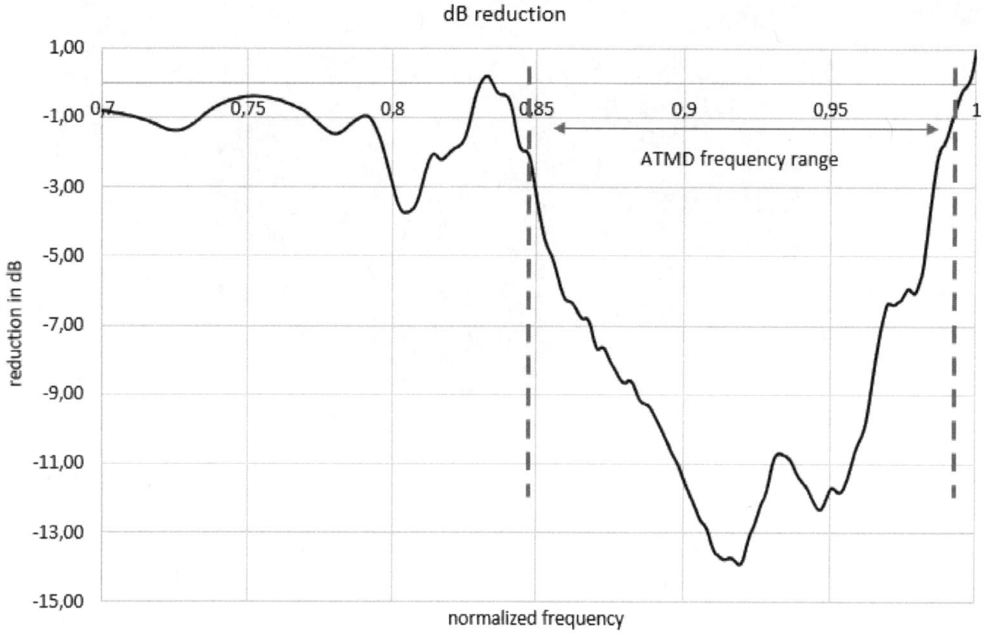

Figure 10: Reduction of the vibration amplitude at ESM shaker table in dB.

The results in *Figure 10* show a significant reduction at the ESM test set-up with 4.7 % mass ratio. For the gear box of a wind turbine, the mass ratio is with < 1 % even lower. According to multibody simulations, considering important application parameters such as frequency, mode shape and mounting position, the reductions caused by the ATMD are still significant even in a < 1 % mass ratio application.

4. Measurements on a gear box test bench

To test the ATMD in a more realistic environment, the ATMD was mounted on a ZF gearbox on a test set-up. In such a test set-up two gearboxes are mounted back-to-back.

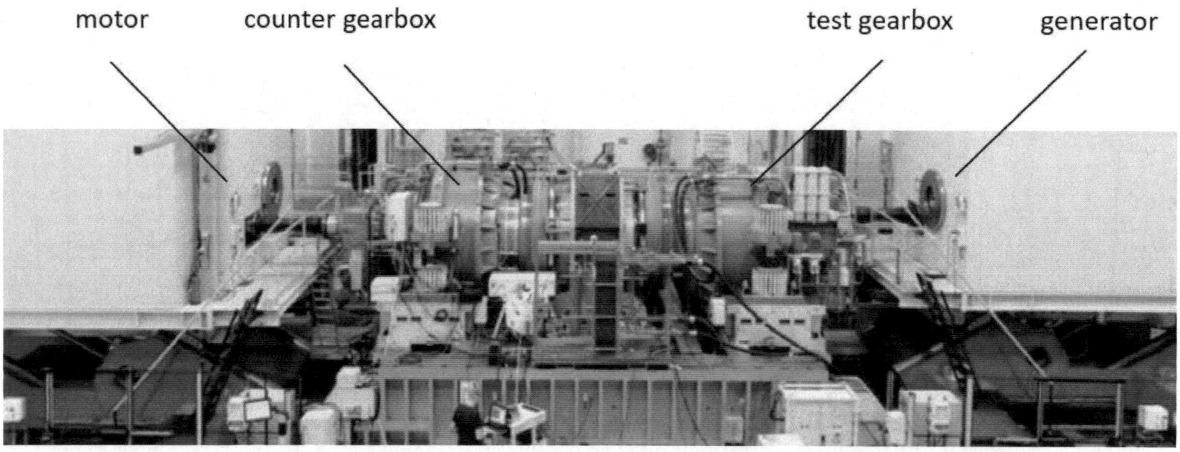

Figure 11: Exemplary ZF gearbox test set-up with back-to-back mounting (13 MW test rig).

The counter gearbox, which is being driven by a motor represents the wind turbine rotor system. The test gearbox is connected to a generator and represents the wind turbine drivetrain system.

The ATMD was mounted on the test gearbox in tangential direction to counter-act torsional eigenmodes. A speed sensor measuring the output shaft rotational speed of the gearbox is used as an input for the ATMD controller unit. With this test set-up, so-called speed run-ups are being performed at different load conditions. The functionality of the ATMD can then be evaluated by tracking the vibration level along the gear mesh orders, responsible for the excitation of dominant eigenmodes, and comparing the situation with and without ATMD.

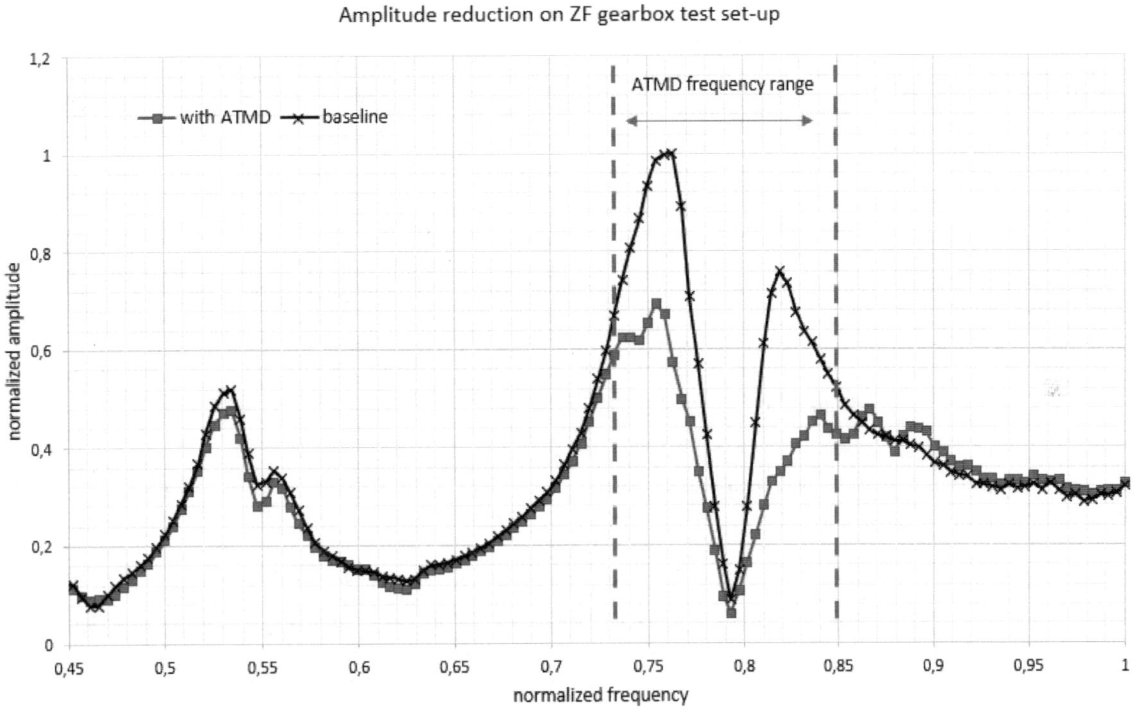

Figure 12: Vibration level on the gearbox with and without the ATMD.

The difference in vibration level for the gearbox in tangential direction with and without an ATMD can be seen in *Figure 12*, where the resonance amplitudes are significantly reduced by the ATMD operation. Outside the ATMD frequency range it can be observed that the ATMD has no effect on the vibration amplitudes. This behaviour is entirely different to regular low damped tuned mass dampers as shown in *Figure 3*. The difference in vibration level reduction converted into dB can be seen in *Figure 13*.

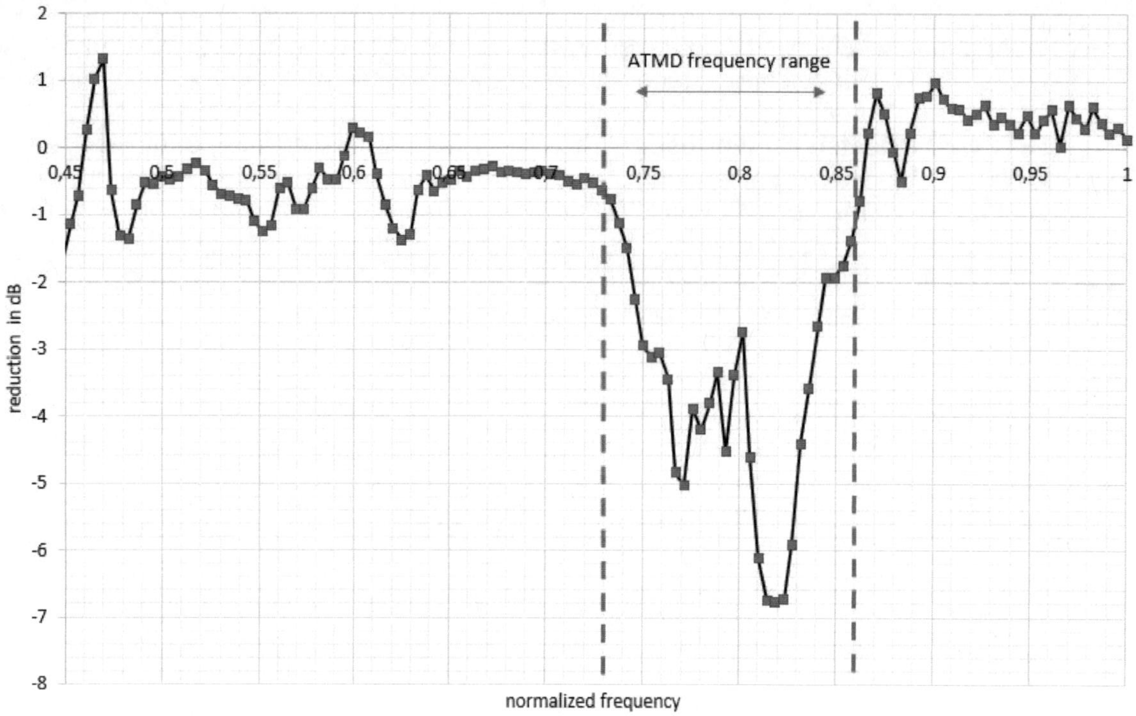

Figure 13: Reduction on the gearbox with and without the ATMD

In comparison to *Figure 10*, *Figure 13* shows a quite good similarity between the test results on ESM shaker test bench and ZF full-size gearbox test bench. Just the achieved maximum reduction level is different (approx. -14dB vs. approx. -7dB), which can be well explained by the different mass ratios applied (approx. 4.7% vs. <1%). However, even with the quite small mass ratio of < 1%, the reductions of the gearbox vibrations achieved by the application of the ATMD are still up to -7 dB.

5. Conclusions

Summarizing the results from this paper, a significant reduction in the vibration amplitudes can be achieved by an ATMD. This applies for a test set-up with shakers, where the excitations are unidirectional and artificially produced as well as for a gearbox testbench with the actual gearbox loads and corresponding excitations as in a wind turbine.

These reductions in the amplitude of structure borne sound can eventually contribute to a reduced turbine tonality level. For this purpose, it is crucial to consider decisive aspects and parameters for the ATMD application. These are at least the frequency range, the transfer path of the structure borne noise towards radiating surfaces in the turbine for a particular mode shape as well as the mode shape of the critical eigenmode itself. Here, also the positioning of the ATMD with respect to oscillation nodes and antinodes needs to fit to the eigenmode.

References

[Gas2013] Gasch, Robert, and Klaus Knothe. Strukturdynamik. Springer-Verlag, 9 Mar. 2013.

[Hal2011] Hall, John F., et al. "Wind Energy Conversion with a Variable-Ratio Gearbox: Design and Analysis." Renewable Energy, vol. 36, no. 3, Mar. 2011, pp. 1075–1080, https://doi.org/10.1016/j.renene.2010.08.037.

[Hill2002] Hill, Simon, et al. An Adaptive Vibration Absorber. 2002.

[Jia2012] Jianu, Ofelia, et al. "Noise Pollution Prevention in Wind Turbines: Status and Recent Advances." Sustainability, vol. 4, no. 6, 29 May 2012, pp. 1104–1117, https://doi.org/10.3390/su4061104.

[Klu2002] Klug, H. (2002, September). Noise from wind turbines: standards and noise reduction procedures. In Forum Acusticum, Sevilla, Spain.

[Oer2007] Oerlemans, S., et al. "Location and Quantification of Noise Sources on a Wind Turbine." Journal of Sound and Vibration, vol. 299, no. 4-5, Feb. 2007, pp. 869–883, https://doi.org/10.1016/j.jsv.2006.07.032.

[Pet2001] Petersen, Christian. Schwingungsdämpfer Im Ingenieurbau. 2001.

[Zel2018] Zeller, P. (2018). Schwingungsminderung. In: Zeller, P. (eds) Handbuch Fahrzeugakustik. ATZ/MTZ-Fachbuch. Springer Vieweg, Wiesbaden.

10th International Conference
on
Wind Turbine Noise
Dublin – 21st to 23rd June 2023

Wind farm compared to road traffic noise onset induced arousal responses during sleep

Peter Catcheside, Flinders University: peter.catcheside@flinders.edu.au
Bastien Lechat, Flinders University: bastien.lechat@flinders.edu.au
Branko Zajamsek, Flinders University: branko.zajamsek@pm.me
Kristy Hansen, Flinders University: kristy.hansen@flinders.edu.au
Gorica Micic, Flinders University: gorica.micic@flinders.edu.au
Duc Phuc Nguyen, Flinders University: ducphuc.nguyen@flinders.edu.au
Claire Dunbar, Flinders University: claire.dunbar@flinders.edu.au
Hannah Scott, Flinders University: hannah.scott@flinders.edu.au
Tessa Liebich, Flinders University: tessa.liebich@flinders.edu.au
Barbara Toson, Flinders University: barbara.toson@flinders.edu.au
Felix Decup, Flinders University: felix.decup@gmail.com
Andrew Vakulin, Flinders University: andrew.vakulin@flinders.edu.au
Nicole Lovato, Flinders University: nicole.lovato@flinders.edu.au
Leon Lack, Flinders University: leon.lack@flinders.edu.au
Colin Hansen: University of Adelaide: chansen@bigpond.net.au
Dorothy Bruck, Victoria University: dorothy.bruck@vu.edu.au
Ching Li Chai-Coetzer, Flinders University: ching.chaicoetzer@flinders.edu.au
Con Doolan, University of New South Wales c.doolan@unsw.edu.au

Summary

Despite rapid growth in wind farms globally, and community concerns regarding noise impacts, potential effects of wind farm noise exposure on sleep remain poorly understood. This study compared wind farm noise (WFN) versus road traffic noise (RTN) effects on electroencephalographically (EEG) defined arousal responses during sleep. Sixty-eight adults underwent in-laboratory sleep studies over 7 nights including, for this study, one night with repeated 20-sec WFN and RTN exposures. Following at least 2 minutes of established sleep and ≥20-sec between noise exposures, pre-recorded WFN or RTN samples were reproduced at sound pressure level (SPLs) of 30, 40 and 50 dBA in random order. The primary outcome was the probability of EEG-defined arousal events (>3 sec EEG shifts to faster frequencies) following the onset of each noise exposure. Awakening responses (≥15 sec EEG frequency shifts) were also evaluated. Noise type, sleep stage and SPL effects on arousal and awakening response probabilities were evaluated using mixed effects logistic regression analyses. Sixty-two participants (mean±SD aged 49±20 years, 35 females and 27 males) had sufficient replicates of noise exposure data for analysis. Arousal response probabilities were low, particularly in deep sleep, but showed a significant noise type-by-SPL interaction (χ^2=13, p=0.001), with marginally but significantly lower WFN compared to RTN arousal probabilities at 40 dBA (mean [95%CI]: 2.1 [1.5, 2.9] vs 3.2 [2.4, 4.2]%, p=0.016) and 50 dBA (5.0 [4.0, 6.2] vs 8.6 [6.9, 10.6]%, p<0.001). Awakenings were infrequent (<4% at 50 dBA) but showed similar effects. These findings support that acute wind farm noise onset events are marginally less sleep disruptive than road traffic noise events of equivalent SPL ≥40 dBA.

1. Introduction

Wind farm noise (WFN) has unique acoustic characteristics that could render it more intrusive, annoying and potentially sleep disruptive compared to other environmental noises. WFN is typically dominated by low frequencies (<200 Hz), which attenuate markedly less over distance and through objects. Consequently, WFN can remain audible at greater distances than higher frequency dominated environmental noises such as road traffic noise (RTN). WFN can also contain an amplitude modulation (AM) component: a regular time variation in SPL that is particularly characteristic of WFN and potentially the most intrusive and annoying noise feature (Muzet, 2007). Modern wind turbines also emit infrasound (Jakobsen, 2016): a very low frequency noise (<20 Hz) below the widely accepted average human audibility threshold in this frequency range (Møller and Pedersen, 2004). WFN is also influenced by other factors, including the number, type and size of turbines, other background noises (or lack of, particularly at night), local topology, wind speed and direction, temperature (and inversions), turbulence conditions, and intervening structures such as neighbouring buildings and the construction properties of residences in which individuals live. Given prominent audible and low frequency dominated features, WFN clearly has the potential to disrupt sleep.

Clearly discernible electroencephalography (EEG) markers of sleep disturbance are well-established for road, rail, and air traffic noise (Basner & McGuire, 2018; Basner, Glatz, Griefahn, Penzel, & Samel, 2008; Basner & McGuire, 2018; Dunbar et al., 2022; Elmenhorst et al., 2012; Smith et al., 2020). These markers include EEG micro-arousals (a shift to faster EEG frequencies for 3 to 15 seconds) and full-awakenings (≥15 second events), which fragment sleep and are associated with poorer daytime functioning and mood (Martin, Wraith, Deary, & Douglas, 1997). EEG responses depend on the type and intensity of the noise stimulus and sleep depth.

Evidence surrounding sleep disturbance from wind farm noise is still emerging (Micic et al., 2018; van Kamp & van den Berg, 2017), and well-controlled studies using objective measures of sleep physiology and WFN are lacking. This study sought to clarify the effects of WFN on sleep compared to RTN, which is already well-known to disrupt sleep and is thus a particularly useful comparator. Therefore, this study aimed to directly compare the dose-response characteristics of WFN versus RTN on the probability of EEG-defined micro-arousals and awakenings from sleep to 20-second noise exposures. Given prominent low frequency characteristics, the primary hypothesis was that EEG arousal responses would be more probable with WFN compared to RTN of equivalent A-weighted SPL.

2. Methods

Ethical and local governance approval was obtained from the Southern Adelaide Clinical Human Research Ethics Committee (Protocol number 343.18) and Flinders Medical Centre.

2.1 Participants

Participants were recruited via advertisements posted on websites, media outlets, and public noticeboards, participant databases from prior studies, and via community presentations and word of mouth. Eligibility criteria included age >18 years, able and willing to consent to the study and to travel and remain at the laboratory for seven consecutive nights. Individuals with self-reported sleep disorders other than insomnia, night shift work (between 10pm - 6am) or recent travel across >2 time zones in the past 2 months were excluded.

Participants were selected into four groups based on residential location and self-reported levels of sleep disturbance to wind turbine or traffic noise. Participants living <10km from a wind turbine who self-reported WFN-related sleep disturbance were classified as "WFN-sleep disturbed". Participants living <10km from a wind turbine who self-reported no WFN-related sleep disturbance were classified as "WFN non-sleep disturbed". Participants living in an urban area who self-reported RTN-related sleep disturbance were classified as "RTN-sleep disturbed". Participants living in a quiet rural area who did not report noise-related sleep disturbance were classified as "Rural control".

2.2 Experimental procedures

All participants completed an adaptation night (no noise) on the first evening of their laboratory stay, followed by six different overnight noise exposure conditions. This study only reports the findings from the 20-second noise battery night, which was specifically designed to examine the acute effects of WFN versus RTN noise exposures during EEG confirmed established sleep.

2.3 Sleep recordings

All participants underwent polysomnography recordings during each sleep opportunity to allow for direct evaluation of sleep disturbance. This included 7-channels of EEG according to the 10-20 electrode placement system (F3, F4, C3, C4, Cz, O1, O2 referenced to M1 or M2), left and right electro-oculograms (EOG), chin electromyogram (EMG), limb movements, electrocardiogram (ECG), and finger pulse oximetry measurements. Electrodes were fitted to achieve impedances <5 kΩ and all EEG signals were amplified and sampled at 512 Hz using Grael 4K Polysomnography and Profusion 4 EEG acquisition software (Compumedics Ltd, Melbourne Australia).

Sleep stages (N1, N2, N3 and REM), arousals (3-15 seconds) and awakenings (≥15 seconds) were scored according to American Academy of Sleep Medicine (AASM) manual sleep scoring guidelines by a single scorer blinded to the noise condition (Iber, Ancoli-Israel, Chesson, & Quan, 2007). An arousal or awakening was considered to have been evoked by the noise stimulus (or control) if it occurred anytime during the noise presentation (i.e., the 20-second window from stimulus onset).

2.4 Noise stimuli

The WFN sample was selected from previous recordings obtained 3.3 km from the nearest wind turbine, and included amplitude modulation with an AM depth representative of median values from year-long wind farm noise data collected at a residence 3.5 km from a wind farm. The RTN noise sample was selected from recordings obtained 700 m from a busy urban road.

On the night of the study, participants were exposed to 20-second samples of WFN and RTN samples in block-randomised order and 20-second background noise (19 dBA) as a control, with an inter-stimulus interval of 20 seconds. Noise samples were continued throughout periods of consolidated N2 or deeper sleep. However, in the event of a full-awakening from sleep (EEG arousals ≥15-seconds), noise stimuli were paused at the end of any currently playing stimulus, and only recommenced after N2 sleep was re-established.

WFN and RTN noise samples were played at three different SPLs: 30, 40, and 50 dBA with a 250 ms ramp-in and ramp-out for all noise samples.

2.5 Noise reproduction

Environmental noises were reproduced using an RME Babyface Pro sound card. Reproduction of the low-frequency and infrasonic components of pre-recorded wind farm and road traffic noise used a Krix KX-4010s loudspeaker (118 (H) x 670 (W) x 410 (D) mm) with closed vent and powered by Crown DC-300 amplifier. Traffic noise was reproduced using a LabGruppen C 10:4X amplifier, and a Krix Pheonix V2.1 loudspeaker (35 Hz to 40 kHz frequency response, 950 (H) x 195(W) x 295(D) mm). Both speakers were placed approximately 3m away from the head of the bed, facing the participant. The noise reproduction was achieved with the 1/3 octave band equalization of the reproduced noise with the original noise samples.

2.6 Synchronization and noise measurement

Noise reproduction and a synchronization timing signal output were controlled by custom software implemented in MATLAB (Version 2018a/b 9.4/9.5, Mathworks, Natick, Massachusetts, USA). Noise reproduction system calibration and equalisation were conducted at the sleeper's head position prior to each study. Noise in the participant's bedroom was also recorded using a PROSIG P8004 24-bit data acquisition system and a GRAS 40AZ microphone placed ~1 meter above the participant's head. This microphone can record noise level as low as 17 dBA (dynamic range: 17 to 132 dB) and from 0.5 Hz to 20 kHz (frequency range ± 2dB). At noise onset and offset, the MATLAB application sent a square pulse (200 msec long) to a separate channel on the RME Babyface Pro sound card which was also re-directed to a DC input channel of the Grael 4K Polysomnography system and sampled at 1024 Hz using Profusion 4 EEG acquisition software (Compumedics Ltd, Melbourne Australia). Trigger pulses were also sent from the MATLAB application to the PROSIG P8004 acquisition system to enable accurate temporal matching of polysomnography to independent noise recordings.

2.7 Statistical analysis

Noise type effects on arousal response probability were tested using mixed effects logistic regression to evaluate the interaction between noise type and SPLs. The association between arousal probability of occurrence and noise SPL and type was examined using mixed effects logistic regression with participant number as a random effect, each with a separate intercept. Effects of sleep stages and study group were investigated using three-way interactions between noise type, noise level and sleep stages or study group. These analyses were repeated with awakenings only (arousals ≥15 sec) in similar models. Results are reported as response probabilities and odds ratios (ORs) with their respective 95% confidence intervals (CI). Summary graphs for each model are presented with marginal probabilities and ORs. All statistical analyses were performed using the computing environment R (R Core Team, 2019) with lme4 (Bates, Machler, Bolker, & Walker, 2015) open-source package for mixed model analysis.

3. Results

3.1 Participant and sleep characteristics

A total of 68 participants consented to participate in this study, from which 62 participants contributed to the final analysis. The first 4 participants received more noise types, resulting in a low number of noise repetitions. Thus, these participants were excluded from analysis and the number of noise types was reduced in subsequent participants to increase overnight noise sample repetition. An additional 2 participants were excluded from analysis due to synchronization pulse failure, precluding accurate noise sample onset timing within polysomnography recordings. The demographics of the final sample of 62 participants are shown in Table 1.

Table 1. Participant demographics and polysomnography characteristics

			Participant Group			
		Overall Sample	**Rural Control**	**RTN-sleep disturbed**	**WFN-sleep disturbed**	**WFN non-sleep disturbed**
n		62	18	18	9	17
Age, years		48.1 (19.8)	46.7 (20.7)	33.5 (15.0)	67.2 (7.2)	54.8 (16.6)
Sex, n(%)	**Males**	27 (43.5%)	4 (22.2%)	10 (55.6%)	4 (44.4%)	9 (52.9%)
	Females	35 (56.5%)	14 (77.8%)	8 (44.4%)	5 (55.6%)	8 (47.1%)
BMI, kg/m^2		27.6 (5.6)	27.9 (5.7)	24.8 (5.9)	28.0 (3.7)	30.1 (5.1)
Total sleep time, min		426 (69)	429 (72)	435 (90)	386 (38)	434 (50)
Wake after sleep onset, min		62 (46)	67.5 (54)	48 (39)	91 (55)	56 (30)
N1, min		47 (25)	45 (27)	44 (24)	43 (12)	57 (29)
N2, min		203 (51)	213 (47)	205 (42)	185 (79)	199 (47)
N3, min		81 (44)	79 (47)	81 (42)	88 (44)	78 (47)
REM, min		95 (36)	92 (38)	105 (40)	70 (34)	99 (26)
AHI, events/hours		8.2 (14.6)	12.3 (22.2)	3.1 (6.2)	8.4 (5.3)	9.3 (13.4)

Data are reported as mean (SD) for continuous variable and n (%) for categorical variable. BMI: Body Mass Index. AHI: Apnoea hypopnea index.

3.2 Noise onset induced arousal responses and awakenings

WFN was less likely to evoke arousals compared to RTN as shown in Figure 1 (SPL-by-noise type interaction, χ^2 = 13, p= 0.001). At 50 dBA, the probability of a noise-evoked arousal response was 5.0 [4.0, 6.2]% (mean [95%CI]) for WFN compared to 8.6 [6.9, 10.6]% for RTN (Figure 1a). There was a similar effect at 40 dBA, although the difference was smaller (approximately 1%, Figure 1a). There was no significant effect of group (SPL-by-noise type-by-group interaction, χ^2 = 4.5, p = 0.610), and no significant three-way interaction between noise type, noise level and sleep stage (overall effect, χ^2 = 0.63, p = 0.730). However, arousal probabilities were much lower in N3 sleep compared to N1/N2 sleep, but were similar in N2/N1 sleep and REM sleep (Figure 2). The probability of full awakenings showed similar effects to arousals with lower probabilities for WFN than for RTN at equivalent SPLs (SPL-by-noise type interaction, χ^2 = 6.9, p = 0.032).

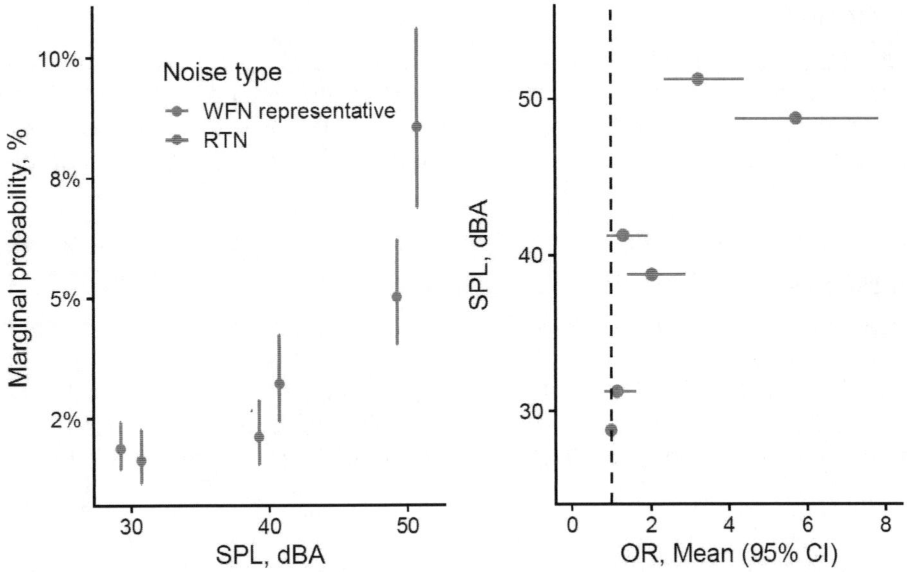

Figure 1. Left: The probability (mean±95% CI) of noise-related arousals or awakenings as a function of sound pressure level (SPL) at 30, 40 and 50 dBA for wind farm noise (WFN) with representative amplitude modulation depth (green) versus road traffic noise (RTN orange). Right: Odds ratio (OR, ±95% CI) of evoking an arousal/awakening compared to traffic noise at 30 dBA. Note that for clarity, WFN and RTN symbols at the same SPL are shown with an offset. n=62.

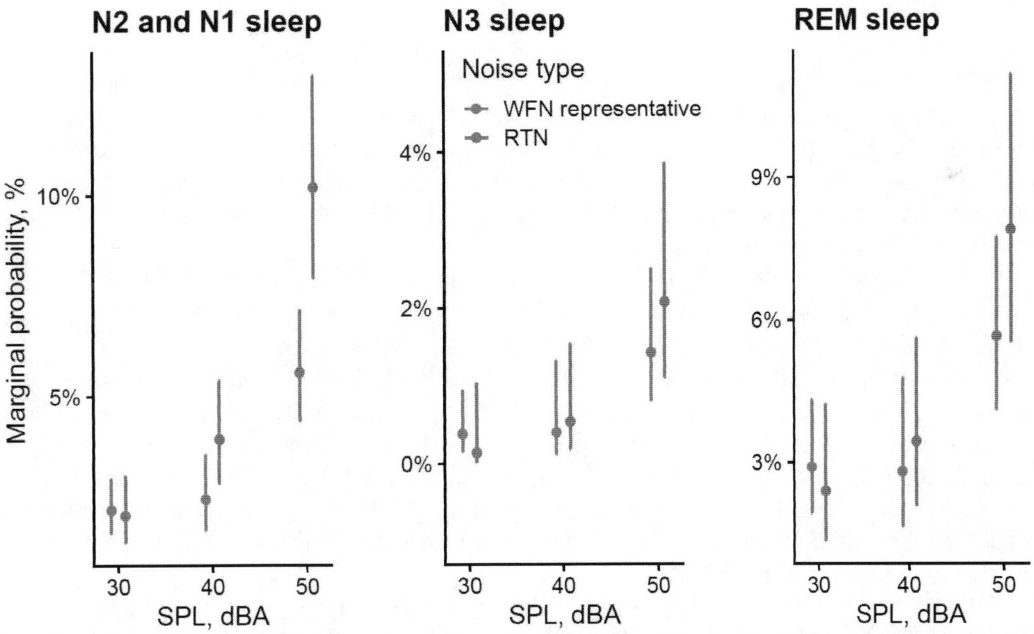

Figure 2. The effect of sound pressure level (SPL) and sleep stage on arousal response probabilities (mean±95% CI) for wind farm noise (WFN) with representative amplitude modulation depth (green) versus road traffic noise (RTN orange). n=62.

4. Conclusions

This study found that acute wind farm noise onset events are marginally less sleep disruptive than road traffic noise events of equivalent SPL ≥40 dBA, and with no significant differences between study groups to support prior noise exposure effects. Overall response probabilities were relatively low, but with a marked rise from 40 and 50 dBA indicating the sleep disruption potential of both noise types.

Acknowledgements

The authors gratefully acknowledge support for this study from study participants, technical support staff, and financial support for this and related work from the National Health and Medical Research Council (Grant 1113571) and the Australian Research Council (Grants DP120102185 and DE180100022 and Fellowship FT120100510). BL also gratefully acknowledges the Flinders Foundation for their support through the Nick Antic Sleep Research Scholarship.

References

Basner, M., Glatz, C., Griefahn, B., Penzel, T., & Samel, A. (2008). Aircraft noise: effects on macro- and microstructure of sleep. *Sleep Med, 9*(4), 382-387. doi:10.1016/j.sleep.2007.07.002

Basner, M., & McGuire, S. (2018). WHO Environmental Noise Guidelines for the European Region: A Systematic Review on Environmental Noise and Effects on Sleep. *Int J Environ Res Public Health, 15*(3). doi:10.3390/ijerph15030519

Bates, D., Machler, M., Bolker, B. M., & Walker, S. C. (2015). Fitting Linear Mixed-Effects Models Using lme4. *Journal of Statistical Software, 67*(1), 1-48. doi:DOI 10.18637/jss.v067.i01

Dunbar, C., Catcheside, P., Lechat, B., Hansen, K., Zajamsek, B., Liebich, T., . . . Micic, G. (2022). EEG power spectral responses to wind farm compared with road traffic noise during sleep: A laboratory study. *J Sleep Res, 31*(3), e13517. doi:10.1111/jsr.13517

Elmenhorst, E. M., Pennig, S., Rolny, V., Quehl, J., Mueller, U., Maass, H., & Basner, M. (2012). Examining nocturnal railway noise and aircraft noise in the field: sleep, psychomotor performance, and annoyance. *Sci Total Environ, 424*, 48-56. doi:10.1016/j.scitotenv.2012.02.024

Hansen, K. L., Nguyen, P., Micic, G., Lechat, B., Catcheside, P., & Zajamsek, B. (2021). Amplitude modulated wind farm noise relationship with annoyance: A year-long field study. *J Acoust Soc Am, 150*(2), 1198. doi:10.1121/10.0005849

Iber, C., Ancoli-Israel, S., Chesson, A. L., & Quan, S. (2007). *AASM - Manual for the Scoring of Sleep and Associated Events*: Westchester, IL: American Academy of Sleep Medicine.

Jakobsen, J. (2016). Infrasound Emission from Wind Turbines. *Journal of Low Frequency Noise, Vibration and Active Control, 24*(3), 145-155. doi:10.1260/026309205775374451

Martin, S. E., Wraith, P. K., Deary, I. J., & Douglas, N. J. (1997). The effect of nonvisible sleep fragmentation on daytime function. *American journal of respiratory and critical care medicine, 155*(5), 1596-1601.

Micic, G., Zajamsek, B., Lack, L., Hansen, K., Doolan, C., Hansen, C., . . . Catcheside, P. (2018). A Review of the Potential Impacts of Wind Farm Noise on Sleep. *Acoustics Australia, 46*, 87-97. doi:10.1007/s40857-017-0120-9

Møller, H. and Pedersen, C.S., 2004. Hearing at low and infrasonic frequencies. Noise & Health, 6(23), pp.37-57.

Muzet, A. (2007). Environmental noise, sleep and health. *Sleep Med Rev, 11*(2), 135-142. doi:10.1016/j.smrv.2006.09.001

R Core Team. (2019). R: A Language and Environment for Statistical Computing. Vienna, Austria: R Foundation for Statistical Computing. Retrieved from https://www.R-project.org/

Smith, M. G., Ogren, M., Thorsson, P., Hussain-Alkhateeb, L., Pedersen, E., Forssen, J., . . . Persson Waye, K. (2020). A laboratory study on the effects of wind turbine noise on sleep: results of the polysomnographic WiTNES study. *Sleep, 43*(9). doi:10.1093/sleep/zsaa046

van Kamp, I., & van den Berg, F. (2017). Health Effects Related to Wind Turbine Sound, Including Low-Frequency Sound and Infrasound. *Acoustics Australia, 46*(1), 31-57. doi:10.1007/s40857-017-0115-6

10th International Conference
on
Wind Turbine Noise
Dublin – 21st to 23rd June 2023

No discernible effect of wind farm infrasound exposure on electroencephalographic markers of sleep disturbance

Peter Catcheside, Flinders University: peter.catcheside@flinders.edu.au
Claire Dunbar, Flinders University: claire.dunbar@flinders.edu.au
Duc Phuc Nguyen, Flinders University: ducphuc.nguyen@flinders.edu.au
Bastien Lechat, Flinders University: bastien.lechat@flinders.edu.au
Hannah Scott, Flinders University: hannah.scott@flinders.edu.au
Kristy Hansen, Flinders University: kristy.hansen@flinders.edu.au
Branko Zajamsek, Flinders University: branko.zajamsek@pm.me
Gorica Micic, Flinders University: gorica.micic@flinders.edu.au
Tessa Liebich, Flinders University: tessa.liebich@flinders.edu.au
Andrew Vakulin, Flinders University: andrew.vakulin@flinders.edu.au
Leon Lack, Flinders University: leon.lack@flinders.edu.au

Summary

There has been protracted community debate arising from unknown impacts of wind farm generated infrasound (<20 Hz noise) on human health and sleep. This study tested for acute wind farm infrasound exposure effects on electroencephalographic (EEG) markers of sleep disturbance. Sixty-eight adults underwent in-laboratory sleep studies over 7 nights, including one night with repeated 3-minute exposures to full-spectrum wind farm noise (WFN) and wind farm infrasound alone, derived via low pass filtering to remove frequencies ≥20 Hz. Noise samples were replayed in random order throughout established sleep to achieve sound pressure levels (SPLs) equivalent to 30 and 35 dBA (without the low pass filter). For infrasound, this corresponded to 71 and 77 dBG, above 90th centile levels previously measured outdoors at residences >1 km from a wind farm. To examine EEG changes relative to the pre-stimulus baseline, sleep disturbance effects with each noise type and SPL were evaluated using logistic regression analyses applied to EEG arousal and K-complex probabilities in each 5-sec window from 20-sec before to 20-sec after each 3-min noise exposure. Changes in EEG spectral power in sleep-related frequency bands over time were also evaluated. Fifty-four participants (mean±SD aged 48±20 years, 31 females and 23 males) had sufficient replicates of WFN and infrasound exposure data for analysis. Full-spectrum WFN exposure produced an abrupt and acute increase in arousal probability within the first 5-sec of noise onset (odds ratio, 2.9 [95%CI 1.6 to 5.4], p= 0.018). In contrast, infrasound exposure produced no discernible changes in arousal or K-complex probabilities, or EEG spectral power. These findings do not support that acute wind farm infrasound exposure produces any discernible EEG changes during sleep.

1. Introduction

Wind farm noise (WFN) contains prominent infrasound, which is conventionally defined as sound below 20 Hz (Leventhall, 2004), and has been speculated to cause sleep disturbance but with little supporting evidence (Bolin et al., 2011; National Health and Medical Research Council 2015). A previous study using gold-standard electroencephalographic (EEG) assessments of brain activity found no discernible EEG changes with overnight exposure to 10 Hz infrasound at 105 dB (Okada & Inaba, 1990). A recent study, using 72 h of simulated wind farm infrasound exposure (1.6–20 Hz at ~90 dB pk re 20 μPa), also found no evidence to support any discernible effects on sleep (Marshall et al., 2023). However, no previous study has specifically tested for acute onset effects of wind farm infrasound on the sleep EEG. Consequently, this study, which commenced around the same time as that of Marshall et al, specifically sought to determine if the onset of pre-recorded wind farm noise, filtered to remove audible components ≥20 Hz, produced any discernible EEG changes during sleep. EEG responses to the unfiltered noise sample retaining audible components were also examined as a positive control.

2. Methods

The study was approved by the local Southern Adelaide Clinical Human Research Ethics Committee (Protocol number 343.18) and governance office and was undertaken in two specially designed sound attenuated bedrooms (overnight background noise 19 dBA) in the Flinders Health and Medical Research Institute: Sleep Health Nick Antic laboratory.

2.1 Study participants

Participants were recruited as part of a larger trial, through advertising on university noticeboards, word of mouth, social media and a computer assisted telephone survey conducted as a part of the wider project. Study eligibility criteria included age >18 years, able and willing to consent to the study and to travel and remain at the laboratory for seven consecutive nights. Individuals with self-reported sleep disorders other than insomnia, night shift work (between 10pm - 6am) or recent travel across >2 time zones in the past 2 months were excluded. Interested individuals were screened for eligibility via an online screening questionnaire (Qualtrics Pty Ltd, Utah, USA) and were provided with comprehensive study information prior to providing informed written consent.

Four groups were recruited; individuals residing <10km from a wind turbine, one group with and another without self-reported WFN related sleep disturbance (WFN sleep disturbed and WFN non-sleep disturbed respectively); individuals living close to a busy road averaging ~40,000 cars per day and with self-reported road traffic noise (RTN) related sleep disturbance (RTN sleep disturbed); and individuals living in a quiet rural area >10 km from any wind turbines and without self-reported environmental noise related sleep disturbance (Rural control).

2.2 Experimental Procedures

All participants completed an adaptation night (no noise) on the first evening of their laboratory stay, followed by six different overnight noise exposure conditions. However, this study only reports the findings from a 3-minute noise battery night during which participants were repeatedly exposed to full-spectrum wind farm noise, infrasound and no noise controls amongst other noise samples presented during established sleep.

2.3 Sleep recordings

On the night of the study, participants attended the sleep laboratory two hours prior to their usual bedtime to allow for sleep study setup and other experimental procedures. Overnight polysomnography recordings included 7 channels of electroencephalography (F3, Fz, F4, C3, C4, O1 and O2 referenced to linked M1 and M2), electrooculography (EOG), electrocardiography (ECG), chin electromyography (EMG), all sampled at 512 Hz, along with leg movement, nasal cannula, oro-nasal thermistor, chest and abdominal motion and finger oximetry. All sleep signals were acquired using Grael 4K hardware and Profusion 4 acquisition software (Compumedics Ltd., Abbotsford, Victoria, Australia).

2.4 Noise stimuli

The WFN sample was selected from previous recordings obtained 3.3 km from the nearest wind turbine. Two different 3-minute noise samples were used in this study. These included pre-recorded full-spectrum WFN, and wind farm infrasound alone, which was derived from the same noise sample by applying a low-pass filter at a cut-off frequency of 15 Hz to ensure that no noise characteristics above 20 Hz would remain audible to participants with normal hearing according to ISO 226:2003.

The noise reproduction system consisted of an RME BabyFace Pro sound card, modified (without vent) Krix KX4010S commercial cinema subwoofer (dimensions 118 x 670 x 410 cm) with 10-inch driver and Crown DC-300 power amplifier with a flat frequency response approaching 0 Hz. The speaker was placed along the bedroom wall approximately one metre from the foot of the participants' bed.

Each noise was played at two sound pressure levels (SPLs) of 30 and 35 dBA (before low pass filtering). These indoor SPLs were chosen to approximate recommended outdoor night-time environmental noise limits of 40 dBA allowing for a median outdoor-indoor SPL difference of approximately 10 dBA (Locher et al., 2018). The reproduced infrasound stimuli were equivalent to (mean±SD) 71±2 and 77±3 dBG, above the 90 percentile of outdoor infrasound levels measured at residences greater than 1 km from a wind farm (Nguyen et al., 2022). Given typical outdoor-indoor attenuation of around 5 dBG (Hansen et al., 2015), these levels were selected to represent realistic near worst-case levels of WF infrasound.

To ensure faithful noise reproduction, noise was also recorded via a PROSIG P8004 24-bit data acquisition system and a GRAS 40AZ microphone, with a flat frequency response down to 0.5 Hz, positioned 1 metre above the participant's head. WFN above 100 Hz was not evident in the selected sample and the reproduced SPL was below the ambient SPL at all frequencies higher than 100 Hz. The reproduced infrasound spectra ≤100 Hz are shown in **Figure 1**.

Overnight technical staff supervising each study used a custom MATLAB interface to commence noise exposure once participants had achieved at least 5 minutes of N2 or deeper (N3 or REM) sleep. Infrasound and full-spectrum noise samples at each SPL were then played throughout the night in randomised order, using a ramp-in of 250 ms and ramp-out of 300 ms to ensure relatively abrupt noise onsets and offsets from which to evaluate EEG response effects. Each 3-minute noise sample was separated by a 20 second quiet (room background noise) period and was accompanied by an onset and offset trigger signal to both acoustics and sleep recording data acquisition systems to enable accurate temporal matching of sleep recordings to each noise stimulus. In the event of a full-awakening from sleep (EEG arousals ≥15-seconds), noise stimuli were suspended at the end of any currently playing stimulus, and only recommenced after N2 or deeper sleep was re-established.

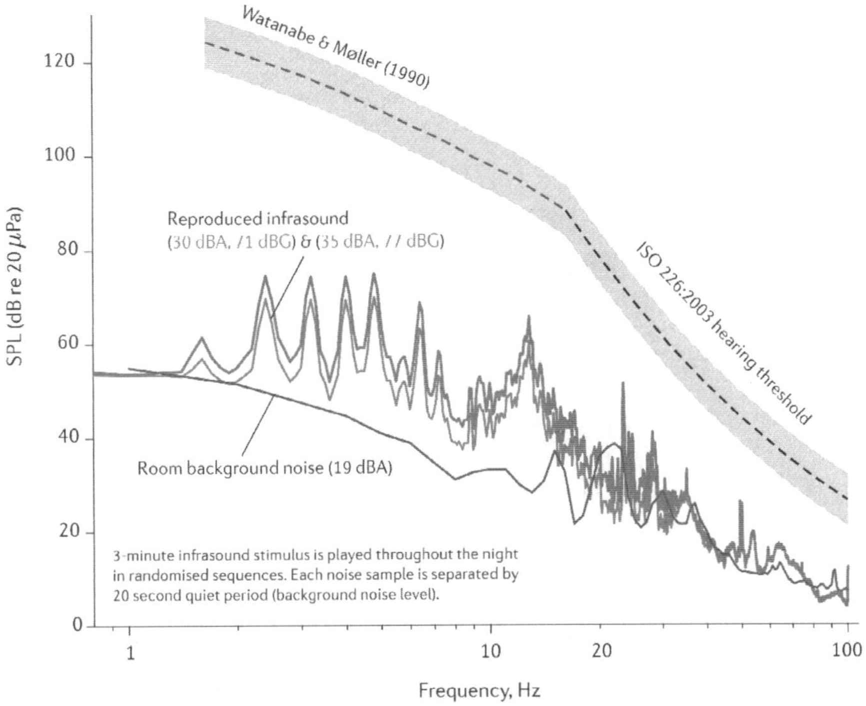

Figure 1. Reproduced wind farm infrasound spectra. The background noise of the sleep laboratory at night was 19 dB(A). The stimulus used during sleep remained well below the infrasound hearing threshold curve (shading indicates ±1 SD) reported by Watanabe and Møller (1990) with no wind farm noise characteristics above 20 Hz within the audible range according to the ISO 226:2003 hearing threshold curve (shading indicates ±1 SD 5.35 dB).

2.5 Analysis

Sleep stages (N1, N2, N3 and REM), arousals (≥3 second shifts in EEG frequencies) were scored according to American Academy of Sleep Medicine (AASM) manual sleep scoring guidelines (Iber et al., 2007) by a single scorer blinded to noise conditions and trigger signals.

The primary analysis examined arousal probability over time in 5-second intervals from 20-seconds before noise onset to 20-seconds after noise offset, using logistic regression models with time and noise type as fixed effects to examine arousal probability changes compared to the pre-stimulus baseline. Models were adjusted for several covariates including noise level, sleep stage, age and gender. Participants and groups were included as random effects to account for repeated measurements and group-level clustering, respectively.

Secondary analyses examined K-complex probability using a previously published K-complex detection algorithm (Lechat et al., 2020), applying a probability cut-off of 50% to define the presence of a K-complex in each 5-second epoch. These analyses were restricted to N2 and N3 to align with data from which this algorithm was trained and validated.

Power spectral analysis in each 5-second epoch was also conducted using a fast Fourier transform multi-taper approach (Dunbar et al., 2021; Lechat et al., 2022) to calculate absolute power in delta (0.5-4Hz), theta (4-8Hz), alpha (8-12Hz), sigma (12-15Hz) and beta (15-30Hz) frequency bands most relevant to sleep EEG.

3. Results

3.1 Participant Characteristics

Sixty-eight participants aged 18-80 years consented to participate and completed the 3-minute noise battery night used in this study. Of these, 14 were excluded due to noise reproduction problems (n=4) and failure of the trigger signal system (n=10), which precluded temporal analysis of EEG responses to noise exposures.

The characteristics of the remaining 54 participants are shown in Table 1. Overall, study participants had normal hearing ability in both ears, although the WFN sleep disturbed group were older and showed poorer hearing.

Table 1

Participant Demographics, hearing and grouping variables

	WFN sleep disturbed	WFN non-sleep disturbed	Rural Controls	RTN sleep disturbed	Overall
N (% Males)	6 (50)	15 (47)	17 (23)	16 (56)	54 (43)
Age (years)	67.3±8.8	55.0±14.7	47.4±21.1	33.8±15.7	47.7±19.6
BMI (kg/m²)	27.1±4.2	29.2±4.9	27.8±5.9	24.1±5.2	27.0±5.5
Weinstein noise sensitivity	64.7±5.2	61.5±20.3	52.7±14.0	64.4±18.2	59.9±17.0
ISI Global	10.7±4.3	8.9±4.9	6.2±3.3	6.4±4.4	7.5±4.4
ESS Global	7.7±2.4	5.3±4.1	4.6±2.8	6.2±5.3	5.6±4.0
PSQI Global	10.5±4.0	8.0±3.8	6.3±3.2	6.0±3.1	7.1±3.7
Self-reported Sleep efficiency (%)	71.0±20.1	76.6±15.4	83.3±12.0	89.2±11.9	81.8±15.0
Average HL 125-1000Hz RIGHT	16.5±16.5	10.9±10.2	7.4±7.8	5.9±5.7	9.1±8.6
Average HL 125-1000Hz LEFT	19.0±10.0	10.3±12.7	8.5±14.4	4.4±5.7	9.2±12.1

Values are mean ± standard deviation. BMI=Body Mass Index. Cut-offs for the Weinstein Noise Sensitivity Scale (range 1-105) >78 indicates high noise sensitivity, scores <26 indicate low noise sensitivity based on upper and lower quartiles of the original study (Weinstein, 1978). ISI (range 0-28, higher scores indicate greater insomnia severity, 0-7= no clinical insomnia, 8-14 = subthreshold insomnia symptoms (>8 indicates clinically relevant insomnia symptoms), 15-21= moderate severity clinical insomnia, 22-28 = severe clinical insomnia (Morin et al., 2011). ESS (range 0-24), scores >=10 indicate excessive daytime sleepiness (Johns, 1991; Johns, 1992). PSQI (range=0-21, >6 = poor sleep quality) (Buysse et al., 1989). PSQI Sleep Efficiency (<85% indicates below normal). HL=Hearing Level. Normal hearing cut off =<20dB.

3.2 Sleep Characteristics

The overall sleep characteristics are presented in Table 2. Overall, participants slept relatively well, for approximately 7 hours and with around 84% sleep efficiency despite noise presentations throughout the night.

Table 2 *Whole night sleep macrostructure metrics*

	WFN sleep disturbed	WFN non-sleep disturbed	Rural Controls	RTN sleep disturbed	Overall
Sleep latency (min)	12.8±11.1	12.1±8.5	27.0±28.7	18.0±28.4	18.6±23.4
REM latency (min)	108.2±80.8	97.4±47.3	39.0±10.3	51.1±9.6	99.4±48.7
Total sleep time (h)	6.7±0.8	7.0±1.0	6.9±1.0	7.0±1.1	6.9±1.0
N1 sleep (%)	11.2±4.7	12.3±8.8	10.3±6.1	9.6±5.1	10.7±6.5
N2 sleep (%)	45.1±14.1	46.3±6.8	46.3±11.1	46.8±6.9	46.3±9.0
N3 sleep (%)	25.2±8.7	19.2±9.2	20.8±8.7	21.1±7.4	20.9±8.4
REM sleep (%)	18.5±9.6	22.2±4.4	22.6±8.6	22.6±3.9	22.0±6.5
Sleep efficiency (%)	79.9±9.7	84.1±9.7	82.1±12.4	86.8±9.6	83.8±10.6
Total arousal index (events/h)	12.3±7.2	14.3±7.9	16.0±16.1	10.2±4.2	13.4±10.5
Apnoea hypopnoea index (events/h)	4.5±5.2	7.5±8.9	12.0±22.6	2.4±4.2	7.1±14.1

Values are mean ± standard deviation.

3.3 EEG response outcomes

In total there were 1328 full-spectrum WFN and 1333 infrasound stimuli presentations available for analysis, with around 25 of each noise type per participant. The probability of an EEG discernible arousal was abruptly and transiently increased in the first 5 seconds after the onset of full-spectrum WFN exposure, from a baseline level of approximately 1%, up to around 3% (Figure 3A). Arousal probability then immediately returned to pre-noise stimulus onset levels, but with some evidence of transiently reduced arousal probability around 20-seconds later. Arousal probability then remained no different from pre-noise onset levels over the remainder of the noise stimulus and following noise offset. In the fully adjusted analysis, the odds of an arousal event within the first 5 seconds of noise were approximately 3-fold higher than in the pre-noise onset baseline (odd ratio, 2.9; 95%CI, 1.6 to 5.4; adj. p= 0.018, Figure 3B). Arousal probability appeared to be higher with 35 compared to 30 dB(A) full-spectrum stimuli (Figure 3A), although the interaction between noise level and time was not statistically significant (χ^2=28.6, df=43, p = 0.95).

In contrast to a clearly discernible abrupt increase in arousal probability with full-spectrum WFN onset, there were no discernible changes in arousal probability with wind farm infrasound exposure at either SPL in unadjusted (Figure 3 C) or adjusted analyses (Figure 3D, time effect χ^2=29.65, df=43, p = 0.94, noise level x time interaction effect, χ^2=29.18, df=43, p = 0.95). A statistical power analysis based on the responses to the audible WFN stimulus showed that an odds ratio as small as 0.5 of the effect size of the positive control (i.e., odds ratio = 0.5×2.9 = 1.5) should have been detectable with 90% power at a significance level of p < 0.05. Thus, a Type II statistical error is unlikely.

There were also no discernible changes in K-complex probability or quantitative EEG power spectral outcomes with infrasound exposure.

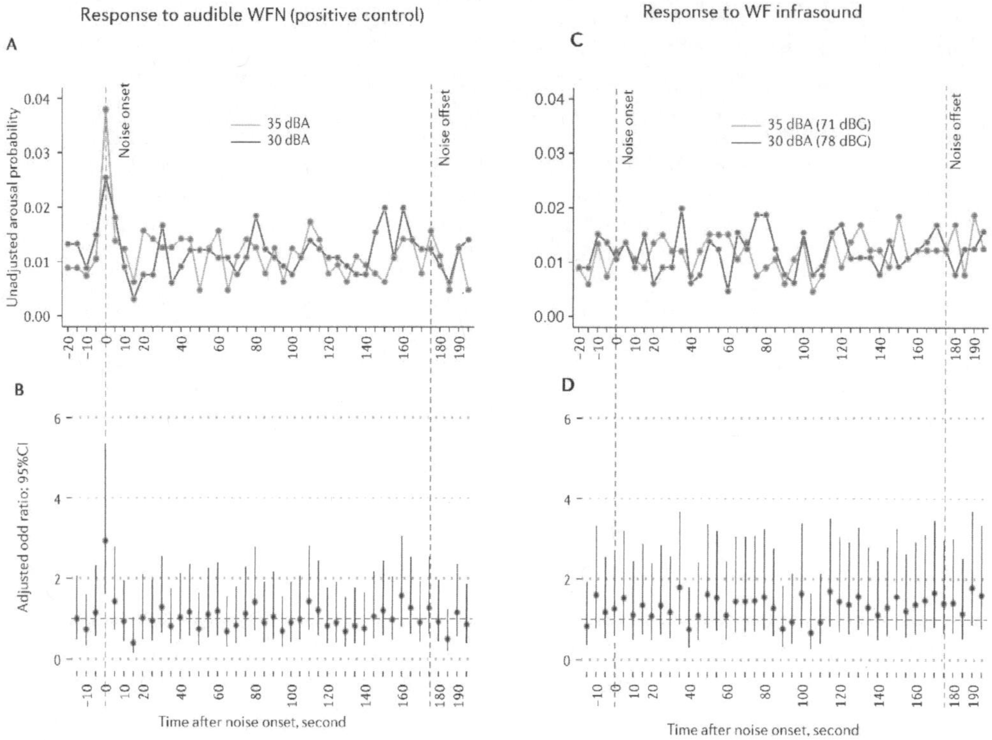

Figure 3. Arousal response to audible WFN and infrasound stimuli. A, Unadjusted arousal probability response to audible WFN (total number of full-spectrum noise stimuli = 1328). B, Odds ratio of the presence of an arousal event in 5-second epochs. C, Unadjusted arousal probability response to infrasound (total number of infrasound stimuli = 1333). D, Odds ratio of the presence of an arousal event in 5-second epochs. The baseline is the odds in the epoch [-20 to -15 s].

4. Conclusions

The findings from this study show clearly discernible, but very brief and transient EEG responses only to the onset of pre-recorded full-spectrum infrasound noise exposure, when replayed at realistic real-world levels during established sleep. In contrast, there were no discernible EEG changes in response to the same noise sample containing only infrasound. These findings add to a growing body of evidence that wind farm noise related sleep disturbance and complaints are not attributable to infrasonic noise features.

References

Bolin, K., Bluhm, G., Eriksson, G., & Nilsson, M. E. (2011). Infrasound and low frequency noise from wind turbines: exposure and health effects. *Environmental research letters, 6*(3), 035103. https://doi.org/doi:10.1088/1748-9326/6/3/035103

Buysse, D. J., Reynolds III, C. F., Monk, T. H., Berman, S. R., & Kupfer, D. J. (1989). The Pittsburgh Sleep Quality Index: a new instrument for psychiatric practice and research. *Psychiatry research, 28*(2), 193-213. https://doi.org/10.1016/0165-1781(89)90047-4

Dunbar, C., Catcheside, P., Lechat, B., Hansen, K., Zajamsek, B., Liebich, T., Nguyen, D. P., Scott, H., Lack, L., Decup, F., Andrew, V., & Gorica, M. (2021). EEG power spectral responses to wind farm compared with road traffic noise during sleep: A laboratory study. *Journal of sleep research*, e13517. https://doi.org/10.1111/jsr.13517

Hansen, K. L., Hansen, C. H., & Zajamsek, B. (2015, Dec). Outdoor to indoor reduction of wind farm noise for rural residences. *Building and Environment, 94*, 764-772. https://doi.org/10.1016/j.buildenv.2015.06.017

Iber, C., Ancoli-Israel, S., Chesson, A. L., & Quan, S. (2007). *AASM - Manual for the Scoring of Sleep and Associated Events*. Westchester, IL: American Academy of Sleep Medicine.

Johns, M. W. (1991). A new method for measuring daytime sleepiness: the Epworth sleepiness scale. *Sleep, 14*(6), 540-545. https://doi.org/10.1093/sleep/14.6.540

Johns, M. W. (1992). Reliability and factor analysis of the Epworth Sleepiness Scale. *Sleep, 15*(4), 376-381. https://doi.org/10.1093/sleep/15.4.376

Lechat, B., Hansen, K., Catcheside, P., & Zajamsek, B. (2020). Beyond K-complex binary scoring during sleep: probabilistic classification using deep learning. *Sleep, 43*(10), zsaa077. https://doi.org/10.1093/sleep/zsaa077

Lechat, B., Hansen, K. L., Melaku, Y. A., Vakulin, A., Micic, G., Adams, R. J., Appleton, S., Eckert, D. J., Catcheside, P., & Zajamsek, B. (2022). A Novel Electroencephalogram-derived Measure of Disrupted Delta Wave Activity during Sleep Predicts All-Cause Mortality Risk. *Annals of the American Thoracic Society, 19*(4), 649-658.

Leventhall, H. G. (2004). Low frequency noise and annoyance. *Noise and Health, 6*(23), 59-72.

Locher, B., Piquerez, A., Habermacher, M., Ragettli, M., Röösli, M., Brink, M., Cajochen, C., Vienneau, D., Foraster, M., Müller, U., & Jean, M., Wunderli. (2018). Differences between outdoor and indoor sound levels for open, tilted, and closed windows. *International journal of environmental research public health 15*(1), 149. https://doi.org/10.3390/ijerph15010149

Marshall, N. S., Cho, G., Toelle, B. G., Tonin, R., Bartlett, D. J., D'Rozario, A. L., Evans, C. A., Cowie, C. T., Janev, O., & Whitfeld, C. R. (2023). The Health Effects of 72 Hours of Simulated Wind Turbine Infrasound: A Double-Blind Randomized Crossover Study in Noise-Sensitive, Healthy Adults. *Environmental Health Perspectives, 131*(3), 037012. https://doi.org/10.1289/EHP10757

Morin, C. M., Belleville, G., Bélanger, L., & Ivers, H. (2011). The Insomnia Severity Index: psychometric indicators to detect insomnia cases and evaluate treatment response. *Sleep, 34*(5), 601-608. https://doi.org/10.1093/sleep/34.5.601

National Health and Medical Research Council. *2015 Information Paper: Evidence on Wind Farms and Human Health. Canberra: National Health and Medical Research Council; 2015* https://www.nhmrc.gov.au/file/7251/download?token=xFcCNj8V

Nguyen, P. D., Hansen, K. L., Lechat, B., Hansen, C., Catcheside, P., & Zajamsek, B. (2022, Dec). Audibility of wind farm infrasound and amplitude modulated tonal noise at long-range locations. *Applied Acoustics, 201*. https://doi.org/ARTN 109106 10.1016/j.apacoust.2022.109106

Okada, A., & Inaba, R. (1990). Comparative study of the effects of infrasound and low-frequency sound with those of audible sound on sleep. *Environment international, 16*(4-6), 483-490. https://doi.org/10.1016/0160-4120(90)90017-Z

Watanabe, T., & Møller, H. (1990). Low frequency hearing thresholds in pressure field and in free field. *Journal of low frequency noise, vibration active control, 9*(3), 106-115. https://doi.org/10.1177/026309239000900030

Weinstein, N. D. (1978). Individual differences in reactions to noise: a longitudinal study in a college dormitory. *Journal of applied psychology, 63*(4), 458-466. https://doi.org/10.1037/0021-9010.63.4.458

INCEEUROPE

10th International Conference
on
Wind Turbine Noise
Dublin - 21st to 23rd June 2023

Exploring the effect of wind farm flow on wind turbine noise propagation through numerical simulations

Jules Colas, Ariane Emmanuelli, Didier Dragna
Univ Lyon, Ecole Centrale de Lyon, CNRS, Univ Claude Bernard Lyon 1, INSA Lyon, LMFA, UMR5509, 69130, Ecully, France

Philippe Blanc-Benon
Univ Lyon, CNRS, Ecole Centrale de Lyon, INSA Lyon, Univ Claude Bernard Lyon 1, LMFA, UMR5509, 69130, Ecully, France

Benjamin Cotté
Institute of Mechanical Sciences and Industrial Applications (IMSIA), ENSTA ParisTech, CNRS, CEA, EDF, Université Paris-Saclay

Richard Stevens
Physics of Fluids Group, Max Planck Center Twente for Complex Fluid Dynamics, J. M. Burgers Center for Fluid Dynamics, University of Twente, Enschede, The Netherlands

Summary

Wind turbine noise is a significant obstacle to the expansion of wind farms. To understand the generation and propagation of wind turbine noise, various models have been developed in recent years. They are generally used to compute the noise generated by a wind turbine according to its geometry and the characteristics of the surrounding flow, and to consider topographic and meteorological effects on sound propagation. The flow around the turbine is known to have a significant effect on wind turbine noise propagation. Although the flow inside a wind farm has been thoroughly studied, insight into the effect of this flow on sound propagation is limited. However, the flow inside and around a wind farm, including the interaction between different wind turbine wakes, can significantly impact wind turbine noise production and propagation.

This study aims to investigate this phenomenon through numerical simulations. A method for coupling several models is employed to predict the noise produced by the wind farm. It uses large eddy simulations to determine the mean flow in and around the wind farm, an extended source model to predict the sound generated by each turbine, and a propagation model to consider the effect of the flow on sound propagation. Two idealized wind farm configurations are studied under neutral atmospheric boundary condition. The wind turbine noise produced and propagated inside a wind farm is compared to an isolated turbine case

1. Introduction

The rapid expansion of onshore wind farms has led to increasing concern over the noise impact on the surrounding areas. Noise from wind turbines comes primarily from the aeroacoustic broadband noise generated by the wind interacting with the rotating blades. This noise is mostly in the low-frequency range and can propagate over several kilometers. Recent measurements have demonstrated the importance of amplitude modulation (AM) in wind farm noise annoyance(van den Berg, 2005; Hansen *et al.*, 2019). Numerical simulations have explored the AM phenomenon, revealing that close to the source, AM is primarily caused by the rotating source and changes in noise production due to flow inhomogeneity (Tian and Cotté, 2016). AM occurring downwind is attributed to the focusing zone generated by the wind turbine's wake. The displacement of these focusing zones due to blade movement creates high variations in downwind sound pressure levels (Barlas *et al.*, 2017a). Therefore, precise sound propagation simulations and source models, that take the atmospheric boundary layer (ABL) into account, are necessary to predict these noise characteristics. Studies have investigated the effects of flow on sound propagation for single wind turbines, including the effects of the wake (Barlas *et al.*, 2017b) and the 3D effects of the flow (Heimann and Englberger, 2018). However, the precise interaction between multiple turbines and its impact on AM or sound pressure level (SPL) has not been fully explored.

The objective of this research is to study the influence of wind turbine interactions on the surrounding noise levels by investigating how the presence of multiple turbines affects the SPL and AM. For this purpose a complete methodology from source to receiver is developed. It includes large eddy simulations (LES) to compute the flow around the turbines, an extended source model based on Amiet's theory and parabolic equation model for sound propagation. Two cases are studied: a baseline case with one turbine and a second case with a pair of turbines arranged in a row.

The paper is organized as follows. First in Sec. 2. the methodology developed for wind turbine noise prediction is described. Then, in Sec 3., the two cases studied are presented. Finally, the SPL and AM predicted for both cases are presented and compared in Sec 4..

2. Methodology

2.1 General methodology layout

The methodology developed to compute wind farm noise is based on three different models as illustrated in Figure 1a. The simulations are performed by first computing a realistic ABL using LES (Gadde *et al.*, 2021b). Different ABL stratification can be simulated while accounting for the interaction between the flow and several wind turbines (see Sec. 2.2). The time averaged 3D velocity fields are then fed into an extended source model and a propagation model. We use the extended source model developed by Tian and Cotté (2016). Each blade is split into uniform segments, that are considered as uncorrelated noise sources. The source model provides the SPL in free field (SPL_{ff})

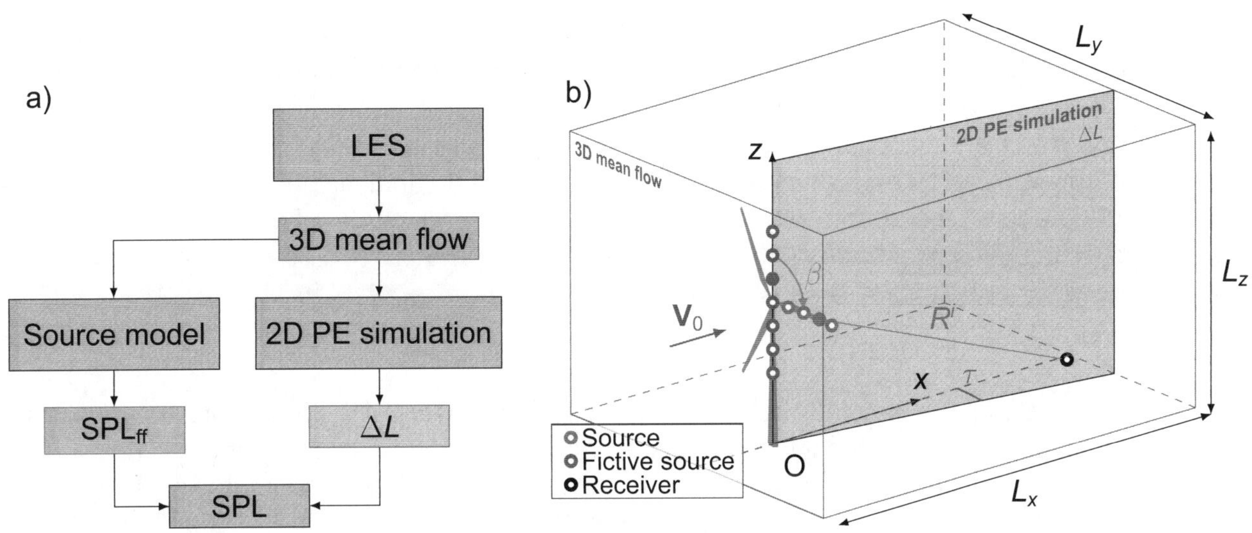

Figure 1: a) Diagram of the complete prediction methodology and b) sketch of the coupling method.

for each blade segment depending on the given wind turbine geometry, incoming velocity profile, and turbulent spectrum (see Sec. 2.3). Finally, the propagation effects, such as ground reflection or atmospheric refraction, are determined by computing the SPL relative to the free field (ΔL) for each source-receiver pair. Practically this is done, as described in Cotté (2019), by performing 2D parabolic equation (PE) simulations for several fictive sources distributed on a vertical line in the rotor plane, as shown in Figure 1b. The corresponding ΔL for each segment position is interpolated from the computed 2D fields. The source and propagation effects are then combined to obtain the total SPL at a receiver produced by one blade segment at a given angular position of the blades:

$$\mathrm{SPL}^i(\mathbf{x}, \omega, \beta) = \mathrm{SPL}^i_{\mathrm{ff}}(\mathbf{x}, \omega, \beta) + \Delta L^i(\mathbf{x}, \omega, \beta) - \alpha(\omega) R^i \tag{1}$$

where i is the index of the considered segment, $\mathbf{x}$ is the receiver position, ω the angular frequency, β the blade angle, $\mathrm{SPL}^i_{\mathrm{ff}}$ the SPL in free field computed for the given segment,ΔL^i the value of ΔL corresponding to the closest fictive source, α the atmospheric absorption coefficient and R^i the distance between the segment center and the receiver.

To take into account the propagation time and compute SPL variations at the receiver, we adapted the methodology from Mascarenhas *et al.* (2022). The propagation time of each source-receiver pair is computed such that:

$$T^i_j = \frac{R^i(\mathbf{x}, \beta_j)}{c_0} + \frac{\beta_j}{\Omega} \tag{2}$$

where $R^i(\mathbf{x}, \beta_j)$ is the distance between the segment i at angular position $\beta_j = j\Delta\beta$ and the receiver at $\mathbf{x}$, $\Delta\beta$ is the angular step between two blade positions, c_0 is the averaged sound speed and Ω is the rotational speed. Therefore, the period of the signal obtained at each receiver is $2\pi/\Omega$. To obtain the signal at each receiver, a full period is sampled at a frequency $f_s = 1/\Delta t$. For every sample at time $t_k = k\Delta t$ the SPL is computed such that

$$\mathrm{SPL}(\mathbf{x}, \omega, t_k) = 10 \log_{10} \left(\sum_{i=1}^{N_s} \sum_{j=1}^{N_\beta} 10^{\,\mathrm{SPL}^i(\mathbf{x},\omega,\beta_j)/10} \right) \quad \text{if} \quad t_k < T^i_j \left[\frac{2\pi}{\Omega} \right] < t_{k+1} \tag{3}$$

The overall SPL (OASPL) are computed by summing the SPL over the frequency. The $\overline{\text{OASPL}}$ corresponds to the OASPL averaged over t, and the AM are computed by taking the difference between the minimum and maximum values of OASPL over one period. One of the main limitations of this method is the use of an average sound speed c_0 for all propagation times. It is worth noting that this methodology allows one to easily modify one of its component: analytical flow profiles can be used instead of LES results, the source model could be extended to include different noise mechanisms and another propagation method could be employed to compute ΔL (Colas *et al.*, 2023).

2.2 Large eddy simulation

In this study the ABL flow is simulated using LES. The largest scales of the flow features are simulated while the smallest eddies are modeled with a Lagrangian average scale-dependent model (Gadde *et al.*, 2021b). The interaction between wind turbines and the flow is simulated with an actuator disk model. Additionally, a concurrent precursor method is implemented in order to generate an unperturbed ABL, which serves as input for the wind farm simulation (Stevens *et al.*, 2014). This code has been extensively validated for neutral, stable and unstable stratification (Gadde *et al.*, 2021b) as well as for a wide variety of wind farm layouts (Stieren and Stevens, 2022; Gadde *et al.*, 2021a).

2.3 Source model

Several noise mechanisms contribute to the overall noise emitted by a turbine. In this study the sound model developed by Tian and Cotté (2016) and later improved by Mascarenhas *et al.* (2022) is used to compute the turbulent inflow noise and trailing edge noise generated by each segment of the turbine blades. This method takes as input the geometry of each segment, the incoming velocity and turbulence spectrum. To compute the trailing edge noise, the wall pressure fluctuation spectrum is determined using Xfoil.

2.4 Propagation with Parabolic equation method

The parabolic equation methods have been widely used for outdoor sound propagation as they provide a good accuracy at long range and have a low computational cost. In this work we implemented a vector wide angle parabolic equation (WAPE_{vec}) proposed by Ostashev *et al.* (2020) that precisely takes into account the mean flow (Kayser *et al.*, 2023; Colas *et al.*, 2023). At the bottom of the domain an impedance boundary condition is used. The WAPE_{vec} is then solved using a Crank-Nicholson algorithm to advance the solution from x to $x + \Delta x$. The derivative in z-direction are computed using second-order finite difference schemes and the starting field is the second order starter from Salomons (2001).

3. Cases investigated

This study focuses on two scenarios presented in Figure 2a and 2b. The first case serves as a baseline and involves a single wind turbine while the second case considers a pair of turbines arranged in a row. The dimension of the domain and position of the turbines can be found in Table 1.

For both cases, we conducted simulations under neutral ABL stratification, where the inflow velocity was set to yield a streamwise wind speed at hub height of 11.4 ms^{-1}, corresponding to the rated

	L_x	L_y	L_z	$\mathbb{T}_0$	$\mathbb{T}_1$
Single-turbine case (baseline)	3.5km	3.2km	0.3km	(0,0)	
Two-turbine case	3.5km	3.2km	0.3km	(0,-400)	(0,400)

Table 1: Domain sizes used for the simulations. $\mathbb{T}_0$ and $\mathbb{T}_1$ corresponds to the turbines' positions.

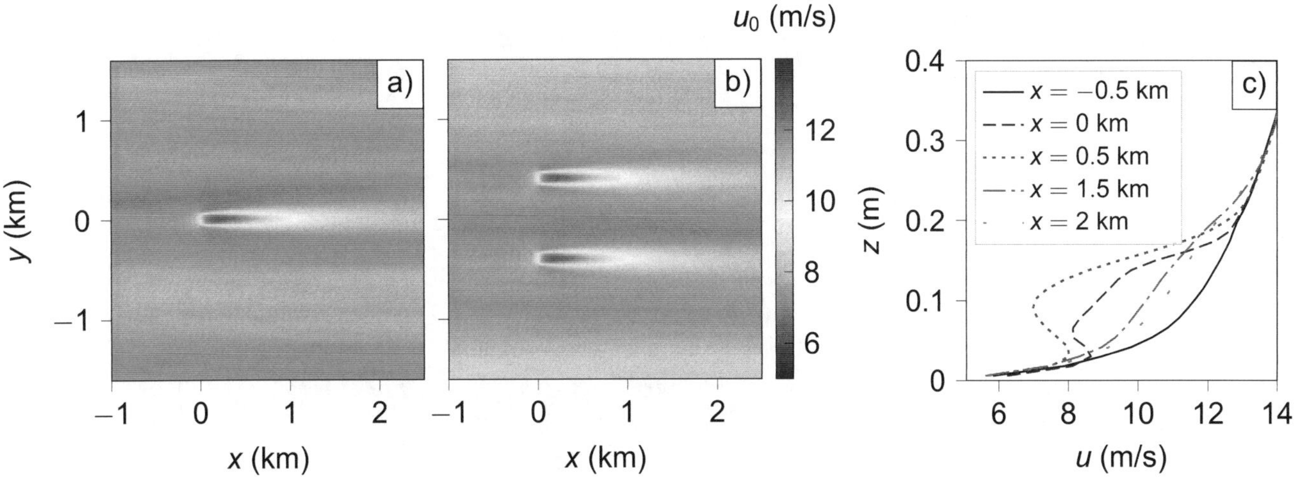

Figure 2: Streamwise velocity component at $z = 100$ m above the ground for a) single-turbine and b) two-turbine case. c) Streamwise velocity component plotted upwind and at 4 downwind locations.

wind speed of the 5MW NREL reference turbine (Jonkman *et al.*, 2009). Figure 2c shows the inflow and perturbed profiles downwind of the turbine. Given the less pronounced downward refraction for neutral stratification compared to stable stratification, we expect a relatively weak focusing effect.

In this work, we slightly modified the wind turbine's geometry defined in Tian and Cotté (2016) by scaling up the rotor diameter to 120 m, corresponding to the 5MW NREL wind turbine, while maintaining a hub height of 100 m. The rotational speed relative to the wind speed at hub height was adjusted based on Jonkman *et al.* (2009). Furthermore, we optimized the pitch angle to achieve an ideal angle of attack of 4° relative to the wind speed at hub height. This implementation of the source model provides results for an idealized wind turbine, disregarding scenarios where the flow detaches and the blade enters a stall regime. The source terms are computed using the incoming wind profile shown in Figure 2c. Therefore, in accordance with the NREL documentation, the rotational speed is set to $\Omega = 12.1$ rpm.

f_c (Hz)	50	63	80	100	125	160	200	250	315	400	500	630	800	1000
f (Hz)	50	63	80	100	125	160	192	241	297	373	467	588	741	926
							208	260	315	391	489	616	770	962
									334	409	512	645	800	1000
										429	536	675	831	1039
													864	1080

Table 2: Frequencies computed (f) and their corresponding third octave band central frequency (f_c).

To compute the ΔL field a set a 2D PE simulations are performed. Simulations are conducted for 35 frequencies presented in Table 2. A third octave band spectrum from 50 Hz to 1 kHz is computed from these frequencies. The grid step is defined such that $\Delta x = \Delta z = \lambda/10$, with λ the wavelength. The ΔL field in the whole domain is interpolated from 360 PE simulations computed in different propagation directions with an angular step of $\Delta\tau = 1°$. For each angle 7 heights of the fictive sources are considered to cover the wind turbine diameter. To simulate a grassy ground, we employed the variable porosity model proposed by Attenborough *et al.* (2011), considering a flow resistivity of 50 $kNsm^{-4}$ and a porosity change rate of 100 m^{-1}.

4. Results

4.1 Single-turbine case

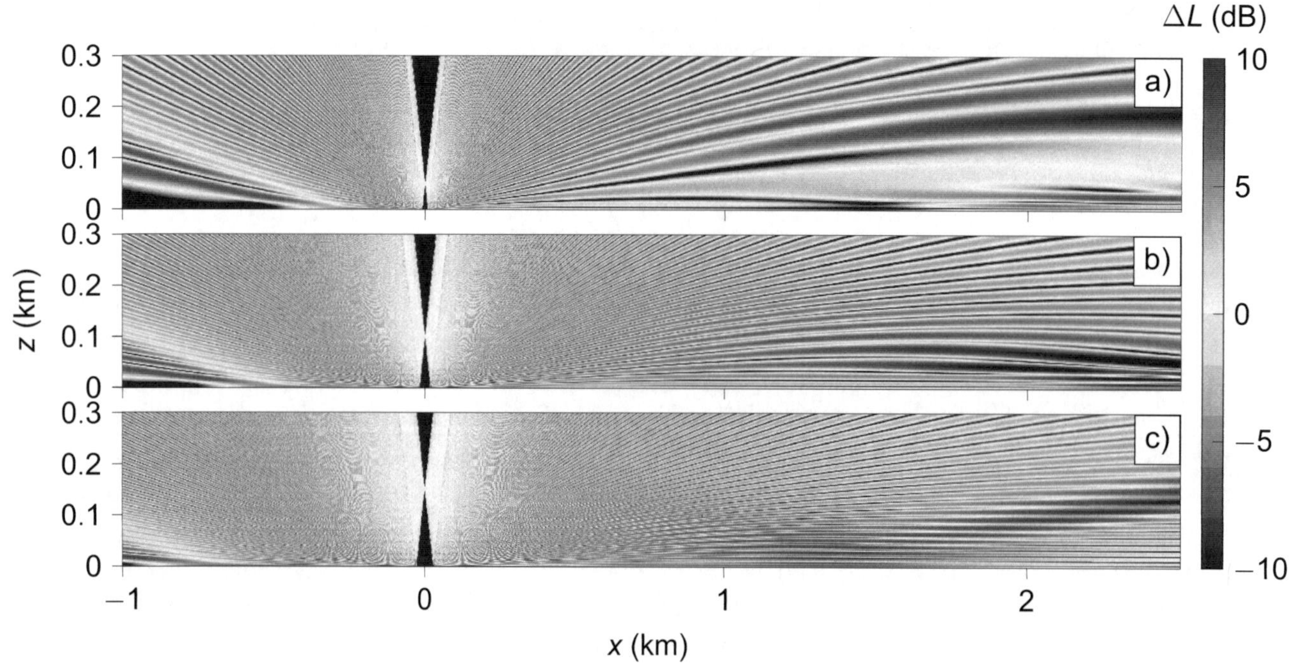

Figure 3: ΔL at $y = 0$ for the single-turbine case for three different source heights: a) 49 m b) 100 m c) 151 m.

Figure 3 displays the ΔL fields obtained from the PE simulation for the single-turbine scenario, showcasing three distinct source heights at a frequency of 100 Hz. This corresponds to the sound pressure relative to the free field solution, hence it shows the effect of the flow and the ground without any source term consideration. The flow effects on the propagation are well visible both downwind ($x > 0$) and upwind ($x < 0$) of the turbine. Upwind a shadow zone is generated by the negative wind gradient. As the source height increases, the boundary of the shadow zone extends farther from the source. An additional effect, not presented in this study to maintain conciseness, is the impact of frequency on the shadow zone. The beginning of the shadow zone is situated closer to the source at higher frequencies. In the downwind direction, a combined impact of the positive wind gradient and the wind turbine wake can be noted. These results reaffirm the previous findings in Barlas *et al.* (2017b) that the downwind focusing pattern is altered by the height of the source, which consequently leads to AM downwind.

Combining the propagation effects with the source terms (see Sec. 2.) allows one to compute the average OASPL and AM, shown in Figure 4 for receivers at 2 m height. Figure 4a clearly shows the dipolar nature of the wind turbine source. Upwind ($x < 0$) the shadow zone becomes discernible from a distance of 800 m away from the source. Closer to the source an SPL increase is noticeable due to convective amplification. Additionally, the figure reveals a slight focusing directly downwind of the turbine. However, due to spatial and frequency averaging, this effect appears less pronounced than in Figure 3.

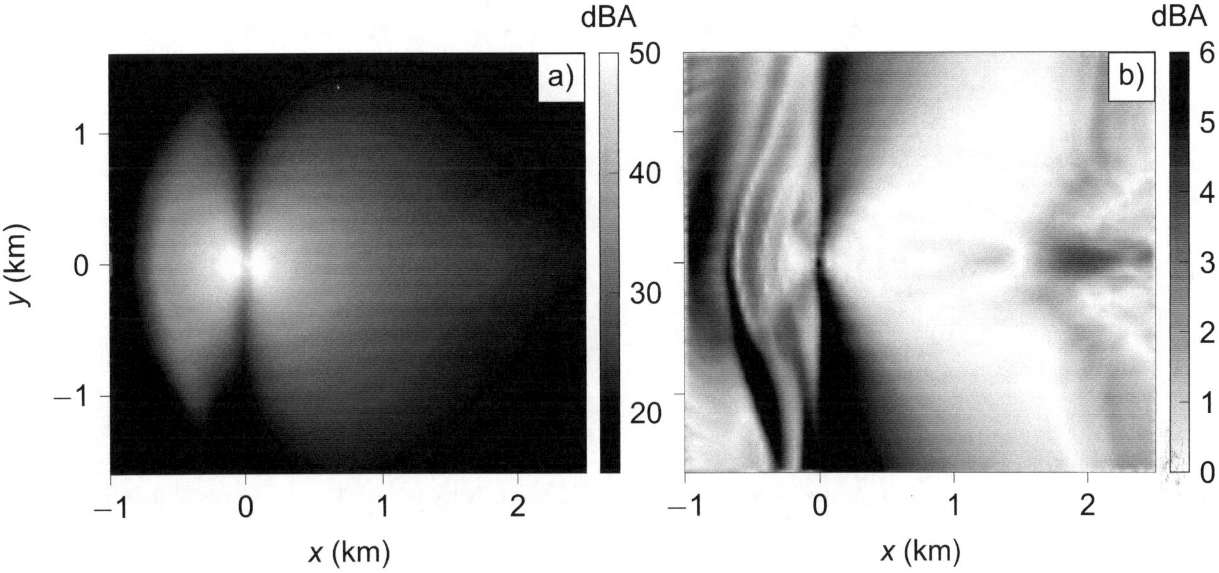

Figure 4: Top view of a) the $\overline{\text{OASPL}}$ and b) AM for the single-turbine case.

The AM field in Figure 4b presents several interesting features. First the AM seems to be concentrated in 3 different zones. A zone upwind corresponds to the moving shadow zone previously described in Cotté (2019). A receiver placed in this zone gets periodically in and out of the shadow zone due to the vertical movement of the sources as shown in Figure 3. The cross wind AM is directly due to the source term as already described by Tian and Cotté (2016). The asymmetric characteristic of the AM along the y axis is probably due to the rotation of the blades, but further investigations are needed as we observed that this asymmetry was also dependent on frequency. This suggests different contributions of the trailing edge or turbulent inflow noise in the final SPL according to the blade position. The third high AM zone is downwind (around $x = 2$ km) and is due to the focusing by the wake. With the sources moving up and down the sound waves undergo different refraction effects leading to the focused zone moving closer or further away from the turbine and hence to variations in SPL at the ground.

4.2 Two-turbine case

To obtain the results for the two turbines we first assume that they both start rotating at the same angular position (with one blade pointing upwards). The two SPL fields are then summed incoherently at each receiver position and at each time step. The resulting SPL are also of period $2\pi/\Omega$ as both turbines are rotating at the same speed. Figure 5 shows the averaged OASPL and AM obtained with two turbines. In Figure 5a, the $\overline{\text{OASPL}}$ corresponds very closely to the superposition of the results for two separate wind turbines. In fact, the interaction between the two turbines is minimal: the incoming

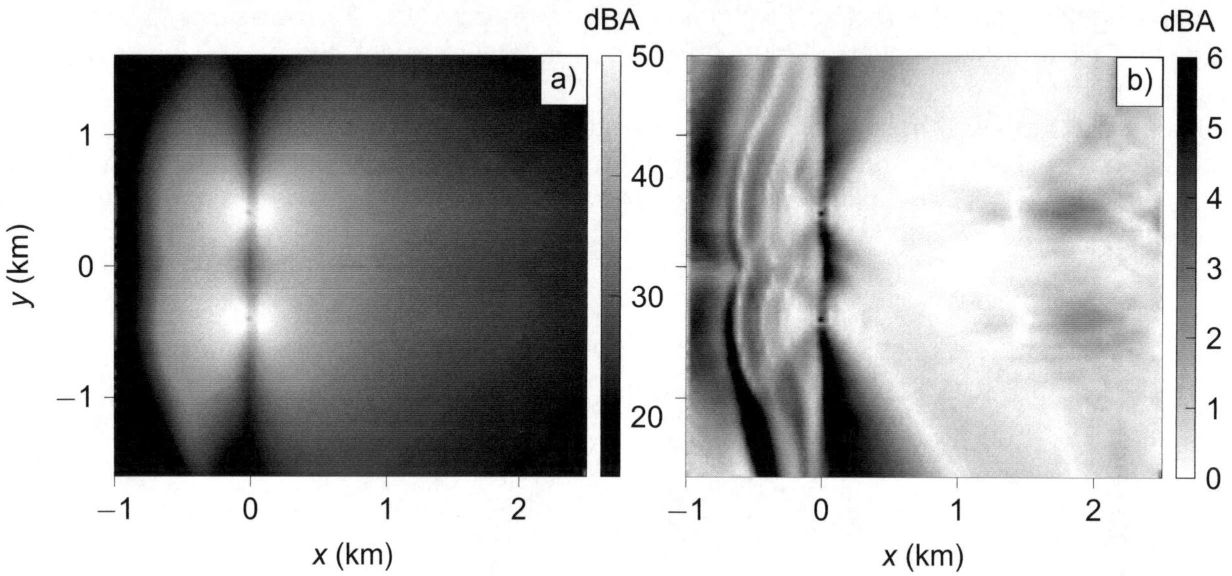

Figure 5: Top view of a) the $\overline{\text{OASPL}}$ and b) AM for the two-turbine case.

flow is similar, and the distance between the turbines is sufficient to ensure that the sound propagation from one is not affected by the neighboring turbine. However, the AM field in Figure 5b, especially downwind, is lower than the AM obtained for an isolated wind turbine. This will be addressed in the next section.

4.3 Effects of the angular offset on the AM

In the previous results, the two turbines were assumed to start rotating at the same angular position, which is a simplified scenario. In reality, the two wind turbines experience different wind conditions, leading to different rotational speeds. For the sake of simplicity, we assume that both turbines rotate at the same speed and consider only the effect of angular offset between the two. We consider two scenarios: one where the turbines are synchronized (the blades start at the same position) and the other where the turbines are desynchronized, with the second turbine starting with an angular offset of 60°. Figure 6 shows the AM obtained for the two scenarios for a line of receivers at $y = 0$. Downwind, between $x = 1.5$ km and $x = 2.5$ km, the AM obtained when summing the contribution of $\mathbb{T}_0$ and $\mathbb{T}_1$ is greatly reduced in the first case (around 0.7 dBA), while in the second case, it is comparable to that of the two separate turbines (around 1.8 dBA). However, this leads to a decrease of AM in other locations, such as upwind at $x = -0.5$ km where AM reduces from 3.5 dB to 1.5 dB.

The time variations of the mean OASPL at $\mathbf{x} = (1.75\text{ km}, 0\text{m}, 2\text{m})$ are also presented for both cases in Figure 6. The mean OASPL increases by 2 dB whether or not the signals from the two turbines are synchronized. However, the AM increases when the turbines are desynchronized. Note that here we logaritmically summed the SPL at different times. Despite both signals looking "out of phase" this does not correspond to pressure fluctuation interferences but to the summation of two uncorrelated signals.

These results indicate that in a realistic situation, where the two turbines would not be rotating at the same speed, the strength of AM would vary as the signals from the two turbines get synchronized and desynchronized. This would create a beating effect, as described in van den Berg (2005).

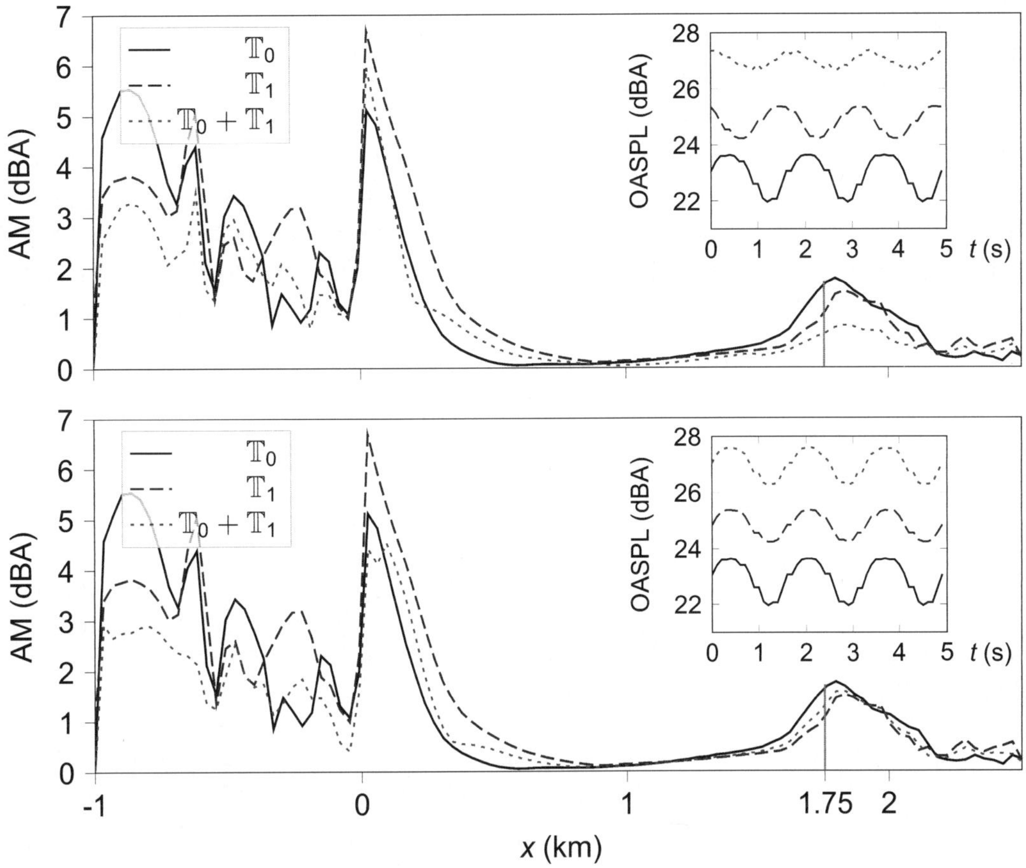

Figure 6: AM for the two-turbine case. On top the turbines are synchronized, at the bottom the second turbine starts with an offset of 60°. The subplots show the OASPL variation at $\mathbf{x} = (1.75\text{ km}, 0\text{ m}, 2\text{ m})$ during one rotation.

5. Conclusion

In this study we presented a methodology to compute wind farm noise. It takes into account flow effects using accurate prediction of the ABL from LES. The source model computes trailing edge and turbulent inflow noise from the flow features and the geometry of the wind turbines. Propagation effects such as atmospheric refraction and ground reflection are computed using a 2D PE method.

SPL and AM obtained for a single wind turbine show some well known features of wind turbine noise such as downwind and upwind AM, dipolar directivity, and focusing zone induced by the wake. We compute the noise from two turbines and we showed that there was limited interaction between the turbines when considering the averaged OASPL. However, by taking into account the propagation time, some interesting behavior of the AM have been highlighted. Angular synchronization of the turbines seems to have an important role on the AM both downwind and upwind.

This study is limited to neutral atmosphere, where the downwind amplification is known to be quite low. These results could be extended to stable and unstable stratification. The scenario treated here, where the two turbines rotate at the same speed, is an interesting boundary case to study the combination of several turbines. Other configurations where the wake of one turbine interacts with a downwind turbine could be interesting, leading to propagation effects and different rotational speeds.

Acknowledgments

This work was performed within the framework of the LABEX CeLyA (ANR-10-LABX-0060) of Université de Lyon, within the program "Investissements d'Avenir" (ANR-16-IDEX-0005) operated by the French National Research Agency (ANR). It was granted access to the HPC resources of PMCS2I (Pôle de Modélisation et de Calcul en Sciences de l'Ingénieur et de l'Information) of Ecole Centrale de Lyon, PSMN (Pôle Scientifique de Modélisation Numérique) of ENS de Lyon and P2CHPD (Pôle de Calcul Hautes Performances Dédiés) of Université Lyon I, members of FLMSN (Fédération Lyonnaise de Modélisation et Sciences Numériques), partner of EQUIPEX EQUIP@MESO. It was supported by the Franco-Dutch Hubert Curien partnership (van Gogh Programme No. 49310UM).

References

Attenborough, K., Bashir, I., and Taherzadeh, S. (2011). *Outdoor ground impedance models,* The Journal of the Acoustical Society of America **129**(5), 2806–2819, doi: 10.1121/1.3569740.

Barlas, E., Zhu, W. J., Shen, W. Z., Dag, K. O., and Moriarty, P. (2017a). *Consistent modelling of wind turbine noise propagation from source to receiver,* The Journal of the Acoustical Society of America **142**(5), 3297–3310, doi: 10.1121/1.5012747.

Barlas, E., Zhu, W. J., Shen, W. Z., Kelly, M., and Andersen, S. J. (2017b). *Effects of wind turbine wake on atmospheric sound propagation,* Applied Acoustics **122**, 51–61, doi: 10.1016/j.apacoust.2017.02.010.

Colas, J., Emmanuelli, A., Dragna, D., Blanc-Benon, P., Cotté, B., and Stevens, R. (2023). *Wind Turbine Sound Propagation: A Comparison of a Linearized Euler Equations model with Parabolic Equation method,* Submitted to J. Acoust. Soc. Am. .

Cotté, B. (2019). *Extended source models for wind turbine noise propagation,* The Journal of the Acoustical Society of America **145**(3), 1363–1371, doi: 10.1121/1.5093307.

Gadde, S. N., Liu, L., and Stevens, R. J. A. M. (2021a). *Effect of low-level jet on turbine aerodynamic blade loading using large-eddy simulations,* Journal of Physics: Conference Series **1934**(1), 012001, doi: 10.1088/1742-6596/1934/1/012001.

Gadde, S. N., Stieren, A., and Stevens, R. J. A. M. (2021b). *Large-Eddy Simulations of Stratified Atmospheric Boundary Layers: Comparison of Different Subgrid Models,* Boundary-Layer Meteorology **178**(3), 363–382, doi: 10.1007/s10546-020-00570-5.

Hansen, K. L., Nguyen, P., Zajamšek, B., Catcheside, P., and Hansen, C. H. (2019). *Prevalence of wind farm amplitude modulation at long-range residential locations,* Journal of Sound and Vibration **455**, 136–149, doi: 10.1016/j.jsv.2019.05.008.

Heimann, D., and Englberger, A. (2018). *3D-simulation of sound propagation through the wake of a wind turbine: Impact of the diurnal variability,* Applied Acoustics **141**, 393–402, doi: 10.1016/j.apacoust.2018.06.005.

Jonkman, J., Butterfield, S., Musial, W., and Scott, G. (2009). *Definition of a 5-MW Reference Wind Turbine for Offshore System Development,* Technical Report NREL/TP-500-38060, 947422 , doi: 10.2172/947422.

Kayser, B., Mascarenhas, D., Cotté, B., Ecotière, D., and Gauvreau, B. (2023). *Validity of the effective sound speed approximation in parabolic equation models for wind turbine noise propagation,* The Journal of the Acoustical Society of America **153**(3), 1846–1854, doi: 10.1121/10.0017653.

Mascarenhas, D., Cotté, B., and Doaré, O. (2022). *Synthesis of wind turbine trailing edge noise in free field,* JASA Express Letters **2**(3), 033601, doi: 10.1121/10.0009658.

Ostashev, V. E., Wilson, D. K., and Muhlestein, M. B. (2020). *Wave and extra-wide-angle parabolic equations for sound propagation in a moving atmosphere,* The Journal of the Acoustical Society of America **147**(6), 3969–3984, doi: 10.1121/10.0001397.

Salomons, E. M. (2001). *Computational Atmospheric Acoustics* (Kluwer Academic, Dordrecht London), doi: 10.1007/978-94-010-0660-6.

Stevens, R. J., Graham, J., and Meneveau, C. (2014). *A concurrent precursor inflow method for Large Eddy Simulations and applications to finite length wind farms,* Renewable Energy **68**, 46–50, doi: 10.1016/j.renene.2014.01.024.

Stieren, A., and Stevens, R. J. A. M. (2022). *Impact of wind farm wakes on flow structures in and around downstream wind farms,* Flow **2**, E21, doi: 10.1017/flo.2022.15.

Tian, Y., and Cotté, B. (2016). *Wind Turbine Noise Modeling Based on Amiet's Theory: Effects of Wind Shear and Atmospheric Turbulence,* Acta Acustica united with Acustica **102**(4), 626–639, doi: 10.3813/AAA.918979.

van den Berg (2005). *The Beat is Getting Stronger: The Effect of Atmospheric Stability on Low Frequency Modulated Sound of Wind Turbines,* Noise Notes **4**(4), 15–40, doi: 10.1260/147547306777009247.

10th International Conference
on
Wind Turbine Noise
Dublin – 21st to 23rd June 2023

Apportionment of wind farm noise limits: Observations from using UK good practice guidance

Alex Dell, TNEI: alex.dell@tneigroup.com
James Mackay, TNEI: james.mackay@tneigroup.com

Summary

Wind farm noise limits in the UK are set in accordance with the guidance contained in ETSU-R-97 'The Assessment and Rating of Noise from Wind Farms'. ESTU-R-97 establishes that noise limits should apply to the combined operation of all wind farms and when multiple wind farms are proposed, it may be necessary to apportion the noise limit.

In 2013, the Institute of Acoustics published 'A good practice guide to the application of ETSU-R-97 for the rating and assessment of wind turbine noise.' (IOA GPG). The IOA GPG provides a number of methods which can be used to consider cumulative noise such that Site Specific Noise Limits can be established for individual schemes. One of the options considered in the IOA GPG is 'apportionment of the ETSU-R-97 limits on an energy basis'.

By assuming apportionment on an energy basis relates to the acoustic energy of the noise predictions for each of the proposed schemes, noise limits would be created where each schemes share a proportionate amount with respect to their predicted noise levels. This approach can work well where predictions indicate that all schemes can coexist within the ETSU-R-97 limits. Where predicted levels exceed the limits apportionment on an acoustic energy basis, all schemes will need to reduce noise emissions (regardless of how acoustically dominant each scheme is).

An alternative approach is that the limits could be apportioned on the basis of the energy yield that will be generated by each scheme. This would result in limits being apportioned in such a way that energy generation is maximised.

Within this paper, example scenarios will be presented where limit apportionment using each method will be undertaken. Commentary will be made on the practicalities of each approach and the potential positive and negative impacts each approach can have on the proposed schemes.

1. Introduction

Wind farm noise limits in the UK are set in accordance with the guidance contained in ETSU-R-97 'The Assessment and Rating of Noise from Wind Farms' [1]. The noise limits established in accordance with ETSU-R-97 should not be exceeded by the combined operation of all wind farms in an area. Where multiple wind farms are proposed, it may be necessary to apportion the noise limit. In 2013, the Institute of Acoustics published 'A good practice guide to the application of ETSU-R-97 for the rating and assessment of wind turbine noise.' (IOA GPG) [2]. The IOA GPG provides several methods which can be used to consider cumulative noise such that separate noise limits can be established for individual schemes ('Site Specific Noise Limits'). This is becoming a consideration of ever-increasing importance due to the continuing proliferation of proposed wind farms.

In Ireland, assessments are currently undertaken in accordance with the Wind Energy Development Guidelines published in 2006 [3] (WEDG 2006). The current guidelines are under review and a draft update was published for consultation in 2019 [4] (WEDG 2019). Cumulative noise is not discussed in the WEDG 2006 and, as a result reference is sometimes made to the guidance in ETSU-R-97 and the IOA GPG. In the absence of specific guidance which applies in Ireland, the approach used to consider cumulative noise for proposed developments varies significantly between applications although the calculation of Site Specific Noise Limits has been included in some more recent wind farm noise assessments.

In 2015, an article was published in the Institute of Acoustics Bulletin titled 'Wind farms cumulative impact assessment' [5]. This document provides guidance on instances where a cumulative assessment is required in accordance with the IOA GPG and ETSU-R-97. Within this document, two approaches are proposed to determine if a cumulative assessment is required. The first suggest creating a 35 dB contour for any exisiting wind farms, plus a 35 dB and 25 dB contour for the proposed development. Any receptors the fall within the 35 dB contour of the proposed development, or both the 25 dB contour of the proposed development and the 35 dB contour of the exisiting wind farms, must be included within the cumulative assessment. Noting that if the total cumulative noise predictions exceed 35 dB at a receptor that fails these either of these criteria then this receptor needs to be included as well. The second method suggests any receptors that both; fall within the total cumulative (including the proposed development) contour of 35 dB, and where the difference between the proposed development and existing developments is less than 10 dB should be considered within the cumulative assessment.

In some instances it is necessary to conduct a cumulative noise assessment for two concurrent applications. In this instance one potential solution included in the IOA GPG to account for cumulative is 'apportionment of the ETSU-R-97 limits on an energy basis'. By assuming apportionment on an energy basis relates to the *acoustic* energy of the noise predictions for each of the proposed schemes, noise limits would be created where each schemes share a proportionate amount with respect to their predicted noise levels. This approach can work well where predictions indicate that all schemes can coexist within the ETSU-R-97 limits. Where predicted levels exceed the limits, through apportionment on an acoustic energy basis, all schemes will need to reduce noise emissions (regardless of how acoustically dominant each scheme is).

Whilst this approach is appropriate for most scenarios where limit apportionment is required, there are instances where this would be unduly restrictive for a scheme which may be less acoustically dominant. Thus, having an impact on total energy yield across the proposed developments. Therefore, a more nuanced approach is sometimes required which would result in limits being apportioned in such a way that energy generation is maximised whilst resulting in different limits than those obtained with the methodology presented within the IOA GPG. This approach could be considered to represent apportionment on an energy *generation* basis.

This Paper is set out as follows; the methodology for noise limit apportionment on an acoustic energy basis will be presented. This method will then be applied to two cumulative wind farm scenarios to highlight the impact this approach can have in differing circumstances. An alternative approach (considering energy generation) will then be introduced which provides limits the total amount of curtailment required when one or more of the proposed developments are not having a meaningful impact at the receptor but need to be included within the cumulative assessment.

2. Apportionment of the ETSU-R-97 limits on an acoustic energy basis

Section 5.4 of the IOA GPG provides an example showing the apportionment of noise limits for two concurrent wind farm applications in the absence of any pre-existing wind farms. By first deriving the Total Noise Limit (TNL) in line with ETSU-R-97 for all receptors in which the proposed schemes will have an impact, wind farm developers can then apportion the noise limit "on an energy basis" (it is assumed this is in reference to *acoustic* energy) such that the cumulative operation of the proposed schemes does not exceed the ETSU-R-97 noise limit. Therefore, the ETSU-R-97 noise limit can only be exceeded if one or more of the wind farms were to operative above its own apportioned noise limits. This is illustrated in Figure 1, which can be found as Figure 7 within the IOA GPG.

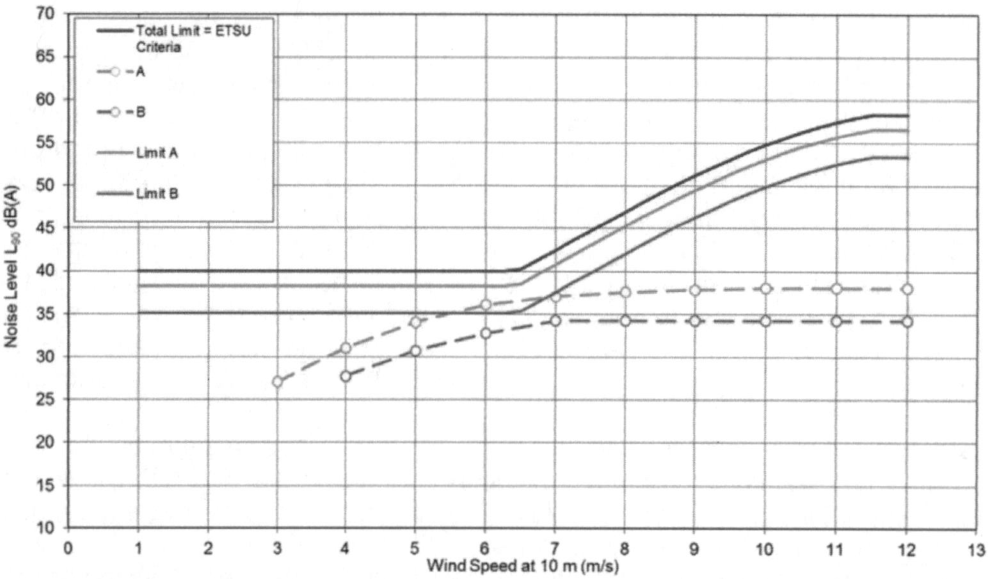

Figure 1: Apportionment of the ETSU-R-97 noise limit between two wind farms (Figure 7 within the IOA GPG).

One exception with this approach is when there are already operational schemes within the area and therefore the full ETSU-R-97 noise limit is not available. In such cases it is necessary to determine the contribution of the operational schemes at each receptor to determine the residual noise limit, which can then be apportioned between the proposed schemes. This is achieved with the following approach.

Consider a receptor in the vicinity of an operational wind farm. At this receptor, there is a TNL which has been determined in accordance with ETSU-R-97. Any apportionment of the TNL must account for the noise immissions from existing developments (ED). Here ED refers to any operational and consented wind farms to be considered within the cumulative assessment. Whilst the ED could in theory use their entire allotted noise limit, in practice this is not usually the case. In cases where a large amount of headroom (margin between predicted levels and the TNL) exists, it can be assumed that a development will not use its entire limit; typically, a minimum of 5 dB headroom is considered necessary, referred to as 'Significant Headroom' within the IOA GPG.

Where Significant Headroom is present, the predicted noise levels for the ED should be logarithmically subtracted from the TNL. The predictions should include an additional buffer (often assumed to be +2 dB), as an appropriate allowance for any potential noise level increases over the development's lifetime. The inclusion of this +2 dB adjustment is a cautious approach, and may over-estimate ED noise levels, however, the assumption that ED's will not exceed their predicted levels may lead to an under-estimation of cumulative noise levels, and ultimately an exceedance of the TNL.

Once the ED predicted levels (including headroom corrections) have been logarithmically subtracted from the TNL, the remaining limit is the Residual Noise Limit (RNL). This is defined in Equation 1, noting that the following expressions are applied for each wind speed across the entire range of standardised wind speeds, typically from 1 m/s to 12 m/s.

$$RNL = 10\text{Log}_{10}\left(10^{\left(\frac{TNL}{10}\right)} - 10^{\left(\frac{ED+2}{10}\right)}\right)$$ **Equation 1**

Where a single wind farm development is proposed, the proposed development can use the entire RNL (subject to Fixed Minimum Limits (FMLs)). Where two or more proposed wind farm developments are going into planning at the same time, the RNL will need to be apportioned, in this instance on an acoustic energy basis.

The Remaining Noise Budget (RNB) is then determined by arithmetically subtracting the combined cumulative noise levels of all proposed developments from the RNL. This is defined in Equation 2 for N proposed wind farms, where WF_i denotes the predicted noise levels for the *ith* wind farm (determined in accordance with ISO9613-2 [6] and the IOA GPG).

$$RNB = RNL - 10\text{Log}_{10}\left(\sum_{i=1}^{N} 10^{\left(\frac{WF_i}{10}\right)}\right)$$ **Equation 2**

When the RNB is positive for a wind speed bin, this indicates that headroom exists between the cumulative predicted noise levels from the proposed developments and the RNL. However, at wind speeds where the RNB is negative, this indicates that there is no headroom available and that the cumulative predicted noise levels exceed the RNL.

Where noise limit apportionment is undertaken on an 'energy' basis (acoustic energy, on an equal basis), each proposed wind farm development should be allotted a Site Specific Noise Limit (SSNL) based on the arithmetic addition of the individual development's predicted noise level and the RNB (subject to SSNL FMLs). This is expressed in Equation 3.

$$SSNL_i = RNB + WF_i \qquad i = \{1, 2, \ldots, N\}$$ **Equation 3**

Using this approach, any curtailment requirements to meet the RNL will be shared across the proposed developments. This is irrespective of how acoustically dominant each development is, which may be unnecessarily restrictive and sub-optimal from a total energy yield perspective.

For instances where noise predictions from a proposed development are 10 dB below the TNL, this development does not require a share of the RNL and can therefore be excluded from the apportionment process.

3. Application of limit apportionment on an acoustic energy basis

3.1 Scenario One

Consider a receptor where a cumulative noise assessment is required for two concurrent wind farm applications (WF1 and WF2). At this receptor, there are existing developments having a meaningful impact such that the total ETSU-R-97 noise limit is not available. Therefore, the RNL must be determined from predictions of the existing developments such that noise limit apportionment can be undertaken for the two proposed developments. The process of limit apportionment is conducted in order of operation for the entire range of wind speeds in Table 1. All windspeeds are the hub height windspeeds standardised to 10 m, as per the IOA GPG, and all noise levels are in L90 dB(A).

Table 1: Scenario One - limit apportionment on an acoustic energy basis.

Windspeed (m/s)	1	2	3	4	5	6	7	8	9	10	11	12
TNL	40.0	40.0	40.0	40.0	40.0	40.0	40.0	40.0	40.0	43.2	47.6	53.0
ED *	-	-	22.0	29.0	33.0	34.3	35.0	35.0	35.0	35.0	35.0	35.0
RNL	40.0	40.0	40.0	40.0	39.0	38.6	38.3	38.3	38.3	42.5	47.6	53.0
WF1	-	-	25.5	32.5	36.5	37.8	38.5	38.5	38.5	38.5	38.5	38.5
WF2	-	-	21.0	28.0	32.0	33.3	34.0	34.0	34.0	34.0	34.0	34.0
WF1+WF2	-	-	26.8	33.8	37.8	39.1	39.8	39.8	39.8	39.8	39.8	39.8
RNB	13.2	13.2	13.2	6.2	1.2	-0.5	-1.5	-1.5	-1.5	2.7	7.8	13.2
$SSNL_{WF1}$	38.7	38.7	38.7	38.7	37.7	37.3	37.0	37.0	37.0	41.2	46.3	51.7
$SSNL_{WF2}$	34.2	34.2	34.2	34.2	33.2	32.8	32.5	32.5	32.5	36.7	41.8	47.2

*As there is significant headroom, values here are ED + 2 dB. This is for all examples presented within the paper.

From Table 1 predictions for the cumulative operation of the proposed wind farms exceed the RNL. This is evident due to the negative quantities for the RNB within the standardised windspeed range 6 m/s to 9 m/s. Because of this exceedance, both schemes are required to curtail by 1.5 dB within this wind speed range. For low wind speeds where the turbines are not operational, the SSNLs have simply been flatlined from the lowest operational windspeed, in this instance at 3 m/s. Plots presenting the limit apportionment of Scenario One from an acoustic energy perspective are available in Figure 2.

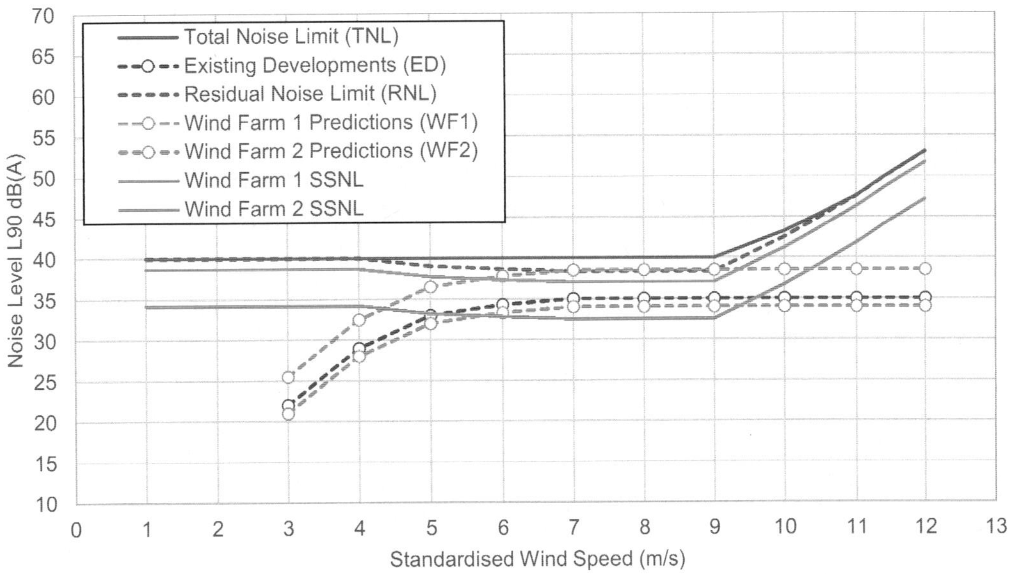

Figure 2: Scenario One - limit apportionment on an acoustic energy basis.

For this scenario, as both schemes are having a material impact, this approach may be an appropriate way to apportion the noise limits by evenly distributing mitigation requirements.

3.2 Scenario Two

Consider a receptor where a cumulative noise assessment is required for two concurrent wind farm applications. At this receptor, there are existing developments having a meaningful impact such that the total ETSU-R-97 noise limit is not available. Therefore, the RNL must be determined from predictions of the existing developments such that noise limit apportionment can be undertaken for the two proposed developments.

Unlike Scenario 1, one of the proposed developments much less acoustically significant than the other but is still within 10 dB of the TNL at this receptor, such that it is only contributing 0.6 dB to the total cumulative noise level. However, noise predictions from both proposed developments are each greater than 25 dB, with cumulative predictions also being greater than 35 dB, meaning both schemes have to be considered within the cumulative assessment. The process of limit apportionment for Scenario Two from an acoustic energy perspective is conducted in order of operation in Table 2.

Table 2: Scenario Two - limit apportionment on an acoustic energy basis

Windspeed (m/s)	1	2	3	4	5	6	7	8	9	10	11	12
TNL	40.0	40.0	40.0	40.0	40.0	40.0	40.0	40.0	40.0	43.2	47.6	53.0
ED	-	-	22.0	29.0	33.0	34.0	35.0	35.0	35.0	35.0	35.0	35.0
RNL	40.0	40.0	40.0	40.0	39.0	38.7	38.3	38.3	38.3	42.5	47.6	53.0
WF1	-	-	26.0	33.0	37.0	38.3	39.0	39.0	39.0	39.0	39.0	39.0
WF2	-	-	18.0	25.0	29.0	30.3	31.0	31.0	31.0	31.0	31.0	31.0
WF1+WF2	-	-	26.6	33.6	37.6	38.9	39.6	39.6	39.6	39.6	39.6	39.6
RNB	-	-	13.4	6.4	1.4	-0.2	-1.3	-1.3	-1.3	2.8	8.0	13.4
$SSNL_{WF1}$	39.4	39.4	39.4	39.4	38.4	38.1	37.7	37.7	37.7	41.8	47.0	52.4
$SSNL_{WF2}$	31.4	31.4	31.4	31.4	30.4	30.1	29.7	29.7	29.7	33.8	39.0	44.4

From Table 2 it is, again, evident that predictions for the cumulative operation of the proposed developments exceeds the RNL within the windspeed range 6 m/s to 9 m/s. Both schemes are required to curtail by 1.3 dB within this range. This is irrespective of the fact that the predictions for Wind Farm 2 are much less acoustically significant at all wind speeds. Plots presenting the limit apportionment of Scenario Two from an acoustic energy perspective are available in Figure 3.

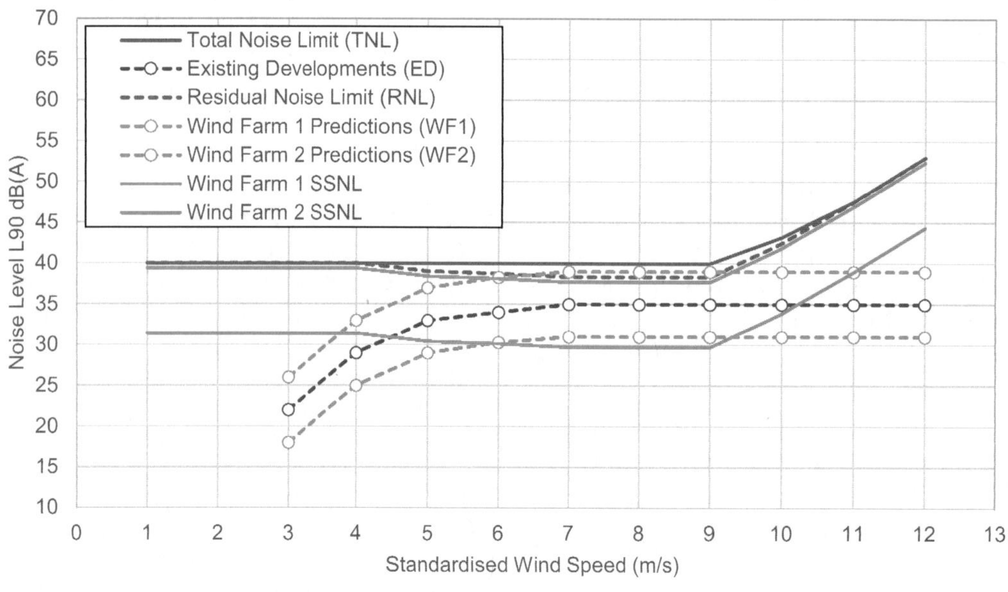

Figure 3: Scenario Two – Limit apportionment on an acoustic energy basis

A limitation of this approach is the requirement for both schemes to curtail, when only one of the schemes is acoustically dominant at the receptor and thus is the primary contributor to cumulative noise issue. Curtailing Wind Farm 2 when it is only contributing a maximum of 0.6 dB to the cumulative noise level is unduly restrictive and would result in an unnecessary reduction in energy yield across the two sites. Therefore, a more nuanced approach to deriving the SSNLs for each of the proposed developments may be warranted.

4. Limit apportionment with respect to the most acoustically dominant wind farms

For instances where limit apportionment between concurrent wind farms is required, but not all schemes are having a material impact at the receptor, limit apportionment on an acoustic energy basis may not be a suitable approach. Instead, it may be more appropriate to derive noise limits based on which schemes are acoustically dominant at the receptor. This approach has the potential to maximise the energy yield between the proposed developments and provides and alternative interpretation to 'apportionment of the ETSU-R-97 limits on an energy basis'.

The following limit apportionment is for Scenario Two, but for instances where the RNB is negative and curtailment is required, the requirement is placed upon Wind Farm 1 to curtail the full amount, with the SSNL for Wind Farm 2 simply defaulting to its predictions within this windspeed range. The resulting limits using this approach are presented in Table 3.

Table 2: Scenario Two - limit apportionment based on the most acoustically dominant wind farm.

Windspeed (m/s)	1	2	3	4	5	6	7	8	9	10	11	12
$SSNL_{WF1}$	39.4	39.4	39.4	39.4	38.4	38.1	37.5	37.5	37.5	41.8	47.0	52.4
$SSNL_{WF2}$	31.4	31.4	31.4	31.4	30.4	30.1	31.0	31.0	31.0	33.8	39.0	44.4

Plots presenting the limit apportionment of Scenario Two based on the most acoustically dominant wind farm are available in Figure 3.

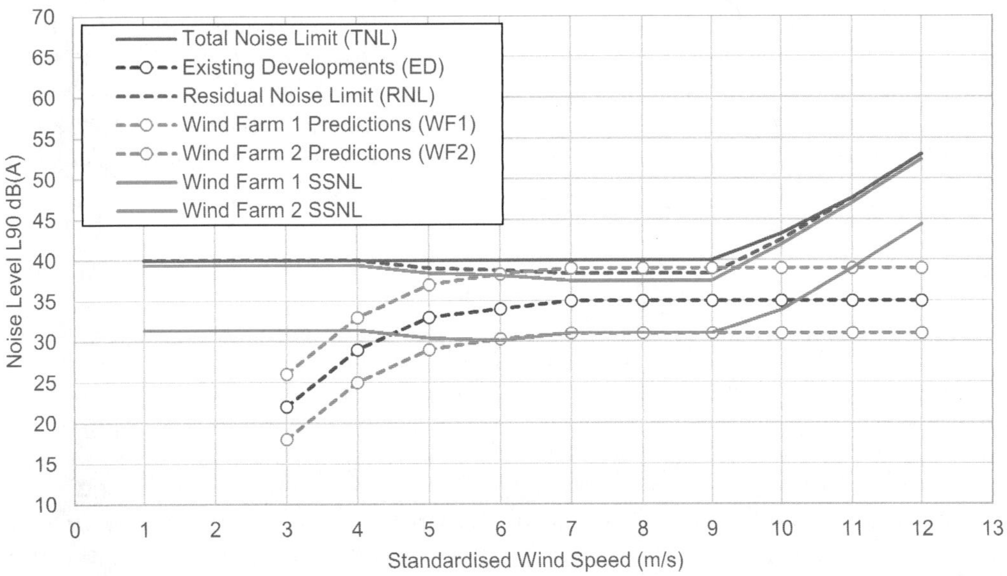

Figure 4: Scenario Two – Limit apportionment based on the most acoustically dominant wind farm.

The benefit of this approach is that the additional curtailment required for Wind Farm 1 (the most acoustically dominant scheme) is negligible in comparison to the amount of curtailment that was required for Wind Farm 2. Therefore, the result is a net positive benefit to the total energy yield across the two proposed developments. This is further exemplified in Table 4 which shows the relative change between the SSNLs between the two approaches. A positive value indicates an increase in the SSNL for limits apportioned with respect to the most acoustically dominant wind farm.

Table 3: Relative change between the SSNLs when deriving limits using apportionment with respect to the most acoustically dominant wind farm and apportionment on an acoustic energy basis.

Windspeed (m/s)	1	2	3	4	5	6	7	8	9	10	11	12
$\Delta SSNL_{WF1}$	0.0	0.0	0.0	0.0	0.0	0.0	-0.2	-0.2	-0.2	0.0	0.0	0.0
$\Delta SSNL_{WF2}$	0.0	0.0	0.0	0.0	0.0	0.0	1.3	1.3	1.3	0.0	0.0	0.0

By using this approach to derive the SSNLs, further curtailment of Wind Farm 1 by 0.2 dB is required and the amount of curtailment for Wind Farm 2 has been reduced by 1.3 dB. Overall, this is a significant reduction in the total amount of curtailment required across the two schemes which could be beneficial from a total energy yield perspective, and this approach can also be deemed a valid alternative approach for limit apportionment when one or more of the schemes involved in the limit apportionment are not having a material impact at the receptor.

One limitation of this approach is there is no consideration to the non-linear relationship between wind farm curtailment for noise mitigation, and the resulting energy yield. To calculate the energy yield impacts due to noise mitigation would be a complex process as it would be a function of the wind direction, the size of the wind farm, the contributions from each turbine to the noise immissions at a receptor, and the availability/ flexibility of low noise modes for the candidate turbine. In some instances, it is feasible that a reduction of 1 dB at a wind farm which is not having a material contribution at a receptor could result in much less energy yield losses than if the acoustically dominant wind farm were to curtail by 0.1 dB. This could simply be due to the size of the schemes where the most acoustically dominant wind farm consists of more turbines than the other wind farm. Consideration of all these factors is only possible where the developers agree to work collaboratively to resolve any cumulative constraints. For instances where this is not possible, apportionment on an acoustic energy basis may be the required apportionment method instead.

5. Conclusion

Within this paper, two interpretations to noise limit 'apportionment of the ETSU-R-97 limits on an energy basis' have been presented. The first approach assumes apportionment on an acoustic energy basis. For instances where schemes can either co-exist or are both having a material impact, this approach can be deemed the most suitable for deriving the apportioned noise limits as it shares the burden evenly between the proposed developments. The second approach assumes on an energy yield basis. This approach can be useful for situations where limit apportionment and curtailment is required, but a not all of the schemes are having a material impact to the total cumulative noise level. This can be useful as it can significantly reduce the total amount of curtailment required for all of the proposed schemes from a noise level perspective.

One benefit to apportionment with respect to the most acoustically dominant wind farms is that this could be beneficial in increasing the total energy yield across all the proposed developments. It is worth noting, however, that there is a non-linear relationship between wind farm curtailment for noise mitigation, and the resulting energy yield. Where energy yield is a function of wind direction, the size of the wind farm, the contributions from each turbine to the noise immissions at a receptor, and the availability/ flexibility of low noise modes for the candidate turbine. Therefore, consideration of all these factors is only possible where the developers agree to work collaboratively to resolve any cumulative constraints. For instances where this is not possible, apportionment on an acoustic energy basis may be the required apportionment method instead.

6. References

[1] ETSU for the DTI (Department of Trade and Industry, "The Working Group on Noise from Wind Turbines ETSU-R-97 The Assessment and Rating of Noise from Wind Farms'," 1996.

[2] Institute of Acoustics, "Good Practice Guidance on the application of ETSU-R-97 for wind turbine noise assessment," 2013.

[3] H. a. L. G. (. Department of Environment, "https://www.gov.ie/en/publication/f449e-wind-energy-development-guidelines-2006/," 2006. [Online]. [Accessed 19 01 2022].

[4] P. a. L. G. Department of Housing, "Draft Revised Wind Energy Development Guidelines," Government of Ireland, 2019.

[5] Bulletin, Institute of Acoustics, "Wind farms cumulative impact assessment," 2016.

[6] International Standards Organisation, "ISO9613:1996 'Acoustics – Attenuation of sound during propagation outdoors' – Part 2: General method of calculation," 1996.

10th International Conference
on
Wind Turbine Noise
Dublin – 21st to 23rd June 2023

Noise curtailment plan optimization and manufacturer constraints

Arthur Finez, ENGIE Green France arthur.finez@engie.com

Summary

To comply with acoustic constraints, wind farms are occasionally operated using predefined curtailment modes. The definition of optimal curtailment plans is a crucial point for wind farm developers and operators. Their implementation by the turbine manufacturer imposes the plan to satisfy some design rules that are not always accounted for in the optimisation process. This constraint introduces complexity in the optimisation process and thus increases significantly computational times. This paper proposes an approach to compute curtailment plans sacrificing a bit of optimality to favour the computation time. The idea is to first compute the optimal curtailment plan without the turbine manufacturer constraints and then fit it to the manufacturer constraints. Some illustrative applications are proposed suggesting that these constraints cannot be ignored within the optimization process, in particular when the number of commands is very limited (below 6) and that the command prioritization option is not possible. These cases are also a reminder that the availability of truly efficient noise reduction operational modes is key.

1. Introduction

As free space to install new wind farms is becoming rare and turbine heights and rotor diameters are dramatically increasing, it is more and more important for wind farm developers and operators to control the noise impacts of wind farms on the neighborhood. Among the tools available to limit these impacts, the possibility to curtail the turbines using predefined reduced operational modes is a common practice. The proper design of acoustic curtailment plans is key to provide a correct balance between the emitted noise levels and the electricity production.
However for the implementation in the turbines to be possible, the curtailment plans have to satisfy some design rules that depend on the turbine manufacturers. For instance, only a maximum number of different modes or commands can be used on a single wind turbine. These rules are generally loose enough to implement simple curtailment plans. However, the more the wind farms are constrained by the regulatory requirements, the more complex the curtailment plans become. In this case, manufacturer constraints may have a severe impact on the production loss and possibly need to be included in the optimization process.
Some efforts to provide curtailment plan optimization algorithms exist in the literature using random search optimization [Nyborg 2023], but without taking into account the manufacturer constraints. Introducing these constraints into the basic optimization problem raises theoretical complications and to the knowledge of the author, attempts to strictly solve the full problem produced an exaggerating increase of computation time. Thus it would probably be beneficial to

define a degraded tool able to compute an "acceptable" curtailment plan in a limited computational time, but taking into account the manufacturer constraints.
In this study a simplified approach is detailed to compute such curtailment plan. It is based on a "degradation approach", i.e. the optimal curtailment plan without manufacturer constraint is first computed and then forced to comply with the constraints. Throughout the study, the computational speed is favored against the solution optimality. As this contribution is a part of an ongoing work, not all the manufacturer constraints are yet included in the existing tool.
The paper is divided in 3 following parts. Part 2 consists in proposing an approach for a quick computation of the free solution (the optimal curtailment plan without the manufacturer constraints). Part 3 deals with the introduction of the manufacturer constraints, its implication in the optimization problem and proposes some tools to compute an approximate solution using the "degradation approach". Finally part 4 provides illustrative examples to apply principles described in the previous sections.

2. Noise curtailment plan optimization without manufacturer constraints

In the following a simplified optimisation problem is set to find curtailment plans maximizing yield while satisfying acoustic targets, not taking into account manufacturer constraints as a first step. Some principles and tools are pointed to tackle efficiently this idealized problem.

2.1 Description of input variables

Choosing a curtailment plan amounts to choosing an operating mode for each turbine and for each external condition (wind speed, wind direction, hour of the day etc...). An acoustic mode is understood as a predefined degraded speed/torque characteristic curve on which the turbine is operated. Compared to the standard full power mode, the rotor speed is reduced for a given incoming wind speed. The blade aerodynamic noise (leading edge and trailing edge noise) which is the main noise source is thus efficiently reduced. The acoustic benefit comes at the price of a lower electricity production. In modern wind turbines at least 4 modes are typically proposed ; they are carefully designed by the turbine manufacturer, not to excite mast resonances or generate tonal noise by the gear box. Each acoustic mode is documented with specific active power curves, thrust coefficients and sound power levels in octave or third octave bands, all three depending on the incoming wind speed. Wind speed bin are generally 1 m/s or 0.5 m/s wide. Each of these three quantities will be used in the following.
The external conditions are usually divided into three groups:

- *Time period*: as local regulation usually depends on the period of the day (night/evening/day), wind turbine curtailments can be defined in local hours. For instance, a wind farm can be curtailed at night between 22h and 7h for noise reasons but operated at full power during the day. It is also possible that a day conditioning is requested. Besides, working days and weekends may fall into distinct categories regarding environmental noise. An additional season dependency is sometimes wanted as background noise is known to significantly vary with seasons (mainly due to foliage seasonality) [Petit 2019],
- *Wind direction*: as wind plays a lead role in sound propagation, it is useful to tune the wind farm operation to the wind direction to mitigate the sound level at most exposed location i.e. usually downwind neighbours. Usually two to four wind direction sectors are chosen to deal with most common situations, for instance the large southwest sector [135°-315°[and the large northeast sector [315°-135°[in Picardie, France,
- *Wind speed range:* turbine sound power levels strongly depend on rotation speed and consequently on incoming wind speed. In countries where the acoustic emergence is regulated, the highest level difference between total noise and background noise is found in the critical medium speed range, typically 5 to 10 m/s at hub height. Highest curtailments are found in this range. The need to use acoustic modes on a small wind speed range is thus often encountered.

In the following a set of abovementioned external conditions (period of the day/wind direction sector/wind speed bin) is called a *situation bin*. According to the site meteorology, each situation bin occur with a known probability and can be associated with a number of occurrence hours in a standard year. As a consequence for each situation bin, the product of the standard occurrence duration by the active power of a candidate mode (at the considered wind speed bin) raises a partial Annual Energy Production (pAEP). The total AEP corresponding to a curtailment plan is the sum of all pAEP. It is thus possible to draw a table of pAEP for each mode in a given situation bin. Note that a distinction between turbines can be introduced in this step using onsite overspeed coefficients. Note also that this production table ignores wake effects, but this will be addressed in a following subsection. Finally it is also possible to take into account energy tariffication in this step. It can be useful if the concerned wind farm is constituted of several parts subjects to different sale prices.
Noise constraints introduce a number of additional inputs : the neighbouring acoustic control points. The acoustic contribution of the whole wind farm is required to be lower than a maximum acceptable level depending both on the control point and the situation bin. The transfer function between a specific wind turbine and the control point can be computed using a propagation model, possibly fitted to onsite measurements. The frequency content of each mode is accounted for by treating separately the frequency bands and summing the obtained quadratic pressure at the control point. Each mode produces an AOSPL at each neighbour point and for each situation bin (as noted previously wind direction and speed are lead actors in the sound propagation).

2.2 Formal statement

A key point is to notice that, as wakes and turbine manufacturer constraints are ignored in this first step, there is no apparent link between situation bins. This is very fortunate since the global problem can be split into many independent small problems; they are easier to solve than the general one. Moreover this operation makes the problem an excellent candidate for core parallelization. The global curtailment plan is obtained by a concatenation of all the smaller plan for each situation bin.
Noting:

- N_{WT} the number of wind turbines concerned by the optimisation process, with corresponding indices i_{WT}
- N_m the total number of candidate acoustic modes, with corresponding indices i_m
- N_Z the number of neighbour acoustic control points, with corresponding indices i_z

The unknown of the simplified optimisation problem is a binary matrix X with dimensions $[N_{WT}, N_m]$. $X[i_{WT}, i_m] = 1$ means that mode i_m is activated on turbine i_{WT}, and 0 else.

- The cost function involves matrix T with dimensions $[N_{WT}, N_m]$, $T[i_{WT}, i_m]$ being the partial AEP (or tariff as discussed in 2.1) of turbine i_{WT} set in mode i_m for the concerned situation bin.

The *objective function* to maximise is

$$f(X) = \sum_{i_{WT}, i_m} T[i_{WT}, i_m] * X[i_{WT}, i_m] \qquad \text{eq.1}$$

Two structural constraints arise from the binary formulation, namely

- There is *one and only one mode activated at a time* on a turbine :

$$\forall i_{WT} \le N_{WT}, \sum_{i_m} X[i_{WT}, i_m] = 1, \qquad \text{eq.2}$$

- The *activated mode must be available on the turbine*. As different wind turbine models can coexist on a windfarm, it is necessary to define an incompatibility matrix C with dimensions $[N_{WT}, N_m]$ stating if $C[i_{WT}, i_m] = 1$, that mode i_m is not available on turbine i_{WT}:

$$X[i_{WT}, i_m] = 0 \; if \; C[i_{WT}, i_m] = 1 \qquad \text{eq.3}$$

- The *acoustic constraint* is expressed using a vector B with dimension $[N_Z]$. $B[i_Z]$ is the maximum bounding value for the quadratic pressure (not in dB) allowed at neighbour location i_Z . Multidimensional array H with dimension $[N_{WT}, N_m, N_Z]$ represent the normalized quadratic pressure contribution of wind turbine i_{WT} set in mode i_m at location i_Z. The final constraints then reads

$$\forall i_Z \leq N_Z, \sum_{i_{WT}, i_m} H[i_{WT}, i_m, i_Z] * X[i_{WT}, i_m] \leq B[i_Z] \quad \text{eq.4}$$

2.3 Solving the simplified problem

Normalized quadratic pressure (p^2/p_{ref}^2) values are preferred to dB to express noise contribution so as to write a full linear problem. This is profitable since it falls in category of *integer linear programming* with binary unknowns for which very efficient solving tools exist and are available. Such problem can indeed be tackled using general solvers like *glpk, symphony* or *cplex.* These broadly distributed tools are able to choose automatically a suited algorithm to solve the provided problem. It is observed that the "Branch and cut" algorithm is always selected in the tested cases. Common data organisation is proposed in the R language by the ROI project. Packages available on CRAN like *ompr* provide an algebraic and intuitive way to model Mixed Integer Linear Programs.

The optimal solution for a standard wind farm of 10 turbines with 10 control points and 50 situation bins is typically found in a few seconds on a desktop computer.

2.4 Wake interaction

Accounting for wake interactions introduces a first degree of complexity in the simple model described above. As turbines in wakes are submitted to lower incoming wind speeds than turbines in clean flow, they harvest a lower energy yield than indicated in the production table depicted in section 2.1. An acoustic mode applied to an upstream wind turbine $X[i_{WTu}, i_m]$ has an influence on the input data of downstream wind turbines $T[i_{WTd}, i_m]$ via its proper thrust coefficient C_t. As a consequence the practical situation slightly departs from the linear model expressed so far. Slightly because the wake alignment usually occur on a part of a wind direction sector and only a small fraction of the energy yield is removed to the downstream turbine table. However the curtailment plan solution $\hat{X}$ of the nonlinear problem is supposed to be also the solution of the linear model when the updated production table $\hat{T}$ is used, including the mode activations in $\hat{X}$. As a consequence an iterative approach is chosen to account for this effect while keeping the benefit from the linear formulation. A first iteration is run with the initial full power production table T_1. The solution $\hat{X}_1$of this first linear problem is used to update the production table in a second version T_2 which is used to produce a second linear model. A new solution $\hat{X}_2$ is obtained which in return is used to compute the update T_3, etc... The process is repeated while a convergence is observed on table T. It is believed that the last solution will be close to the non linear problem solution if the convergence is indeed reached.

In practice a small number of iterations is usually necessary to reach convergence, say around 3. However in some cases, it is observed that the iterative produce enters a cyclic scheme and is necessary to stop the process before a complete convergence. The effect of the iteration process on production will be briefly illustrated in section 4.

In the following the proposed solution is referred to as the "free solution" (FS) since it does not take into account turbine manufacturer constraints.

3. Adding manufacturer constraints

3.1 Review of manufacturer constraints

The operational implementation of acoustic modes into the turbines depends on the firmware architecture of the control system and greatly vary from a manufacturer to another. For instance certain manufacturers provide a small number of memory slots linked to a single acoustic mode with application on a lot of time intervals, several wind sectors but a single wind speed range.

Other manufacturers indicate no limitation in the number of wind speed bins and direction criteria but require the acoustic power not to vary more than a threshold value between adjacent situation bins.
Another example concerns the "stop" state which can be acoustically considered as a drastic reduction mode. For some manufacturers it can be deported in an external system (often bats curtailment system) allowing to exclude it of the limitation in number of modes, while this is not possible to others.
In the following turbine manufacturer constraints will be abbreviated "TMCs". All in all it is possible to draw a general list of constraints for a curtailment plan to be implementable in turbines, gathering most of the requirements of the manufacturers:

- *TMC 1: Limitation in the number of modes :* a maximum number of different acoustic modes is allowed to be used by a single turbine over all the encountered conditions. The standard mode do generally not fall in this category. This parameter ranges from 1 for old turbine systems to 20,
- *TMC 2: Limitation in the number of commands :* a command can be defined as the addition of a range of wind direction, of wind speed and a time range condition associated with a single application mode. This parameter ranges to 4 for small systems to no higher limit
- *TMC 3: Maximum number of commands in the same time period and wind sector situation*
- *TMC 4: Minimal length of a wind speed range:* To prevent the turbine to constantly switch between several mode due to windspeed natural variability, it can be required that the wind speed range inside a command does not fall below to a certain value, often 2 m/s,
- *TMC 5: Maximum Offset in Sound Power Levels between two adjacent situation bins.*

Moreover manufacturer description of these implementation constraints introduces variations in applications of these TMC:

- *The STOP mode counted in the mode number limitation:* some manufacturer do not count the stops as a reduction mode since it can be implemented elsewhere,
- *The STOP mode counted in the command number limitation*
- *Command prioritization:* For some manufacturer it is possible to overlap commands. For instance if a turbine is curtailed with mode A from 5 m/s to 10 m/s, and with mode B from 7 m/s to 8 m/s with a higher priority, it will result in the following table:

Speed (m/s)	5	6	7	8	9	10
Mode applied	A	A	B	B	A	A

It provides some flexibility, in particular when the number of commands is highly restricted. For other manufacturers, this is not possible and command must meet excluding conditions.
Constraints TMC 1 to TMC 3 are purely imposed by firmware structure while constraints TMC 4 and TMC 5 are intended to prevent undesired stress to the wind turbine because a change in acoustic modes often induces a change in load on the rotor. Constraints on the number of modes (TMC 1) and on the number of commands (TMC 2) are probably the most important to address as they are shared by the vast majority of manufacturers.
Requirements TMC 1 to TMC 5 are not applicable at the same time since some of them are specific to isolated turbine manufacturers (namely TMC 3, TMC 4, TMC 5) but a wind turbine operators endeavour to take all of them into account to compute the best implementable curtailment plan, i.e. the curtailment plan that produces the higher AEP while satisfying all acoustic and manufacturer constraints.

3.2 Implications on the problem reformulation

Adding constraints of section 3.1 into the optimisation problem of section 2.2 forms the "global optimisation problem"; it has several implications.
The most obvious one is that any curtailment plan satisfying both TMCs *and* natural constraints (eq. 2-4) will of course comply with natural constraints only. So if it exists, the "global solution" will have a production value ($f(X)$ of equation 1) lower or equal to the FS. Adding constraints is thus detrimental to the electricity production; the quantification of this effect will be carried out on some examples in section 4.

A second implication is that the TMCs introduce a link between the initially separated problems of distinct situations bins. For instance the *limitation in the number of modes* (TMC 1) clearly raises a dependency between several bins because the counting of the mode usage has to be done over all curtailed bins (but separately over turbines). As a consequence, it would be necessary to add an ending dimension to variable $X[i_{WT}, i_m, i_s]$ with index $i_s \leq N_s$ referring to the numbering of situation bins where curtailment is needed. Then constraints 1 and 2 would read:

$$\forall i_{WT} \leq N_{WT}, \sum_{i_m, i_s} X[i_{WT}, i_m, i_s] * \delta[i_m, i_{WT}] \leq N_m[i_{WT}] \quad \text{eq.5}$$

Where $N_m[i_{WT}]$ is the maximum number of modes allowed on turbine i_{WT} and $\delta[i_m, i_{WT}] = 1$ if mode i_m must not be counted for turbine i_{WT} and 0 else. Matrix δ is an input parameter of the problem and is useful because the STOP mode can occasionally be omitted in the counting. This property is turbine dependant as several turbine types can take part in the same optimisation problem, especially when wind farm extensions are studied.
The constraint limiting the number of commands TMC 2 supposed to define the notion of "commands" as a group of adjacent situation bins of the same activated mode (for the case where prioritization is not possible). Adjacent situation bins refer to wind speed bins separated by the wind speed step (usually 1m/s), but also to related wind sectors and time periods. For instance the northeast $[0° - 90°[$ and southeast $[90° - 180°[$ wind sectors can be joined to form the east sector$[0° - 180°[$. Following the same idea the evening period $19{:}00 - 22{:}00$ and the night period $22{:}00 - 7{:}00$ can be joined to form the 12 hour period $19{:}00 - 7{:}00$. If the same acoustic mode is employed in the evening and night period on the same (possibly joined) wind sector and on the same wind speed range, then it is possible to count a single command line. While the counting of commands is accessible from a numerical point of view, its analytical description is probably out of the scope of this paper regarding the choices made in the following section. Suffice it to say that this important constraints TMC 2 introduces most probably strong non linearities in the optimisation problem.

3.3 Two solving strategies : global genetic algorithm and degradation

Adding a dimension to X with equation 5 and introducing dependencies between situation bins prevent us from using the bin splitting trick of section 2.2. As a consequence instead of dealing with a myriad of small independent problems, one has to face a single bigger problem. This is detrimental from the computational point of view but still tractable with already cited optimisation tools.
A more significant problem comes with constraint TMC 2 because it introduces nonlinearity in the problem. Some types of nonlinearities can be dealt with adequate tools listed for instance in the CRAN Task View: "optimisation and mathematical programming"[1] but an efficient solver able to find the global optimum of the full problem (including acoustic and TMC) in a manageable amount of time is not known to the author.
From this point several strategies can be adopted. A first one is to resort to stochastic methods like evolutionary computation such as genetic algorithms. The power of this class of algorithms lies in its high flexibility [Sivanandam 2007]. A candidate curtailment plan is called an individual which is evaluated by a fitness function. This function can include any type of objectives, including nonlinear additional constraints, simply integrated as penalization terms. The process of natural selection is mimicked by an iterative process involving populations of individuals. Best individuals are selected to the next generation; "genetic" crossovers and random mutations are applied to favour positive evolution before next fitness evaluations. As a counterpart of this flexibility, the convergence and solution optimality is not guaranteed. It comes with a rather high computational time when compared to the branch and cut algorithm in simple linear cases.
The full constrained optimisation problem can be implemented as a single genetic algorithm introducing equations 1-4 and TMCs in the fitness function. The definition of acoustic modes in all situation bins for all wind turbines are sought at once. This approach is highly flexible and

[1] https://cran.r-project.org/web/views/Optimization.html

typically proposes a solution in one hour on a desktop computer for a medium size wind farm, but the optimality of the proposed curtailment is not guaranteed.
Another strategy relies on the *degradation* of the FS of section 2.4 to force TMC compliance. This approach reproduces the historical manual procedure where the proposed curtailment plan is usually modified for each turbine by the manufacturer to comply with its own implementation rules. For instance, if the number of used modes is higher than requested by TMC 1, some modes can be replaced by "lower" modes that are already in use i.e. modes with lower acoustic contributions to all control point on all frequency bands. As many replacement as necessary are made to comply with the constraint. This strategy can be adapted to the other TMCs; it is referred to as the "degradation approach" and detailed in the next section.

3.4 Degrading the free curtailment plan

In the following, a methodology is proposed to compute a degraded curtailment plan complying with some TMC, starting from the FS. As one of the objectives is to keep computational time tractable, a *turbine splitting* strategy is adopted. For each turbine, a curtailment plan degradation is proposed separately to comply with acoustic constraints and TMC. This choice is a trade-off in favour of computational speed (since the problem size is considerably reduced) and against the electricity production (since mutual benefits of curtailment between turbines cannot be exploited). Successive degradation is realized starting with the compliance to TMC 1 using mode replacement and then TMC 2 (command number limitation). TMC 3 to 5 are not addressed in this contribution but this could theoretically be included in the previous steps.

3.4.1 Mode replacement (TMC 1)

For each turbine, a new binary programming problem is formulated where the main unknown is now a binary replacement matrix Y with dimensions $[N_m, N_m]$ where $Y[i_m, j_m] = 1$ means that mode i_m will be replaced by mode $j_m$and 0 elsewhere. This is a global replacement which does not depend on the situation bin. Let $\tilde{X}[i_m, i_s]$ be the binary representation of the FS computed in the previous step for the concerned turbine (for mode i_m and situation bin i_s). The pAEP value for mode $j_m \leq N_m$ and situation bin $i_s \leq N_s$ is noted $\tilde{T}[j_m, i_s]$. The objective function for the mode replacement problem in the turbine in question reads

$$g(Y) = \sum_{\substack{i_m \leq N_m \\ j_m \leq N_m \\ i_s \leq N_s}} \tilde{X}\,[i_m, i_s] * \tilde{T}[j_m, i_s] * Y[i_m, j_m] \qquad \text{eq.6}$$

As in section 2.2, there is a structural constraint to satisfy which is the fact that a mode has to be replace by one and only one other mode (possibly itself). This reads

$$\forall i_m \leq N_m \sum_{j_m \leq N_m} Y[i_m, j_m] = 1 \qquad \text{eq.7}$$

One could think of another constraint to consider : a replacing mode shall not be replaced itself by another mode. There should not be any circular replacement. Another way of saying it is that the replacement matrix Y should be a projection ($Y * Y = Y$). Adding this constraint would transform the problem into Quadratic Programming. However in practice there is no need to implement it since a full permutation can not have a better cost function value than the FS (by definition of the FS). An exception would be if the true FS has not been found because of the non optimal wake iteration procedure. As a consequence, only a check on the compliance to this constraint is done, but a violation has not been detected on any example up to now.
The acoustic constraint is again formulated in terms of quadratic pressure to keep linearity. For all used situation bins i_s, all control points $i_z \leq N_z$ the replacing mode $j_m$should have a lower acoustic contribution $H[j_m, i_s, i_z]$ than the replaced mode i_m : $H[i_m, i_s, i_z]$. Because of equation 7, this can be written in a linear sum form :

$$\forall i_m \leq N_m, i_s \leq N_s, i_z \leq N_z, \sum_{j_m \leq N_m} H[j_m, i_s, i_z] * \tilde{X}\,[i_m, i_s] * Y[i_m, j_m] \leq H[i_m, i_s, i_z] \qquad \text{eq.8}$$

Finally TMC1 can be coded by limiting the number of modes to a maximum value $N \geq 1$ using a mode counting binary vector $M[j_m]$ of dimension $N_m$which is a new unknown. $M[j_m] = 1$ if and only if mode j_m is finally used at least once in matrix Y. TMC1 can be written

$$\sum_{j_m \leq N_m} M[j_m] \leq N \quad \text{eq.9}$$

The link between M and Y can be encoded following [Bradley 1977] using two additional constraint inequalities

$$\forall j_m \leq N_m, \sum_{i_m \leq N_m} Y[i_m, j_m] \geq M[j_m] \quad \text{eq.10}$$

$$\forall j_m \leq N_m, \sum_{i_m \leq N_m} Y[i_m, j_m] \leq K * M[j_m] \quad \text{eq.11}$$

Where K is a constant with a sufficiently high value, the total number of modes being sufficient. For each j_m, equations 10 and 11 are equivalent to the sentence "$M[j_m] = 0$ if and only if corresponding column in Y is full of zeros".
Including the STOP mode in the list of possible modes ensure the existence of a solution since at least the trivial candidate where it replaces all used operational modes satisfy equations 7 to 11.
This new Binary Programming problem being linear, the maximization of $g(Y)$ can again be realized very efficiently using the tools mentioned in 2.3. Because the number of binary unknown is very small $(N_m * (N_m + 1))$ the computation time does not usually exceeds a few seconds on a desktop computer.

3.4.2 Reducing the number of commands (TMC 2)

Once the mode number reduction is done, the compliance to TMC 2 is reevaluated. If a violation is found, it is necessary to degrade further the curtailment plan of each turbine. Given the nonlinearities discussed in section 3.2, it has been estimated that a more flexible tool is needed than the linear programming techniques used up to here. A genetic algorithm is used to encode the limitation in the number of commands (TMC 2) separately for each turbine. Beside this allows to include the prioritization possibility in a rather natural way.
To this end a binary genetic algorithm is set up. An individual is defined as a table where each row is constituted of a command line defined by

- the starting wind speed bin of application,
- the length of the application wind speed range,
- the acoustic mode to be applied,
- the (possibly joined) time period/wind direction sector identifier.

All variables in the unknown table are indices which can be binary coded to benefit from standard tools. The number of rows is the maximum number of commands allowed by TMC 2, so that this constraint is naturally satisfied by each individual. From the command table, a turbine curtailment plan is deduced, accounting for the possibility of prioritization. If prioritization is not allowed, a violation is noted and an overlap penalization term will be computed.
Each individual is evaluated against production objective and constraints through a fitness function. It is constituted of the sum of the electricity production losses with respect to the standard mode diminished by the overlap penalization and an acoustic penalization term. The acoustic penalization is activated each time the noise contribution by the turbine on a control point is higher than its contribution according to the FS. This ensures that the farm total noise at all control points is lower than the FS total noise.
The implementation of this genetic algorithm is done using powerful and widespread R language package *ga* [Scrucca 2013][Scrucca 2017]. This allows to benefit from already developed functions to tune crossovers and selection functions and efficient parallelization tools. The fine choice of these functions is still a matter of questions so that the degradation is not too harsh and

will be partially explored in section 4. Typically stops should not be extended more than necessary.
The degradation approach has the advantage that the compliance to TMC is evaluated after the computation of the free solution. If by any chance the turbine control system is flexible enough or the computed FS simple enough, the compliance is already satisfied and an optimal solution has been found in a very small computation time. If on the contrary TMC violation are found, it is the job of degradation algorithms to increase production losses as little as possible. But as TMC are getting looser thanks to the efforts of the turbine manufacturers, this case will occur less and less often. Another advantage of using a genetic algorithm for degradation purposes is that the including of TMC 3 to 5 could theoretically be included with limited efforts.

4. Some applications

In this section some illustrative examples are provided to support some of the assertions of the previous sections.

4.1 Wake interactions

As mentioned in section 2.4, a specific iterative procedure is set up to deal with non linearities due to turbine wake interactions. Applications are proposed here on some real field examples, with limited site and turbine information however for confidentiality reasons. Announced figures of yield losses are computed using site long term wind rose statistics and turbine power curves provided by the manufacturers.
The first typical example is an existing wind farm (noted wind farm 1) constituted of 6 turbines in France. The turbines are relatively close to each other (mean spacing 5 rotor diameters) and the turbines are organized on a north/south row while the long term wind rose shows a predominance of southwest winds. As a consequence wakes are found to partially impinge turbines on a limited but significant part of encountered wind conditions. The wind farm is heavily curtailed for noise reasons (10% of yield loss).

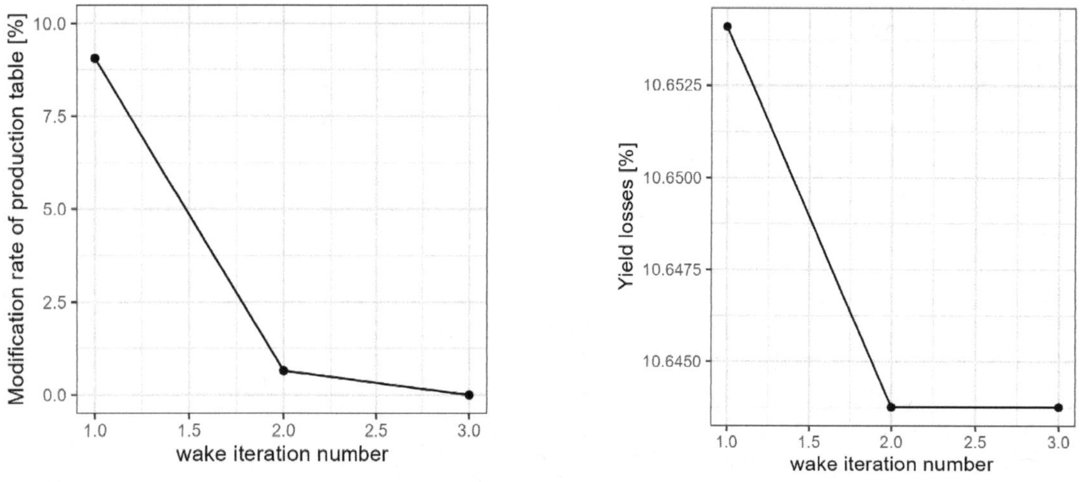

Figure 1: Effects of the iterative wake procedure on windfarm 1. Left : rate of change of the production table T at the end of each iteration. Right : implied yield loss found on each iteration.

The iterative wake procedure is applied on this wind farm to compute the free solution (FS, without manufacturer constraints). The procedure is found to converge in 3 iterations. Figure 1 shows the modification proportion of the production table T along the procedure and the corresponding yield loss. At the end of the first iteration, the computed curtailment plan induces a modification of 9% of the production table. The second iteration produces a new curtailment plan which necessitates less that 1% adjustment of the production table. The curtailment plan optimized in iteration 3 is the same that the one in iteration 2, which closes the convergence procedure. Although beneficial, the implication on yield losses is very limited with a loss reduction of less than 0.01% AEP difference between the 2 computed curtailment plans. A general quantification of the optimization of the curtailment plans including wake effects has not been conducted but all the tested example shows a very limited implication on yield performances.

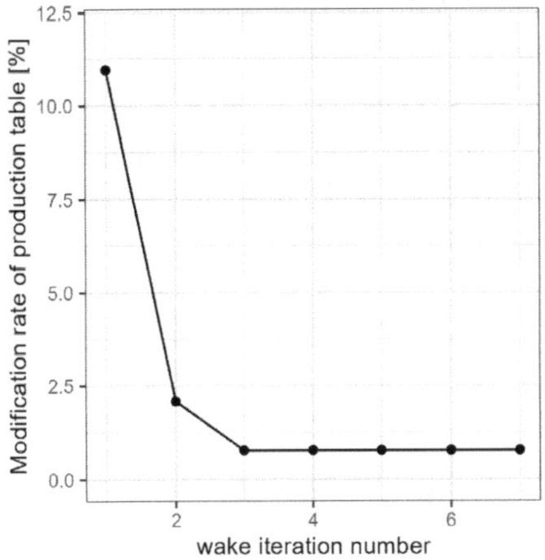

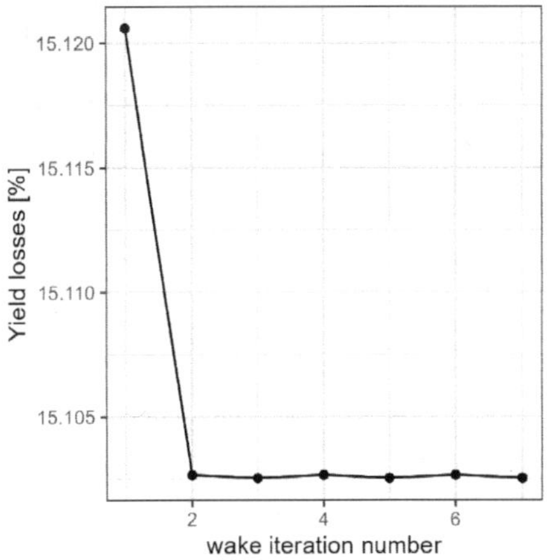

Figure 2: Effects of the iterative wake procedure showing a loop patten. 1. Left : rate of change of the production table T at the end of each iteration. Right : implied yield loss found on each iteration.

A second example is provided with the cyclic scheme mentioned in section 2.4. It occurs in a study concerning the same site in the framework of its repowering with 6 turbines of a different manufacturer and a 15% higher hub height. Figure 2 shows that the procedure in this case is not convergent since the change rate of T is stuck on a very small value of 0.7%. The corresponding production oscillates between two very close values. Comparing the optimized curtailment plans proposed in two successive iterations shows that they are actually very close and differ only on a single situation bin. The curtailment plan computed in iteration 3 induces a production table for which the optimized solution is the curtailment plan of iteration 4. This solution generates in return the already computed production table at the beginning of iteration 3. This results in the oscillating curve shown in Figure 2 right. As the yield figures are very close, these solution are considered equivalent on a practical point of view and for simplicity it is chosen to interrupt the iterative procedure as soon a the modification rate of T strictly stops decreasing. In this example the retained solution is thus the forth computed curtailment plan which has slightly higher yield losses than the preceding one.

4.2 Mode number reduction

In this section the mode replacement technique proposed in section 3.4.1 is illustrated by means of another real life example. The mode number degradation procedure is applied and a small parametric study is provided against the maximum allowed mode number. To exhibit specific effects of the mode number limitation, the number of commands is not limited in this section. The STOP mode is not integrated in the mode count limitation, assuming that the commands can be deported in another part of the firmware.

The studied wind farm is composed of 4 modern turbines delivered with 5 acoustic modes. Severe acoustic constraints imply that the wind farm needs to be curtailed to comply with local regulations.

The FS involves all provided acoustic modes plus the STOP mode on the turbines. Corresponding yield losses are evaluated to 8.2 %. From the FS situation using 5 modes, we progressively reinforce the constraint on the maximum number of modes down to a single mode authorized (in addition to the STOP mode and the standard full power mode). The single mode constraint can be found in old turbine softwares. Figure 3 illustrates that as expected the yield loss regularly increases as the number of modes is reduced. It is first to be noted that a even a (small) reduction of 5 to 4 modes can be detrimental to the production as an increase of 0.2% is observed which is considered moderate but meaningful. On the extreme opposite a severe increase is found between values 2 and 1 of the parameter, where loss reaches more than 33%. As shown on Figure 3 right, this is linked to an extensive usage of stops which can not be avoided on more that 10% of the situation bins.

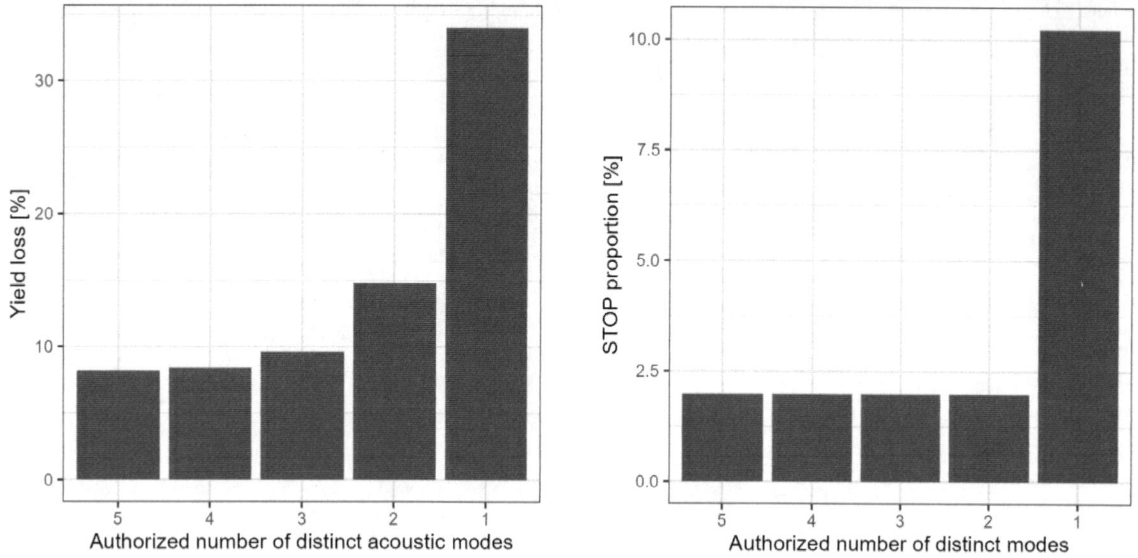

Figure 3:Effect of the mode number constraint on an example windfarm. Left : production losses as the function of the main constraint parameter. Right : proportion of usage of the STOP modes in the situation bins.

An interesting comparison is the analysis of the same wind farm candidate with another model of turbine. The previous turbine (model A) was delivered with 5 predefined acoustic modes while the new one (mode B) is delivered with 12 distinct modes. A comparison between the sound power specifications is provided in Figure 4 showing that model B is significantly quieter than mode A (with the same rated electrical power). This translates into lower production losses since the FS computed using model B for all turbines is linked 0.6% production losses. It uses up to 5 different acoustic modes on a same turbines. Now reducing progressively the allowed number of acoustic modes produces Figure 4 right. It is interesting to see that the loss increase is very limited event for a single mode used by turbine (losses of 1.5%). This is first because the wide variety of available acoustic modes makes it possible to use a suited mode in each turbine case. Another reason is that the STOP mode is not necessary for the model B case, even in the single mode situation. A more detailed investigation of the model A case showed that all STOP occurring in the $N = 1$ situation for model A were observed between 8 m/s and 13 m/s, a wind speed range where Model B provides quieter acoustic modes. The availability of efficient quiet acoustic modes has a direct impact on production losses in particular in constrained situations.

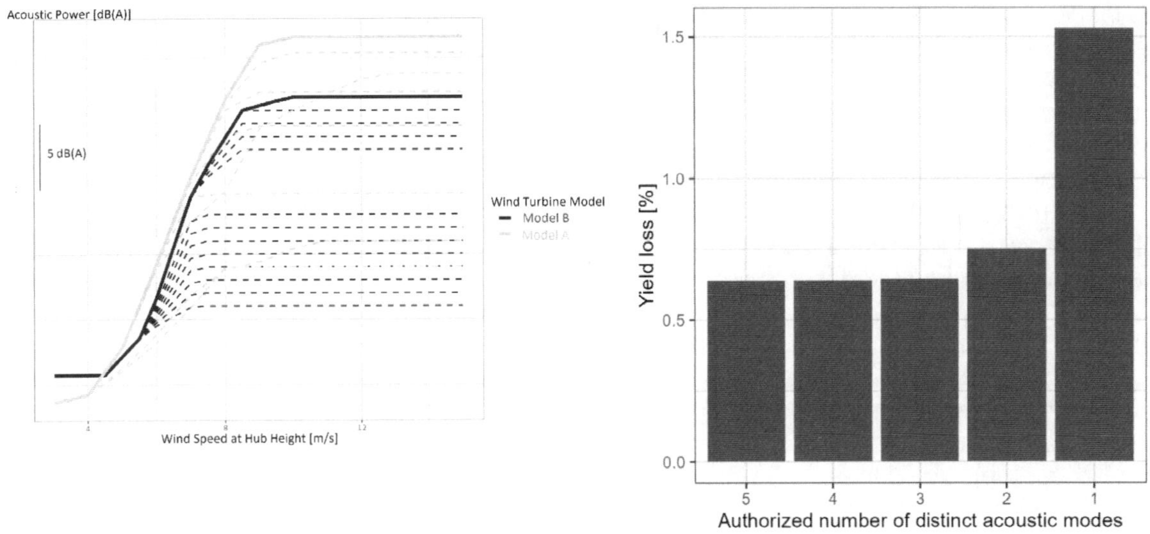

Figure 4 : Acoustic power of both turbine type studied in section 4.2

4.3 Command number reduction

The influence of the reduction of command number is illustrated in this section on the examples used so far. It is simpler to start with Model B than Model A since the noise overshoots are limited and a limited number of commands (actually 6) is needed to completely encode the FS (Figure 5, mode names have been changed). The command overlap and prioritization is not necessary for FS encoding.

CHB	WT	8 m/s	9 m/s	10 m/s	11 m/s	12 m/s	13 m/s	14 m/s
Nuit_NE	E_1		Mode 2	Mode 2	Mode 2	Mode 5	Mode 5	
	E_2		Mode 4		Mode 2		Mode 4	
	E_3		Mode 4					
	E_4		Mode 5	Mode 6	Mode 1			
Nuit_SO	E_1		Mode 4	Mode 4	Mode 5	Mode 5	Mode 5	
	E_2		Mode 4	Mode 3		Mode 1	Mode 5	
	E_3		Mode 2	Mode 2			Mode 4	
	E_4		Mode 2	Mode 5				

Figure 5: FS curtailment plan computed for Model B wind turbine. The curtailment occurs only at night (stringiest regulations) for northeasting and southwesting winds. Green spaces means that the standard mode is applied (no acoustic curtailment).

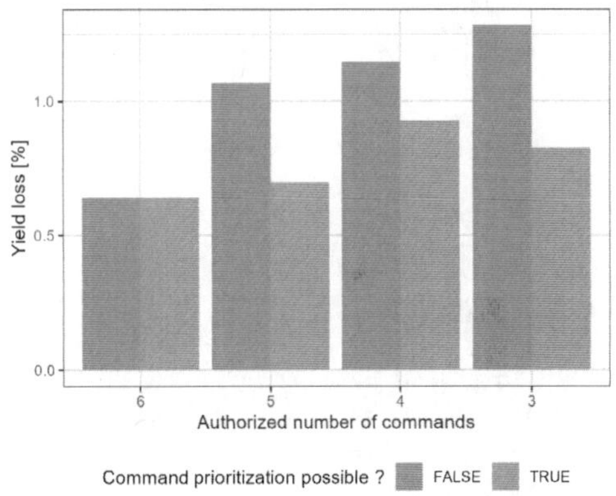

Figure 6: Influence of the number of commands on the Model B example.

Starting from this point, the same methodology than in section 4.2 is applied to progressively reduce the authorized number of commands (without limiting the number of modes). As can be seen in Figure 6, on this case the increase of production losses is limited (below 1% of increase), first because the FS is not complicated but also because the variety and the large number of modes allows to concatenate mode commands without strong loss damages. Two computations are run with and without the possibility to overlap and prioritize commands. It is seen that the prioritization offers a simple way to limit the loss increase on this case.

It can also be noted that the proposed curtailment plan with prioritization with 4 commands is less productive than the one with 3 commands. It illustrates the fact that the degradation methodology with a genetic algorithm does not always provide the optimal solution. Actually in some cases a manual degradation of the curtailment plan is more efficient than the algorithm output. Obviously there is room for improvement in the parametrization of the genetic algorithm.

The last example concerns the turbine of Model A for which the FS is more complex than Model B as illustrated on Figure 7, with 11 command lines. It should be noted intuitively that the impact of TMC is all the more important than the number of curtailed situation type (wind direction and time period class) is high. Figure 7 shows that 4 situation types are curtailed while only two were curtailed in Figure 5.
The influence of the command number limitation in this case on the production loss is shown in Figure 8. In this more complex case, the increase of production loss due to command number limitation is very significant and can reach 5% increase for realistic cases (4 commands maximum is an active requirement for some turbine models). Again the possibility to use command prioritization allows to limit the loss increase (up to 4 commands at least). Note that the no command prioritization case seems more complex to handle by the genetic algorithm than the prioritization case. Indeed the red bars are not steadily increasing while the green bars appears much more stable. In some case ($N_c = 7$) it even fails to provide an acceptable solution. This is again a manifestation of the inability of the genetic algorithm (as parametrized today) to provide a truly acceptable solution.

CHB	WT	7 m/s	8 m/s	9 m/s	10 m/s	11 m/s	12 m/s	13 m/s	14 m/s
Jour_NE	E_1			Mode B	Mode B	Mode B	Mode B	Mode B	
	E_2				Mode A				
	E_3								
	E_4		Mode C	Mode C	Mode A				
Jour_SO	E_1			Mode C	Mode E	Mode B	Mode B	Mode B	
	E_2			Mode A	Mode B	Mode B	Mode B	Mode B	
	E_3				Mode A				
	E_4			Mode B					
Nuit_NE	E_1		Mode C	Mode D	Mode D	Mode D	STOP	STOP	
	E_2		Mode E	Mode D	Mode C	Mode B		Mode B	
	E_3		Mode C	Mode D	Mode C	Mode C	Mode B	Mode B	
	E_4		Mode E	Mode D	STOP	Mode D	Mode B		
Nuit_SO	E_1		Mode E	Mode D	Mode D	STOP	STOP	STOP	
	E_2		Mode E	Mode D	Mode D	Mode B	Mode B	Mode C	
	E_3		Mode E	Mode D	Mode E	Mode B		Mode B	
	E_4		Mode C	Mode D	STOP	Mode D	Mode B		

Figure 7 FS curtailment plan computed for Model A wind turbine. The curtailment occurs nights and days for northeasting and southwesting winds

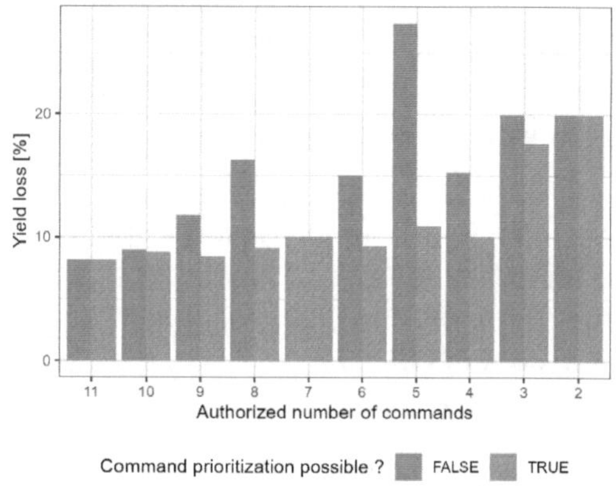

Figure 8 Influence of the number commands on the Model A example

5. Conclusions

Defining an acoustic curtailment plan for a wind farm is a matter of balance between emitted noise and electricity production. A general approach to find an optimal solution can be defined in setting noise limits that have to be strictly respected and maximizing the electricity production in this framework. This basic problem can actually be handled relatively easily using standard and efficient binary linear optimization tools.
A great deal of complexity is introduced as manufacturer constraints are taken into account, mostly because they add dependency between wind speed bins and introduce non linearities in the optimisation problem. As a consequence more flexible tools have been deployed. This comes with either a prohibitive increase in computational time or the abandon of solution optimality.
This contribution aims at proposing a degraded approach (non optimal curtailment plans) terminating in manageable timescales. The main idea is to compute the solution without manufacturer constraints (noted FS) and progressively transforms each wind turbine curtailment plan separately to force the manufacturer constraint compliance. This provides the advantage that degradation only operates when needed. Thanks to manufacturer efforts, this will be less and less often in the near future.
The approach is largely perfectible since it only partially takes into account the manufacturer constraints (only the mode number and the command number limitations). Moreover the proposed solution appears in complex cases to be largely suboptimal when compared to a human expert degradation, and production losses are not always an increasing function of the constraint reinforcement. However the flexibility of the approach gives hope to progressively improve the

quality of the proposed solution and include more constraints. Moreover the limited computational time makes it already interesting when gross estimates of production loss are required.
Application to field examples shows that turbine manufacturer can be the source of dramatic loss increase (which can be discriminant in the model competition for a wind farm equipment). These very partial results suggest that turbine manufacturer constraints cannot be ignored in particular when the number of commands is very limited (below 6) and that the prioritization option is not possible. These cases are also a reminder that the availability of efficient noise reduction operational modes is key.

References

[Nyborg 2023] Nyborg, C. M.; Fischer, A., Réthoré, P.-E. , Feng, J. *Optimization of wind farm operation with a noise constraint* WES, 8, 255–276, 2023
[Petit 2019] Petit Arthur, Le Bourdat Colin (2019) *Background noise seasonality and its influence on acoustic curtailments*, Wind Turbine noise 2019
[Scrucca 2013] Scrucca L. (2013). GA: A Package for Genetic Algorithms in R. Journal of Statistical Software, 53(4), 1-37, doi: 10.18637/jss.v053.i04.
[Scrucca 2017] Scrucca, L. (2017) On some extensions to GA package: hybrid optimisation, parallelisation and islands evolution. The R Journal, 9/1, 187-206, doi: 10.32614/RJ-2017-008.
[Sivanandam 2007] Sivanandam S., Deepa S. (2007). Introduction to Genetic Algorithms. Springer-Verlag, Berlin Heidelberg
[Bradley 1977] Bradley, Hax, and Magnanti *Applied Mathematical Programming*, Addison-Wesley, 1977 chapter 9

INCE EUROPE

10th International Conference
on
Wind Turbine Noise
Dublin – 21st to 23rd June 2023

Do measured immission data support the accepted norms on wind turbine propagation directivity as reported by the IoA GPG and is wind speed a significant variable?

Merlin Garnett, Green Cat Renewables Ltd: merlin@greencatrenewables.co.uk
Kira Marshall, Green Cat Renewables Ltd: kira.marshall@greencatrenewables.co.uk
Cameron Sutherland, Green Cat Renewables Ltd: cameron@greencatrenewables.co.uk

Summary

This paper provides analysis of directivity effects on wind turbine sound propagation from a far field perspective, and assesses the influence of wind speed. The study draws from more than 8 years' worth of post-completion measurements conducted by Green Cat Renewables (GCR) since the current Institute of Acoustics (2013), 'A Good Practice Guide to the Application of ETSU-R-97 for the Assessment and Rating of Wind Turbine Noise' guidance[1] (IoA GPG) was issued in full. The study highlights where results support or contradict the studies referenced in the IoA GPG; particularly: Wind Turbine Acoustics, NASA Technical Paper 3057, (1990). The 'NASA' results reproduced in the IoA GPG indicate that, for distances of up to 5.25 tip heights, attenuation of 2dB relative to downwind propagation can be expected for directions ±10° from cross-wind but with no attenuation shown for upwind conditions.

1. Introduction

The ETSU-R-97 guidance followed by all noise impact assessments in the UK require turbine immission to be assessed cumulatively considering all nearby wind projects. As the number of approved wind developments has increased, so has the complexity and scope of cumulative impact assessments. Many 'in-fill' sites (located between existing projects) that would otherwise be suitable for development are constrained on the basis that propagation modelling suggests that cumulative immission limits would be exceeded. One of the limiting factors is the assumption of simultaneous un-attenuated propagation from disparate turbines under all wind directions save a narrow crosswind sector of ±10 degrees (as per the NASA results).

[1] https://www.ioa.org.uk/sites/default/files/IOA%20Good%20Practice%20Guide%20on%20Wind%20Turbine%20Noise%20-%20May%202013.pdf

The data that support this study are pooled from 27 individual measurements selected from an archive of 58 post-completion measurement campaigns made by GCR during 8 years of fieldwork. The 27 measurements were selected because they were conducted closer to turbines than typical receptor distances (proxy locations). The majority of these proxy locations fall with 5.25 tip height distances of turbines.

The turbines measured ranged from 50m to 130m tip height; 20 of the 27 were in the 70m to 100m range. For all proxy locations, an individual turbine was the dominant contributor of measured immission levels. This analysis incudes standardised v_{10} wind speed as a variable with the aim of quantifying any associated refraction effects.

Subjective impressions of immission levels experienced while carrying out post-completion fieldwork suggested that levels of attenuation during upwind and cross-wind conditions may be greater than those detailed in the charts shown on p22 of the IoA GPG. Curiosity as to whether the measured data would support these observations is the motivation for this study.

A meta-analysis of these 27 measurement results has been carried out to establish whether the key questions above could be answered.

2. Methodology

2.1 Overview

This study aggregates a number of small datasets to both mitigate the measurement uncertainty inherent in contributing results and derive any overall trends present. The following approach was adopted to increase the data available for analysis: The study assesses twenty-four 15-degree sectors from downwind but assumes directional symmetry with reference to a line from upwind through downwind of the turbine. Therefore sectors 1&24 are pooled as are 2&23 etc, resulting in 12 data bins; as given in Figure 1.

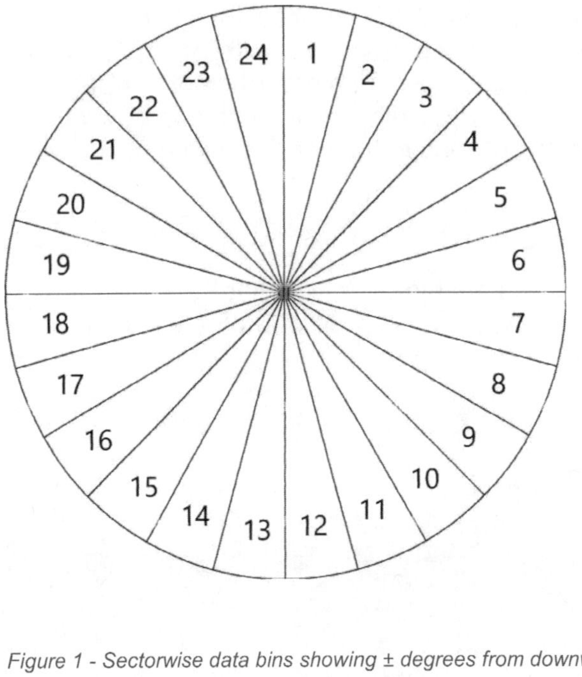

direction bin	sector	min	max	sector	min	max
1	1	0	15	24	-15	0
2	2	15	30	23	-30	-15
3	3	30	45	22	-45	-30
4	4	45	60	21	-60	-45
5	5	60	75	20	-75	-60
6	6	75	90	19	-90	-75
7	7	90	105	18	-105	-90
8	8	105	120	17	-120	-105
9	9	120	135	16	-135	-120
10	10	135	150	15	-150	-135
11	11	150	165	14	-165	-150
12	12	165	180	13	-180	-165

Figure 1 - Sectorwise data bins showing ± degrees from downwind

2.2 Investigation Carried Out

The investigation carried out considered a total of 58 wind turbine noise compliance assessment data sets that GCR have measured in Scotland between January 2015 and December 2022 inclusive. 27 of these data sets were captured at proxy locations, across a total of 23 projects, and form the basis of this study.

A standardised 10m wind speed range of $4ms^{-1}$ to $10ms^{-1}$ was considered, as this is the wind speed range typically of most relevance to wind turbine planning requirements.

For each assessment and each wind speed bin, average level differences (LDs) between T-on and T-off data (measured as $L_{A90,10min}$) were tabulated along with the comparison of resulting turbine L_{A90} with the IoA GPG prediction at the proxy location – this comparison with prediction is referred to as ΔP.

Although each individual proxy location dataset was subject to IoA GPG minimum data requirements, a minimum data count of 3 was also used to ensure that no ΔP was assumed to be representative unless based on results from at least 3 contributing results and was derived from an LD of 3dB or greater.

2.3 Measurement Details

Each of the measurements undertaken followed the IoA SGN5 approach. The following sections note key points.

2.3.1 Overview

Typical background noise levels at the 23 sites of interest would reach an L_{A90} of around 40dB(A) at a 10m wind speed of $10ms^{-1}$, approximately corresponding to the turbine(s) reaching rated sound power levels.

Locations for proxy noise monitoring positions were therefore chosen to ensure a level difference between operational and shut-down sound of around 3dB at rated sound power. Using the assumptions above, for example, the proxy position prediction would correspond to a turbine L_{A90} (=$L_{A,eq}$ – 2dB, as per IoA GPG recommendation) minimum prediction of 41dB(A) such that measured L_{A90} with turbine(s) operational could reasonably be expected to be around 43.5dB(A) at 10m wind speed of $10ms^{-1}$, ie, more than 3dB above the corresponding background L_{A90} at that wind speed.

2.3.2 Turbine shut-downs

The turbines were scheduled to shut down periodically to enable operational noise to be adjusted to remove the influence of background noise. A typical shut-down program used is shown in Table 1.

Table 1 – shutdown schedule

Period	Stop	Restart
1	20:00	21:00
2	22:00	23:00
3	00:00	01:00
4	02:00	03:00

2.3.3 Data reduction

Typically, data from the eight hours between 20:00 and 04:00 was assessed to ensure the quietest times of day were targeted.

2.3.4 Data exclusions

Analysis of operational (T-on) levels excluded periods when the turbines were starting up, shut down for maintenance, or not performing to their full capacity.

Under ETSU-R-97 guidelines[2], data must be removed if they are likely to have been affected by rain or are: *'considered atypical of the noise environment which normally prevails at the property'*. Exclusions were made corresponding to logged rainfall or any noise logger calibration drift.

2.3.5 Regression analysis

Time synchronisation between all data sets was confirmed using correlations and time series plots.

Noise levels were plotted against wind speeds and their relationship was established using polynomial trend lines.

Turbine noise levels were derived for integer wind speeds using the procedure outlined by the ETSU-R-97 guidelines.

2.3.6 Calculation of wind speed from SCADA

Wind speeds at the hub height of the turbine were calculated using the methodology described in BS EN 61400-11 'Wind turbine generator systems - Acoustic noise measurement techniques', third edition. The method defines a portion of the turbine's power curve (called the "allowed range") that can reliably be used to determine wind speed. The accuracy or tolerance of the power curve has been deemed to be +/- 3% of maximum power output (suggested value). This produces an allowed range of the power curve typically covering just above cut-in to around 90% of rated power output. A linear correlation of power curve derived wind speeds within the allowed range and hub anemometry was obtained and used to correct hub anemometry wind speed measurements above the allowed range for rotor shadow.

2.3.7 Wind Shear

Wind shears were calculated for each ten minute interval between the turbine hub height and a 10m onsite met-mast measurement, considering only the allowed range of wind speeds as defined above. Additionally, 10 minute periods when the nacelle anemometer logged an average wind speed of less than $5ms^{-1}$ or when 10m mast measured wind speeds were less than $2ms^{-1}$ were excluded as these low wind speed periods tended to produce atypically high wind shear values.

2.3.8 Wind speed adjustments to 10m mast

For the assessments that underly this study, turbine noise was calculated by logarithmic subtraction of shut-down measurements from operational data (as per SGN05). In this context, a consistent v_{10} wind speed reference was required that can be applied to data associated with both the shut-down and operational status of the turbine.

BS EN 61400-11 (Edition 3) includes a method for the standardisation of wind speeds measured by a temporary met-mast during periods when the turbine was shut-down: *section 8.2.2 - Wind speed measurements during background noise measurements.*

---

2 https://assets.publishing.service.gov.uk/government/uploads/system/uploads/attachment_data/file/49869/ETSU_Full_copy__Searchable_.pdf

This standardisation method was applied to all wind mast measurements but was expanded to correct for wind shear by wind speed bin as well as direction sector; incorporating the influence of wind speed on wind shear improved the consistency of the resulting v_{10} wind speed reference.

Figure 2 shows an example comparison of the v_{10} wind speeds derived by the bin-wise method described above with one that is derived directly from SCADA on a 10-minute basis during operational periods.

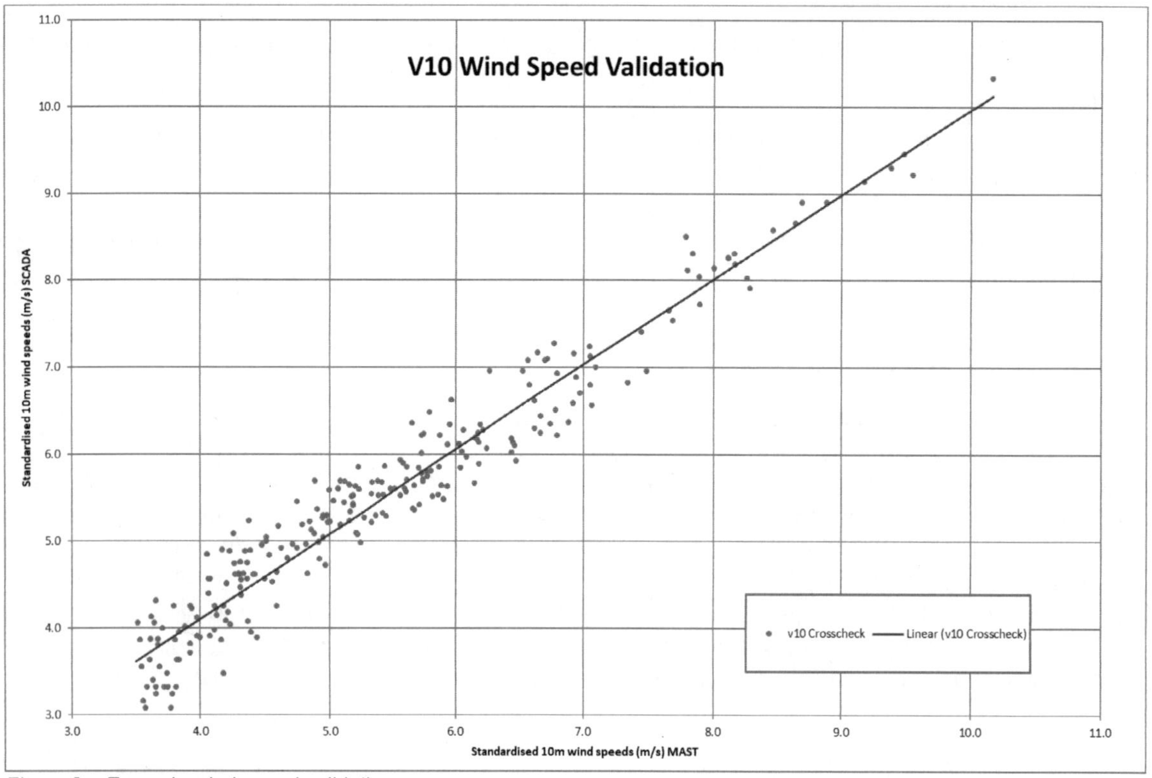

Figure 2 – Example wind speed validation

2.3.9 Data Collection

A 10m met-mast was used to log on-site wind speed and direction consistently with turbine(s) switched on and off. Wind direction and rainfall data were logged at 10 minute intervals.

Wind speed and direction data was also retrieved from SCADA data logged at the wind turbine(s) for the duration of the survey.

Noise data was logged at 10min intervals for each proxy monitoring location using an IEC 61672-1 Class 1 sound level meter. The proxy location was chosen to be representative of the receptor location, with most locations being slightly more exposed and less close to nearby vegetation or buildings.

Data collection would typically last 2 to 3 weeks.

2.3.10 Typical Equipment Used

Details of the typical equipment used are shown in Table 2.

Table 2 – Typical survey equipment

Equipment List	
Sound Level Meter (IEC 61672-1 Class 1):	Rion NL-52
Acoustic Calibrator (IEC 60942 Class 1):	Rion NC-74
Microphone:	Instrument standard
Tripod:	Single integrated pole
Wind Shield:	Rion WS-15 double skinned wind shield
10m met mast	Various manufacturers
10m anemometer	NRG#40C
10m wind vane	NRG#200P
Rain Gauge	Davis II

2.11 Normalisation

Individual measurements were normalised such that the maximum ΔP within the results was adjusted to zero. By the same method, the aggregated results for each integer v_{10} wind speed were then normalised such that attenuation factors (AF), relative to downwind for each speed and direction bin, could be compared.

3. Results and Discussion

The tabulated results show data count, LD, and normalised AF results per direction bin.

Table 3 – data count (#)

	Standardised 10m wind speed (m/s)						
Direction Bin	**4**	**5**	**6**	**7**	**8**	**9**	**10**
1	23	20	16	16	13	6	4
2	22	23	22	18	16	6	2
3	22	21	20	17	11	6	3
4	23	20	19	12	9	6	3
5	13	18	18	13	9	4	2
6	15	15	16	17	14	6	3
7	15	17	15	11	9	4	5
8	16	16	17	13	11	7	3
9	18	15	19	13	10	3	0
10	14	13	13	10	6	3	2
11	7	9	8	7	5	5	4
12	9	9	8	7	6	2	1

Table 4 – LD (dB)

	Standardised 10m wind speed (m/s)						
Direction Bin	**4**	**5**	**6**	**7**	**8**	**9**	**10**
1	5.2	6.2	5.2	4.8	4.1	2.3	
2	5.0	5.6	5.3	5.2	4.8	3.7	
3	5.3	5.7	5.7	5.8	4.6	2.8	
4	4.8	5.8	6.6	5.4	4.5	4.0	2.3
5	3.3	4.8	5.1	4.5	4.2	3.6	
6	3.8	4.5	4.3	3.9	3.5	3.7	3.6
7	3.0	4.1	3.8	3.7	3.9	4.4	3.0
8	4.6	3.8	5.0	5.6	5.2	4.7	4.1
9	4.4	5.2	5.6	5.9	6.4	8.0	
10	5.5	6.3	5.8	6.5	6.5	7.1	
11	7.3	6.6	5.8	6.3	4.8	3.6	
12	3.6	5.0	5.3	5.2	5.1		

Table 5 – normalised attenuation factors (AF) for measurements made at up to 5.25 tip height distance

	Standardised 10m wind speed (m/s)						
Direction Bin	**4**	**5**	**6**	**7**	**8**	**9**	**10**
1	0.0	0.0	-0.2	-0.4	-0.4		
2	-1.3	-0.2	-0.1	0.0	0.0		
3	-0.9	0.0	0.0	-0.3			
4	-0.5	-0.7	-0.4	-1.5	-0.8		
5	-1.8	-0.9	-1.0	-1.3	-1.8	-0.5	
6	-3.1	-2.2	-2.2	-2.2	-1.9	0.0	
7	-3.7	-2.1	-2.2	-2.9	-1.9		
8	-2.1	-3.1	-1.7	-1.8	-1.4	-0.6	
9	-0.7	-0.5	-0.1	-0.3	-0.3		
10	-0.7	-0.3	-0.6	-0.8	-0.6		
11		-1.3	-1.4	-0.9	-0.6		
12		-2.3	-0.8	-1.6	-1.1		

The data available for analysis decreases as the angle from downwind and v_{10} speed increase (Table 3). LD of results contributing to the study maintain a margin of 3dB for v_{10} speeds up to 9m/s where LDs become more variable (Table 4).

Direction bins 6 and 7 each represent a 15 degree sector of crosswind directions and for v_{10} speeds of 5m/s to 8m/s, Table 5 results show close agreement with the NASA results, given in the IoA GPG, typically showing ~2dB of attenuation. However, the results above indicate that a wider sector of crosswind directions are typically attenuated. The influence of v_{10} speed did not produce a significant trend in the results other than where higher AFs are typically shown for 4m/s than for higher speeds.

Results for upwind directions (bins 11 and 12) show higher than typical variance but indicate marginal attenuation of around 1dB relative to downwind propagation. This result indicates that the onset of upwind attenuation may occur before the 5.25 tip height distances suggested by the NASA results.

The results shown in Table 6 include only measurements made between 3 and 5.25 tip height distances to further investigate the onset of upwind attenuation.

Table 6 - normalised attenuation factors (AF) for measurements made at between 3 and 5.25 tip height distance

	Standardised 10m wind speed (m/s)						
Direction Bin	**4**	**5**	**6**	**7**	**8**	**9**	**10**
1	-0.3	-0.4	-0.9	-0.5	-0.3		
2	-0.8	0.0	-0.5	0.0	0.0		
3	-0.6	-0.2	0.0	0.0			
4	0.0	-0.8	-0.6	-1.5	-0.2		
5	-2.0	-1.1	-0.5	-0.7	-1.9		
6	-3.4	-2.2	-2.1	-1.8	-2.3	-1.1	
7	-3.4	-2.4	-2.7	-2.6	-1.6		
8	-3.1	-4.3	-2.3	-1.5	-1.2	-1.9	
9	-1.0	-1.3	-0.2	-0.3			
10	-1.2	-0.8	-0.3	-0.1	-0.7		
11		-2.5		-2.5			
12		-2.6	-1.4	-2.4	-1.3		

The results in Table 6 suggest that an attenuation of around 2dB can be expected for locations at a distance of at least 3 tip heights upwind (bins 11 and 12) from the turbine. Crosswind AFs remain similar to Table 5. Table 7 shows the results for measurement distances of between 4 and 5.25 tip heights.

Table 7 - normalised attenuation factors (AF) for measurements made at between 4 and 5.25 tip height distance

	Standardised 10m wind speed (m/s)						
Direction Bin	**4**	**5**	**6**	**7**	**8**	**9**	**10**
1	-0.6	-1.4	-3.0				
2	-1.5	-0.4	-0.9	-0.1	0.0	0.0	
3		0.0	-0.6	-0.1	0.0	0.0	
4	-0.1	-1.2	-1.5	-0.4	0.0		
5	0.0	-1.0	-1.2	-0.8	-0.3		
6	-3.7	-3.2	-5.3	-4.7	-3.0		
7			-7.0	-5.2	-2.1		
8	-2.8	-4.7	-4.4		-0.7		
9	-2.5	-4.2	-1.5	0.0			
10	-2.3	-2.9	0.0	-0.3	-1.8		
11		-6.0	-6.1	-5.1			
12		-6.4	-6.4	-5.0	-2.8	-0.8	0.0

Whereas the previous results were aggregated from at least three measurements, the results shown in Table 7 for bins 11 and 12 are from a single measurement. Nevertheless, the results suggest significant levels of upwind attenuation for measurement locations closer than 5.25 tip heights. These results differ from those shown in the NASA data; that, for topographically complex sites such as those in this study, upwind attenuation may be limited to 2dB for measurement locations as distant as 7.5 tip heights.

4. Conclusions

A meta-analysis of 27 measurements were undertaken to determine whether the assumptions for turbine directivity reported in the IoA GPG are supported or contradicted by analysis of measured immission data. The preceding results show close agreement with IoA GPG data in terms of the level of crosswind attenuation but differ in terms of the range of crosswind directions that are attenuated; this study suggests a significantly wider range of crosswind directions are attenuated.

The study also found that the onset of upwind attenuation occurred at shorter distances than suggested in the IoA GPG i.e at between 3 and 4 tip height distances rather than 5.25, though this finding was informed by fewer data than were available for the crosswind analysis.

For v_{10} speeds of greater than 4m/s, wind speed was not found to be a significant influence.

In summary, the IoA GPG directivity norms are significantly more conservative in defining crosswind directions and upwind attenuation onset distance, than the results of this study.

References

IOA Good Practice Guide on Wind Turbine Noise – (IoA 2013)
Shepherd, K. P.; Hubbard, H. H. Wind Turbine Acoustics, NASA Technical Paper (TP) 3057, Dec 1990.
BS EN 61400-11 (Edition 3)

10th International Conference
on
Wind Turbine Noise
Dublin – 21st to 23rd June 2023

The use of proxy monitoring locations to assess planning compliance at receptor distances.

Merlin Garnett, Green Cat Renewables Ltd: merlin@greencatrenewables.co.uk
Kira Marshall, Green Cat Renewables Ltd: kira.marshall@greencatrenewables.co.uk
Cameron Sutherland, Green Cat Renewables Ltd: cameron@greencatrenewables.co.uk

Summary

This paper reports an investigation to establish the ideal position for a proxy measurement, relative to a wind turbine or wind farm, from which the results could be extrapolated to receptor distances for the purposes of establishing planning compliance in the UK. The study draws from more than 8 years' worth of post-completion measurements conducted by Green Cat Renewables (GCR) since the current Institute of Acoustics (2013), 'A Good Practice Guide to the Application of ETSU-R-97 for the Assessment and Rating of Wind Turbine Noise' guidance (IoA GPG) was issued in full, in particular 'Supplementary Guidance Note 5: Post Completion Measurements' (SGN5)[1]. Data collected at several distances from a variety of different turbines was compared with expected immissions according to SGN5 assuming the turbines were operating to manufacturers' expectations based on IEC-61400-11 measurements. Level difference (LD) between operational turbine noise and background noise was considered in relation to the validity of turbine immissions derived from measurements.

1. Introduction

In the UK, noise limits are based on turbine sound immissions at receptors, rather than total noise as is the case in some other jurisdictions. Compliance with turbine noise requirements has to be assessed by measuring the difference between sound levels at a resident with and without turbine sound.

Using the methodology detailed by IoA GPG, it can be problematic to establish turbine sound levels at receptor distances because the level difference between noise measurements taken at properties with turbines switched on (T-on) or off (T-off) can be small, particularly at higher wind speeds where wind generated sound is more significant. At receptor locations, a level difference (LD) between operational turbine sound and background is unlikely to be 3dB(A), resulting in calculation uncertainty for turbine sound. This issue continues to be important at all

[1] https://www.ioa.org.uk/sites/default/files/IOA%20GPG%20SGN%20No%205%20Final%20July%202014.pdf

wind speeds because of the number of sites where 'apportioned' limits are used, particularly in cumulative situations.

In order to ensure that level differences between T-on and T-off are sufficient to robustly establish turbine noise levels, SGN5 supports the use of proxy locations closer to the turbine(s) of interest. If the proxy location is too far from the turbine(s), the level difference between turbines being switched on and off may not be sufficient to obtain robust results.

Therefore, two key questions arise concerning the use of proxy locations and obtaining a robust assessment of turbine sound immission: what is the minimum level difference required between T-on and T-off data to obtain reasonable results; and is the optimal distance for the proxy location an absolute distance or a function of the turbine size?

In the period since SGN5 was released, GCR has made 58 compliance measurements at wind turbine sites in Scotland, most of which have turbines that are between 400m and 1000m from the nearest properties. Even at these relatively modest receptor distances, it has often been considered necessary to employ proxy measurement locations to improve the probability of obtaining robust compliance assessment results.

These measurements were undertaken close to turbines that would now be considered small by today's standards (50-130m tip height). With far larger turbines routinely being deployed with lower A-weighed sound power levels per MW capacity, it was important that this analysis considered whether proxy measurements should be located at distances proportional to the size of turbines or whether an absolute range of distances was likely to be most appropriate.

A meta analysis of these measurements has been carried out to establish whether the key questions above could be answered.

2. Methodology

For the purposes of this study, all data taken within +/-45 degrees downwind of a turbine was considered to represent downwind conditions. Such data could be compared directly with turbine sound level predictions using the requirements of the IoA GPG, subject to GPG minimum data requirements.

2.1 Investigation Carried Out

The investigation carried out considered a total of 58 wind turbine noise compliance assessments from wind turbine sites in Scotland between January 2015 and December 2022. 27 of these data sets were at proxy locations across a total of 23 projects. Of these, 4 had no downwind data suitable for this analysis. The remaining 23 data sets were used in this analysis.

A standardised 10m wind speed range of $4ms^{-1}$ to $10ms^{-1}$ was considered, as this is the wind speed range typically of most relevance to wind turbine planning requirements in the UK.

The key variables considered in this analysis were: standardised 10m wind speed; average level differences (LDs) between T-on and T-off data (measured as $L_{A90,10m}$); A comparison of resulting turbine L_{A90} with the IoA GPG prediction at the proxy location – this comparison with prediction is referred to as ΔP; the number of datasets leading to the results; and a standard deviation of ΔPs (SD) was also calculated.

With LDs, number of data sets, ΔPs and standard deviations of ΔPs collated, three filters were used to investigate the results further: Minimum LD; Maximum ΔP; and minimum data count.

Minimum LD was investigated to understand what a sensible minimum would be that still allowed credible turbine noise results to be derived from datasets where some of the results had a low LD.

Maximum ΔP was used as a filter to help understand whether large ΔP was associated with higher standard deviation in results.

Although each individual proxy location dataset was subject to IoA GPG minimum data requirements, a minimum data count of 3 was used to ensure that no ΔP was assumed to be representative unless based on at least 3 results.

2.2 Measurement Details

Each of the measurements undertaken followed the SGN5 approach. The following sections note the key points.

2.2.1 Proxy Noise Monitoring Locations

Typical background noise levels at the sites of interest would reach an L_{A90} of around 40dB(A) at a 10m wind speed of 10ms^{-1}, approximately corresponding to the turbine(s) reaching rated sound power levels.

Locations for proxy noise monitoring positions were therefore chosen in the hope of ensuring a level difference (LD) between operational and shut-down sound of around 3dB at rated sound power. Using the assumptions above, for example, the proxy position prediction would correspond to a turbine L_{A90} (=$L_{A,eq}$ – 2dB, as per IoA GPG recommendation)[2] prediction of 41dB(A) such that measured L_{A90} with turbine(s) operational could reasonably be expected to be around 43.5dB(A) at 10m wind speed of 10ms^{-1}, ie, around 3dB above the corresponding background L_{A90} at that wind speed.

2.2.2 Turbine shut-downs

The turbines were scheduled to shut down periodically to enable operational noise to be adjusted to remove the influence of background noise. A typical shut-down program used is shown in Table 1:

Table 1 – Shutdown schedule

Period	Stop	Restart
1	20:00	21:00
2	22:00	23:00
3	00:00	01:00
4	02:00	03:00

2.2.3 Data reduction

Typically, data from the eight hours between 20:00 and 04:00 was assessed to ensure the quietest times of day were targeted.

[2] IoA GPG Sect 4.2.5

2.2.4 Data exclusions

Analysis of T-on levels excluded periods when the turbines were starting up, shut down for maintenance, or not performing to their full capacity.

Under ETSU-R-97 guidelines[3], data must be removed if they are likely to have been affected by rain or are: *'considered atypical of the noise environment which normally prevails at the property'*.

Exclusions were made corresponding to logged rainfall or any noise logger calibration drift.

2.2.5 Regression analysis

Time synchronisation between all data sets was confirmed using correlations and time series plots.

Noise levels were plotted against wind speeds and their relationship was established using polynomial trend lines.

Turbine noise levels were derived for integer wind speeds using the procedure outlined by the ETSU-R-97 guidelines.

2.2.6 Calculation of wind speed from SCADA

Wind speeds at the hub height of the turbine were calculated using the methodology described in BS EN 61400-11 'Wind turbine generator systems - Acoustic noise measurement techniques', third edition. The method defines a portion of the turbine's power curve (called the "allowed range") that can reliably be used to determine wind speed. The accuracy or tolerance of the power curve has been deemed to be +/- 3% of maximum power output (suggested value). This produces an allowed range of the power curve typically covering just above cut-in to around 90% of rated power output. A linear correlation of power curve derived wind speeds within the allowed range and hub anemometry was obtained and used to correct hub anemometry wind speed measurements above the allowed range for rotor shadow.

2.2.7 Wind Shear

Wind shears were calculated for each ten minute interval between the turbine hub height and a 10m onsite met-mast measurement, considering only the allowed range of wind speeds as defined above. Additionally, 10 minute periods when the nacelle anemometer logged an average wind speed of less than $5ms^{-1}$ or when 10m mast measured wind speeds were less than $2ms^{-1}$ were excluded as these low wind speed periods tended to produce atypically high wind shear values.

2.2.8 Wind speed adjustments to 10m mast

For the assessments that underly this study, turbine noise was calculated by logarithmic subtraction of shut-down measurements from operational data (as per SGN05). In this context, a consistent v_{10} wind speed reference was required that can be applied to data associated with both the shut-down and operational status of the turbine.

---

3 https://assets.publishing.service.gov.uk/government/uploads/system/uploads/attachment_data/file/49869/ETSU_Full_copy__Searchable_.pdf

BS EN 61400-11 (Edition 3) includes a method for the standardisation of wind speeds measured by a temporary met-mast during periods when the turbine was shut-down: *section 8.2.2 - Wind speed measurements during background noise measurements.*

This standardisation method was applied to all wind mast measurements but was expanded to correct for wind shear by wind speed bin as well as direction sector; incorporating the influence of wind speed on wind shear improved the consistency of the resulting v_{10} wind speed reference.

Figure 1 shows an example comparison of the v_{10} wind speeds derived by the bin-wise method described above with one that is derived directly from SCADA on a 10-minute basis during operational periods.

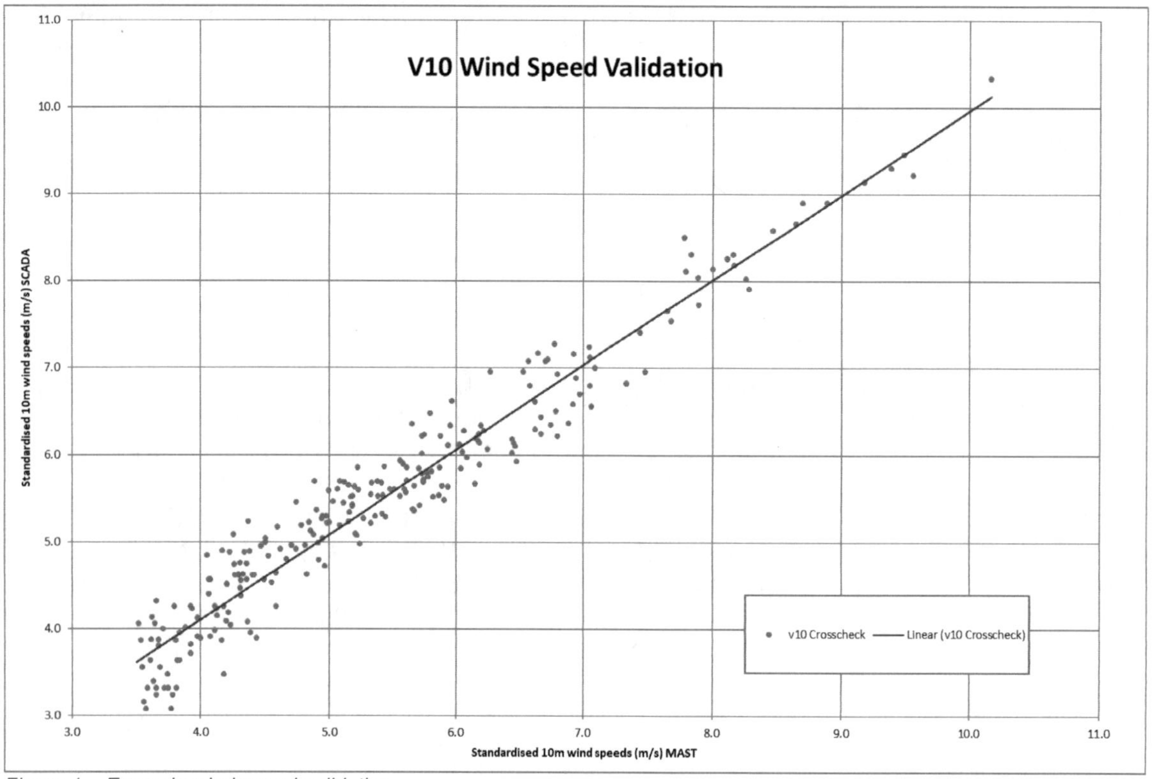

Figure 1 – Example wind speed validation

For this analysis, the IEC method was applied to wind mast measurements made for T-on and T-off periods. This was corrected for wind shear by wind speed bin, direction sector and time of day.

2.2.9 Data Collection

A 10m met-mast was used to log on-site wind speed and direction consistently with turbine(s) switched on and off. Wind direction and rainfall data were logged at 10 minute intervals.

Wind speed and direction data was retrieved from SCADA data logged at the wind turbine(s) for the duration of the survey.

Noise data was logged at 10min intervals for each proxy monitoring location using an IEC 61672-1 Class 1 sound level meter. The proxy location was chosen to be representative of the receptor location, with most locations being slightly more exposed and less close to nearby vegetation or buildings.

Data collection would typically last 2 to 3 weeks.

2.2.10 Typical Equipment Used

Details of the typical equipment used are shown in Table 2.

Table 2 – typical survey equipment

Equipment List	
Sound Level Meter (IEC 61672-1 Class 1):	Rion NL-52
Acoustic Calibrator (IEC 60942 Class 1):	Rion NC-74
Microphone:	Instrument standard
Tripod:	Single integrated pole
Wind Shield:	Rion WS-15 double skinned wind shield
10m met mast	Various manufacturers
10m anemometer	NRG#40C
10m wind vane	NRG#200P
Rain Gauge	Davis II

3. Results and Discussion

3.1 Baseline Data

All the data from the 23 data sets used was tabulated to form the baseline results subject to a minimum LD of 0dB(A).

Table 3 - Level difference, Data number, ΔP and standard deviation from all data

Standardised 10m wind speed							
	4	**5**	**6**	**7**	**8**	**9**	**10**
LD [dB(A)]	5.4	6.0	6.2	5.7	4.2	2.7	1.7
Data #	22	23	23	21	18	13	7
ΔP [dB(A)]	-1.3	-1.6	-1.3	-0.2	0.1	-0.2	-1.9
SD [dB]	2.8	2.8	2.2	2.2	3.0	3.3	4.4

The table shows that, on average, ΔP was low and below zero at all but one wind speed. Where ΔP was positive it was 0.1dB(A), which is insignificant. Given that GPG predictions include uncertainty, these figures were broadly as expected.

However, as the minimum LD was 0dB(A), it can be seen that standard deviation in ΔP was more than 2dB(A), which suggests a high variability in the results. Indeed, the highest standard deviations were at 9 and 10ms^{-1} where the average LDs were the smallest and the number of data sets was the lowest.

It was noted, also, that the standard deviation was lowest at 6 and 7ms^{-1}, which had amongst the highest average LDs. However, the standard deviation was higher at low wind speeds of 4 and 5ms^{-1} where the average LDs were also high.

Data collected at 4 and 10ms^{-1} tended to be at the wind speed extremes for these type of measurements. Therefore, results in these wind speed bins may be the result of a trendline fitted through data that 'saturates' in sound level, leading to higher uncertainty.

3.2 Sensitivity of Results to Level Difference

Standard deviation was plotted against LD based on filtering datasets by minimum LD to assess how important having a minimum LD at a given wind speed is to having confidence in assessment results. The baseline data were filtered to include on those where the calculated turbine level was no more than 3dB(A) above prediction, in order to prevent outlying results from skewing the analysis.

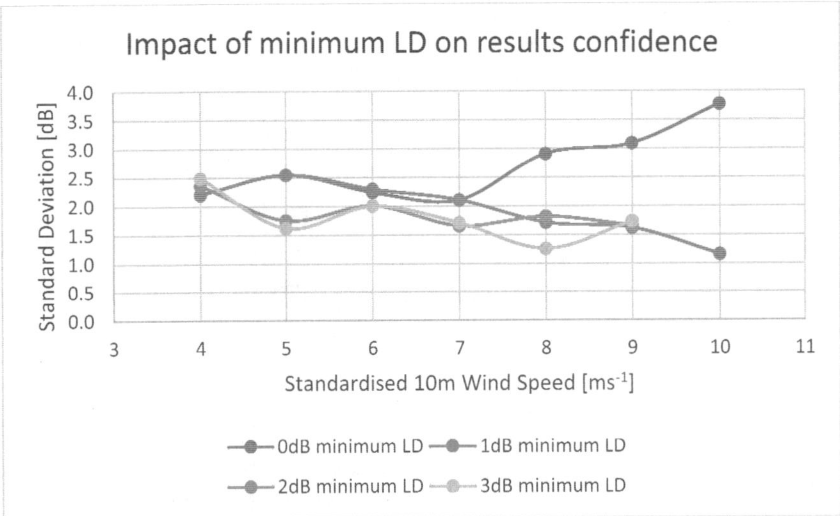

Figure 2 - Impact of minimum LD on the confidence in results.

As can be seen, the variation of results is fairly consistent at wind speeds up to 7ms^{-1}, reflective of the generally higher LDs observed at these wind speeds.

With no minimum LD, uncertainties are high at wind speeds of 8ms^{-1} and above, as expected. By contrast, any minimum LD of 1dB or more seemed to give rise to similar levels of variation, all of which were lower than with no minimum. Moreover, the uncertainties were generally smaller at these higher wind speeds.

This was a surprising result for 1dB or 2dB minimum LDs, which were a lot more common at 8 and 9ms^{-1} than when using a 3dB minimum. For example at 8ms^{-1}, there were 7 datasets with a 3dB minimum LD, whereas this increased to 13 for a 1dB minimum LD.

3.3 The Influence of Distance on Results

To compare the influence of distance on results, all baseline results were considered with no minimum LD and no maximum ΔP, in order to maximise the number of studies that could be compared.

Average LD, average ΔP, and SD values were plotted against distance for each wind speed bin. Distance was 'measured' in 3 different ways: Number of tip heights; Number of rotor diameters; and absolute horizontal distance in m. These 3 sets of data are plotted below.

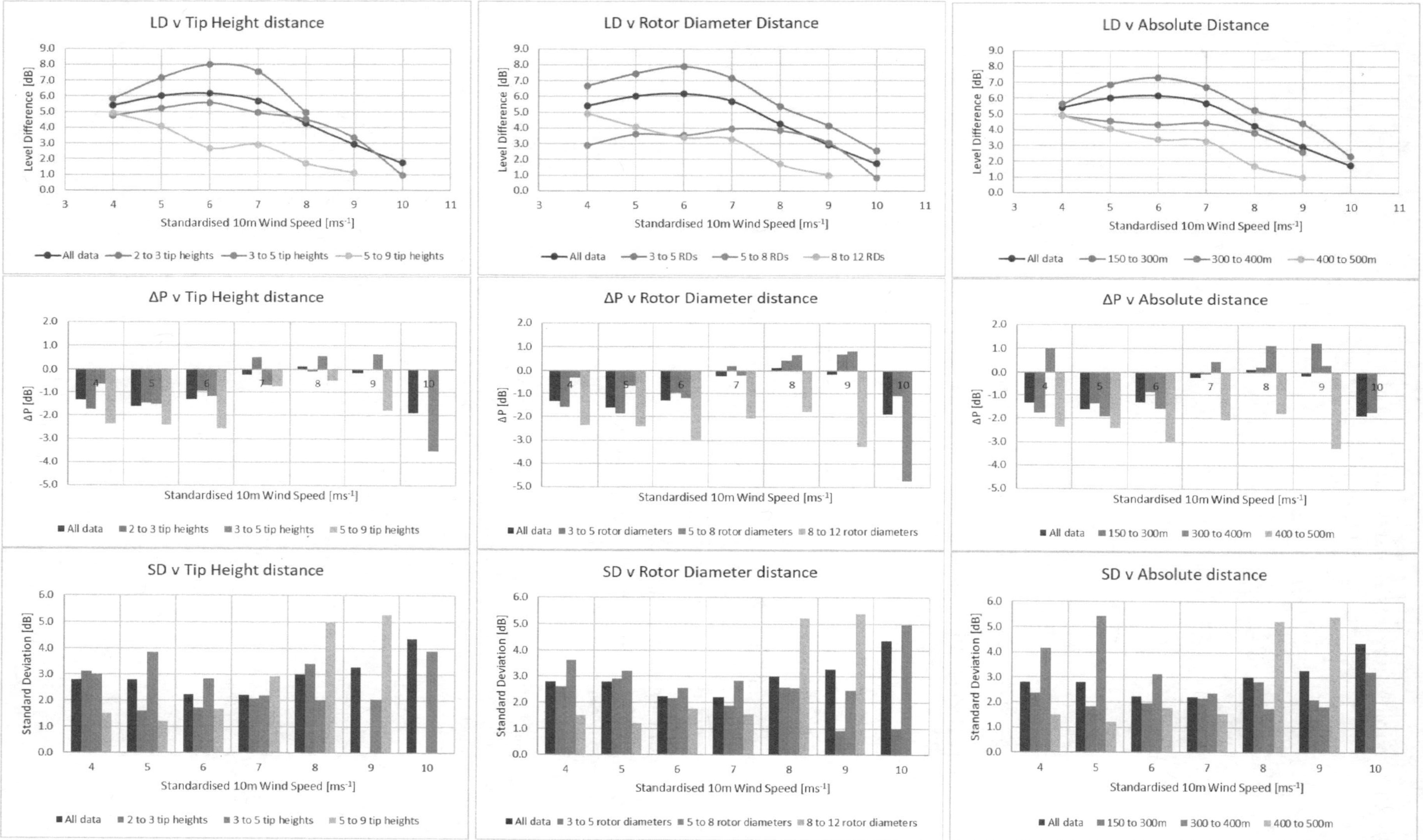

Figure 3 - Impact of distance on: LD, ΔP and SD.

A theoretical LD for each absolute distance was also compared with measurements from all data.

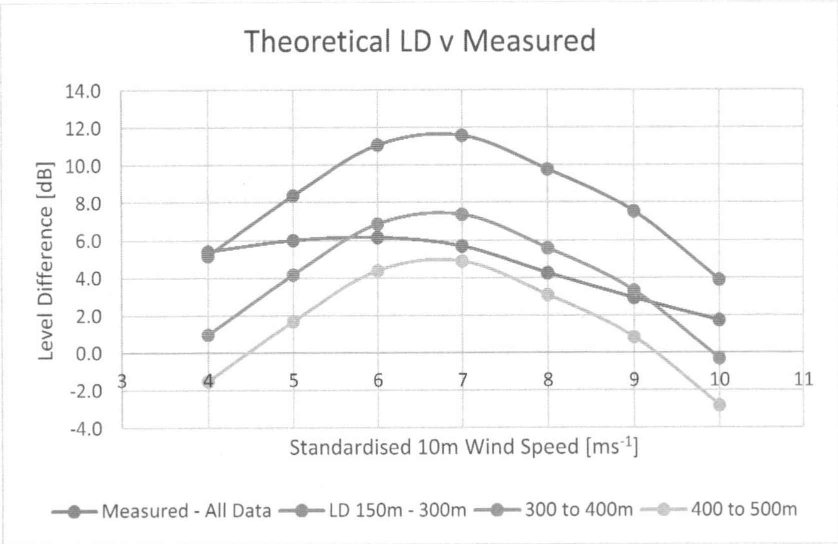

Figure 4 - Theoretical LDs for each absolute distance category compared with measured LD

3.3.1 LD Variation

The most consistent trend of how LD varied with distance was with respect to absolute distance. Maximum average LDs were observed in the wind speed range 4 – 7ms^{-1} at all distances. However, there were some interesting features:

The measured average LDs were quite different from what would theoretically be expected, even allowing for different background environments.

6ms^{-1} was where there was the greatest difference between measured LDs with distance. The theoretical curves suggest that the greatest differences should be at higher wind speeds. However, this could simply illustrate that, on average, background noise levels measured have been higher at 7 and 8ms^{-1} than a theoretical background noise curve would suggest. This study has not aggregated the background noise measured into an 'average background noise' curve which may have assisted in interpreting results.

The LDs at 4ms^{-1} were similar and consistently averaged around 5dB regardless of distance (though when judged against rotor diameter there was more variation). This was surprising firstly because at 4ms^{-1} wind speeds there was the least potential for wind induced microphone noise to affect results. It was also surprising because, as the theoretical curve illustrates, there is a relative lack of sound energy coming from turbines at this 10m wind speed, particularly given that the largest size of turbine measured was 2.3MW, 100m to tip height. And thirdly, the background sound level measured should be consistent with distance, therefore, the T-on sound level measured should decline with distance without an equivalent decline in T-off level, leading to greater LD.

3.3.2 ΔP and SD Trends

Trends in ΔP and SD values were a little less consistent when considered as a function of absolute distance away from a turbine than when looked at against tip height or rotor diameter. The comments below are referenced to distance in rotor diameters but are also applicable to distance in tip heights unless otherwise specified.

Data collected at 10ms^{-1} was typically high in uncertainty and gave results below GPG prediction. This made sense in that the average LDs were below 3dB for this wind speed, even at 3 to 5 rotor diameters from the turbine(s).

ΔP was consistently greater in magnitude at the greatest distances and, perhaps surprisingly, resulted in average predictions consistently below GPG prediction.

Another surprise was that the projects at greatest distance gave results with lowest SD for windspeeds of 4 to 7ms^{-1}. Therefore, there is higher confidence that those turbine sound levels were below the GPG prediction than for the results at lower rotor diameter distances.

In contrast to this, at 8 and 9ms^{-1}, the data showed very large variations at the 8 to 12 rotor diameters distance, which was as expected and would justify measuring at proxy locations less than 8 rotor diameters, 5 tip heights and 400m away.

At distances closer to the turbine, there was a consistent trend of ΔP being negative at wind speeds from 4 to 6ms^{-1} and ΔP having a magnitude of less than 1dB at wind speeds in the range 7 to 9ms^{-1}.

4. Conclusions

This meta analysis was able to directly compare 23 compliance measurements using proxy measurement locations taken between 2015 and 2022, following publication of SGN5.

The analysis suggested that filtering the data to measurements with minimum LDs of greater than 1dB was sufficient to remove the most uncertain high wind speed results. Filtering on higher LDs of 2 and 3dB did not appear to improve confidence in the results.

It was found that a 3dB LD was unlikely at windspeeds of 10ms^{-1} even at only 3 rotor diameters or 2 tip heights away from the turbine.

It was found that a 3dB LD was unlikely at windspeeds of 8ms^{-1} and above beyond 400m away from a turbine.

For proxy locations less than 400m away or 8 rotor diameters, it was found that prediction averaged within 1dB of theoretical between 7 to 9ms^{-1} and were generally lower than theoretical at lower wind speeds. However, even at these distances, the SD of results was ~2-3dB. This highlights the potential for ongoing challenge in verifying apportioned limits.

Finally, this analysis did not find a clear difference between absolute distance and size of turbine in determining the ideal location for a proxy measurement. Therefore, more analysis would be required to assess whether such differences do exist and, therefore, whether advice could be offered of relevance to the measurement of larger turbines.

References

IOA Good Practice Guide on Wind Turbine Noise – (IoA 2013)

IOA Good Practice Guide on Wind Turbine Noise; SGN5 – (IoA 2013)

BS EN 61400-11 ed 3

Jones, A and McKenna T (2010) *A new method for statistical analysis: the two part reverse interrogation of sparse data* Jnl Reverse Statistics 55(2), 65-219

INCE EUROPE

10th International Conference on Wind Turbine Noise Dublin – 21st to 23rd June 2023

Tone propagation and receptor levels

Mark Hempel, Graduate: mahml@vestas.com
Kaj Dam Madsen, Senior Product Functional Lead: kadma@vestas.com
Mranal Gupta, Product Functional Lead: mrgup@vestas.com

Summary

A frequency component sticking out from the overall frequency spectrum shape leads to tonal characteristic. This is a concern as such noise could be annoying. There are standards which define how a tone is assessed and evaluated.

The current paper describes tone propagation to evaluate the consequences at receptor. The various considerations that are needed for advanced evaluation of the typical conditions at receptors are presented. The major parameters that influence the perceived tonal response at receptors are highlighted.

1. Introduction

Wind turbines keep expanding their role in a world with rapidly increasing demand for clean energy and therefore the individual turbines are optimized for providing more energy production to lower the cost and utilize the wind resources to a maximum - also in more densely populated areas. This together drives bigger turbines, larger rotors to capture the wind and potentially more noise emission from the individual turbine.

When wind turbines are operated, a critical aspect to the local environment is the noise performance of the turbines. The noise experienced from turbines is evaluated by considering the overall noise as well as any noticeable tones.
This paper is concerned with the tonal part of the turbine noise.

If assessment of tonal performance at neighbour locations, from wind turbines, is not considered prior to setting up turbines, it is a risk that the noise performance, due to the tonal signature of the turbines, becomes noncompliant leading to neighbour complaints, shutdowns, decreased power output and/or component exchanges to reach compliance.

Avoiding these adverse effects to ensure the general acceptance of wind turbines and to the business case of operating a wind turbine make up the case for creation of a tool capable of assessing the tonal response from wind turbines at arbitrary sites using known turbines.

It is of interest to create a tool capable of representing different turbine configurations, both regarding architecture and tonal performance, as well as arbitrary site setups, allowing for receptor positions to have varying distances from the turbine and height from the ground.

The IEC standard 61400-11 [1] is used for tonality assessment at turbine level. Other tonality assessment methods exist such as FGW [2].

2. Background

Turbine siting is generally done considering the neighbours, often referred to as receptors. The siting is normally governed by applicable regional regulations. While some areas have turbines sited based on noise emitted at fixed wind speeds, other regions have more advanced requirements accounting for background noise and/ or time of day, etc.
Various tools such as e.g., WindPro holding different propagation models for the general noise are used to site turbines. During planning, the 'immission' levels at neighbour positions or noise ISO lines are calculated depending on turbine noise emission.

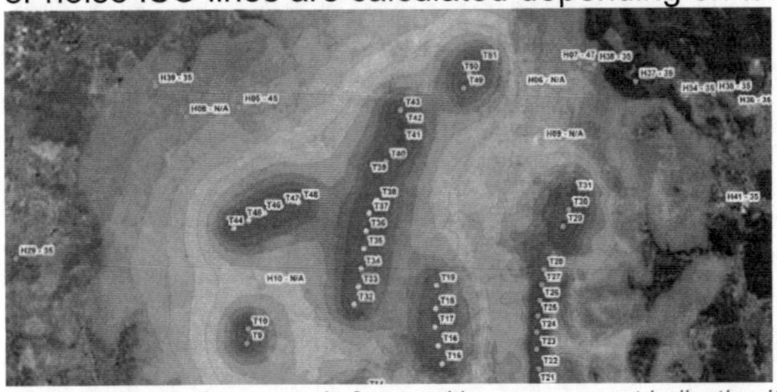

Figure 1. Illustrative example from a siting assessment indicating ISO lines for sound pressure.

These tools mainly work on the overall sound pressure level. To account for tones, a special consideration regarding reflection from ground needs to be made as the sound waves from the turbine will interfere which can lead to an amplification or cancellation.
Thus, the presented tool is created to predict tonal audibility more effectively at, for example, residences neighbouring wind turbines or other noise sensitive locations, in which noise tonality is critical to control due to annoyance and health concerns.
The tool is intended to be used with readily available information about the turbine and site of interest, meaning prediction of tonality is meant to be possible without need for customized measurements specific to each simulation case. The information used as inputs for simulation is therefore chosen based on measurements already taken by following the IEC 61400-11 standard [1].

Noise and tonality requirements vary on a case-by-case basis, in terms of both allowed levels and the location at which the noise is evaluated (receptor distance and height from ground). The tool therefore needs to be able to assess tonality at arbitrary receptor locations, allowing for receptors lifted from the ground.

By lifting the receptor from the ground, ground effects caused by interference of the incoming and reflected sound waves from the turbine become frequency dependent, which thus must be considered.
This is done by utilizing the Nord2000 model [3,4] for propagation of noise to the receptor position, as it provides the needed features.

2.1 Additional functionality

While the tool takes offset in assessing tonality at a receptor for a single turbine, additional functionality for assessing a site with multiple turbines is included. In doing so, careful considerations regarding how to handle turbine interdependencies are needed. The interdependencies include, but are not limited to, wake effects, varying turbine tonal configurations, windspeed variations across turbines, whether addition of contribution from additional turbines would add to total tonal energy in the critical band or could be below threshold and add masking, among others.

3. Process

The far field tonality tool makes use of user inputs to define site and turbine configuration. The validation of the tool is carried out based on field measurements.

3.1 Calculation method

Parameters of hub height, rotor diameter, windspeed, receptor distance and height, tone frequency, IEC tone or masking energy and IEC tonal audibility represent the simulation setup of turbine and site.
As turbines, generally, are measured at the IEC position at varying windspeeds, the tonality information needed for representing turbines in simulations are based on these measurements.

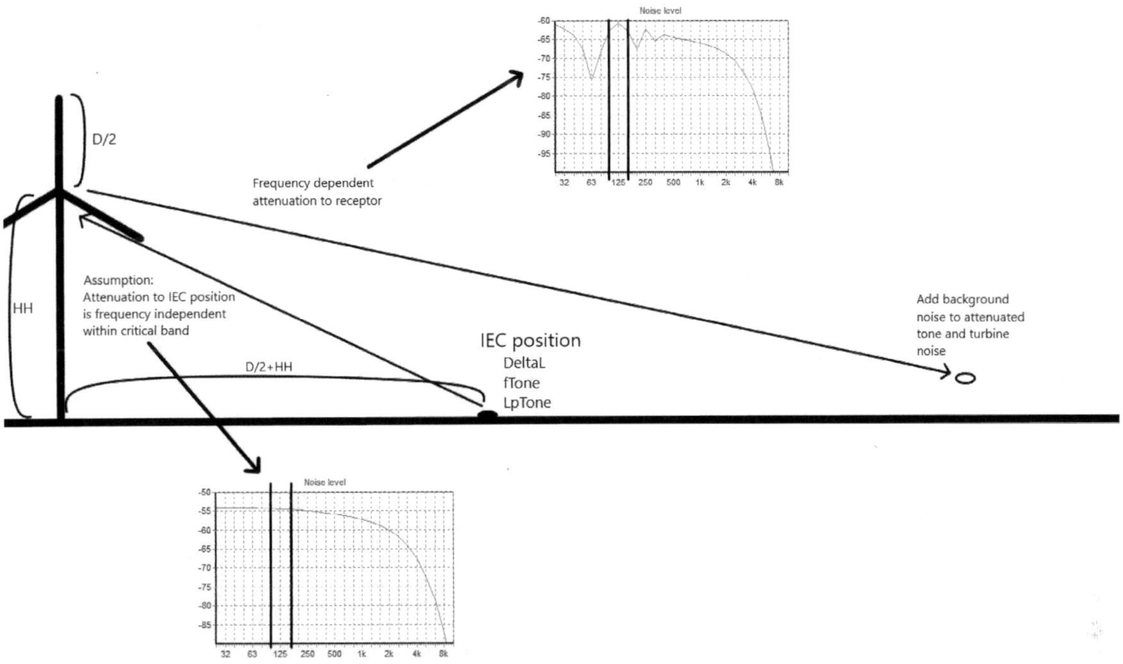

Figure 2. Sketch of simulation setup for a single turbine. Turbine tonal information from IEC position is transformed to far field through transformation to receptor position by propagation to turbine and then to receptor.

The first step is making sure the turbine noise, the critical frequency band, is accurately portrayed. As the IEC measurements include both noise energy from the turbine and from the environmental background, the environmental background noise is energetically subtracted from the critical band. This is ideally done using an environmental noise measurement from the site where the IEC measurement is done, representing the conditions used in the simulation (windspeeds, time-of-day, surroundings…). Alternatively, if such measurements are not available, a lookup table of fitting background noise spectra from other sites can be used. Thus, the background-corrected IEC tonality information represents just noise from the turbine.

The subtraction of the environmental background noise at the IEC position generally leads to increases in tonality. This is due to the masking energy having some of its energy level stemming from the environmental background noise. Additionally, the tone is influenced relatively less by the environmental background noise than the masking.

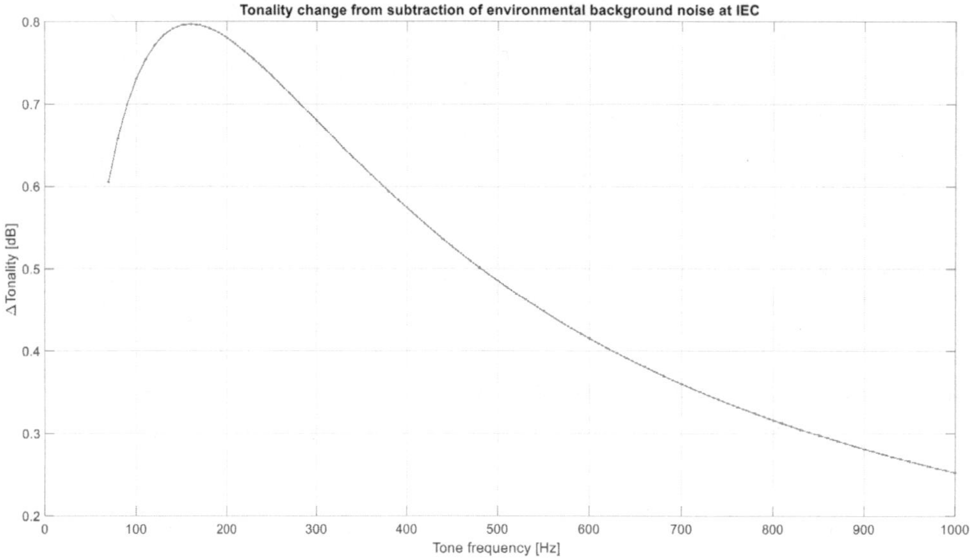

Figure 3. Tone frequency solution space of tonality change from subtraction of IEC background noise. IEC TA = 2 dB. IEC masking = 38 dB. Windspeed = 11.5 m/s. Receptor height = 1.5 m

To represent the noise from a turbine at an arbitrary receptor distance and height, a transformation of the background-corrected IEC tonality information of masking noise and tone level to the receptor position is performed.

The background-corrected IEC masking and tone level is first propagated to the turbine accounting for spherical divergence, after which the Nord2000 model [3,4] is used to further transform the tonality information from the turbine to the receptor position. The Nord2000 calculation to transform from the turbine to the far field receptor assumes flat ground.

With this transformation spectrum created, it is possible to assess the change to the tonality stemming purely from transforming the turbine noise from the IEC to the far field receptor position.

In general, the tonality change from transformation of the IEC tonality information can lead to slight increases or decreases in the tonality at distances similar to the IEC distance. The change is greatest at close distances when the receptor is lifted from the ground owing to the frequency dependent ground effect caused by interference of the directly incoming and ground-reflected sound wave. At far receptor distances, the ground effect diminishes.

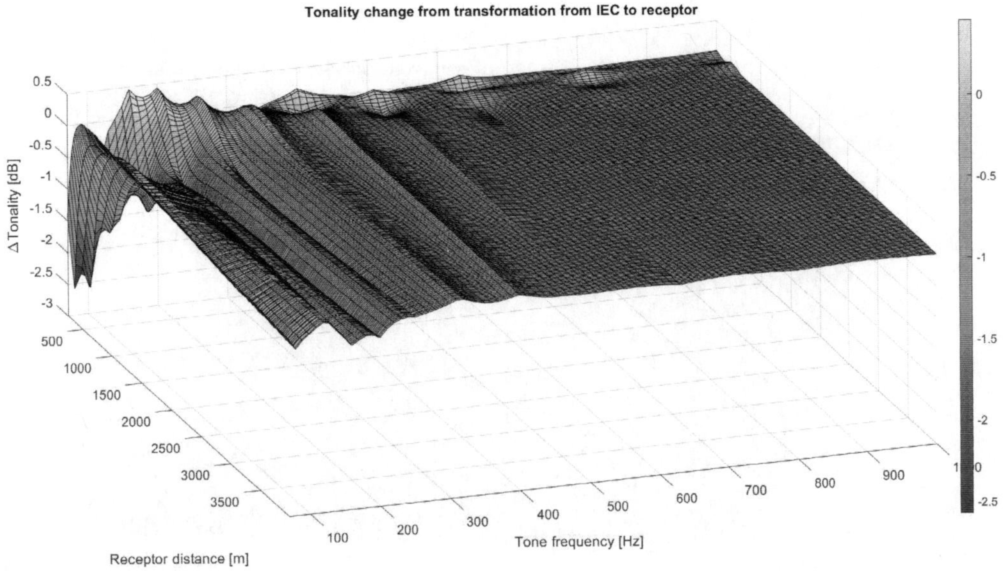

Figure 4. Tone frequency and receptor distance solution space of tonality change from transformation from IEC to the receptor. IEC TA = 2 dB. IEC masking = 38 dB. Windspeed = 11.5 m/s. Receptor height = 1.5 m. IEC horizontal distance = 247 m

With the turbine noise tonal information, determined at the receptor position, a final step of adding back the environmental background noise at the receptor is required. Just as for the IEC position it is of preference to use a measured spectrum from the location of interest. Alternatively, if one is not available, a lookup table can also be used here.

Addition of the environmental background noise to the transformed turbine noise at the receptor location, allows for representation of the overall expected noise at the receptor.

The addition of the environmental background noise at the receptor position generally leads to decreases in tonality. This is due to the masking energy having sizeable contribution at the receptor stemming from the environmental background noise. Expectedly, the masking is influenced relatively more by the background than the tone. Furthermore, due to the far field receptor position generally being at a further distance than the IEC position, the overall relative energy impact is greater at the receptor position. This leads to the decrease in tonality due to addition of receptor environmental background noise generally outweighing any increase to tonality due to subtraction of IEC environmental background noise.

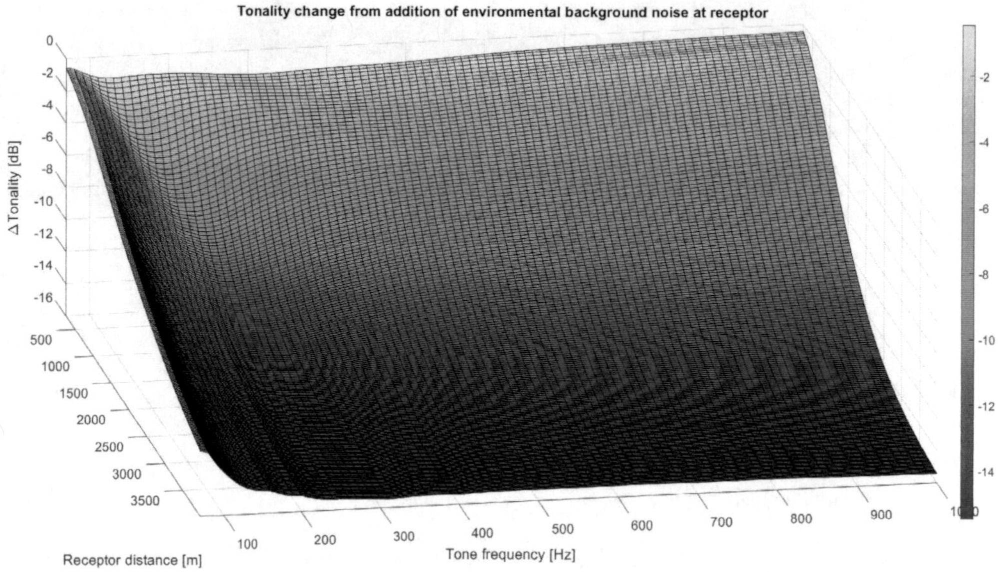

Figure 5. Tone frequency and receptor distance solution space of tonality change from addition of receptor background noise. IEC TA = 2 dB. IEC masking = 38 dB. Windspeed = 11.5 m/s. Receptor height = 1.5 m. IEC horizontal distance = 247 m

From the masking and tone level, it is then possible to assess the overall tonal audibility at the far field receptor stemming from tonality change due to each of the parts of the simulation (subtraction of IEC environmental noise, transformation from IEC to receptor position, addition of receptor environmental noise)

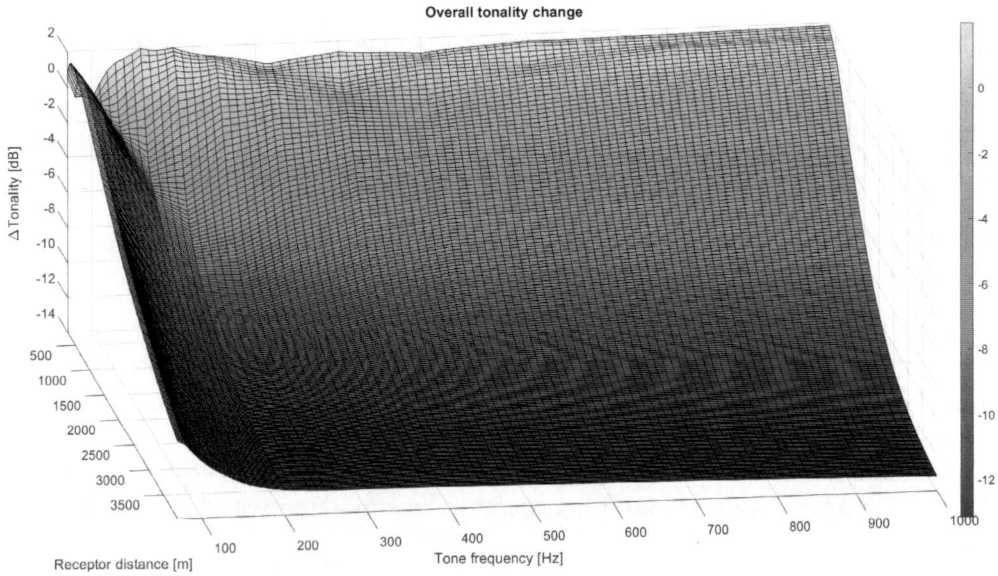

Figure 6. Tone frequency and receptor distance solution space of overall tonality change. IEC TA = 2 dB. IEC masking = 38 dB. Windspeed = 11.5 m/s. Receptor height = 1.5 m. IEC horizontal distance = 247 m

Additionally, it is possible to add up noise from multiple turbines to evaluate more complicated setups such as wind farms.

In this case each of the separate turbines would have their background-corrected IEC tonality information transformed to the receptor. The masking and tone levels of each of these turbines would then be combined prior to adding back the environmental background noise at the receptor. Here it is assumed that the noise from each turbine is independent of each other.

In this way the environmental background noise is only included once, while the turbine noise from a multi-turbine site setup is considered when determining the overall tonal audibility, making it a conservative approach.

In a wind farm, due to the different components and slight differences in operation the tone frequency from individual turbines will not be identical. The amplitudes from tones would also vary based on the distances of the turbines from the receptor. In practice this could lead to situations where the different turbines are resulting in reduced annoyance by increasing the masking.

The multiple turbine setup is overly simplified in the tool to maintain conservatism in estimations, compared to a real wind farm setup, in which individual turbines are not independent of each other due to for example wake effects. These effects can be considered by configuring the individual turbines differently in the simulation inputs, allowing them to have different tonal information. However, this is just a work-around and the wake effect would in this case be accounted for externally to the program.

3.2 Validation

Validation efforts are planned to be done in a data-based manner, evaluating many points across the large input-space to ensure general model accuracy.

In validating, concurrent measurements at the IEC position and at the receptor point of interest need to be performed. The measurements at the IEC position are used as turbine configuration input and environmental background noise at IEC, while the receptor position measurements are used as output to be used for comparison with the model results and environmental background noise at receptor.

For a given turbine, a set of concurrent 10 second averaged measurements at the IEC position and at a far field receptor positioned at 500-meter horizontal distance and 4 meter height are performed across a 24-hour period. The 10 second averages are grouped according to windspeed and tone. Using each of the 10 second IEC measurements as inputs for far field tonality calculations, allows for accompanying calculated far field values for each 10 second average to be computed and used for comparison with the measured far field values.

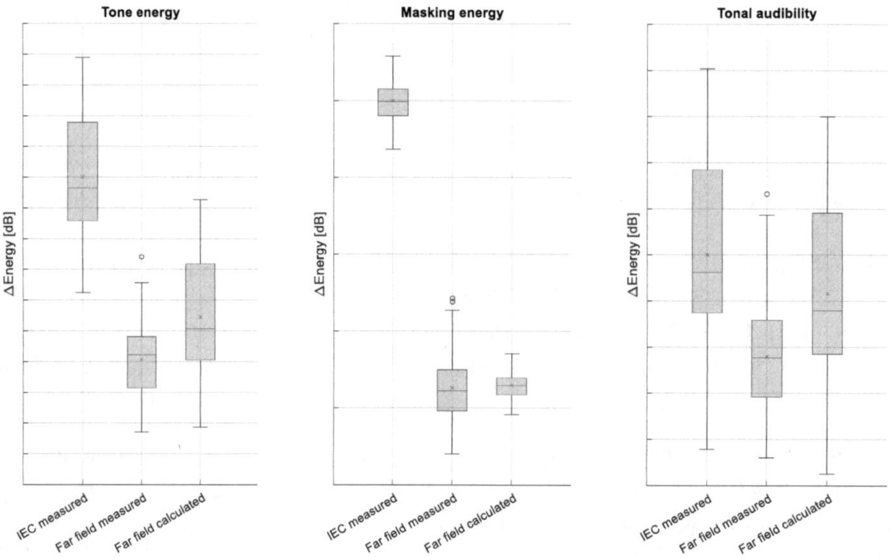

Figure 7. Boxplots of measured and calculated tonal energy, masking energy and tonal audibility. 142 10-second averages and tone frequencies from 140 to 165 Hz at 6.5 m/s windspeed. Ticks indicate 2 dB change in energy.

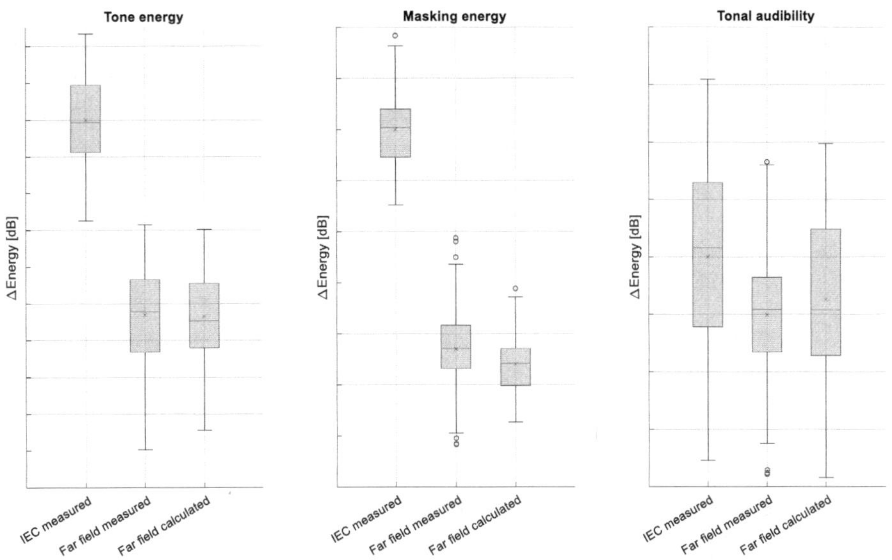

Figure 8. Boxplots of measured and calculated tonal energy, masking energy and tonal audibility. 141 10-second averages and tone frequencies from 90 to 110 Hz at 8.5 m/s windspeed. Ticks indicate 2 dB change in energy.

The boxplots show the spread in measured values of tonal energies and masking energies at both the IEC and far field receptor, leading to a similar large spread in the measured tonal audibilities. As the 10 second averaged IEC measurements are used as inputs, it is expected that the calculated far field energies show a similar spread to the IEC measurements.

Interestingly, it is seen that the spread between the measured IEC masking and calculated far field masking decreases more than for the tonal energy. This is due to the background environmental energy making up a larger part of the masking energy compared to the tone energy, decreasing the overall energy spread for masking more than for the tone after calculation.

The change in energy from the IEC to far field position is well represented in the calculation for the masking energy and tonal energies. The far field masking energy leans towards slight underestimation, while the far field tonal energy leans towards slight overestimation. This leads to a conservative estimate of the far field tonal audibility. This tendency could partly stem from the conservative propagation model used from the IEC to the turbine.

More stringent validation efforts are to be performed, allowing for tool verification across larger input parameter space spanning turbine and site configurations.

4. Conclusions

The tool created can be used to predict and assess tonality noise performance of turbines at set receptor locations using turbine noise information known from standard measurements at the IEC location. Corrections are included for environmental background noise.

The receptor location can be chosen at arbitrary distance from the turbine and height from the ground, with the ground effect considered using the Nord2000 propagation model.

The turbine input configuration allows for representation and assessment of turbine tones across varying frequency and energy levels, as well as hub height and rotor diameter.

A brief validation of the tool has been performed using concurrent measurements taken at IEC and far field positions. Using the 10 second averaged IEC measurements, for binned windspeed and tone frequency, as inputs for turbine tonal performance, allowed for calculation of far field tonal performance. This calculated far field tonal performance was compared with far field tonal measurements for binned windspeed and tone frequency. The comparison between measured and calculated far field tonal performance showed the tool to be slightly conservative.

Additionally, simplified functionality for assessing multiple turbines is included, allowing for representation of wind farms. This functionality is made from assuming independence of individual turbine performance.

This feature of the tool could be used to assess the impact of setting up or changing the tonal performance of a turbine, both individually and as part of a wind farm. This could be used to ensure tonal performance is controlled and allows for decisions to be made to ensure regulations are met and turbines are accepted by the neighbourhood.

As an example, a wind farm with different source frequencies, arising out of different suppliers, could be made to mask any potential tones or turbines could be operated differently through sector management depending on the wind direction.

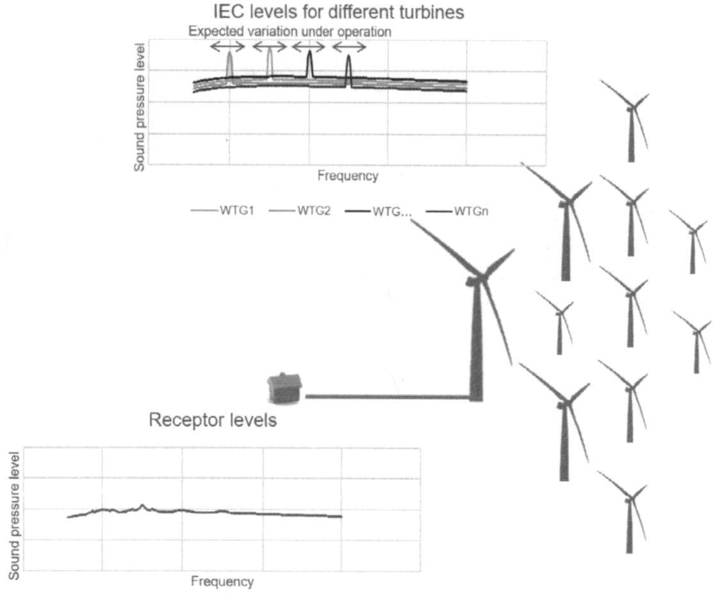

Figure 9. Illustrative example of multiple turbines operating to keep annoyance levels below the limits.

Noise tonality is a concern as we care about neighbours, customers, regulations, etc. We have good understanding of our turbines and tonality, as we take it to be part of our core competence. Vestas has extensive experience from field noise measurements and simulations. Thus, we are well positioned to leverage the potential out of the data we acquire from measurements and simulations using advanced analysis.
At Vestas, we ensure integration of all the complex modules (and components) from multiple suppliers to have the turbines and wind farm operation compliant with our tonality requirements, in addition to the components primary function. We also leverage the operation of the turbine to avoid tonality.
We aim to work in a focused and simplified communicable way to achieve desired objective of tonality free turbines and wind farms without unnecessarily complicating or avoiding the issue but finding solutions where most optimal.

References

[1] IEC 61400-11. Edition 3.0 2012-11. INTERNATIONAL STANDARD. *Wind turbines – Part 11: Acoustic noise measurement techniques*

[2] FGW. Technical Guidelines for Wind Turbines PART 1 Determination of Noise Emission Values, FGW, Rev. 19, 2021.

[3] B. Plovsing, Nord2000. Comprehensive Outdoor Sound Propagation Model. Part 1: Propagation in an Atmosphere without Significant Refraction, DELTA Acoustics & Vibration, Report AV 1849/00 (revised), Hørsholm 2006.

[4] B. Plovsing, Nord2000. Comprehensive Outdoor Sound Propagation Model. Part 2: Propagation in an Atmosphere with Refraction, DELTA Acoustics & Vibration, Report AV 1851/00 (revised), Hørsholm 2006.

10th International Conference
on
Wind Turbine Noise
Dublin – 21st to 23rd June 2023

Study on the Capabilities of a TNO-Based Trailing Edge Noise Prediction Tool Applied to Boundary Layer Suction on a NACA 64-418 Airfoil

S. Haubold, F. Seel, T. Lutz, A. Beck, E. Krämer
University of Stuttgart,Institute of Aerodynamics and Gas Dynamics, Germany
Corresponding Author: sabrina.haubold@iag.uni-stuttgart.de

Summary
Recent improvements to the TNO-based IAGNoise+ trailing edge noise prediction tool better capture the noise emissions for angles of attack up to moderate flow separation. This improved prediction capability is of particular use in studies on the efficiency of boundary layer suction as a trailing edge noise mitigation measure. As a method of active flow control this can delay the onset of flow separation along the airfoil surface and reduce the turbulence kinetic energy and turbulent length scales of the flow when passing the trailing edge, where turbulent energy is then scattered into sound. In order to predict the decrease in noise achieved by this kind of boundary layer control, an accurate prediction of the baseline sound emissions without suction is essential. Wind tunnel measurements are available for a NACA 64_3-418 airfoil with and without boundary layer suction through a porous plate for angles of attack of 0°, 3° and 6° and can be used for validation of the predictions. Based on steady Reynolds-Averaged Navier Stokes computations, the previous as well as the improved IAGNoise+ implementation will be compared to these wind tunnel results. First, the boundary layer parameters related to trailing edge noise generation will be examined, followed by the resulting sound spectra.

1 Introduction
The expansion of wind energy as a renewable energy resource is severely limited by the availability of suitable sites especially for onshore wind parks. The noise generated by the wind turbines contributes to this limitation, as strict regulations are required to be met, in order to protect human populations as well as wildlife from any harmful effects that may result from these noise emissions. A reduction of the primary noise sources on wind turbines would free up additional locations for onshore turbines, as the radius around the turbine within which the noise was above tolerable levels would decrease. The dominant noise sources on modern wind turbines are the aerodynamic noise caused by inflow turbulence, which is influenced by many environmental factors and thus complex to predict and mitigate, and airfoil self-noise generated mostly at the trailing edge of the rotor blades [1]. The trailing edge noise (TEN) source is easier to influence by both passive and active measures. The most common methods for passive

noise reduction are porous trailing edges [2], trailing edge serrations [3, 4, 5], brushes [6, 7], as well as airfoil design for reduced boundary layer thickness [8, 9]. Active measures offer the advantage of being adjustable to changing flow conditions, so as to not incur penalties when run off-design. Active flow control (AFC) in the form of tangential blowing or boundary layer suction is commonly used to stabilize the boundary layer and prevent flow separation, which also results in a drag reduction [10]. The aspect of TEN reduction was examined in [11], [12] and [13] and specifically for wind turbine noise applications in [14] and [15]. Faszer et al. [12] compared tangential blowing and boundary layer suction both through slits of different size and using a porous plate on a NACA0012 airfoil. While both blowing and suction through a slit resulted in either little change or even an increase in TEN, the noise reduction achieved by suction through a porous plate was shown to be quite promising [12]. Investigations on a flat plate by Szoke and Azarpeyvand [13] showed similar results. Uniform perpendicular suction through a porous plate served to decrease velocity fluctuations, boundary layer thickness, displacement thickness, momentum thickness and the turbulence kinetic energy content within the boundary layer downstream of the plate towards the trailing edge, resulting in lower TEN emissions [13]. The achieved noise reduction was most pronounced in the lower frequencies within the considered frequency range. Wolf et al. [14, 16] examined the influence of boundary layer suction through a porous plate in the main pressure recovery region of a NACA 64_3-418 airfoil, commonly employed in the outboard section of wind turbine blades. The wind tunnel measurements were conducted at the Laminar Wind Tunnel of the Institute of Aerodynamics and Gas Dynamics (IAG) at the University of Stuttgart. Similar to the previous studies, the boundary layer suction showed a promising TEN reduction and was thus applied in a following study by Arnold et al. [15] to the outboard part of the blades of both a generic NREL 5 MW turbine, as well as a modern M117 wind turbine. Arnold identified an optimum spanwise extent of the suction region between $0.7 \leq r/R \leq 1$ and showed that both the overall noise signature as well as the aerodynamic efficiency could be improved with the optimal suction configuration [15]. The identification of these optimal configurations for various different wind turbines requires time-efficient and accurate prediction tools. The in-house TNO-based prediction tool IAGNoise+ [17] is built to provide fast results based on steady Reynolds-Averaged Navier-Stokes simulations. The original code was called Rnoise and developed by Kamruzzaman [17]. The newest implementation by Hornung [18], applies a modified model for the wall pressure fluctuations (WPF) based on Blake [19], which does not neglect turbulence-turbulence interaction (TTI). It is thus able to predict the increase in noise seen with increasing angles of attack due to moderate flow separation [18]. This increased prediction quality should allow for better quantification of the achieved decrease in trailing edge noise through boundary layer suction, as the delayed separation would be reflected specifically in a decrease of the TTI component. An evaluation of the prediction accuracy of both the previous (MK) as well as the new IAGNoise+ model (CH) is performed by comparing results for various angles of attack and various degrees of suction to the available wind tunnel results for the NACA 64_3-418 airfoil by Wolf [14]. The aim is specifically to quantify the improvements in prediction accuracy for cases with boundary layer suction gained by the inclusion of the TTI term and the modified anisotropy formulation in the new CH model.

In the first section of this work, the measurements by Wolf [14] providing the validation data are explained, followed by the numerical methods, i.e. the CFD-RANS using the flow solver FLOWer and the IAGNoise+ prediction. Once the methods are established, a discussion of the results will follow. The CFD results in terms of turbulent boundary layer data are examined first, as they form the basis for the IAGNoise+ evaluation. Finally, the noise predictions are validated against the measured noise spectra.

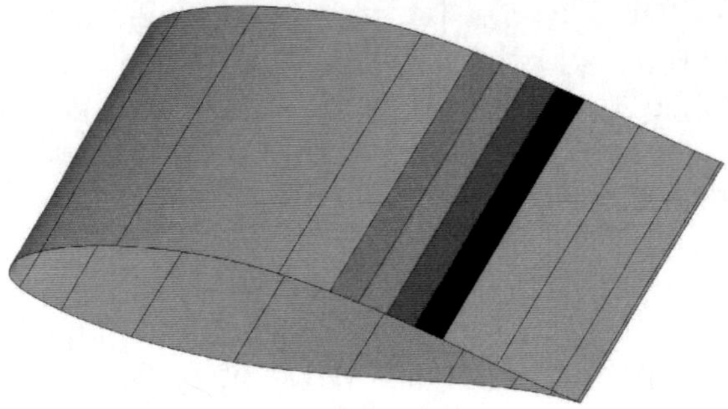

Figure 1. Positions of suction panels (colored) on NACA 64_3-418 airfoil (C1=orange, C2=green, C3=blue, C4=black).

2 Experimental Data

All the experimental data used for the validation of the predictions stems from the measurement campaign by Wolf [16] conducted at the Laminar Wind Tunnel (LWT) of the IAG. The LWT is a very low turbulence (0.02 %) open return wind tunnel with a closed test section [20]. Laminar to turbulent transition was forced using turbulators at 5 % chord on both suction and pressure side. The aforementioned NACA 64_3-418 airfoil was equipped with porous suction panels from 55 % to 75 % chord along the suction side, as shown in Fig. 1. The panels had a porousity of 25 %, meaning the holes made up 25 % of the entire panel surface. The hole diameter was 250 µm. The panels were connected to 4 different chambers, allowing suction to be switched on and off independently for the 4 sections. For the investigations, either all 4 sections seen in Fig. 1 (C1234), only the two upstream sections from 55 % to 65 % chord (C12, orange and green) or only the two downstream sections from 65 % to 75 % chord (C34, blue and black) were switched on. Two different suction rates were tested, resulting in a total massflow of either $\dot{m}_s = 0.04$ kg/s (for C12 or C34) or $\dot{m}_s = 0.08$ kg/s (for C1234) at the lower rate of $C_Q = 0.0114$ and $\dot{m}_s = 0.06$ kg/s or $\dot{m}_s = 0.12$ kg/s at the higher rate of $C_Q = 0.0171$. The dimensionless suction rate is defined as:

$$C_Q = \frac{\dot{m}_s}{\rho \cdot U_\infty \cdot A_s} \tag{1}$$

with $\dot{m}_s$ being the suction massflow, A_s the area of the suction surface, the density ρ and the freestream velocity U_∞. The following table shows all of the available measurement data recorded by Wolf in his measurement campaign at the LWT.

The Coherent Particle Velocimetry (CPV) method developed by Herrig et al. [21] at the IAG was employed for the measurement of trailing edge noise. In this method, velocity fluctuations are measured by a hot-wire above and below the trailing edge just outside the boundary layer. Background noise is mostly eliminated via cross correlation of the signals from both wires. The accuracy of the overall method is given as +/- 1 dB [20]. Above 5 kHz the measurements are affected by electronic noise. For frequencies below about 1 kHz the measurements are usually considered to be less reliable, as a phase shift in the cross correlation spectrum is seen and the background noise covers up the trailing edge noise signals. The exact frequency at which the phase shift occurs is dependent on the specific airfoil. Frequencies between 500 Hz and 5 kHz were measured by Wolf for the various configurations [14]. A phase shift at 800 Hz was encountered in the experiment, therefore data at frequencies around 800 Hz had to be interpolated [14]. This is especially relevant when comparing the overall sound pressure levels (OASPL) for frequencies between 500 Hz and 5 kHz, as these will also be affected by possible inaccuracies with the lower frequencies ($f < 800$ Hz). The boundary layer profile and velocity fluctuations within the boundary layer (BL) 1 mm behind the trailing edge were recorded using a single hot-wire. The turbulence kinetic energy(tke) was calculated using Eq. 2 from the measured velocity fluctuation $\langle u_1^2 \rangle$ in only one direction by applying the anisotropy assumption

$\langle u_1^2 \rangle : \langle u_2^2 \rangle : \langle u_3^2 \rangle = 4:2:3$ by Kamruzzaman [17].

$$k_t(y) = \frac{9}{8} \langle u_1^2 \rangle (y) \quad (2)$$

configuration	$\dot{m}_s[kg/s]$	C_Q	AoA [°]	acoustic measurement	BL measurement
baseline	-	-	0	Yes	Yes
C12	0.04	0.0114	0	Yes	Yes
C34	0.04	0.0114	0	Yes	Yes
C12	0.06	0.0171	0	Yes	No
C34	0.06	0.0171	0	Yes	No
C1234	0.08	0.0114	0	No	Yes
C1234	0.12	0.0171	0	Yes	Yes
baseline	-	-	3	Yes	Yes
C12	0.04	0.0114	3	Yes	Yes
C34	0.04	0.0114	3	Yes	Yes
C12	0.06	0.0171	3	Yes	No
C34	0.06	0.0171	3	Yes	No
C1234	0.08	0.0114	3	Yes	Yes
C1234	0.12	0.0171	3	Yes	Yes
baseline	-	-	6	Yes	Yes
C12	0.04	0.0114	6	Yes	Yes
C34	0.04	0.0114	6	Yes	Yes
C12	0.06	0.0171	6	Yes	No
C34	0.06	0.0171	6	Yes	No
C1234	0.08	0.0114	6	Yes	Yes
C1234	0.12	0.0171	6	Yes	Yes

Table 1. Test matrix of the wind tunnel measurements ($U_\infty = 70$ m/s, $c = 0.6$ m, Re_c =2.5 Mio, $x_{tran}/c = 0.05$) performed by Wolf [14].

3 Numerical Methods

In accordance with the wind tunnel measurements, a NACA $64_3 - 418$ airfoil with a chord length of $c = 0.6$ m at a chord based Reynolds number of $Re_c = 2\,500\,000$ and an inflow velocity of $U_\infty =$ 70 m/s was investigated numerically in a two-step process. Steady 2.5D RANS calculations were run using the flow solver FLOWer and the CFD results were then used as input for the IAGNoise+ predictions using both the new (CH) as well as the previous (MK) model.

3.1 CFD Computations

The flow was calculated using the structured grids finite volume based CFD solver FLOWer developed at the DLR (German Aerospace Center) [22] and extended for specific applications at the IAG. The two-equation k-ω RANS turbulence Model was chosen. Transition was forced at $x_{tran}/c = 0.05$ on both the suction and pressure side.

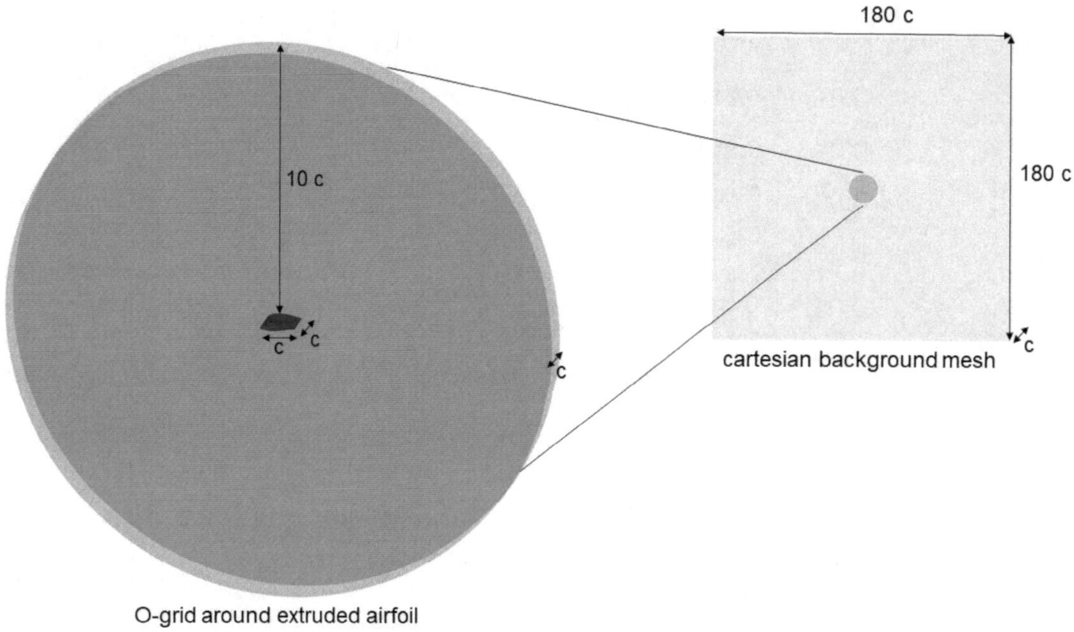

Figure 2. Schematic of the computational domain.

3.2 CFD Setup

For the numerical setup, the NACA 64_3-418 airfoil was extruded by one chord length in spanwise direction. Symmetry conditions were applied to the spanwise boundaries of the computational domain. The resolution of the boundary layer (BL) was chosen to maintain a y^+ of less than 1 even in the case of suction, which leads to steeper velocity gradients near the wall. The entire height of the boundary layer was covered by 56 cells. The four separate suction areas, according to the wind tunnel tests (see Fig. 1), were each resolved by 21 points in streamwise direction. Suction and pressure side were each resolved by 312 cells in total. The various angles of attack were applied using the Chimera overset grid technique, where an O-grid around the wing, containing about 700.000 cells, was rotated on a stationary Cartesian background mesh of about 800.000 cells. The full schematic is depicted in Fig. 2. A multigrid W-cycle with 3 refinement levels was used to accelerate convergence of the simulations. The suction boundary condition in FLOWer was implemented at the IAG [23] and is based on the actuation boundary condition contained in the flow solver for hybrid meshes TAU, which was also developed at the DLR [24]. The boundary condition, based on the method of characteristics, applies a suction pressure at a ghost cell beneath the panel surface, calculates the flow response and adjusts that pressure iteratively to reach a specified target massflow chosen in accordance to measurements [23]. The thus iterated suction pressure is constant across the boundary section, as can be seen in the surface pressure distribution in Fig. 3. This leads to a suction profile marked by low or even negative suction velocities (blowing) at the upstream border and the maximum velocity at the downstream edge of the suction surface. This point of maximum suction velocity and the end of the panel is then followed by a non-slip wall, where the wall nor-

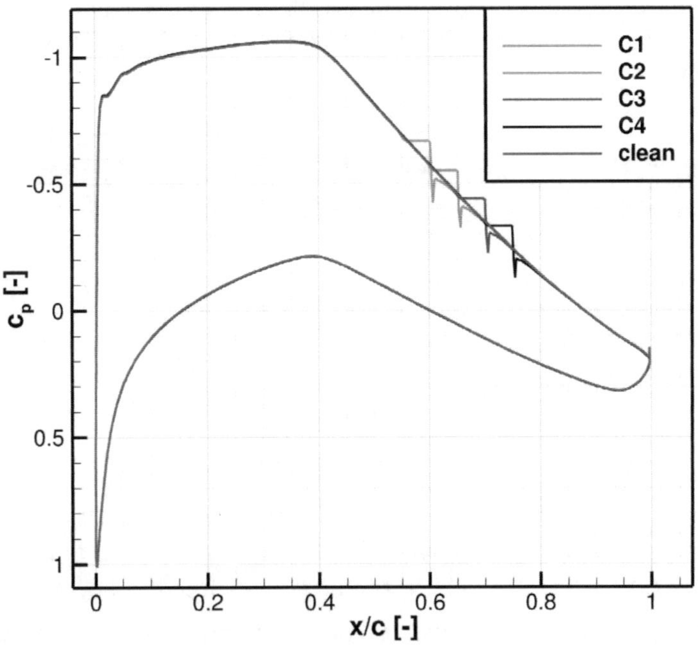

Figure 3. Pressure distribution with suction at panels C1, C2, C3 and C4 seperately with $C_Q = 0.0114$ compared to the clean case at $\alpha = 3°$.

mal velocity needs to be zero, resulting in a strong deceleration and a sharp surface pressure increase in relation to the total massflow, as is visible in Fig 3, before it reverts to normal levels. A slightly more sophisticated version of this suction model which considers pressure losses through the porous plate and thus achieves a more realistic distribution of the suction pressure is available. However, the pressure iteration is already quite sensitive in terms of stability and the more complicated version of the boundary condition adds to this issue, while it was shown in [23] that the effect on the trailing edge noise predictions is negligible. Therefore, the simpler and more stable version was chosen for this investigation.

3.3 Noise Prediction Tool

The IAGNoise+ prediction tool is based on the TNO model by Blake [19] and Parchen [25] [26], a semi-analytical model for trailing edge noise. The original implementation, also known as Rnoise, was created by M. Kamruzzaman [17]. For later comparisons, this version will be referred to as the MK model, to differentiate it from the new "CH model", created by C. Hornung [18]. The main difference between this and the former model is that the last term, associated with turbulence-turbulence interaction (TTI), in the source equation for the Wall Pressure Fluctuations (WPF) is not neglected, as it is in the original model:

$$\frac{\partial^2 p'}{\partial x_i^2} = -\rho \left(\underbrace{2 \frac{\partial \overline{U_i}}{\partial x_j} \frac{\partial u'_j}{\partial x_i}}_{MTI} + \underbrace{\frac{\partial^2}{\partial x_i \partial x_j} \left(u'_i u'_j - \overline{u'_i u'_j} \right)}_{TTI} \right) \tag{3}$$

Neglecting the TTI part of the equation was proposed by Blake, as the ratio of Mean-shear Turbulence Interaction (MTI) to TTI was assumed to be greater than 10 [19]. However, this assumption is no longer justified when regarding highly loaded boundary layers at higher angles of attack, where the wall normal mean velocity gradient diminishes and thus the ratio of MTI to TTI decreases [18]. Therefore, the new model was modified to include this part of the WPF and thus produce more accurate results especially for slightly to moderately separated flows. Additionally, the original anisotropy formulation by Kamruzzaman was extended to include adverse pressure gradient effects [18].

$$f_{22,aniso,CH} = \underbrace{Re_{\lambda}^{-0.09}}_{f_{22,aniso,MK}} \cdot e^{\left(\frac{u_p}{0.09U_e}\right)} \text{ with } Re_{\lambda} = \frac{\sigma\lambda_g}{\nu} \text{ and } u_p = \left(\frac{\nu}{\rho}\frac{dp}{dx_1}\right)^{\frac{1}{3}} \tag{4}$$

Re_{λ} represents the Reynolds number based on the Taylor micro scale λ_g, u_p is the so called pressure velocity and U_e the flow velocity at the outer edge of the boundary layer. The resulting power spectral density of the WPF is calculated according to Eq. (5) from statistical turbulence quantities provided by the RANS flow solution [18].

$$P(k_1,k_3,\omega) = 4\rho^2\left(\frac{1}{k_1^2+k_3^2}\right)\cdot\int_0^{\infty}\left(k_1^2\frac{\partial\overline{U_1^2}}{\partial x_2}+\frac{2(k_1^2+k_3^2)}{15\nu}\varepsilon\right)\Lambda_2\Phi_{22}\cdot\langle u_2'^2\rangle\Phi_m e^{-2|k|x_2}dx_2 \tag{5}$$

Within the equation, k_1 and k_3 represent the wave number in streamwise and spanwise direction respectively, ω corresponds to the angular frequency, ρ is the density and ν the kinematic viscosity. The velocity gradient in wall-normal direction is given by $\frac{\partial\overline{U_1^2}}{\partial x_2}$ the dissipation by ε and the turbulent integral length scale in vertical direction by Λ_2. The variable Φ_{22} is the normalized spectrum of the vertical velocity fluctuations $\langle u_2'^2\rangle$ (i.e. Reynolds stresses), Φ_m the moving axis spectrum and k the wavenumber vector.
The resulting Wall Pressure Fluctuations are then propagated using the following far field model derived by Chase [27] and Brooks and Hodgson [28].

$$G_{ff}(f) = \frac{L\overline{D}}{2\pi R^2}\int_0^{\infty}\frac{\omega}{c_0|k_1|}\frac{P(k_1,k_3=0,\omega)}{1-\frac{\omega}{c_0|k_1|}}dk_1 \tag{6}$$

$$SPL(f) = 10\cdot\log\left[\frac{G_{ff}(f)\cdot df}{(2\cdot 10^{-5})^2}\right] \tag{7}$$

The model includes a directivity function $\overline{D}$ for trailing edge noise. The parameter L represents the wetted length in spanwise direction, c_0 the speed of sound and R the observer distance.
The IAGNoise+ evaluation point was set as $x/c = 0.995$ in accordance with the hot-wire location for the coherent particle velocimetry (CPV) [14] within the wind tunnel measurements. The observer location was chosen to be $R = 1$ m above the suction side of the airfoil.

4 Results and Discussion

4.1 Boundary Layer Analyses at Trailing Edge

The prediction accuracy of the IAGNoise+ results is highly dependent on the quality of the boundary layer data provided by the CFD RANS simulations. Both the velocity profiles as well as the turbulence kinetic energy were measured by Wolf [16] in the wind tunnel tests and are available for comparison with the CFD data. The following figures 4 to 6 show the boundary layer data at the chordwise evaluation point of $x/c = 0.995$ for different angles of attack and suction configurations. It is important to note that the boundary layer data in the wind tunnel was measured slightly further downstream, 1 mm past the trailing edge, resulting in a distance of 4 mm between the evaluation positions. A small difference in boundary layer thickness and profile would therefore be expected. However, a rough estimation based on the theory of Schlichting for a flat plate [29] according to Eq. (8) suggests that the difference in turbulent boundary layer thickness would be only about 0.06 mm for an inflow velocity of 70 m/s.

$$\delta(x) = 0.37x\frac{U_\infty x}{\nu} \tag{8}$$

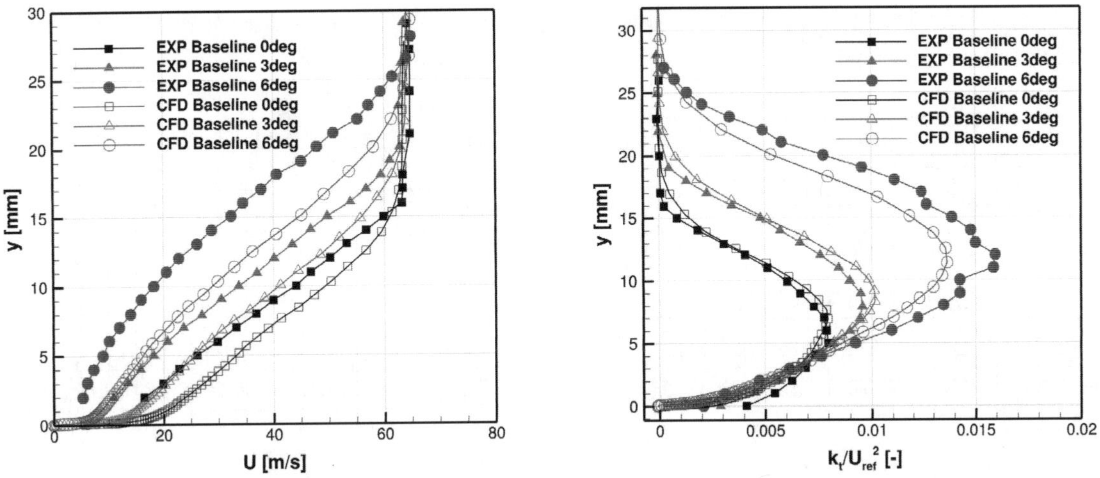

(a) Velocity profiles. (b) Turbulence kinetic energy distribution.

Figure 4. Comparison of wall-normal distribution of streamwise velocity and turbulence kinetic energy near the trailing edge (suction side) between CFD-simulation and experiment [14] at different angles of attack.

Figure 4 shows the boundary layer data in the baseline case, without any suction applied. The boundary layer at $x/c = 0.995$ extracted from the CFD simulations is marked by higher velocities and a lower boundary layer thickness at all three angles of attack, than the measured boundary layer data 1 mm past the trailing edge, as would be expected. The difference however appears to be larger than the estimation based on Schlichting would suggest. The turbulence kinetic energy matches the wind tunnel data quite well. This is in agreement with the frozen turbulence hypothesis by Taylor [30], according to which the properties of turbulent eddies do not change as they are advected by the mean streamwise velocity. In terms of the relation between the angle of attack and the boundary layer, the increase in boundary layer thickness, the velocity decrease and the increase in the turbulence kinetic energy with increasing angle of attack are all reflected in the CFD results. Overall, the effects of beginning separation for increasing angles of attack occurred earlier in the wind tunnel tests than in the CFD RANS simulations, which was also noted by Wolf [14] and might be related to the two equation turbulence model used in the RANS simulation. A decrease of k_t in the lower region of the boundary layer is seen in the experimental results but not in the CFD data. Wolf [14] attributes this to the measurement position 1 mm behind the trailing edge, where some mixing between suction and pressure side could have already occurred.

In the case of suction, the impact on the boundary layer is not fully captured by the simulations. In Fig. 5 the $\alpha = 3°$ case was arbitrarily chosen to show the differences, as the trends were the same for all angles of attack. The velocity profile for the suction cases in the CFD simulations appears to match the measusurements quite well. The simulations predict a higher boundary layer thickness in the suction case in spite of the more upstream evaluation position. The change in boundary thickness is clearly underpredicted. In the measurements, the velocities at the outer edge of the boundary layer appear to be slightly reduced in the suction cases.

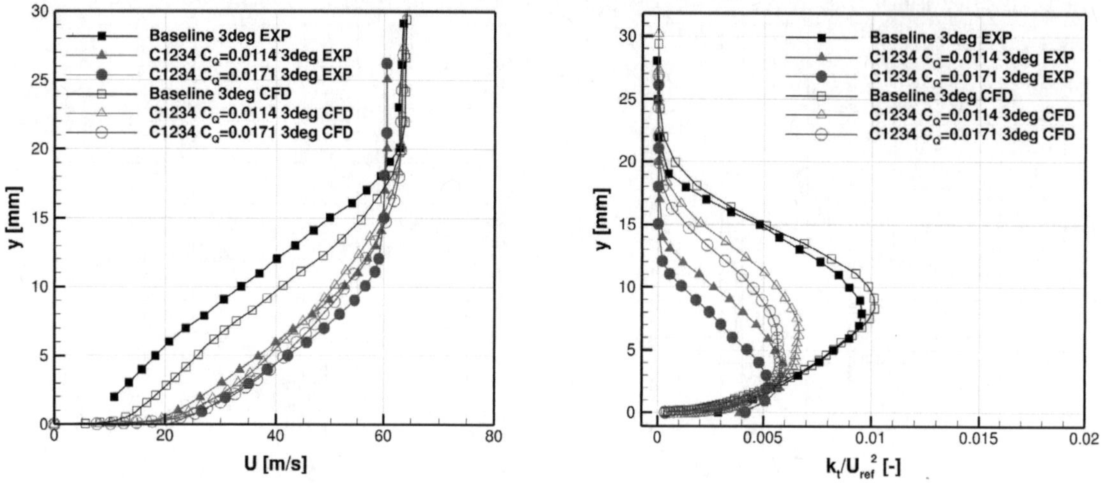

(a) Velocity profiles. (b) Turbulence kinetic energy distribution.

Figure 5. Comparison of boundary layer profiles near the trailing edge (suction side) for $\alpha = 3°$ when all suction panels are active at $C_Q = 0.0114$ or $C_Q = 0.0171$ to the baseline case (Measurement data from [14]).

This effect is not seen in the CFD results. The figure on the right (Fig. 5b)) also shows a less significant decrease in turbulence kinetic energy in the upper region of the boundary layer predicted by the simulation. The CFD-RANS using a two-equation model does not exhibit as strong of an effect of the boundary layer suction on the airfoil boundary layer as was seen in the measurements. This could be related to the fact that the simulations predicted less separation than was seen in the experiment [14], in which case the impact of boundary layer suction would be expected to be somewhat lower.

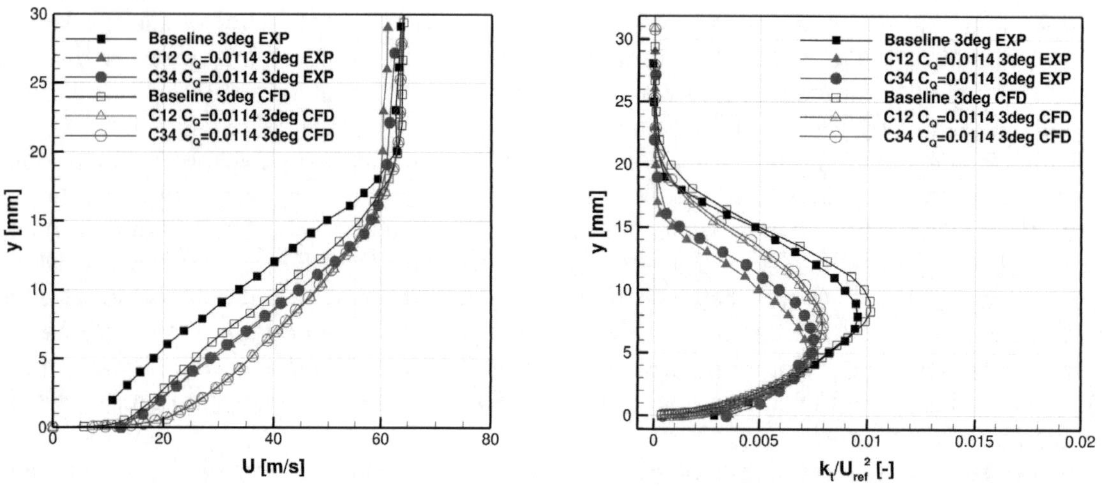

(a) Velocity profiles. (b) Turbulence kinetic energy distribution.

Figure 6. Comparison of boundary layer profiles near the trailing edge (suction side) for $\alpha = 3°$ when only two suction panels are active at $C_Q = 0.0114$ (C12 or C34)(Measurement data from [14]).

The simplified implementation of the suction boundary condition may also influence the effect on the boundary layer. Inaccuracies at this stage would be difficult to mitigate in the later IAG-Noise+ calculation. The improvements gained by resolving turbulence better in the CFD simulation approach or adjusting the suction boundary condition should be investigated, to evaluate whether the more accurate boundary layer results would merit the increase in computational cost.
Figure 6 shows that similar effects are present in the case of smaller suction regions (C12 and C34) and a thus lower massflow. The boundary layer thickness in the case of suction predicted by the simulations is higher than in the measurements. The simulations show a more significant change in the velocity gradient, while in the experiment the gradient in the case of suction is still quite similar to baseline. The steeper gradient in the simulations could be related to the implementation of the suction boundary condition. The turbulence kinetic energy calculated from the measurements also shows a slightly stronger dependence on the position of the suction region than the tke extracted from the simulations.

4.2 Trailing Edge Noise

In this section, the trailing edge noise predictions produced by the previous (MK) and the new (CH) IAGNoise+ model are compared to the sound pressure levels measured in the experiments to gauge the improvement gained by the inclusion of the TTI component and the modified anisotropy factor. The sound pressure spectra for the baseline cases shown in Fig. 7 reflect the shift in the spectrum towards lower frequencies with increasing angle of attack. The CH model predictions match the experiment quite well, both in terms of absolute values, as well as the slope of the spectra. The spectra predicted by the MK model appear to be shifted towards lower frequencies, while the CH model based spectra are shifted slightly towards higher frequencies, when compared to the measured spectra. The pressure side boundary layer is usually thinner at positive angles of attack, resulting in slightly smaller vortices, which contribute more to the higher frequencies, while the suction side contribution is seen more in the lower frequencies due to the higher boundary layer thickness. This would suggest that the new CH model captures more of the suction side contribution and even slightly overestimates it, while both suction and pressure side contributions appear to be slightly underestimated by the previous MK model.

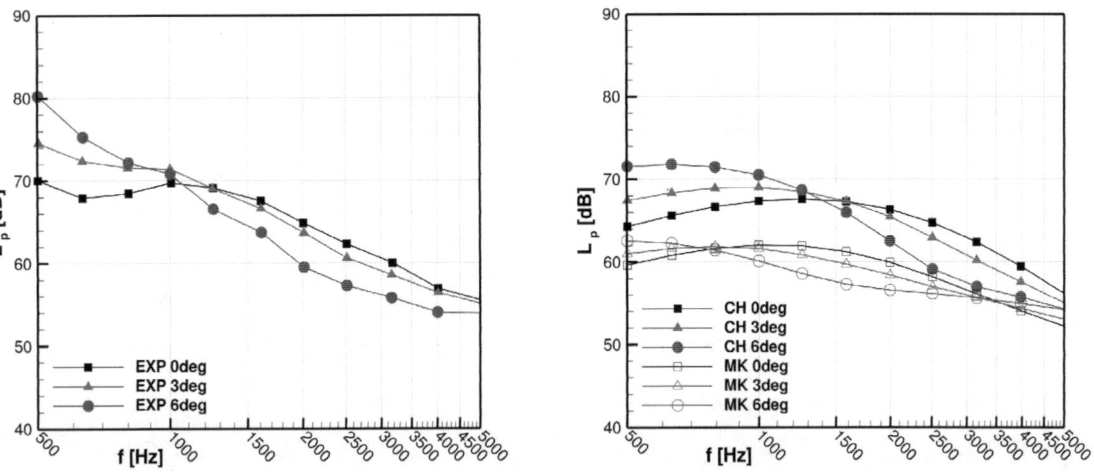

(a) Measured spectra [14]. (b) Predicted spectra.

Figure 7. Comparison of baseline 1/3 octave sound pressure spectra at different angles of attack.

It is also important to remember at this point that the phase shift in the CPV measurement and subsequent interpolation of values at 800 Hz [14] makes the experimental results below this frequency less reliable. An artifact of this can be seen in the experimental spectrum in Fig. 7a) for $\alpha = 0°$, where a new peak is visible at 500 Hz in addition to the main hump at around 1 kHz. For this reason, the following discussions will focus mainly on frequencies above 1 kHz.
As Fig. 8 shows, the slope of the measured spectra is still slightly higher than the predictions based on the CH model. The model appears to predict a higher contribution by the suction side. This is consistent with the results shown in [18]. However, the overall improvement gained compared to the MK model prediction is clearly visible, not only for the higher angles of attack but even at a symmetrical inflow.

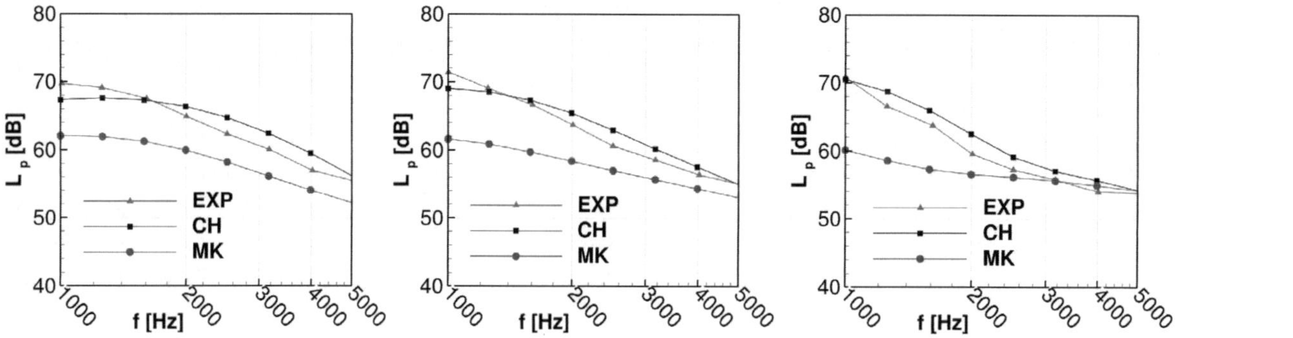

(a) $\alpha = 0°$. (b) $\alpha = 3°$. (c) $\alpha = 6°$.

Figure 8. Comparison of baseline 1/3 octave sound pressure spectra produced by each sound prediction model to wind tunnel data [14] for different angles of attack.

As the overall sound pressure levels (OASPL) in the measurement were calculated for a frequency range from 500 Hz to 5 kHz, some inaccuracies are to by expected, due to the inclusion of the less reliable data at the lower frequencies. However, the OASPL predicted by the CH model are close to the experimental results with deviations of less than 2 dB, as can be seen in table 2.

-	$\alpha = 0°$	$\alpha = 3°$	$\alpha = 6°$
experiment [14]	76.91 dB	79.16 dB	81.37 dB
CH model	77.21 dB	78.3 dB	79.71 dB
deviation	+0.3dB	-0.86 dB	-1.66 dB
MK Model	70.6 dB	70.45 dB	70.6 dB
deviation	-6.31 dB	-8.71 dB	-10.77 dB

Table 2. Comparison of baseline OASPL for frequencies from 500 Hz to 5 kHz.

After examining the influence of the angle of attack on the sound pressure spectra in the baseline case, cases with active boundary layer suction will now be investigated. The effect of boundary layer suction on the sound spectrum is opposite to that of increasing angles of attack. As separation is delayed and the boundary layer thickness near the trailing edge is decreased, the lower frequencies, generally related to the larger turbulent scales, are reduced and an increase is seen in the higher frequency components, indicating a higher contribution by the smaller scales of turbulence.

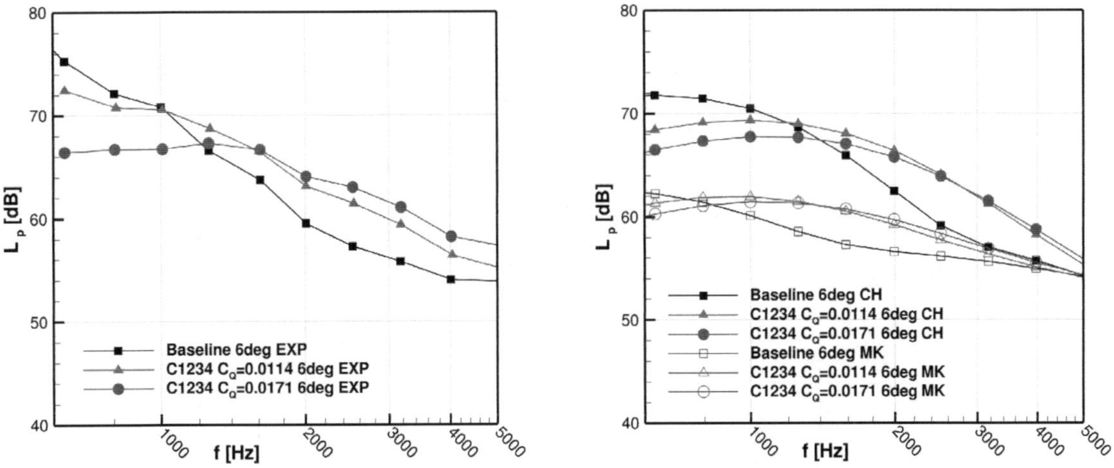

(a) Measured spectra [14]. (b) Predicted spectra.

Figure 9. Change in 1/3 octave sound pressure spectra with suction at all 4 panels (C1234) for $\alpha = 6°$.

In spite of the inaccuracies within the RANS-based input data shown in the previous chapter, the predictions based on the CH model show good alignment with the experiment (see Fig. 9). The MK model predictions reflect the shift of the spectral peak towards higher frequencies as well as the influence of the suction rate qualitatively, however the changes are not as significant as in the measurement or the CH model prediction. This can be explained by the relation between separation and the TTI fluctuation term. As mentioned in section 3.3, accurately capturing the change in trailing edge noise seen with increasing angle of attack and beginning separation requires an inclusion of TTI in the noise prediction. As boundary layer suction reduces boundary layer thickness and delays separation, it largely affects the TTI component. This effect is not captured in the previous model (MK), which neglects the term completely and thus predicts less change in the noise spectrum.

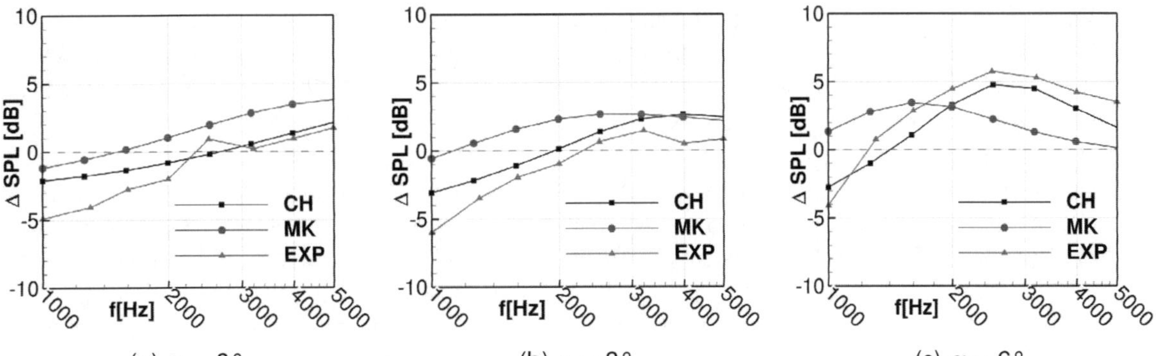

(a) $\alpha = 0°$. (b) $\alpha = 3°$. (c) $\alpha = 6°$.

Figure 10. Difference in sound pressure level due to suction at all 4 panels with $C_Q = 0.0171$, $\Delta SPL = SPL_{C1234,C_Q=0.0171} - SPL_{Baseline}$ (Measurement data from [14]).

In Fig. 10, a comparison of the ΔSPL resulting from suction at all 4 sections (C1234) with $C_Q = 0.0171$ is shown for the various angles of attack. The CH prediction matches the measurement slightly better with deviations of only up to 3 dB, but overall the relative impact of suction on the spectrum appears to be reasonably similar for lower angles of attack. At $\alpha = 6°$ the MK model shows the most significant differences. Figure 10c) also shows that the increase in SPL for the higher frequencies is quite significant in the $\alpha = 6°$ case, which would be reflected in the OASPL as well, decreasing the total noise reduction across the frequency range of 500 Hz to 5 kHz. An increase in the higher frequencies would be related to a higher contribution by smaller vortices within the turbulent boundary layer or a higher meanflow velocity at the trailing edge.

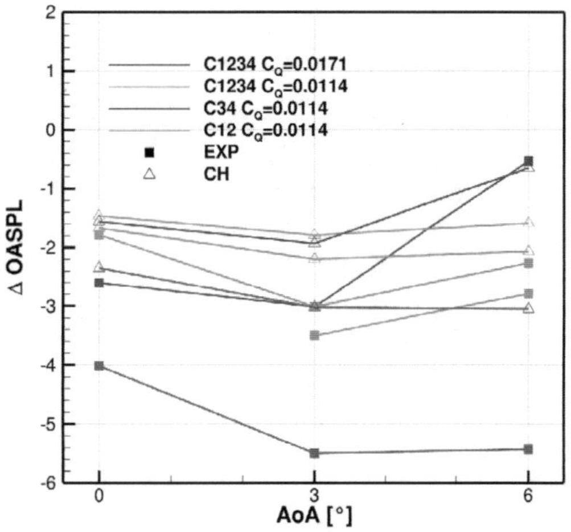

(a) Noise reduction (difference in OASPL).

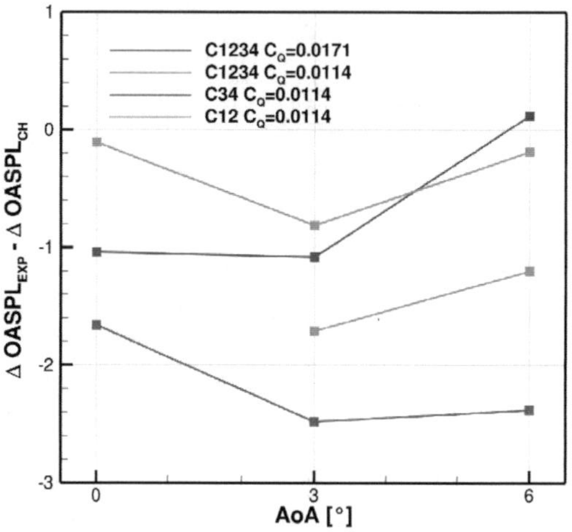

(b) Difference between prediction and measurement.

Figure 11. Comparison between predicted and measured noise reduction in terms of OASPL (500 Hz to 5 kHz) at different angles of attack and suction configurations ($\Delta OASPL = OASPL_{suction} - OASPL_{baseline}$).

The total noise decrease in terms of the overall sound pressure levels (OASPL) is compared in Fig. 11 for different angles of attack and suction configurations. Both the experiment and the CH model predictions show the highest noise reductions at $\alpha = 3°$. The lower noise reduction due to the aforementioned increase in higher frequency noise in the $\alpha = 6°$ case is also visible, though the difference appears to become smaller for higher suction massflows (i.e. C1234 $C_Q = 0.0171$). At $\alpha = 6°$ the position of the suction panels also seems to become more relevant, as the difference between C12 and C34 is much more significant is this case. For an angle of attack of $\alpha = 0°$ the boundary layer suction still results in a decrease in overall noise, although less significant than at higher angles of attack. These results suggest that the relation between the angle of attack and the suction rate in terms of the effectiveness of the boundary layer suction is not simply linear and a more detailed investigation would be required to derive design guidelines for these devices. With the new CH model, the predicted noise decrease is still generally lower than that measured in the wind tunnel. It is important to remember at this point that the reliability of the total sound pressure levels from the measurements is somewhat questionable, due to the inclusion of the lower frequency data, which could not be measured properly and had to be interpolated. The difference in noise reduction between different angles of attack is also lower in the IAGNoise+ predictions than in the measured data, where differences are much more pronounced. Subtracting the predicted noise decrease from the measurement data leads to Fig. 11b) on the right, which shows the deviations. The figure makes it clear that the difference between prediction and experiment increases with higher suction massflows, which would suggest that the effect of the boundary layer suction on the overall noise is not yet fully captured by the predictions. To further verify this and identify the relevant mechanisms, it could be useful to perform direct noise computations, which more accurately capture the internal phenomena within the boundary layer and the effect on the noise signature.

5 Conclusion and Outlook

In this work, the noise predictions by two different TNO-based model variations within the IAGNoise+ code were validated against wind tunnel measurement data for a 2D airfoil with and without boundary layer suction and different angles of attack between $\alpha = 0°$ and $\alpha = 6°$. The CFD-RANS based boundary layer data in terms of the velocity profile and the wall-normal distribution of the turbulence kinetic energy was compared to hot-wire measurements. It was found that for the baseline case without suction, the boundary layer thickness in the simulations was considerably lower and the velocities slightly higher than in the experiment and the turbulence kinetic energy in the $\alpha = 6°$ was slightly underestimated. The reduction in the turbulence kinetic energy resulting from boundary layer suction was also underpredicted by the RANS simulations. However, when the IAGNoise+ predictions based on the RANS data were compared to the sound pressure spectra measured in the wind tunnel using the CPV method, the newest IAGNoise+ "CH" model by Hornung showed good alignment both for the baseline as well as the suction cases. The improvement compared to the previous "MK" model by Kamruzzaman was clearly visible, especially at higher angles of attack. This was attributed to the inclusion of turbulence-turbulence interaction (TTI) within the new CH model, which widened the range of applicability to higher angles of attack and up to moderate separation, while additionally improving the prediction quality at lower angles of attack as well. The improved alignment in the predictions for cases with boundary layer suction and the resulting change in the spectral distribution suggests that this trailing edge noise mitigation measure largely affects the TTI contribution to the sound pressure spectrum. The previous MK model only captures the effect of boundary layer suction on the meanflow-turbulence interaction (MTI), which for the lower angles of attack still allowed for a reasonably accurate prediction of the spectral changes for the lower angles of attack, but failed in the prediction of the spectral shift at $\alpha = 6°$. Overall, the predictions based on the new CH model provide a good basis for the investigation of different suction configurations and their noise mitigation potential upon various airfoils as well as rotor blade geometries. However, a deeper understanding of the underlying mechanisms gained through Direct Noise Computations (DNC) would aid in discovering ways to further improve upon the modeling and predictions specifically for cases with boundary layer suction. Large Eddy Simulations (LES) would offer a good balance between accuracy and computational cost for this purpose.

Acknowledgements

The research presented in this paper was funded by the Friedrich und Elisabeth Boysen Foundation as part of the project BOY-170. The authors gratefully acknowledge the High-Performance Computing Center Stuttgart (HLRS) for providing computational resources.

References

[1] S. Wagner, R. Bareiß, and G. Guidati, *Wind Turbine Noise.* Springer, 1996. [Online]. Available: http://link.springer.com/10.1007/978-3-642-88710-9

[2] M. R. Khorrami and M. M. Choudhari, “Application of passive porous treatment to slat trailing edge noise,” 2003, NTRS Author Affiliations: NASA Langley Research Center NTRS Document ID: 20030059013 NTRS Research Center: Langley Research Center (LaRC). [Online]. Available: https://ntrs.nasa.gov/citations/20030059013

[3] M. S. Howe, “Noise produced by a sawtooth trailing edge,” *The Journal of the Acoustical*

Society of America, vol. 90, no. 1, pp. 482–487, 1991, publisher: Acoustical Society of America. [Online]. Available: https://asa.scitation.org/doi/abs/10.1121/1.401273

[4] S. Oerlemans, M. Fisher, T. Maeder, and K. Kögler, "Reduction of wind turbine noise using optimized airfoils and trailing-edge serrations," *AIAA Journal*, vol. 47, no. 6, pp. 1470–1481, 2009, publisher: American Institute of Aeronautics and Astronautics _eprint: https://doi.org/10.2514/1.38888. [Online]. Available: https://doi.org/10.2514/1.38888

[5] A. Finez, M. Jacob, M. Roger, and E. Jondeau, "Broadband noise reduction of linear cascades with trailing edge serrations," *17th AIAA/CEAS Aeroacoustics Conference (32nd AIAA Aeroacoustics Conference)*, 2011, _eprint: https://arc.aiaa.org/doi/pdf/10.2514/6.2011-2874. [Online]. Available: https://arc.aiaa.org/doi/abs/10.2514/6.2011-2874

[6] M. Herr, "Design criteria for low-noise trailing-edges," *13th AIAA/CEAS Aeroacoustics Conference (28th AIAA Aeroacoustics Conference)*, 2007, _eprint: https://arc.aiaa.org/doi/pdf/10.2514/6.2007-3470. [Online]. Available: https://arc.aiaa.org/doi/abs/10.2514/6.2007-3470

[7] A. Finez, M. Jacob, E. Jondeau, and M. Roger, "Broadband noise reduction with trailing edge brushes," *16th AIAA/CEAS Aeroacoustics Conference*, 2010, _eprint: https://arc.aiaa.org/doi/pdf/10.2514/6.2010-3980. [Online]. Available: https://arc.aiaa.org/doi/abs/10.2514/6.2010-3980

[8] F. Bertagnolio, H. Madsen, and C. Bak, "Experimental validation of TNO trailing edge noise model and application to airfoil optimization," *European Wind Energy Conference and Exhibition*, 2009.

[9] T. Lutz, A. Herrig, W. Würz, M. Kamruzzaman, and E. Krämer, "Design and wind-tunnel verification of low-noise airfoils for wind turbines," *AIAA Journal*, vol. 45, pp. 779–785, 4 2007.

[10] F. Collier, "An overview of recent subsonic laminar flow control flight experiments," *23rd Fluid Dynamics, Plasmadynamics, and Lasers Conference*, 1993, _eprint: https://arc.aiaa.org/doi/pdf/10.2514/6.1993-2987. [Online]. Available: https://arc.aiaa.org/doi/abs/10.2514/6.1993-2987

[11] S. R. Koh, W. Schröder, and M. Meinke, "Numerical study of noise reduction via wall turbulence control," *16th AIAA/CEAS Aeroacoustics Conference*, 2010, _eprint: https://arc.aiaa.org/doi/pdf/10.2514/6.2010-3990. [Online]. Available: https://arc.aiaa.org/doi/abs/10.2514/6.2010-3990

[12] A. Faszer, T. Hynes, C. Blaabjerg, and H.-C. Shin, "Acoustic beamforming and holography measurements of modified boundary layer trailing-edge noise," *12th AIAA/CEAS Aeroacoustics Conference (27th AIAA Aeroacoustics Conference)*, 2006, _eprint: https://arc.aiaa.org/doi/pdf/10.2514/6.2006-2566. [Online]. Available: https://arc.aiaa.org/doi/abs/10.2514/6.2006-2566

[13] M. Szoke and M. Azarpeyvand, "Active flow control methods for the reduction of trailing edge noise," *23rd AIAA/CEAS Aeroacoustics Conference*, 2017.

[14] A. Wolf, "Reduktion des Hinterkantenlärms von Windenergieanlagen mittels aktiver Strömungskontrolle," Ph.D. dissertation, Universität Stuttgart, 2016.

[15] B. Arnold, "Wind turbine noise reduction by means of boundary layer suction," Ph.D. dissertation, 2019.

[16] A. Wolf, T. Lutz, W. Würz, E. Krämer, O. Stalnov, and A. Seifert, "Trailing edge noise reduction of wind turbine blades by active flow control," *Wind Energy*, vol. 18, pp. 909–923, 5 2015.

[17] M. Kamruzzaman, "Study of turbulence anisotropy and its impact on flow induced noise emission," Ph.D. dissertation, University of Stuttgart, 2012.

[18] C. Hornung, T. Lutz, and E. Krämer, "A model to include turbulence-turbulence interaction in the prediction of trailing edge far field noise for high angles of attack or slightly separated flow," *Renewable Energy*, vol. 136, pp. 945–954, 6 2019.

[19] W. K. Blake, *Mechanics of Flow-Induced Sound and Vibration Volumes I and II.* Academic Press Inc., 1986.

[20] A. Herrig, "Validation and application of a hot-wire based method for trailing-edge noise measurements on airfoils," Ph.D. dissertation, Universität Stuttgart, 2012.

[21] A. Herrig, W. Würz, T. Lutz, and E. Krämer, "Trailing-edge noise measurements using a hot-wire based coherent particle velocity method," *Collection of Technical Papers - AIAA Applied Aerodynamics Conference*, vol. 4, pp. 2605–2616, 2006.

[22] N. Kroll, C.-C. Rossow, D. Schwamborn, K. Becker, and G. Heller, "Megaflow - a numerical flow simulation tool for transport aircraft design," *ICAS Congress*, 2002.

[23] A. Hatzipanagiotou, "Implementierung einer Aktuatorrandbedingung in den Strömungslöser FLOWer," p. 86, 2012.

[24] T. Knopp, "The actuation boundary condition for flow control in the DLR TAU Code," 2010.

[25] R. Parchen and K. Looijmans, "Progress Report DRAW," 1996.

[26] R. Parchen, "Progress Report DRAW: a prediction scheme for trailing edge noise based on detailed boundary layer characteristics," 1998.

[27] D. M. Chase, "Noise radiated from an edge in turbulent flow," *AIAA Journal*, vol. 13, pp. 1041–1047, 1975. [Online]. Available: https://doi.org/10.2514/3.60502

[28] T. Brooks and T. Hodgson, "Trailing edge noise prediction from measured surface pressures," *J. Sound Vib.*, vol. 78, pp. 69–117, 1981.

[29] H. Schlichting, *Boundary-Layer Theory.* McGraw Hill, 1979. [Online]. Available: https://royalsocietypublishing.org/doi/10.1098/rspa.1938.0032

[30] G. I. Taylor, "The spectrum of turbulence," *Proceedings of the Royal Society of London. Series A - Mathematical and Physical Sciences*, vol. 164, pp. 476–490, 2 1938. [Online]. Available: https://royalsocietypublishing.org/doi/10.1098/rspa.1938.0032

10th International Conference
on
Wind Turbine Noise
Dublin – 21st to 23rd June 2023

Analysis of Mitigation Measures for Wind Turbine Noise Annoyance – A field experiment in the interdisciplinary project Inter-Wind

Hübner, Gundula – Martin-Luther-University Halle-Wittenberg, MSH Medical School Hamburg: gundula.huebner@psych.uni-halle.de
Müller, Florian Johannes Yanic – MSH Medical School Hamburg: florian.mueller@medicalschool-hamburg.de
Pohl, Johannes – Martin-Luther-University Halle-Wittenberg: johannes.pohl@psych.uni-halle.de

Summary

Due to the growing distribution of renewable energies, increasingly, people live in the vicinity of wind turbines (WTs). In this context, acceptance conflicts can arise due to annoyance caused by WT immissions. In the interdisciplinary project Inter-Wind on WT noise, objective measurement data were collected synchronously with subjective assessments of residents at a wind farm with three WTs in southern Germany. After strongly annoyed residents were identified in a survey, measurements were carried out at the WTs and in the municipality. Meteorological, acoustic, and ground motion measurement data as well as operational parameters of the wind farm were collected, while residents (n = 46) reported and described annoying WT noise using an app. Constantly high rotational speeds (rpm) as well as high variability in rpm, were associated with annoyance. To address these rotational patterns, mitigation measures with three different operational modes were tested in a following field experiment. During the experimental period, 36 residents used the app to log their noise perception every night. Compared to periods of normal operation (one WT noise reduced) noise reduction of all three WTs did not result in lower levels of sound perception and noise annoyance.

1. Wind turbine sounds – Annoyance and the need for tailored mitigation measures

Wind energy seems to be one of the most important pillars of the energy transition. Annoyance of wind turbines (WTs) noise, whether expected or experienced, is one of the major acceptance factors of residents' acceptance (Hübner et al., 2019; Pawlaczyk-Łuszczyńska et al., 2018; Pohl et al., 2021). Worries about the effects of noise immissions can initiate negative social dynamics early on and can result in long-term opposition (Baxter et al., 2013; Songsore & Buzzelli, 2014). Consequently, the understanding of how WT sounds become annoying noise is of high relevance. Over the last decade, this relevance

is mirrored in the increasing research on WT immissions and their impact on residents (e.g., Hübner et al., 2019; Pohl et al., 2021; Bakker et al., 2021; Jalali et al., 2016; Hansen et al., 2021; Michaud et al., 2016a,b,c; Poulsen et al, 2018a,b; Turunen et al., 2021a). Furthermore, to help residents, who are or may be negatively affected in the future, the understanding of annoyance needs to be applied to implement and evaluate measures that are able to reduce annoyance. Additionally, if specific noise annoying situations can be detected, tailored mitigation measures can be developed. A win-win situation could result: less annoyance for residents, higher WT output due to less general reductions.
To detect noise impacts of WTs, epidemiological, laboratory and field studies are conducted. Epidemiological studies analyse if WTs do have negative effects on residents, investigating if diseases, e.g., cardiovascular diseases or diabetes, occur more often in the vicinity of WTs (e.g., Poulsen et al, 2018a,b; Turunen et al., 2021b). These studies are important to understand stress effects of WTs, yet they do not allow conclusions about the circumstances, which lead to noise annoyance and symptoms. To better understand specific situation which lead to noise perceptions, recent studies combined written noise diaries with recordings of sound pressure levels (SPL), wind speed and direction (e.g., Hansen et al., 2021) or a smartphone app with calculated SPL, simulated meteorological data and operating data of the WTs (e.g., Søndergaard et al., 2021). These studies find associations between outdoor SPL and annoyance, but with large variations of SPL for similar levels of annoyance. Other factors, like power output or wind speed are added, in order to explain annoyance better. Pohl et al. (2020) simultaneously assessed acoustic, ground motion and meteorological measurements. They did not find a correlation between SPL and annoyance, rather they found that annoyance was primarily correlated with residents' attitude to wind energy or their perception of fairness of the planning process. Hübner et al. (2022) provided residents an app where they could report annoying sounds, while, simultaneously, measurements of acoustics, ground motions, and meteorological conditions were performed as well as operational parameters of the wind farm were assessed. They found that noise reports predominantly coincided with two rotational patterns: a constantly high rotation rate (10–12.5 rpm) and a high variability of the rotation rate. Based on this finding, the joint research group of the interdisciplinary project Inter-Wind derived mitigation measures–tailored to the specific annoying situations. These mitigation measures were implemented in a field experiment at a wind farm in southern Germany. Three different operation conditions of the wind farms' three WT were run. The present report presents the subjective assessments–how strongly annoyed residents (Müller et al., 2023) perceived the wind turbine sounds depending on the three mitigation operation conditions.

2. Method

2.1 Location and Participants

The location was the municipality Kuchen, in a rural, mountainous area of southern Germany. The local wind farm Tegelberg consists of three WTs (hub height 139 m, rotor diameter 120 m, 2.78 MW rated power). Overall, 148 residents of the wind farm were recruited for interviews in 2020 (mean age M = 62.55 years; SD = 11.72, range = 24–83 years, see Müller et al., 2023). A larger proportion of males participated (68.6 %). None of the participants received financial benefits from the local wind farm or was working in the wind energy industry. Out of the total sample, a smaller proportion of 36 participants agreed to use a real-time noise report app for the present study, an even smaller group (n = 20) actually did so. While, descriptively, these app users had higher levels of general WT noise annoyance (M = 2.45, SEM = 0.28) than the other interviewees (M = 1.83, SEM = 0.15, n = 128), this difference was not statistically significant ($t(30.71) = 1.99$, $p = .055$, Cohen's d = 0.27, small effect size).

2.2 Procedure and Assessments

The chosen mitigation measures were based on the earlier finding that residents were predominantly annoyed when at least one WT ran at a constant high rotational speed (10–12.5 rpm) or there was a high variability of rotational speed (Hübner et al., 2022). To address these patterns, during a period of 12 weeks from November 2022 to February 2023 mitigation measures with three different modes of operation were tested at the wind farm. During normal operation (NO) one WT ran in a noise reduced mode (2.64 MW) between 10 pm and 6 am, while the other two WTs were operated with a rated power of 2.78 MW. The first mitigation measure (noise reduced operation 1, NRO1) reduced the variability between the three WTs by reducing the rated power of all three WTs to 2.64 MW for the whole day. The second measure reduced the rated power even further down to 2.28 MW for all three WTs for the whole day (noise reduced operation 2, NRO2). All three modes of operation were implemented for two 2-week time periods, each. See Table 1 for the order of operational modes.

Table 1. Order of operational modes during the field experiment. NO = normal operation, NRO1 noise reduced operation 1, NRO2 = noise reduced operation 2, week = experimental week.

Week (year)	1/2 (2022)	3/4 (2022)	5–8 (2022(23)	9/10 (2023)	11/12 (2023)
Mode	NRO2	NRO1	NO	NRO2	NRO1

To evaluate the effectiveness of these measures, participants were provided an app and asked to report every night before they went to bed whether they heard WT sounds, and, if so, how annoying they perceived them to be (on a 5-point scale from 0 "not at all annoyed" to 4 "very annoyed"). Additionally, they could report at any other time when they heard WT sounds. For each noise report six 10 min intervals of the operational parameters were analysed, in order to describe the hour leading up to the report.

3. Results

During the test period, 21 app users made 417 reports. In order to ensure accuracy of the reports, any report for which the time of app usage differed more than two hours from the time the report is referring to, were filtered out, resulting in 345 reports by 20 users. Almost half of the reports (47.8 %) were about situations when WT sounds were heard. When WT sounds were heard, residents were on average somewhat annoyed (M = 2.21, SEM = 1.07).

3.1 Comparison of operational modes

As expected, overall, WT sounds were perceived significantly less frequently under the experimental noise reduced conditions (NRO1: 47.9 %, NRO2: 31.2 %) in comparison to the normal operation (NO: 74.4 %). Also, when WT sounds were perceived the average level of noise annoyance was significantly lower in the most rpm reduced condition NRO2 (M = 1.79, SEM = 0.16) in comparison to the NRO1 (M = 2.29, SEM = 0.14; p = 0.009, d = 0.46, small effect size) and the NO (M = 2.41, SEM = 0.13; p =.002, d = 0.60, medium effect size). The NRO1 and NO conditions appeared to be similar.

However, when differentiating between the test periods in 2022 and 2023, inconsistent results appeared. For the NRO1 in 2023 and the NO conditions comparable high levels of annoyance were reported. In contrast, during the NRO1 period in 2022 and both NRO2 periods (2022/23) similarly lower annoyance levels were observed–each significantly lower in comparison to the NO (Figure 1; p = .002 to p = .041, d = 0.44 to d = 0.76, small to medium effect sizes).

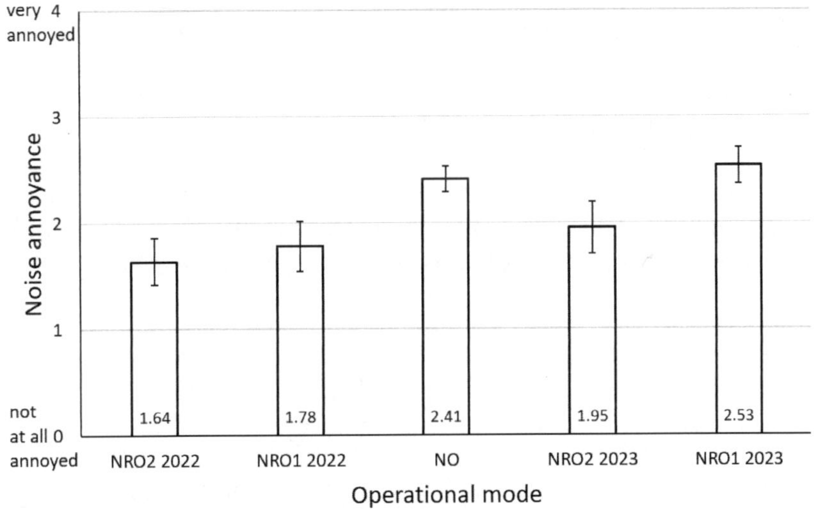

Figure 1. Average level of noise annoyance across the different test periods or operational modes for app reports when WT sounds were heard. NO = normal operation, NRO = noise reduced operation.

Additionally, during the NRO1 in 2023 and the NO a comparatively high percentage of reports with audible WT sounds were observed (Table 2).

Table 2. Number of reports with and without WT sound, across the different test periods of operational modes. NO = normal operation, NRO1 noise reduced operation 1, NRO2 = noise reduced operation 2.

	NO 2022/23	NRO1 2022	NRO1 2023	NRO2 2022	NRO2 2023	Total
No sound	22 (25.6 %)	45 (71.4 %)	18 (31.0 %)	64 (74.4 %)	31 (59.6 %)	180 (52.2 %)
WT sound	64 (74.4 %)	18 (28.6 %)	40 (69.0 %)	22 (25.6 %)	21 (40.4 %)	165 (47.8 %)

To analyse the inconsistent finding we checked the rotation rates of the rotor blades (rotations per minute, rpm) and wind speed. First, the daily rotation rates fluctuate more strongly across the test periods than could be explained by the power reduction alone during the mitigation conditions (Figure 2). Specifically, the NO reached full-load operation regularly, but most NRO periods (reduction to approximately 12 rpm in NRO1, and 11 rpm in NRO2) did not. Only during the first week of NRO1 in 2023 high wind speeds led to a period with (reduced) full-load operation–and comparable annoyance patterns as in the NO (Figure 1).

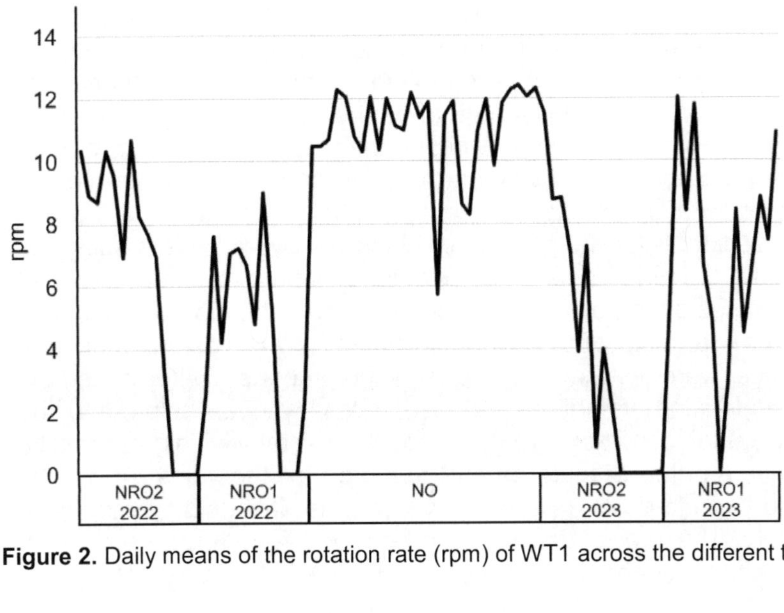

Figure 2. Daily means of the rotation rate (rpm) of WT1 across the different test periods.

Second, rpm and wind speed were analysed for the 60 minutes preceding app reports of WT sounds (based on six intervals of ten minutes). We found that the mean rotation rate was higher during NO than in all NRO periods (Table 3, all $p < 0.0001$, all $d \geq 0.97$, large effect sizes). The rotation rate of NRO1 in 2023 was also significantly higher than in the other three NRO conditions ($p < .0001$ or $p = .044$, $d = 0.45$ to $d = 1.27$, small to large effect sizes). At first glance, that the two periods with higher rpm showed the highest average annoyance seemed to corroborate the assumption of a relevant relation between rpm and annoyance. However, a small correlation between the two measures contradicted the assumption–being too small to be relevant ($r = .27$). No relevant correlations with annoyance were found for pitch angle ($r = –.24$) and wind direction ($r = .22$) either.

Even more, when considering the wind speed, we found empirical evidence for a substantial relation between wind speed and annoyance ($r = .47$). Further, counterintuitively in the NRO1 in 2023 the highest wind speeds at nacelle height were observed in comparison to all other conditions, even to the NO condition (Table 3; all $p \leq .001$, $d = 0.65$ to $d = 1.93$, medium to large effect sizes). This difference led us to control the influence of wind speed as a covariate in an analysis of variance. When controlling for the impact of wind speed on annoyance, we could not confirm the assumed main effect of the NRO conditions: The average annoyance in each test condition converged towards a range between $M = 2.04$ to $M = 2.31$, without significant differences (pairwise comparisons). In other words, when controlling for wind speed the differences between the test conditions disappeared.

Finally, we checked for a relation between annoyance and set back distance. The correlation observed was irrelevant ($r = .02$), corroborating earlier findings (e.g., Hübner et al., 2019). Overall, the findings provided evidence for the importance of wind speed over the operation conditions to explain WT noise annoyance in the observed mountainous test area.

Table 3. Mean rotation rate (rpm) and mean wind speed of all 3 WTs during 60 minutes preceding the annoyance time when hearing WT sound (M, SEM). NO = normal operation, NRO1 noise reduced operation 1, NRO2 = noise reduced operation 2.

	NO 2022/23	NRO1 2022	NRO1 2023	NRO2 2022	NRO2 2023	Total
rpm	11.95 (0.32)	5.73 (0.60)	9.65 (0.40)	8.28 (0.54)	5.32 (0.56)	9.38 (0.28)
wind speed [m/s]	9.20 (0.33)	6.45 (0.63)	10.97 (0.42)	5.74 (0.57)	6.38 (0.58)	8.51 (0.25)

4. Summary

In the present study a field experiment to evaluate the effectiveness of two modes of noise reduced operation on residents' noise annoyance was described. These two modes aimed to reduce the variability of rotational speeds between the three WTs of the wind farm by aligning the reduction of the rated power across all WTs. Unfortunately, over long periods of the test period low wind speeds led to rotation rates often not reach full-load operation. The times with fast and slow rotation rates were distributed unequally, so that during periods of noise reduced operation the reduction rarely came into effect. One test period with noise reduced operation was, however, characterised by fast wind speeds and higher rotation rates. Comparing this test period with the normal mode of operation (NO) did not reveal any differences with regard to the frequency of sound perception by residents or their levels of noise annoyance. This suggests that reducing the power output of all WTs did not yield any additional benefit over NO. This NO included a power reduction at night times of one of the WTs, already before the start of this study. Instead, the wind speed explained the amount of annoyance, partly. So far, it remains unclear which effects related to annoyance were evoked by higher wind speeds. Different sound characteristics (e.g., amplitude modulation) might be induced, possibly even related to the landscape morphology. Specific weather conditions (e.g., atmospheric stability, high humidity, fog, wind differences between valley and top of escarpment) may add explanations. Additional analyses as well as better tailored experimental conditions, allowing to fit reduction modes to weather conditions more accurately. The present research offers an interesting approach to reduce annoyance. But, further research is needed before deriving generalizable results.

5. Acknowledgements

We thank the local authorities of Kuchen for their support, as well as the Stadtwerke Schwäbisch Hall, the KWA Contracting AG for WT access and data. Furthermore, we acknowledge the support of local residents, and thank Esther Blumendeller, Laura Gaßner, and Maayen Wigger for support in conducting the experiment and data analysis. The presented data are part of the interdisciplinary project Inter-Wind, supported by the Federal Ministry for Economic Affairs and Climate Action on the basis of a decision by the German Bundestag (grant 03EE2023A-D).

References

Ascone, L., Kling, C., Wieczorek, J., Koch, C., & Kühn, S. (2021). A longitudinal, randomized experimental pilot study to investigate the effects of airborne infrasound on human mental health, cognition, and brain structure. *Scientific Reports*, 11(1), 1–9.

Bakker, R. H., Pedersen, E., van den Berg, G. P., Stewart, R. E., Lok, W., & Bouma, J. (2012). Impact of wind turbine sound on annoyance, self-reported sleep disturbance and psychological distress. *Science of the Total Environment*, 425, 42–51.

Baxter, J., Morzaria, R., & Hirsch, R. (2013). A case-control study of support/opposition to wind turbines: Perceptions of health risk, economic benefits, and community conflict. *Energy Policy*, 61, 931–943.

Cooper, S. E. (2014). *The results of an acoustic testing program – Cape Bridgewater wind farm*. Melbourne.

Gaßner, L., Blumendeller, E., Müller, F. J. Y., Wigger, M., Rettenmeier, A., Cheng, P. W., ... Pohl, J. (2022). Joint analysis of resident complaints, meteorological, acoustic, and ground motion data to establish a robust annoyance evaluation of wind turbine emissions. *Renewable Energy*, 188, 1072-1093. doi:10.1016/j.renene.2022.02.081

Hansen, K. L., Nguyen, P., Micic, G., Lechat, B., Catcheside, P., & Zajamšek, B. (2021). Amplitude modulated wind farm noise relationship with annoyance: A year-long field study. *The Journal of the Acoustical Society of America*, 150(2), 1198–1208.

Hübner, G., Müller, F. J. Y., & Pohl, J. (2022, October 24–28). Inter-Wind - an Interdisciplinary Analysis of Wind Turbine Noise. Paper presented at the 24th International Conference on Acoustics, Gyeongju, Korea.

Hübner, G., Pohl, J., Hoen, B., Firestone, J., Rand, J., Elliott, D., & Haac, R. (2019). Monitoring annoyance and stress effects of wind turbines on nearby residents: A comparison of US and European samples. *Environment International*, 132, 105090. doi:10.1016/j.envint.2019.105090

Jalali, L., Bigelow, P., Nezhad-Ahmadi, M.-R., Gohari, M., Williams, D., & McColl, S. (2016). Before–after field study of effects of wind turbine noise on polysomnographic sleep parameters. *Noise & Health*, 18(83), 194–205.

Michaud, D. S., Feder, K., Keith, S. E., Voicescu, S. A., Marro, L., Than, J., ... Villeneuve, P. J. (2016a). Self-reported and measured stress related responses associated with exposure to wind turbine noise. *The Journal of the Acoustical Society of America*, 139(3), 1467–1479.

Michaud, D. S., Feder, K., Keith, S. E., Voicescu, S. A., Marro, L., Than, J., ... Bower, T. (2016b). Exposure to wind turbine noise: Perceptual responses and reported health effects. *The Journal of the Acoustical Society of America*, 139(3), 1443–1454.

Michaud, D. S., Keith, S. E., Feder, K., Voicescu, S. A., Marro, L., Than, J., ... Lavigne, E. (2016c). Personal and situational variables associated with wind turbine noise annoyance. *The Journal of the Acoustical Society of America*, 139(3), 1455–1466.

Müller, F. J. Y., Leschinger, V., Hübner, G., & Pohl, J. (2023). Understanding subjective and situational factors of wind turbine noise annoyance. *Energy Policy*, 173, 113361. doi:10.1016/j.enpol.2022.113361

Pawlaczyk-Łuszczyńska, M., Zaborowski, K., Dudarewicz, A., Zamojska-Daniszewska, M., & Waszkowska, M. (2018). Response to noise emitted by wind farms in people living in nearby areas. *International Journal of Environmental Research and Public Health*, 15(8), 1575.

Pedersen, E., Van Den Berg, F., Bakker, R., & Bouma, J. (2009). Response to noise from modern wind farms in The Netherlands. *The Journal of the Acoustical Society of America*, 126(2), 634–643.

Pohl, J., Gabriel, J. & Hübner, G. (2014). *Untersuchung der Beeinträchtigung von Anwohnern durch Geräuschemissionen von Windenergieanlagen und Ableitung übertragbarer Interventionsstrategien zur Verminderung dieser.* Halle (Saale): Institut für Psychologie der Martin-Luther-Universität Halle-Wittenberg.

Pohl, J., Gabriel, J., & Hübner, G. (2018). Understanding stress effects of wind turbine noise–The integrated approach. *Energy Policy*, 112, 119–128.

Pohl, J., Hübner, G., Liebig-Gonglach, M., & Hornberg, C. (2020). *Verbundvorhaben: Objektive Kriterien zu Erschütterungs- und Schallemissionen durch Windenergieanlagen im Binnenland (TremAc); Schlussbericht zu den Teilvorhaben Umweltpsychologische Analyse der Windenergie-Immissionswirkungen auf Akzeptanz und Wohlbefinden der Anwohner (MLU) und Umweltmedizinische Analyse der Wirkung von Windenergieanlagen auf Gesundheit und Wohlbefinden von Anwohnern/innen (UBI)*. Halle (Saale) und Bielefeld: Institut für Psychologie, Arbeitsgruppe Umwelt- und Gesundheitspsychologie, Martin-Luther-Universität Halle-Wittenberg und Fakultät für Gesundheitswissenschaften, Arbeitsgruppe 7 Umwelt & Gesundheit, Universität Bielefeld.

Pohl, J., Rudolph, D., Lyhne, I., Clausen, N.-E., Aaen, S. B., Hübner, G., ... Kirkegaard, J. K. (2021). Annoyance of residents induced by wind turbine obstruction lights: A cross-country comparison of impact factors. *Energy Policy*, 156, 112437.

Poulsen, A. H., Raaschou-Nielsen, O., Peña, A., Hahmann, A. N., Nordsborg, R. B., Ketzel, M., ... Sørensen, M. (2018a). Long-term exposure to wind turbine noise at night and risk for diabetes: A nationwide cohort study. *Environmental Research*, 165, 40–45.

Poulsen, A. H., Raaschou-Nielsen, O., Peña, A., Hahmann, A. N., Nordsborg, R. B., Ketzel, M., ... Sørensen, M. (2018b). Short-term nighttime wind turbine noise and cardiovascular events: A nationwide case-crossover study from Denmark. *Environment International*, 114, 160–166.

Schäffer, B., Pieren, R., Schlittmeier, S. J., & Brink, M. (2018). Effects of different spectral shapes and amplitude modulation of broadband noise on annoyance reactions in a controlled listening experiment. *International Journal of Environmental Research and Public Health*, 15(5), 1029. doi:10.3390/ijerph15051029

Schäffer, B., Pieren, R., Hayek, U. W., Biver, N., & Grêt-Regamey, A. (2019). Influence of visibility of wind farms on noise annoyance–A laboratory experiment with audio-visual simulations. *Landscape and Urban Planning*, 186, 67-78. doi:10.1016/j.landurbplan.2019.01.014

Schäffer, B., Schlittmeier, S. J., Pieren, R., Heutschi, K., Brink, M., Graf, R., & Hellbrück, J. (2016). Short-term annoyance reactions to stationary and time-varying wind turbine and road traffic noise: A laboratory study. *The Journal of the Acoustical Society of America*, 139(5), 2949-2963. doi:10.1121/1.4949566

Smith, M. G., Ögren, M., Thorsson, P., Hussain-Alkhateeb, L., Pedersen, E., Forssén, J., ... Persson Waye, K. (2020). A laboratory study on the effects of wind turbine noise on sleep: results of the polysomnographic WiTNES study. *Sleep*. doi:10.1093/sleep/zsaa046

Søndergaard, L. S., Volk, C. P., Hansen, T. R., Enggaard, L., Sørensen, T., & Peña, A. (2021). Wind farm neighbourship investigated by a daily app questionnaire combining weather, noise, and annoyance. Paper presented at the 9th International Conference on Wind Turbine Noise, Remote from Europe.

Songsore, E., & Buzzelli, M. (2014). Social responses to wind energy development in Ontario: The influence of health risk perceptions and associated concerns. *Energy Policy*, 69, 285–296.

Turunen, A. W., Tiittanen, P., Yli-Tuomi, T., Taimisto, P., & Lanki, T. (2021a). Self-reported health in the vicinity of five wind power production areas in Finland. *Environment International*, 151, 106419.
Turunen, A. W., Tiittanen, P., Yli-Tuomi, T., Taimisto, P., & Lanki, T. (2021b). Symptoms intuitively associated with wind turbine infrasound. *Environmental Research*, 192, 110360.

10th International Conference
on
Wind Turbine Noise
Dublin – 21st to 23rd June 2023

Practical application of standard 61400-11 : lessons learned on a case study

Fabrice JUNKER, EDF-Renewables, France: fabrice.junker@edf-re.fr
Thomas QUILLET, EDF-SA, France: thomas.quillet@edf.fr

Summary

This paper describes a measurement campaign to test the sound power level of a wind turbine following precisely the IEC 61400-11 standard. In a first step, an IEC compliant analysis is done. This is followed by a discussion on the practical implications and choices of the IEC methodology.

1. Introduction

Developing competitive wind power projects needs to master every kind of uncertainties and margins. Among them, the margin taken by wind turbine supplier on the sound power level data can be an optimization cursor. Especially, post-installation sound power testing can reveal margins that can lead to revise the curtailment plans.

IEC 61400-11 [1] is the international standard for wind turbines Sound Power Level (SPL) measurement and is used world-wide. Though, very few optimization tests concerning the Wind Turbines SPL are done during the commissioning of a new project and there isn't much feedback available about the pros and cons of strictly following the standard guidance to first check the acoustical emission of the turbines.

After a summary of the IEC requirements, the measurement campaign is described and the results of the tests are shown. A discussion on the results and methodology is presented in the last part of the paper.

2. Main requirements of the IEC standard

A summary of the main requirements (non-exhaustive list) of the IEC standard is given below:

1. One acoustic measurement is a 10 second recording, measured downwind at a distance equal to the tip height of the turbine. The sound level meter has to be located downwind of the turbine, with a tolerance of +/- 15° of the incident wind direction angle. This requires the use of several measuring devices to cover a wide range of incident winds;
2. For each 1 m/s wind speed bin, the sound level (L_{Aeq}) is defined as the energetic average of at least 10 measurements recorded at this wind speed. This is then used as an input to calculate the sound power Lw of the wind turbine for this wind speed bin;
3. A noise level L_{Aeq} measured during the operation of the wind turbine is only valid when it is at least 3 dB above the background noise (including all the noise around the tested turbine). A result with an emergence of only 3 to 6 dB is considered

degraded but still tolerable. The emergence is calculated as the difference between L_{Aeq} and the background noise, provided that the latter is measured immediately after or before the measurement, under similar operating and wind conditions;

4. During the test, the wind speed is determined:
    a. Using the supplier's power curves when the wind turbine is in operation and the supplier's power curve has sufficient resolution for the relevant wind range. If the second condition is not met, the anemometer in the nacelle of the wind turbine can be used;
    b. By direct measurement on a weather mast of at least 10 metres for residual noise measurements. This mast should be located outside the area disturbed by the wind turbine.

3. Measurement campaign

The measurements were done on one single turbine on a six turbine site. The selection of the tested turbine is a compromise between the representativeness of the turbine and its location that ensure a lower contribution of the other turbines at the recording point.

3.1 Measurement locations

Seven sound level meters were located on a semicircle opposite to the prevailing south west (SW) wind direction, at a distance of 179.5m (tip height) from the wind tested turbine.
In order to avoid the disturbance area due to the wind turbine for all wind directions covered by the instrumentation, the weather mast was placed outside the area of influence of the wind turbine at a distance of 131m, at an angle of 22.5° S-SW towards the wind turbine.
However, due to the changes of the meteorological conditions during the test, three measurement points had to be moved to the E-NE sector. The wind speed measurements for the 345-65° wind direction during the shutdown phases do not comply with the IEC requirement as the measuring mast is located in the area of influence of the wind turbine. For those samples, the measurements from the nacelle anemometer are used instead.
Figure 1 below shows the location of the various measurements. The relative angle of the acoustic measurements is given according to the incident wind (for example, the BK10 sound level meter is positioned downwind of the wind turbine for an incident wind of 308°)

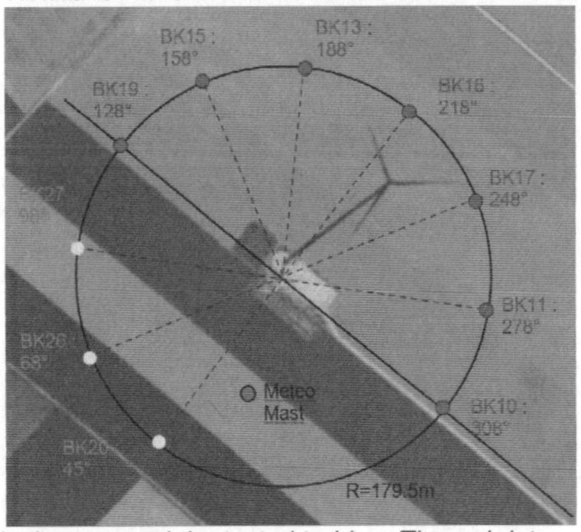

Figure 1: Location of the recording devices around the tested turbine. The red dots are the 7 initially positioned measuring devices considering a SW prevailing wind direction. The yellows dots are the measuring devices moved during the campaigns as the main wind direction as changed to E-NE. The blue dot is the meteo mast.

3.2 Background noise

Background noise measurements already carried out on site together with Code_TYMPAN [2] calculations are used to reach an optimum between the signal-to-noise ratio predicted for the tests and the loss of production associated with the necessary machine downtime. Thus, all the measurements have been done only during nights and only one of the closest turbine of the tested one was stopped during the tests. The closest turbine which was stopped during the recording was the one located upwind the tested one. This pre-calculation stage therefore made it possible to optimise the quality of the measurements and the impact of the shutdown of disturbing turbines on the production loss.

3.3 Measurement campaign duration

In order to ensure that a sufficient number of measurement samples covering sufficient wind speed ranges were obtained, the installation was left in place for 10 days, due to the quite low wind speed conditions observed during the campaign.
The number of sample recorded in each wind bin for both total and background noise measurement is given in Table 1.

4. Data analysis

4.1 Recorded samples

Figure 2 shows the $L_{Aeq,10s}$ recorded both on operation and downtime periods.

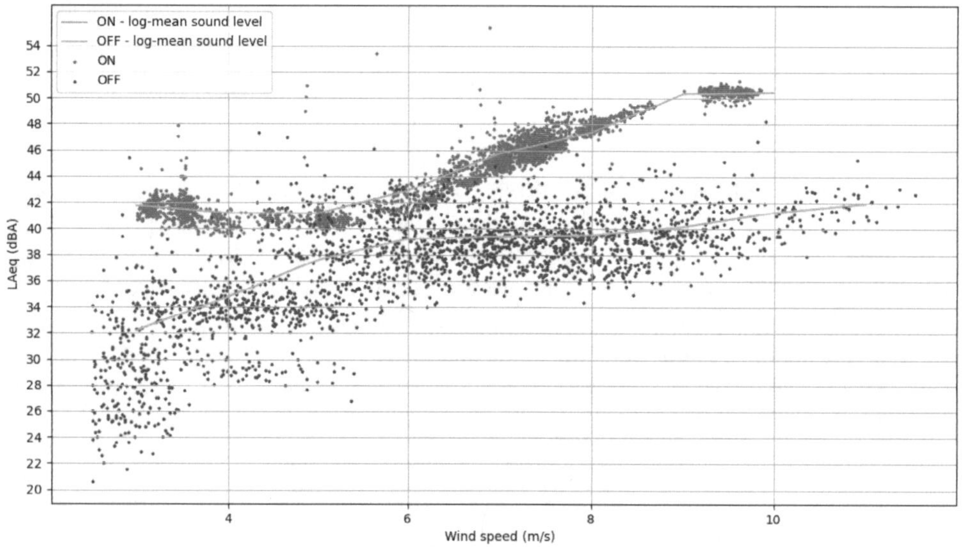

Figure 2: $L_{Aeq,10s}$ recorded sample for both turbine and background noise

4.2 Particular Noise Calculation

The samples shown on Figure 2 are grouped into consistent wind classes.

Table 1 shows the Total Noise, Background Noise and Particular Noise (background corrected noise level), obtained after the following calculations:

- Offset on the spectral data to get the global noise level (eq. (8) of the standard [1]);
- Logarithmic mean of the samples per wind band (eq. (9) of standard [1]) :
- Alignment of the acoustic measurements to the centre value of each wind band, for each frequency band (eq. (20) of the standard [1]);
- Calculation of the particular noise or background corrected noise level, for each frequency band, according to eq. (23) of the standard [1];
- Calculation of the overall levels by summation of the sound levels per frequency band;

- Verification of the emergence of the Total Noise with respect to the Residual Noise (> 6 dB or > 3 dB at least, (c.f. Figure 3);
- Calculation of the associated uncertainties, for each frequency band, according to equations (22), (24), (25) of [1]). The standard uncertainties of the measuring device u_B2, as defined in Annex B of standard [1], are taken from Table 3 of standard [3]. The standard uncertainty of the wind measurement via the power curve is 0.2 m/s according to Annex C of the standard [1].

The results are shown in Table 1.

Wind Speed [m/s]@hub height	Number of sample (ON/OFF)	Total Noise [dBA]	Residual Noise [dBA]	Sound emergence [dBA]	Particular Noise [dBA]	Uncertainty [dBA]
3	295 / 321	41.6	32.3	9.3	41.1	1.1
4	335 / 289	41.3	34.8	6.5	40.2	1.3
5	247 / 309	41.2	37.5	3.7	39.2 (*)	2.0 (*)
6	251 / 388	43.2	39.2	4	41.2 (*)	1.8 (*)
7	946 / 454	45.9	39.4	6.5	44.8	1.4
8	518 / 393	47.0	39.5	7.5	46.2	1.2
9	164 / 228	50.4	40.1	10.3	49.9	1.0
10	156 / 80	50.4	41.2	9.2	49.9	1.0

Table 1: Particular Noise or Background corrected Noise Level and their associated uncertainties. The (*) indicates the measures where the sound emergence is lower than 6 dB

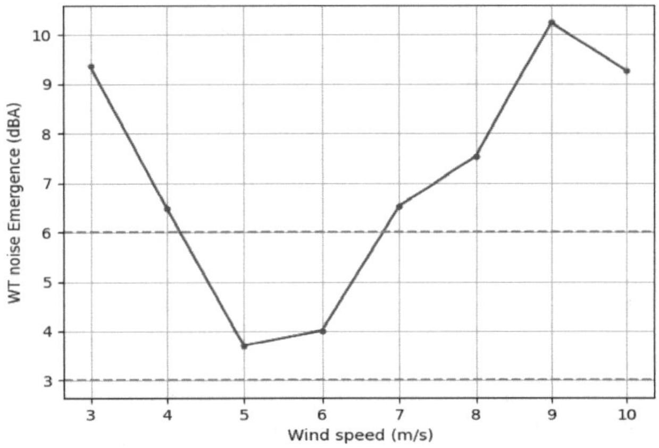

Figure 3 : Sound emergence is below 6 dB at 5 and 6 m/s hub height. This will increase the uncertainty on the particular noise.

4.3 Sound Power Level Calculation

Equation (26) of [1] allows a correction for geometric divergence for the calculation of apparent sound power level (L_{WA}) from the particular noise results. We propose a new correction taking into account geometric divergence and air absorption (see section 5.4 for details). The apparent sound power level resulting of this new correction is labelled "L_{WA} aa". The uncertainty of the apparent sound power level is calculated according to equation (28) of [1].

The final results are given in Table 2 and Figure 4 below.

Wind Speed @10m [m/s]	Manufacturer Data [dBA]	L_{WA} [dBA]	L_{WA_aa} [dBA]	Uncertainty [dBA]
2	/	92.5	*93.9*	1.0
3	94.0	91.1 (*)	*92.6 (*)*	1.7 (*)
4	94.9	92.3 (*)	*93.5 (*)*	1.9 (*)
5	101.2	96.8	*97.8*	1.5
6	104.7	100.5	*101.4*	1.3
7	104.9	101.1	*101.9*	1.1

Table 2: L_{WA} comparison with manufacturer data

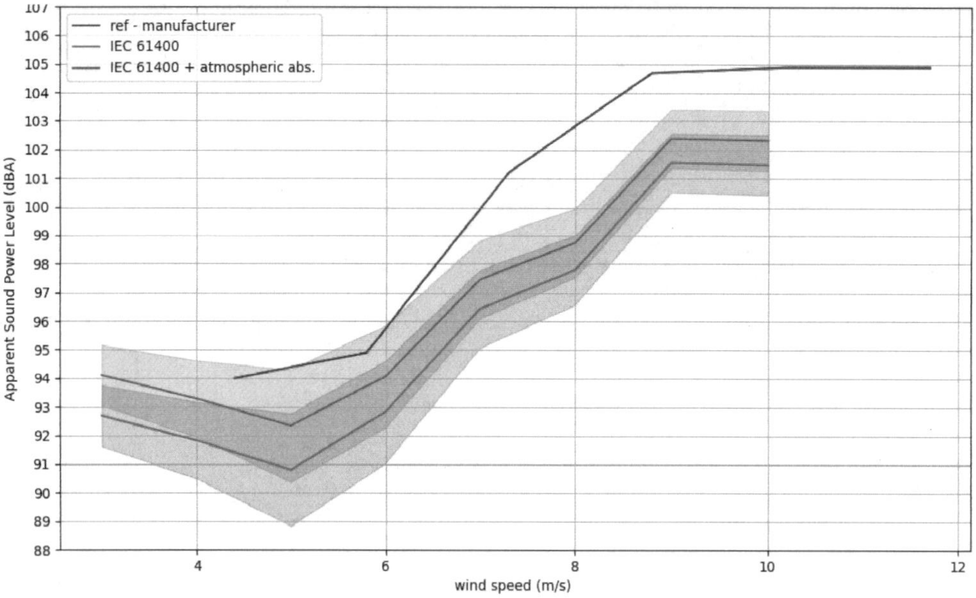

Figure 4 : L_{WA} results : Manufacturer data (blue). IEC compliant L_{WA} value (red). L_{WA} aa using atmospheric absorption correction. Wind speed is given at hub height.

The results show that whatever the assumption on the methodology, the measured maximum L_{WA} is at least 3 dBA lower than the specification of the manufacturer.

5. Discussion

5.1 Practical aspects

The in-situ test has provided sufficient data to calculate the apparent L_{WA} of the tested turbine. Nevertheless, we faced some difficulties in fully complying with the IEC standard, in particular regarding the collection of meteorological data. We propose the following improvements:

1. The test could use meteorological data from the site mast (if available) or from nacelle anemometers of the closest turbine in operation when the tested wind turbine is off. This would avoid the need for 2 to 3 people to set up a meteorological mast. The whole test could be carried out by a smaller team.
2. Over two weeks, only a few well-chosen hours of measurement would have been sufficient. It would be wise to have one person coming at each wind changes to carry out a few short (10 minute) on/off cycles in real time, in agreement with the operation center. The operator would then only need a single sound level meter to be repositioned downwind of the turbine and the duration of the test could be reduced to one day (or night). Moreover, since the measurement plates do not allow the use of rain kits, the test can only be carried out on a clean weather window : reducing the test duration makes it easier to achieve.

3. In carrying out the present test, we have taken care not to generate too much downtime for the operator. In particular, the wind turbine adjacent to the tested one and not located upwind was left in operation. The financial gain of this operation amounts to approximately 2 k€ over a little more than 3 nights of testing. As this value is relatively low, we recommend that all adjacent turbines be shut down to maximise turbine emergence and prioritise the quality of the measurement over a small gain in production, especially if the optimisation of the test duration mentioned in the previous point is implemented.

4. For sites with low emergence, measurements should take place at night in order to maximise the quality of the measurement.

5.2 Particular noise for low signal-to-noise ratio spectral data

For a given third octave band, if the total noise is not at least 3 dB higher than the background noise, the particular Noise calculated as the total noise minus 3 dB. This method can lead to overestimate the particular noise.

Figure 5 below shows the effect of the IEC methodology to calculate the particular noise. The effect here is about +0,4 dBA on the global sound power level and up to 2,6 dB at 500 Hz.

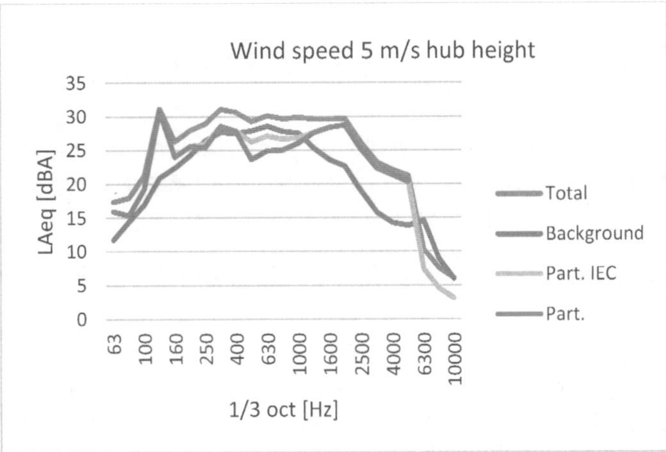

Figure 5: L_{Aeq} spectral data shows the effect of the IEC calculation (Part. IEC) compared to the usual calculation (Part.).

5.3 Logarithmic averaging

The use of a logarithmic average to calculate the average noise per wind block is questionable. This averaging method is frequently used to calculate the average noise, L_{Aeq}, of contiguous samples. However, it no longer makes physical sense when the samples are from disparate recordings. Furthermore, this method tends to favour extreme levels that are often due to background noise variations. On the specific case under study, the difference between the use of L_{Aeq} instead of L_{50A} would decrease the global L_{WA} by a maximum of 0.5dBA which indicates that the global measurements are not especially perturbated by strong extraneous noises sources.

Though, at certain frequencies, inconsistent measurements are easily detected, as shown in Figure 6. The affected noise spectra are over-estimated at these frequencies, resulting in possible errors on the sound power spectra – even if it doesn't affect the global SPL. The use of L_{A50} or at least a statistical filter would solve the problem efficiently.

Figure 6 below shows the measured $L_{Aeq,10s}$ data in the third octave band 6.3 kHz to illustrate the well-known effect of "outliers" on the calculation of L_{Aeq}.

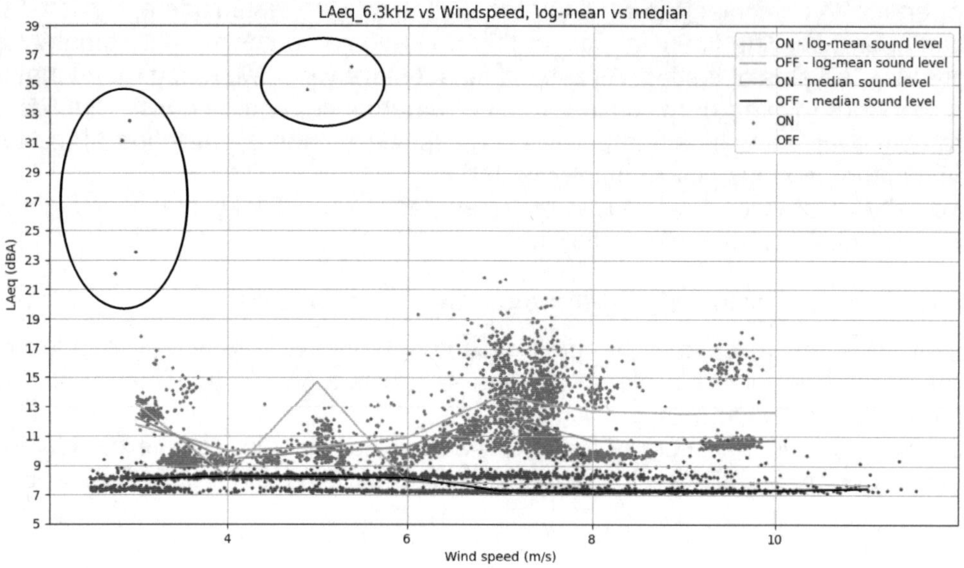

Figure 6: Scatter plot of the collected data at 6,3 kHz showing how outliers can hardly affect the L_{Aeq} values.

The use of the median indicator instead of the logarithmic mean could avoid outliers measurements often due to extraneous noise to increase artificially the noise indicator.

5.4 Simplified propagation model used to calculate the apparent sound power level

Equation (26) of [1] describes the geometric divergence correction to be applied for the calculation of apparent sound power level. The proposed formula is consistent with the standard [4] for the geometrical divergence. However, equation (26) does not includes any correction for atmospheric absorption [5]. As the average distance between the source and the receiver is more than 200m, this attenuation reaches 25 dB for high frequencies (see Figure 7 below).

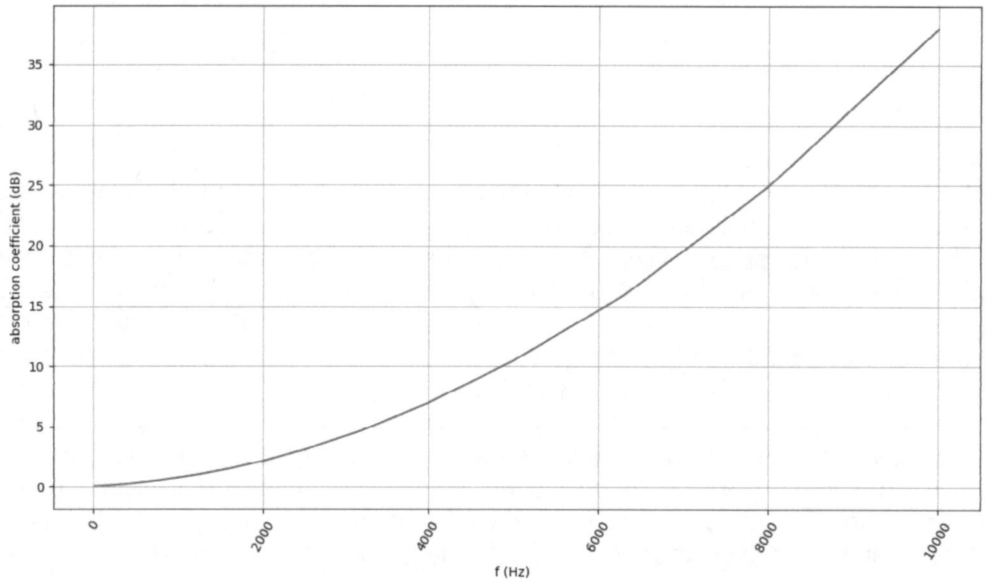

Figure 7: Atmospheric absorption at T=15°C and hr=50%

Below shows the effect of the atmospheric absorption correction on the apparent sound power together with the effect of overestimation due to the correction forced to -3 dBA for low signal-to-noise ratio measurements. The values at frequencies greater than 5 kHz show a non-physical behaviour due to the fact that the correction is applied when particular noise is overestimated as the background noise is too high (see section 5.2).

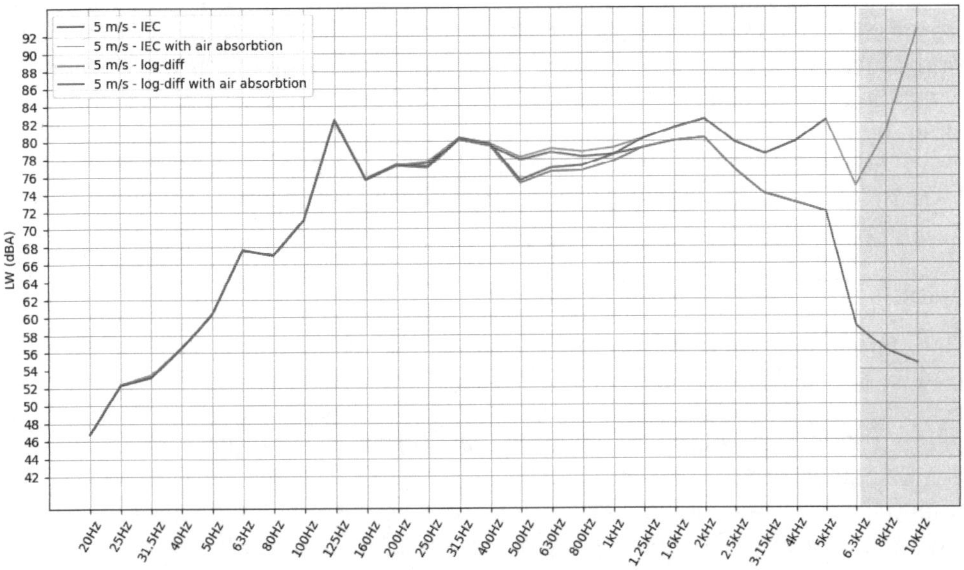

Figure 8: Apparent Sound Power Level spectrum at 5 m/s hub height.
With and without taking the air absorption into account and with and without applying the -3dB IEC correction for low signal-to-noise ration measurements.

The apparent sound power level will be better estimated if the atmospheric absorption correction is used to calculated the apparent sound power level.

5.5 Uncertainties

Due to the large number of samples collected, the statistical uncertainty is very small and the standard uncertainties of the instruments constitute the major part of the calculated uncertainty. More than the bias explained in section 5.2, the flat-rate correction principle is also applied in the uncertainty calculation by fixing the residual noise uncertainty to 3 dBA and logarithmically combining this 3 dBA to the total noise uncertainty ([1] Equation (25)) . This leads to a misestimation uncertainty for L_{WA} for these low signal-to-noise ratio measurements.

Wind Speed @10m [m/s]	Uncertainty IEC [dBA]	Uncertainty No flat-rate [dBA]
2	1.0	1.0
3	1.7 (*)	1.7
4	1.9 (*)	2.1
5	1.5	1.5
6	1.3	1.3
7	1.1	1.1

Table 3: Uncertainties on the L_{WA} depending on the use or not of a fix value of the uncertainty

In this specific study, the difference between the IEC compliant uncertainties calculation and a no flat-rate calculation is not significant : 1.7 and 1.9 dBA respectively at 3 and 4 m/s instead of 1.7 and 2.1 dBA. This is because the biggest uncertainties occur at some frequencies at low levels and because the total uncertainty is obtained by adding the sound level weighted uncertainties in each frequency band (equation 25 of [1]). But, if one imagines a situation where large uncertainties arise at higher noise levels, then the IEC method may underestimate the uncertainty.

5.6 Margins allowed by the measurement results

The curtailment plan designated for this site on the basis of the SPL data specified by the manufacturer results in a loss of production during 3% of the year. A new optimisation can be done allowing considering the new SPL value so that all noise curtailments are now unnecessary. Even if the resulting gain on this case is not so huge, it shows that the cost of the measuring test is largely offset by the gains.
More generally, sound power gains of the order of 2 to 3 dBA have been estimated from the tests conducted. On many machines, these gains make it possible to increase the production capacity of a machine by about 15%.

6. Conclusion

The feedback of this measurement campaign provides some practical lessons learnt to reduce the time and the cost of the sound power level testing of wind turbines.

There are a few shortcoming in the IEC method that can be avoided by:

- using L_{50} instead of L_{Aeq}
- forgetting the flate-rate noise correction for low signal-to-noise ratio measurements ;
- taking into account the air absorption for the calculation of L_{wA}.

Away from the methodology aspects, this example of testing a wind turbine on a specific site as shown that it is worth doing tests in order to evaluate the margins of a project. Even if the IEC standard has some requirements and some methodological simplified assumptions, it can be applied with some adaptation which is enough to get good measurements and evaluate the performance of a turbine on site. This will be done more often on the new EDF-R windfarm projects.

References

[1] *IEC 61400-11 : Acoustic noise measurement techniques,* June 2018.

[2] *Code_TYMPAN- https://sourceforge.net/projects/code-tympan/.*

[3] *Norme CEI 61672-1: « Électroacoustique – Sonomètres – Partie 1: Spécifications »*, 2013.

[4] *ISO 9613-2: Attenuation of sound during propagation outdoors — Part 2: General method of calculation,* 1996.

[5] *ISO 9613-1: Attenuation of sound during propagation outdoors — Part 1: Calculation of the absorption of sound,* 1993.

10th International Conference
on
Wind Turbine Noise
Dublin – 21st to 23rd June 2023

Wind turbine tonality - A holistic approach to its prediction and mitigation

E. Kanpolat, ZF Wind Power, emre.kanpolat@zf.com
S. Schmidt , ZF Wind Power

Summary

Tonal noise in wind turbines is a result of vibrations at discrete frequencies which can be excited by different sources such as gearbox, generator, etc. This structure-borne sound propagates through the drivetrain to finally be radiated from blades, tower or large surfaces of the drivetrain housings. IEC61400-11 defines how to measure such narrow band sounds relative to the broadband noise emitted by the entire wind turbine. If the narrow band noise exceeds the wind turbine masking noise, this is referred to as a tonality and becoming increasingly a hot topic due to certain trends in the wind energy market. On the other hand, reducing the structure-borne noise is becoming increasingly difficult due to a cost pressure, increasing torque density and new integrated powertrain architectures.
This situation requires scanning of a wide design space at an early phase of the project to create risk maps for the powertrain and optimize the components thereof according to wind turbine operation and sensitivities. This paper aims to explain an wholistic approach to evaluating and optimizing tonality performance of its products throughout the design process. Based on example powertrain, a tonality risk assessment is carried out and optimizations on powertrain structures are done to reduce tonality risks.

1. Introduction

With the increased need of sustainable energy solutions and the hence increasing amount of installed wind turbines, wind farms inevitably get closer to inhabited areas. This trend and the necessity of acceptance by the local population lead legislative authorities to release laws limiting the allowed emitted noise from a wind turbine and its sub-components, like for example the powertrain. In wind turbine powertrains vibrations are generated at discrete frequencies e.g. by the meshing of the gears. They propagate through drivetrain and wind turbine and are radiated from the blades, tower or large surfaces of the drivetrain housings. IEC61400-11 is the basis for many national legislations and defines how to measure such narrow band sounds relative to the broadband noise emitted by the entire wind turbine. If the narrow band noise exceeds the wind turbine masking noise, this is referred to as tonality. The continuing trend to decrease the overall rotor size specific sound emitted by wind turbines makes narrow band sounds in relative terms more prominent and hence increases the tonality risk. At the same

time, reducing structure-borne sound originating in the powertrain is becoming increasingly difficult due to a cost pressure, the requirement to increase the torque density, new powertrain architectures and turbine operation control trends (like noise reduced operating curves).
This situation requires scanning of a wide design space at an early phase of the project to create risk maps for the powertrain and optimize the components thereof according to wind turbine operation and sensitivities. This paper aims to explain the holistic approach at ZFWP of evaluating and optimizing tonality performance of its products throughout the design process. For this task, ZF Wind Power is developing a Python based framework (NOVA – Noise and Vibration Analytics) to run automated model building, baseline evaluations, DOEs and tailoring of its products.

2. Tonality free wind turbine and power train design

2.1 Importance of a holistic approach

Certification of onshore wind turbines includes successfully passing a tonality measurement according IEC61400-11. The main drivers that determine the level of tonality are:

- masking noise (mainly determined by the aeroacoustics of the rotor)
- wind turbine torque-speed curves (operating modes)
- gear excitation
- transfer path from gear excitation to the IEC microphone which can be split into:
    - powertrain transfer path
    - transfer paths over tower and blades
- direct airborne radiation of powertrain components

Designing a tonality-free wind turbine requires very close co-operation between all project partners as a precise assessment of the wind turbine characteristics must be done. An approach that links all relevant information into an overall tonality prediction is indispensable. For this purpose, ZF Wind Power has developed a Tonality Free Wind Turbine (**TFWT**) design approach which bases on a close co-operation between development partners [Fig.1]. A tonality prediction is carried out by incorporating

- the gear excitation
- structural transfer path between excitation source and sound emitter
- radiation from tower and blades
- masking noise
- wind turbine operation characteristics

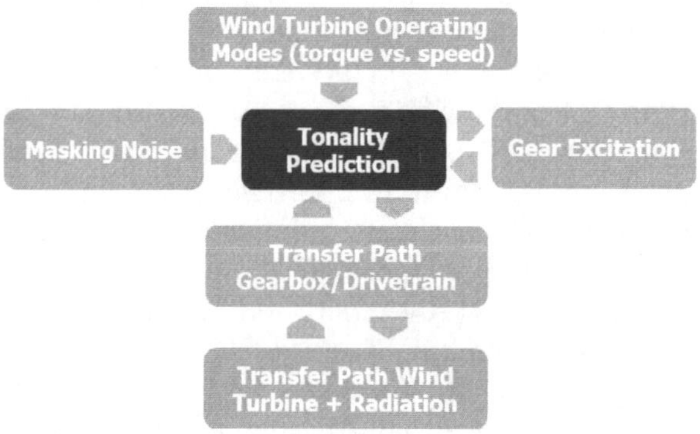

Figure 1 Tonality Prediction - Building Blocks

For each building block, optimization loops are run during the concept and detailed design phases. [SCH19] describes how gear excitations can be optimized with respect to tonalities. Tonality predictions and structural optimizations focusing on powertrain transfer path are carried out using NOVA,.

2.2 Noise and vibration analytics framework (NOVA)

Different computer-aided engineering (CAE) software is used during the TFWT process. Each of the software packages has its own purpose and strength, but the inputs and outputs are not standardized, which makes it difficult to automate the tool chain. To define a fully automated process the inputs and outputs of the individual software packages need to be controlled. For that purpose, the NOVA framework was developed. The framework is Python based and connects all required software packages (see Figure 2). NOVA fetches the required inputs and triggers the requested calculations. The output of the calculations is stored in a centralized database.

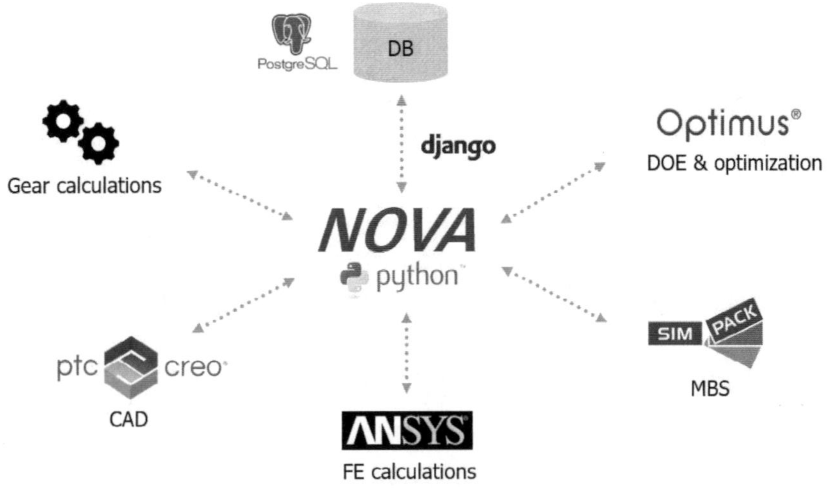

Figure 2 Structure of NOVA Framework

2.2.1 Fully parametric multi-body simulation models

An important part of the tonality prediction is the wind turbine powertrain transfer path analysis. Extensive effort has been put in the validation of multi-body simulation of wind turbine powertrains in the past decades. [PEE04] did first sensitivity studies of the dynamics of a wind turbine powertrain using multi-body simulation. [SCH13] validated a flexible multi-body simulation of a wind turbine powertrain with respect to dynamic loads and structure borne sound. [VAN15] used a SIMPACK model with flexible bodies of a wind turbine gearbox in a test rig environment to validate the gearbox transfer path.

For precise validation previous research did focus on detailed models with modal condensation of all structural components according [CRA68]. Whereas this approach provides a precise representation of the structural dynamics, it also has two main drawbacks in design studies and optimization of the transfer path. Firstly, local stiffness variation of flexible bodies cannot be assessed and studied directly in the multi-body simulation. Secondly, design studies and optimizations require updating the flexible bodies, which requires high effort in modelling and FE calculations in each iteration.

Therefore, TFWT process makes use of a fully discretised MBS model of the wind turbine powertrain with torsional and axial degrees of freedom during concept and design phases of the project. This approach enables:

- accessing the local structural stiffness, mass and inertia
- scanning large design spaces with low effort
- running optimizations on the transfer path characteristics
- low effort implementation of local design changes and design updates

Such models consist of blades, hub, main shaft, main bearing, base frame or main bearing housing assembly, elastomer elements, gearbox, high speed coupling and generator. The discretization is done based on CAD and FE models of components. The complete powertrain assembly is discretized to ~100 point masses / inertias and torsional / axial stiffness'.

2.2.2 Tonality prediction approach

Tonality prediction can be summarized in two steps. First one is calculation of the mechanical excitation profile of the powertrain, which consists of:

- **the excitation**, a torque dependent excitation force represented by the transmission error (T.E.) and gear mesh stiffness (C_{mesh}). This excitation force is also dependent on gearbox stage and harmonic (order), which can be regarded as a frequency dependency.
- **the powertrain system behaviour**, represented by the frequency response function (FRF) of the powertrain model, which is calculated between the excitation source and tower / blade interface of the powertrain.

The first step can be shown in mathematical form as follows:

$$PTR\_excite(f,T) = T.E.(f,T) * C_{mesh}(f,T) * FRF(f,T) \qquad \textbf{Eq. 1}$$

Second step is to model how this excitation is transferred to the IEC microphone, which consists of calculating a **transfer function** (TF) between a certain type of unit interface load and the sound pressure at the IEC microphone. This second step can be represented in its simplified, mathematical form as follows:

$$Tonality(f,T) = PTR\_excite(f,T) * TF - masking\ noise\ (f) \qquad \textbf{Eq. 2}$$

Aim of the process is to create a heat map of excitations from powertrain at turbine interfaces and from there deduct a tonality prediction combining this heat map with turbine transfer path, radiation characteristics and masking energy. The turbine operation envelopes and/or power curves can be overlaid on such graphs to evaluate the performance for individual modes of operation. Examples of those result are shown in the following figures.

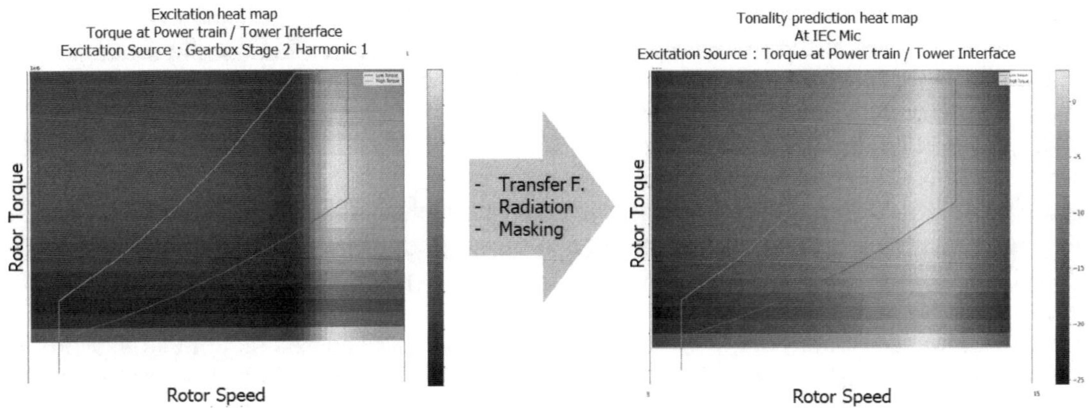

Figure 3 Example of PTR and WTG heat maps

2.2.3 DOEs, surrogate models and design optimization

With fully parametric simulation models parameters like powertrain stiffnesses can be systematically varied by performing a Design of Experiments (DOE) analysis. Based on the outcome of the DOE an optimization with the objective of minimal tonality/response can be run. Two main designs for DOEs are utilized:

- **Latin-Hypercube Design:** random distribution of a pre-defined number of samples within the design space. Good for searching for interactions between input and output parameters within a large design space.
- **Adaptive DOE design:** this iterative algorithm doesn't set all experiments at once, but it starts from a learning population and works like swarm optimization algorithms, deciding on the next population based on the current population. At the end of each population internal interpolation models are built to decide on the next population. Thereby the focus can be put on maximizing the space filling or feature learning capabilities. This control over DOE design quality, makes it more appropriate for using it as an input to surrogate model building. [OPT1]

The optimization can be either performed directly on the results of the DOE or on a surrogate model (*Response Surface Model*, RSM). A surrogate model is an approximation model which defines an analytical function between the parameters of a DOE (inputs) and the system responses (outputs). A key advantage of a surrogate-based optimization compared to a direct optimization is a very fast computing time. [CAV13]
Common methods for creating a surrogate model can be grouped in parametric and non-parametric methods. Parametric methods like *Polynomial Regression* (PRG) or *Kriging* (KRG) [KRI51] aim to approximate the system by a global function. To achieve models with an adequate model quality a-priori knowledge about dependencies between the factors and the maximum degree of the system are required. Non-parametric methods like *Radial Basis Function* (RBF) [BAX92] or *Deep Neural Network* (DNN) [AGG18] are independent of system knowledge and can be used for large multi-dimensional DOEs (>500 experiments).
According [CAV13] optimization algorithms can be classified as deterministic, stochastic or hybrid algorithms. Commonly used deterministic algorithms are *Steepest Descent* [CUR44] *or Sequential Quadratic Programming* (SQP) [STO85]. State-of-the-art algorithms like the *Nonlinear Programming Quadratic Line Search* (NLPQL) [SCH85] combine the SQP approach with a *Line Search* for a very reliable and efficient optimization.
All deterministic approaches follow a strict mathematical process, which results in fast convergence and a reproducible solution. As most of the deterministic algorithms are gradient-based, the results of the optimization is highly dependent on the choice of the starting point. For

optimization problems with more than one local optimum multiple optimizations with starting points equally spread over the design space are required in order to find the global optimum. Stochastic algorithms are based on the presence of randomness and are often inspired by principles of the nature. Common stochastic optimization algorithms are evolutionary and genetic algorithms like the *Differential Evolution Algorithm* (DEA) [STO97] or *Particle Swarm Optimization* (PSO) [KEN95]. These algorithms mimic the concept of evolution or the interaction of a bird swarm to find an optimum. Due to the stochastic approach these algorithms are more robust in finding the global optimum compared to deterministic algorithms. On the other hand, the computing time can be by a factor of 100 higher than for deterministic approaches [WÜN17].
Hybrid optimization combines the advantages of both deterministic and stochastic algorithms. Often a two-step approach is followed. In the first step a stochastic optimization algorithm is used to randomly explore the design space and find areas with local optima. After that, the optimum is used as a starting point for a deterministic algorithm. With this approach a global optimum can be found reliably and efficiently.

3. Optimizing the tonal characteristics of a wind turbine powertrain

The following summarizes results of optimization of a wind turbine powertrain for a certain operation mode. First a baseline evaluation is carried out and tonality risks are determined. Secondly, optimization objectives are defined based on those risks. Thirdly, DOEs are run to scan the design space and identify important parameters. In a fourth step DOEs are repeated with only the important parameters and surrogate models are built based on those. At the last step an optimization is carried out.

3.1 Wind turbine and gearbox under investigation

Some specifications of the wind turbine subject to this study is shown in the below table:

Table 1 WTG Specifications

Rated Power	approx. 6MW
Rotor Size	approx. 150m
Hub Height	approx. 110m
Powertrain type	Medium-Speed Integrated
Gearbox type	3 planetary stages

The operation data used for investigations are shown in the below figure. Data is modified to protect the OEM's IP.

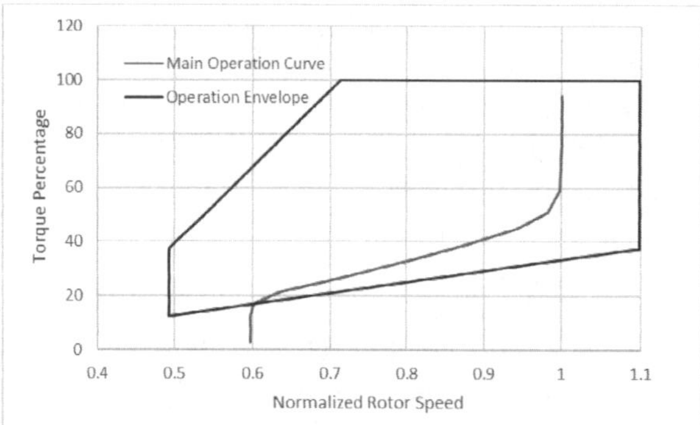

Figure 4 Operation curve and envelope

To evaluate tonal behaviour, heat map for response limits at tower interface is calculated from IEC measurement results (tonality and permissible tone line levels) and transfer function between the tower interface and IEC microphone position. For purposes of this paper noise emission from blades are neglected. Figure below shows this heat map along with the order tracks for gear mesh excitations from 2nd and 3rd stages of the gearbox.

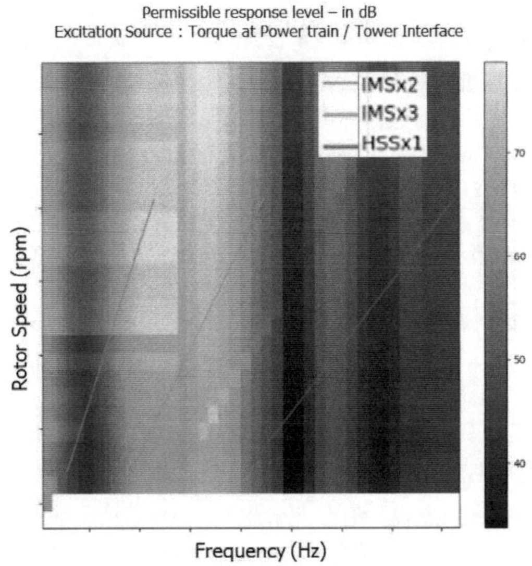

Figure 5 Permissible response level at tower interface

3.2 Baseline assessment

The baseline assessment focuses on the responses to excitations from 2nd planetary stage, especially in the 50 – 150 Hz range where we observe excitation of several powertrain eigen modes. Figure 6 shows the normalized response levels for stage 2 harmonics 2 and 3.

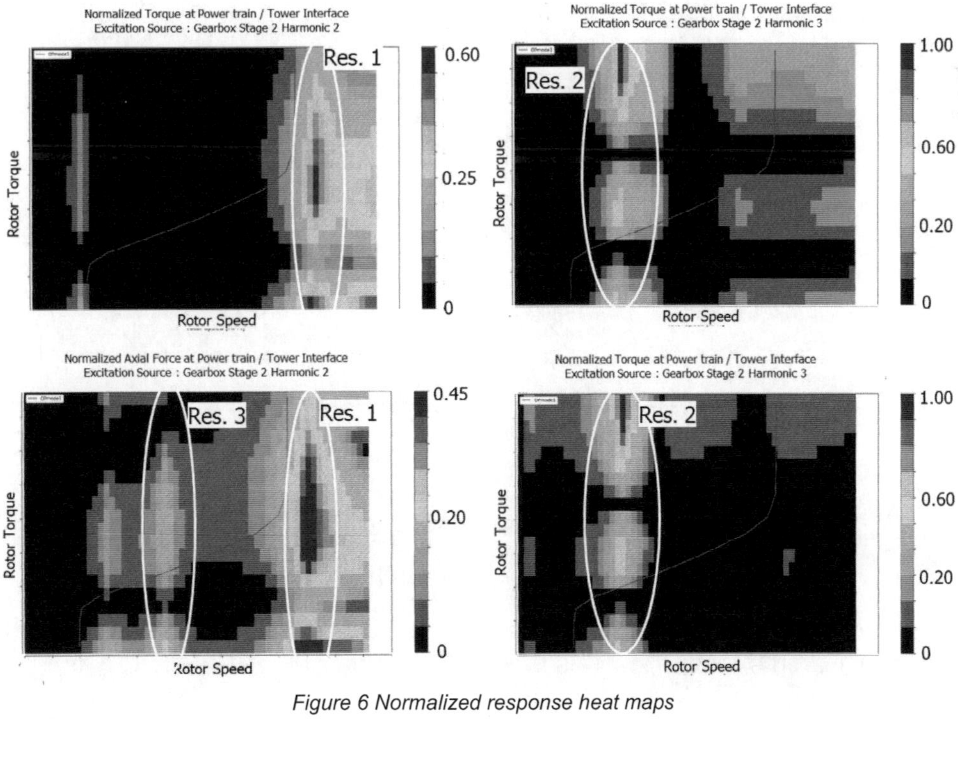

Figure 6 Normalized response heat maps

The current paper will focus on the resonance #3. This resonance is due to excitation of an eigen-mode featuring an out-of-phase axial pumping of the planet carriers of stage 2 and 3. The operation mode (red curve in Figure 6) drives through this resonance at 30% torque and mid-rotor speed, with relatively low masking energy. The combined influence of axial and torque response at that speed is calculated to exceed the response limit by ~7dB.

3.3 Definition of the optimization objectives

An optimization scenario is built under following boundaries:

- **Objective:** minimize the integral of axial force response at PTR / TWR interface due to excitation from stage 2 harmonic 2

  Constraints: don't exceed nominal values for the integral of axial force response at PTR / TWR interface due to excitation from stage 2 harmonic 3

3.4 Design of experiments and surrogate model

Design of experiments and optimizations are done using Optimus by Noesis software solutions [Opt2].

3.4.1 Scan of the design space

First step is to determine the dependencies between inputs and outputs of the black box and identify the important input parameters. This first step includes 100 input variables and 20 output vectors. Design space is divided into 3 with respect to input variables : stiffnesses, masses and inertias of the powertrain discretised model. This is mainly done to be able to run DOEs in parallel.
Vector outputs used throughout the investigation base on an operation slice of the response heat maps along the operation curve (Figure 6). The peak amplitude, integral and rotor speed at peak amplitude are extracted as scalar outputs from the operation slice vector (Figure 7). Those scalar outputs are used to evaluate DOEs, built RSMs and optimize the powertrain structure.

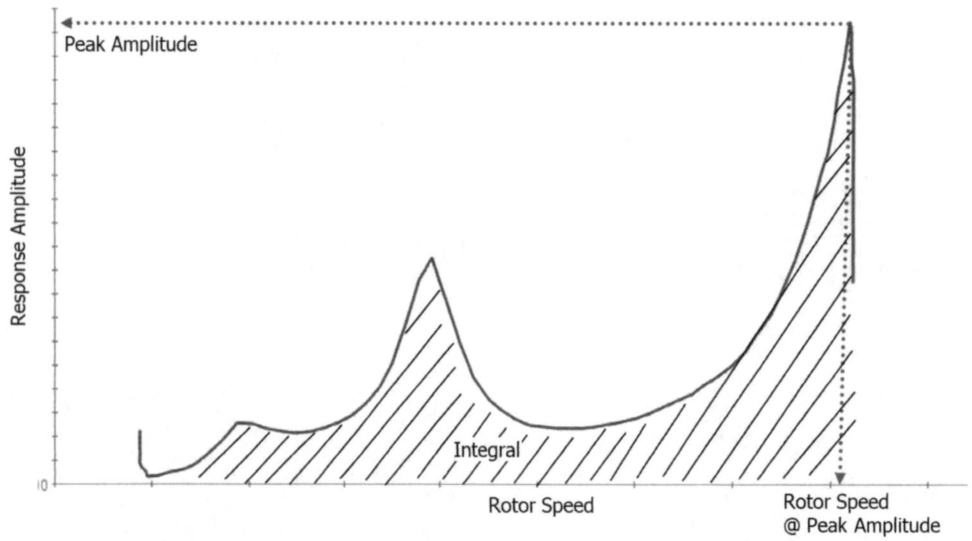

Figure 7 Representation of scalar outputs used in the study

In the first step Latin-Hypercube design are used and 3 DOEs with a total of 1800 experiments are run. The total CPU time for this stage is around 30 hours.

The correlation between inputs and outputs are evaluated based on the Pearson coefficients [TUR22]. It is important to mention that although Pearson coefficients give a good impression of correlations on relative scale, it doesn't give an indication on how important the absolute influence is. As a result of this first scan 7 axial stiffness parameters are found to have important influence on scalar outputs. 5 of those stiffness belong to gearbox housing, one is a bearing (param #7) and last is representing axial stiffness of a planet carrier (param #6).

	Axial Response							
	Stage 2							
	Main Frame				Main Shaft			
	Stg2_MFR_Ax_2_Int	Stg2_MFR_Ax_2_YMax	Stg2_MFR_Ax_3_Int	Stg2_MFR_Ax_3_YMax	Stg2_MSH_Ax_2_Int	Stg2_MSH_Ax_2_YMax	Stg2_MSH_Ax_3_Int	Stg2_MSH_Ax_3_YMax
Param_ax 1	0.044 (0.041)	0.041 (0.037)	0.029 (0.029)	0.016 (0.017)	0.039 (0.032)	0.030 (0.025)	0.011 (0.008)	-0.001 (0.001)
Param_ax 2	-0.381 (-0.259)	-0.432 (-0.350)	-0.244 (-0.213)	-0.595 (-0.588)	-0.367 (-0.252)	-0.461 (-0.419)	-0.102 (-0.091)	-0.379 (-0.335)
Param_ax 3	0.672 (0.645)	0.593 (0.544)	0.932 (0.945)	0.645 (0.644)	0.668 (0.652)	0.586 (0.566)	0.857 (0.879)	0.460 (0.499)
Param_ax 4	0.001 (0.006)	-0.014 (-0.015)	0.005 (0.007)	-0.009 (-0.013)	0.001 (0.002)	-0.009 (-0.021)	0.007 (0.006)	0.006 (-0.002)
Param_ax 5	-0.067 (-0.050)	-0.068 (-0.046)	-0.117 (-0.120)	-0.006 (-0.008)	-0.065 (-0.051)	-0.053 (-0.042)	-0.150 (-0.146)	-0.242 (-0.372)
Param_ax 6	-0.141 (-0.152)	-0.221 (-0.294)	-0.126 (-0.105)	-0.421 (-0.402)	-0.254 (-0.262)	-0.423 (-0.461)	-0.416 (-0.372)	-0.427 (-0.335)
Param_ax 7	0.007 (0.019)	0.009 (0.035)	-0.107 (-0.104)	0.027 (0.029)	0.010 (0.020)	0.009 (0.020)	-0.139 (-0.130)	-0.057 (0.040)

	Torsional Response							
	Stage 2							
	Main Frame				Main Shaft			
	Stg2_MFR_Tors_3_Int	Stg2_MFR_Tors_3_YMax	Stg2_MFR_Tors_2_Int	Stg2_MFR_Tors_2_YMax	Stg2_MSH_Tors_2_Int	Stg2_MSH_Tors_2_YMax	Stg2_MSH_Tors_3_Int	Stg2_MSH_Tors_3_YMax
Param_ax 1	-0.019 (-0.017)	-0.015 (-0.015)	-0.005 (-0.019)	-0.009 (-0.016)	0.010 (0.003)	-0.010 (-0.021)	-0.015 (-0.015)	-0.018 (-0.019)
Param_ax 2	0.158 (0.149)	0.463 (0.449)	-0.409 (-0.358)	-0.399 (-0.342)	-0.698 (-0.673)	-0.436 (-0.390)	0.234 (0.228)	0.539 (0.529)
Param_ax 3	-0.510 (-0.488)	-0.763 (-0.778)	-0.830 (-0.893)	-0.853 (-0.888)	-0.575 (-0.578)	-0.807 (-0.837)	-0.768 (-0.768)	-0.705 (-0.720)
Param_ax 4	0.017 (0.027)	0.016 (0.019)	0.007 (-0.017)	-0.001 (-0.014)	-0.013 (-0.017)	-0.001 (-0.018)	0.031 (0.030)	0.018 (0.020)
Param_ax 5	0.470 (0.468)	0.152 (0.146)	0.029 (0.042)	0.018 (0.029)	0.026 (0.040)	0.019 (0.029)	0.328 (0.309)	0.173 (0.176)
Param_ax 6	0.137 (0.113)	0.328 (0.299)	-0.119 (-0.104)	-0.166 (-0.146)	-0.183 (-0.153)	-0.245 (-0.227)	0.156 (0.137)	0.285 (0.250)
Param_ax 7	0.565 (0.565)	0.118 (0.111)	0.021 (0.019)	0.016 (0.016)	0.035 (0.037)	0.022 (0.021)	0.396 (0.380)	0.143 (0.147)

Figure 8 Example showing Pearson correlation coefficients for axial stiffness inputs and scalar outputs

3.4.2 Building of remote surface models

With a design space that is narrowed down to 7 input variables, new DOEs are run with the target of building response surface models between inputs and scalar outputs. At this stage an adaptive DOE algorithm is used. A total of 600 experiments are run and RSMs are built using 3 approaches:

1. Best approach: Optimus software internally choses the best response surface model
2. Deep neural network (DNN)
3. Shallow neural network (SNN)

The quality of models are measured against a control set consisting of 100 experiments. Normalized sum of squared errors is used as a measure as shown in the following formula.

$$SSE_{norm} = \frac{\sum_{i=1}^{n}(y_i - y_i^{predicted})^2}{y_{nominal}^2} \qquad \textbf{Eq. 3}$$

Following table shows an example of RSM errors and the chosen models for the selected scalar outputs. All models were calculated from the same experiment set using different settings.

Table 2 Normalized SSE for axial response

	Response		Normalized SSE Best	DNN	SNN
Stage 2 Harm. 2	PTR / TWR Axial	Peak Amp.	0.29	**0.15**	0.16
		Integral	0.24	**0.16**	0.17
	PTR / Hub Axial	Peak Amp.	0.13	**0.09**	0.10
		Integral	0.24	0.19	**0.19**

It is observed that the error values for integral output is less compared to that of peak amplitude. RSM calculations for the 3rd scalar output (rotor speed at peak amplitude) fail since this result has a discrete character. Percentage errors can be evaluated with model residual plots. An example is shown for PTR/TWR interface and integral of axial force response vector. The bars represent percentage error for each experiment in the control set. Except some outliers, the errors lie in +/-5% range.

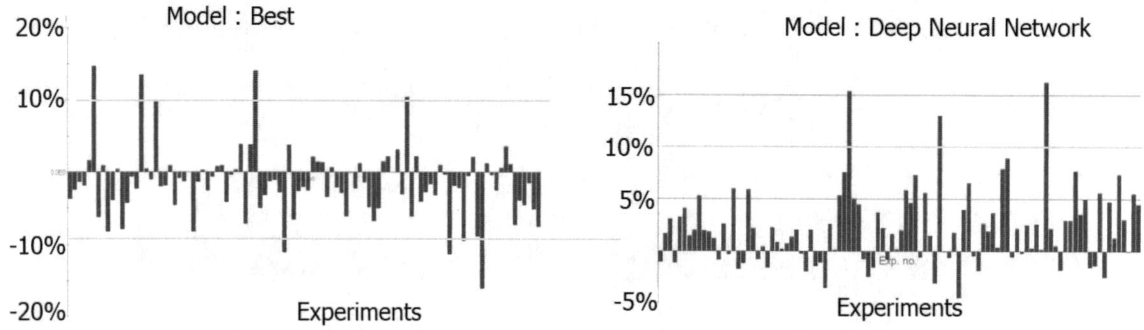

Figure 9 Model residual plots comparing two RSMs to a DOE of 100 experiments

3.4.3 Optimization of the axial stiffness

A model based differential evolution algorithm is utilized and the convergence is reached after 26 iterations and 4050 experiments in under 3 minutes. The resulting design is shown in below tables. After the optimum design was found, an analysis is done for this design. Table 3 shows the delta changes for scalar outputs between the nominal and optimum designs based on analysis. The percent difference between the results of the analysis and values calculated by the RSM is included as a measure of the model precision. As can be taken from the table, the objective function was reduced by 8dB. Moreover, there are ~3dB reductions for peak amplitudes of axial scalar outputs.

Figure 10 shows a comparison of normalized response vectors for stage 2 harmonics 2 and 3. It is important to note that the resonance 3 is shifted to higher rotor speed where it would benefit from a higher limit on response level. This is not only due to an increase in masking energy but also due to a lower amplification within the turbine transfer path. All effects combined, the predicted tonality level is reduced from +5dB to -6dB at the optimum point.

Table 3 Delta between optimum and baseline because of optimization of axial stiffness

	Response		Delta in dB	Model Error in %
Stage 2 Harm. 2	PTR/TWR Interface Axial Force	Peak Amp.	-3.50	2.54
		Integral	-8.10	13.52
	PTR/TWR Interface Torque	Peak Amp.	0.28	0.00
		Integral	0.25	0.06
Stage 2 Harm. 3	PTR/TWR Interface Axial Force	Peak Amp.	-2.65	-0.04
		Integral	-1.62	-0.14
	PTR/TWR Interface Torque	Peak Amp.	0.30	0.01
		Integral	0.09	0.00
Stage 3 Harm. 1	PTR/TWR Interface Axial Force	Peak Amp.	-0.18	0.71
		Integral	-0.01	0.00
	PTR/TWR Interface Torque	Peak Amp.	-3.44	0.08
		Integral	-1.89	0.08

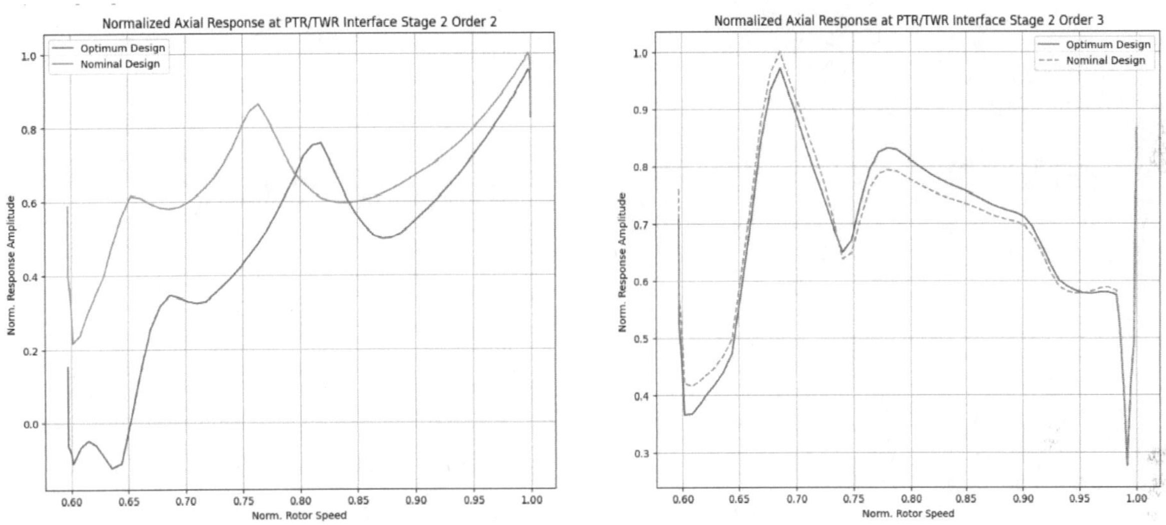

Figure 10 Normalized response - Nominal and Optimum

The changes in design parameters (i.e. cost) required to realize the numerical optimum and robustness of the output scalars should also be discussed. In the optimum design, the parameters were changed as shown in table 4. As can be seen all parameters were changed significantly, however it can be taken from figure 7 not all parameters have dominant effect on the output scalars. Moreover, the aim should be to decide on a robust range for design parameters to give the designer enough freedom in achieving the design and account for production spread. Thus, a design space around optimum should be explored.

Table 4 Change in input variables - Nominal to Optimum

	Param_ax_1	Param_ax_2	Param_ax_3	Param_ax_4	Param_ax_5	Param_ax_6	Param_ax_7
Delta in %	-30%	-26%	-30%	-30%	38%	50%	50%

Figure 11 shows such an exploration. Following can be concluded:

- Parameters 1 and 4 have no correlation with output scalars and can be kept at their nominal designs.
- Parameters 3, 6 and 7 are the most important influencers and they are also set to the lower or higher limits of the design space. Hence the optimization can benefit from a larger design space if feasible.
- It is possible to define a range of +/- 5% around the optimum where the output scalars don't change greatly.

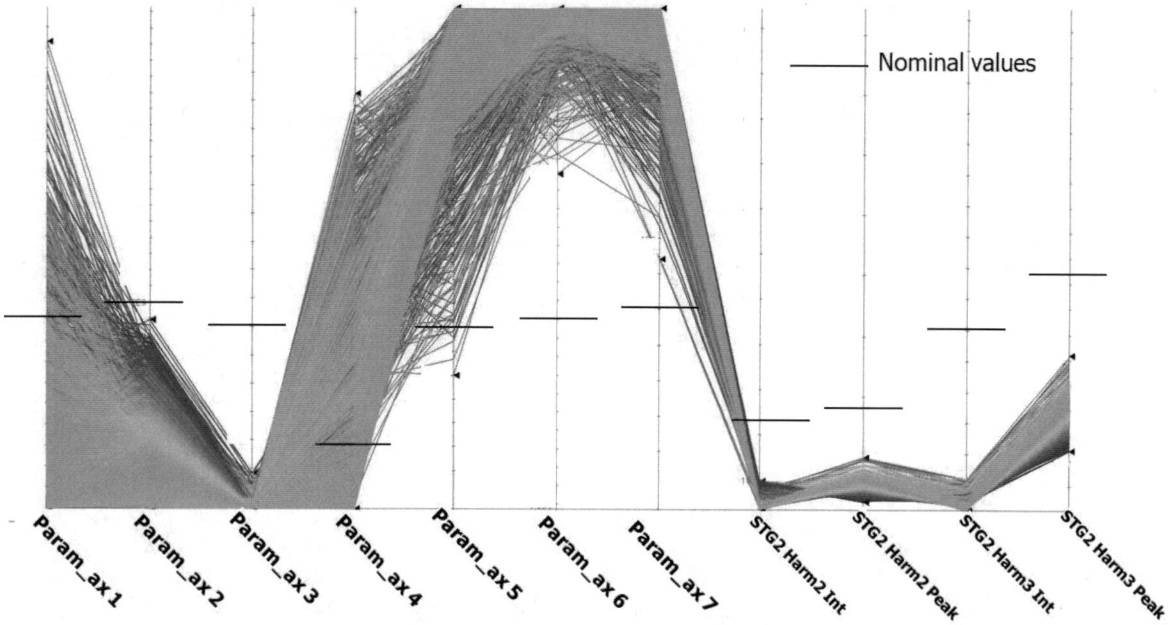

Figure 11 Design space around optimum point

4. Conclusions

The investigation has shown that:

- Although the optimization converges at the borders of the design space, this local optimum satisfies defined limits and provide good reductions in the objective function. With the aid of increase in masking and decrease in amplification on the turbine transfer path, the predicted tonality level is reduced by approximately 10dB.
- A design space and tendencies can easily be extracted from optimization iterations as a guidance to the designer. Meeting exact values of stiffness is not required to achieve a design that is satisfactory from a dynamic point of view.
- Model based optimizations are very swift and efficient. High RSM qualities can be achieved from relatively small sample sizes depending on the number of variables and the degree of nonlinearity. Effort required for model building increase exponentially with number of samples, therefore the sample sizes should be kept as small as possible and as large as necessary.

References

[PEE04] Peeters, J., Vandepitte, D., Sas, P.: Flexible multibody model of a three-stage planetary gearbox in a wind turbine. In: Proceedings ISMA. 2004.

[SCH13] Schmidt, S., Vath A.: Optimization of Windturbine Drive Train Dynamics through Flexible Multi-Body Simulation. In: Proceedings of Conference for Wind Power Drives, 2013

[VAN15] Vanhollebeke, F.: Dynamic analysis of a wind turbine gearbox – Towards prediction of mechanical tonalities. Dissertation, KU Leuven, 2015

[CRA68] Craig Jr, R. R., Bampton, M. C.: Coupling of substructures for dynamic analyses. In: AIAA journal, 6(7), pp. 1313-1319. 1968.

[OPT1] Optimus 2022.2 – Theoretical Background for Optimus 2022.2, Noesis Solutions N.V.

[OPT2] Optimus 2022.2 – Users' Manual, 2022.2. Noesis Solutions N.V.

[CAV13] Cavazzuti, M.: Optimization methods: From theory to design. First edition, Springer-Verlag, Berlin, New York, 2013

[KRI51] Krige, D. G.: A statistical approach to some basic mine valuation problems on the witwatersrand. In: Journal of the Chemical, Metallurgical and Mining Society of South Africa, vol. 52(6), pp. 119–139, 1951

[BAX92] Baxter, B. J. C.: The interpolation theory of radial basis functions. Trinity College, Cambridge University, 1992

[AGG18] Aggarwal, C. C.: Neural networks and deep learning.nSpringer, 2018

[STO85] Stoer, J.: Foundations of Recursive Quadratic Programming Methods for Solving Nonlinear Programs. In: Computational Mathematical Programming, 15, 1989

[SCH85] Schittkowski, K.: NLPQL: A Fortran subroutine solving constrained nonlinear programming problems. In: Annals of Operations Research, 5, pp. 485–500, 1985

[STO97] Storn, R., Price, K.: Differential evolution – a simple and efficient heuristic for global optimization over continuous spaces. In: Journal of Global Optimization, 11(4), pp. 341-359, 1997

[KEN95] Kennedy, J., & Eberhart, R. C.: Particle swarm optimization. In: Proc. IEEE International Conference on Neural Networks, 4, pp. 1942-1948. Perth, 1995

[WÜN17] Wünsch, A.: Effizienter Einsatz von Optimierungsmethoden in der Produktentwicklung durch dynamische Parallelisierung. Otto-von-Guericke-Universität, Magdeburg, 2017

[TUR22] Turney, S. (2022, December 05). Pearson Correlation Coefficient (r) | Guide & Examples. Scribbr.

INCEEUROPE

10th International Conference
on
Wind Turbine Noise
Dublin – 21st to 23rd June 2023

A Study on Sound Propagation of Mid-Speed Drivetrains and how to avoid Tonality Issues

Alexander Kari, Geislinger GmbH: kari@geislinger.com
Dr. Christof Sigle: sigle@geislinger.com
Karina Windhofer: windhofer@geislinger.com
Martin Cardaun, CWD: martin.cardaun@cwd.rwth-aachen.de
Dr. Ralf Schelenz, CWD: ralf.schelenz@cwd.rwth-aachen.de
Wilhelm Schünemann, CWD: wilhelm.schuenemann@cwd.rwth-aachen.de

Summary

Advances in aerodynamic noise over the recent years led to comparatively low sound emissions of wind turbines (WT). But, at low wind speeds or, in low-noise operating modes, reduced aeroacoustics masking energy makes mechanical sounds, noise originating from the drivetrain, audible. Excitations stem from the gearbox, predominately from the gear mesh of the second planetary carrier. The structure-borne sound propagates further through the drivetrain and finally to the surfaces of the wind turbine and is radiated to the ambient air.
Two basic ways of addressing tonalities are described by this paper, all drivetrain integrated. Tonalities can efficiently be mitigated by systematically decoupling excitations from sound emitting surfaces, like with a low-speed shaft coupling (LSSC). The load reducing effect of a LSSC on the gearbox had been well studied already (Kari, A., et al.). The work of this shall paper investigate the influence of a LSSC on the structure-borne sound propagation within an integrated drivetrain concept. Detailed numerical investigations by means of a modified generic model were performed to understand and to quantify the effect and the value of a LSSC to lower sound power levels of a wind turbine. The second part of this paper examines two other decoupling elements, but also a torsional damper, all integrated to the gearbox. Other than with the LSSC, these investigations were limited to torsional vibration analysis (TVA) to provide a good indication of their effect on tonality mitigation. This paper shows how different powertrain elements allow to tune the system in such a way to achieve tonality-free drivetrains.

1. Introduction, Motivation

In addition to advances in aerodynamic noise and therefore, the reduction of masking energy, there are a couple of other trends in wind turbine design which can result in serious noise and tonality concerns. To maintain competitiveness, but also to limit dimensions of growing onshore structures for road transportation, ever-increasing power density is more important than ever.

Thus, a shrinking ratio of excitation energy versus damping effect of the drivetrain mass does make the system prone to mechanical sounds. Growing power densities entail another challenge: Wind turbine and gearbox manufacturers understood that the architecture of future powertrains needs to be rethought, leading to a need for a higher degree of drivetrain component integration. That is a fully or at a least semi-integrated medium-speed drivetrain (mid-speed). Directly coupled arrangements of main bearings, gearbox, and generator will lead to new questions regarding gearbox input loads and noise, vibration, and harshness (NVH) behaviour. Sources of vibrations, like gearbox and generator, can no longer be separated regarding loads and structure-borne sound. This leads to tonalities inevitably, which needs to be handled before reaching the sound radiation surfaces of the tower and the blades. Vibrations, which can lead to mechanical sounds and tonalities can be mitigated or even neutralised by many ways. Two effective attempts based on integrated powertrain components will be presented with this paper.
First, a systematically decoupling of excitations from sound emitting surfaces. A LSSC made of advanced composites already is a well-proven solution to eliminate parasitic gearbox forces. This paper will show, how such a coupling is able to cut the transfer path of structure-borne sound, radiating into components of WT structures. For this purpose, a full flexible state-of-the-art-model was developed in a multibody simulation environment. A reference model is compared to a coupling model, with a LSSC between the main shaft and the gearbox. Transient simulations, dynamic analysis, and transfer path analysis (TPA) were carried out, followed by a calculation of the radiated sound power of the major structures, tower, and blades.
Second, other decoupling elements but also a torsional damper, all integrated to the gearbox, were considered too. What all these options have in common is, getting back to the root cause of the problem: Torsional excitations are being mitigated at its source to avoid the formation of structure born sound, rather than trying to dampen the transfer of noise at any location of the transfer path throughout the drivetrain structure. In WTs, one of the main sources of drivetrain noise is the excitation caused by gear mesh in the second planetary stage. These torsional excitations put the planetary carrier in resonance with the gearbox housing, inducing torsional vibrations to the entire drivetrain ("torsional mode").
The results section of this paper describes and summarizes how different powertrain elements allow to tune the system in such a way to achieve tonality-free drivetrains and to understand the effect of every individual solution on the systems dynamic of the powertrain.

2. Modelling Approaches

To investigate on the effect of a LSSC, multi body simulations (MBS) by means of a state-of-the-art medium-speed onshore wind turbine model with a rated power of 6 MW and a rotor diameter with 164 m were applied, in combination with transfer path analysis (TPA) and the calculation of the radiated sound power.
As the gearbox integrated solutions are designed for a different working principle than the LSSC, a mitigation of the torsional mode, investigations were carried out by applying torsional vibration analysis (TVA) with a software capable of transient calculations. A complex gearbox model of real 4.X MW high-speed drivetrain was transferred into a simplified model.
As both model approaches had been described by other publications already, the MBS model in 2023 (Cardaun, et al.), the TVA model in 2021 (Windhofer, et al.), the descriptions in chapters 2.1 and 2.3 are limited to summaries only.

2.1 MBS Model

Experts from the Center for Wind Power Drives took CAD data of their generic 6 MW model with a 4-point suspension (*Figure 1*) and finite element (FE) models of tower and rotor blades as a basis to convert it to a medium-speed drivetrain for the intended investigations on the LSSC. This architecture of currently developed integrated drivetrain designs with the main bearing unit composes one tubular shaped drivetrain housing together with the gearbox and the generator. Drivetrain components are not individually suspended and are directly coupled and attached by the joint housing to the yaw bearing.

Figure 1. 3D-models of the original 4-point support (left) and the modified medium-speed integrated drivetrain (right)

The rigid arrangement of the main bearing unit which is attached to the mainframe requires to a change over to two tapered roller element bearings. The integrated design increases the loading of the main bearings resulting in bigger and robust bearings in combination with an optimal distance between rotor-sided and non-rotor-sided bearings.
The parallel gear stage, as well as the high-speed shaft and the related housing were removed and the generator rotor and the stator including housing needed a fully new design. A housing was designed to be mountable directly to the gearbox housing. The generator rotor diameter needed to be substantially increased to reach the desired power characteristics while working with much lower rotational speeds and higher torques. Instead of generator bearings, the rotor is attached to the gearbox output shaft which is suspended by two double-row tapered bearings.
Dynamic models were derived from this design with all drivetrain components including gear wheels considered as flexible structures during the time domain simulation, to allow identifying effects of a LSSC on the structure-borne sound distribution in the drivetrain. Allowing to introduce loads at the respective locations, selected degrees of freedom (DOF) were identified as Master-DOF. Structural eigenmodes were investigated up to 200 Hz. As load distribution has an influence on structure-borne sound transfer, advanced bearing models were applied, allowing to take individual roller element loads and displacements into account.
Figure 2 illustrates the differences of the 'reference model' (left) and the 'coupling model'(right): The reference model is the modified version of the original generic model, a fully integrated drivetrain with the main shaft coupled directly to the first planet carrier, whereas the coupling model is semi-integrated with the LSSC integrated between the main shaft and first planetary carrier, which is supported by two roller element bearings.

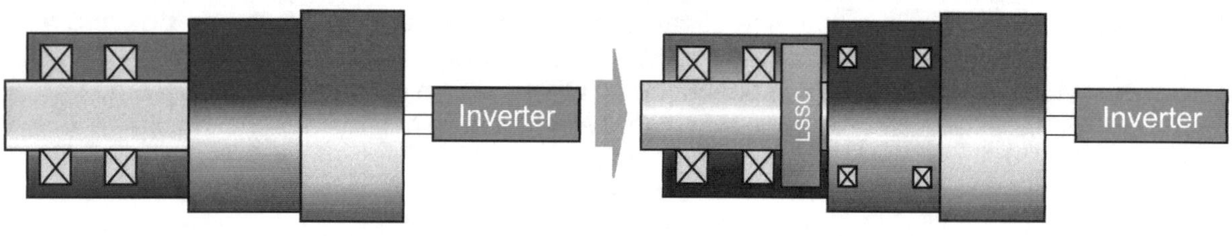

Figure 2. Models of the fully integrated reference model (left) and the semi-integrated drivetrain with the LSSC and planet carrier bearings (right)

2.2 Analysis Methods

Raw data generated by dynamic simulations, consisting of forces, displacements and modal parameters of all model states was plot as raw data into the time domain but also processed via a Fast-Fourier-Transformation (FFT) into the frequency domain.

In a second step, data was analysed further using a TPA to study the transmission of vibrations from an active source to a passive structure. With the classic TPA applied for this investigation, the FRF is determined for the assembled system, excitations are measured with the interface forces at the transfer path from the dynamic simulation.

Individual path contributions are calculated by multiplying the FRF by the interface forces, the exciting amplitude. Graphical representation of the path contributions was done using a Partial Path Contribution Plot (PPCP), allowing to plot the amplitude of the path contributions for each path on top of each other, depending on the speed, frequency at an operating point or order (Schüneman, et al.).

To maintain a reasonable extent of this paper, results of the PPCP are not part of it. The results section focuses on the FRF and the radiation of the sound power of the sound emitting structures. The calculation of the actual sound radiation was achieved by recovering the surface displacements of the respective structures using the modal parameters of the dynamic simulation (Cardaun, et al.).

The picture to the left of *Figure 3* illustrates the evaluation of the radiated power by the main structures, the tower, and the blades, with the local measurement spots of tower head and hub accelerations. The graph to the right shows wind and generator speeds of a transient laminar wind profile, starting from cut-in to rated wind speed and remaining for 30 seconds, to ensure all operational points of the structural dynamics can be detected.

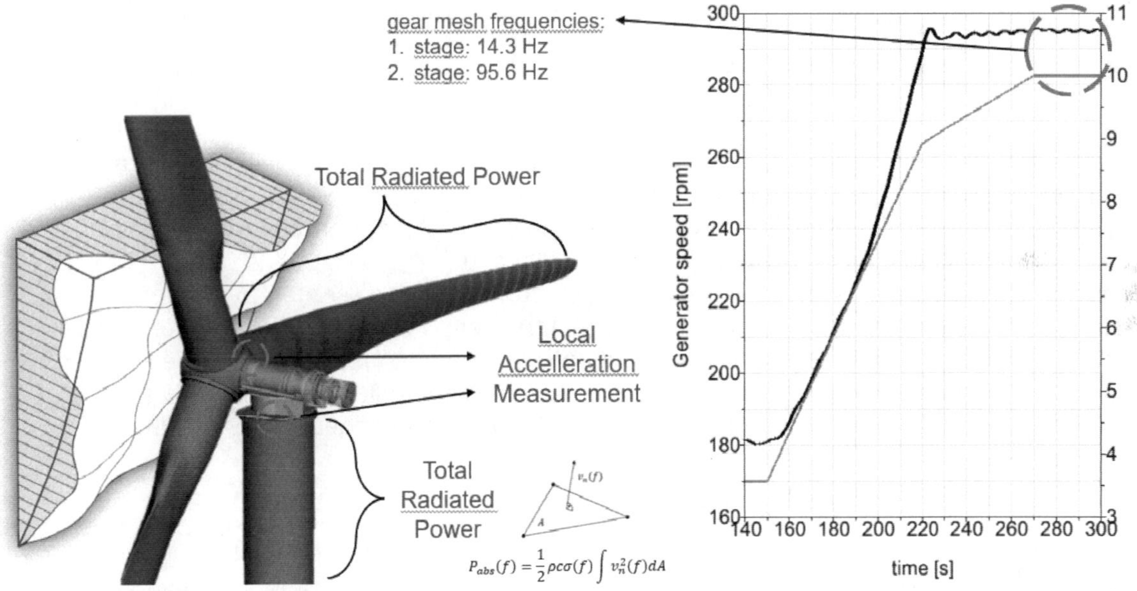

Figure 3. Evaluation method with interface points and sound emitting structures (left) and wind speed – generator speed profile (right).

2.3 TVA Model

In a first step, the complex OEM gearbox model was reduced to a minimum, *Figure 4* shows the complex model (top) and the simplified one (below).

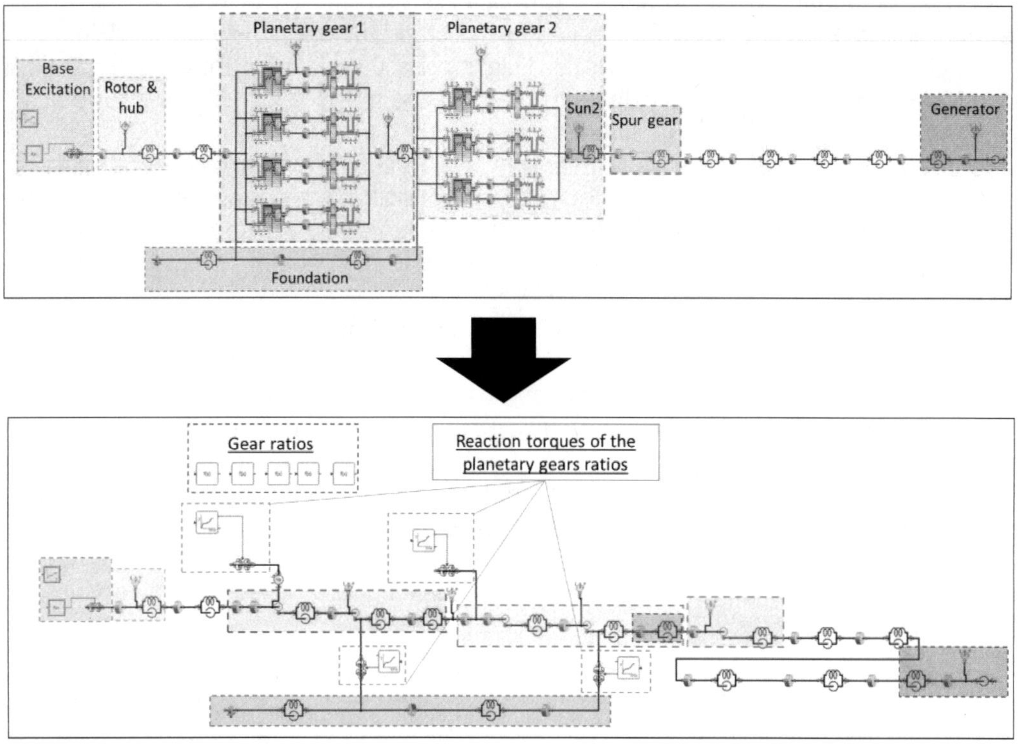

Figure 4. Detailed model (top) and simplified model (bottom) of a three-stepped gearbox (Simulation-X)

First complexity reduction was achieved by simplifying the planet gears i.e., by dividing the gear ratios of the ring gear to the planetary gears and the planet gears to the sun gear. After the subdivision into the individual translations, the inertias of the planet gears were merged into a single inertia and, stiffnesses were reduced to a single value.
The challenge of fluctuating tooth stiffnesses (which cannot be represented by a normal stiffness element) was tackled by describing the stiffness as the sum of static stiffness and dynamic stiffness by two parts for better approximation: A static tooth stiffness variation of ± 5% for both planetary gears was assumed. Rotor and hub were modelled as simplified rigid inertias, while constant torque with a first harmonic oscillation described by a sine curve of ± 2.5% of static torque was applied as basic excitation.

3. Results

As described in chapter 2., different models were applied to achieve results of all relevant powertrain options. For the LSSC, which results are depicted in chapter 3.1, a full-flexible model with dynamic simulation allowed to calculate the radiated sound for proper quantification. While for the gearbox-integrated options, only a relative comparison was possible by performing a TVA and giving an indication by comparing the torque levels, the amplitude, of each powertrain solution against the reference system i.e., without an additional coupling or damper. These results are depicted in chapters 3.1 to 3.3. Anyway, two out of these three options already were validated on tower, enabling a comparison of the TVA results against measured values.

3.1 Low-Speed Shaft Coupling

The LSSC considered for these investigations is a Geislinger Compowind® coupling, consisting of two straight, flexible elements made of advanced composites with a steel intermediate shaft as connecting element to allow for misalignment also in radial direction (vertical and horizontal related to the wind turbine axis). Decoupling the rotor shaft from the gearbox input shaft has been a proven solution to virtually eliminate parasitic gearbox forces. A load study, comparing a reference model, a generic model of a 6 MW offshore wind turbine with four-point-support, against the coupling model, with the gearbox mounted rigidly and a LSSC integrated between rotor shaft and gearbox input shaft. These investigations were carried out by the Center for Wind Power Drives and confirmed a reduction of bending loads by more than 90% (*Figure 5, Kari, A., et al.)*.

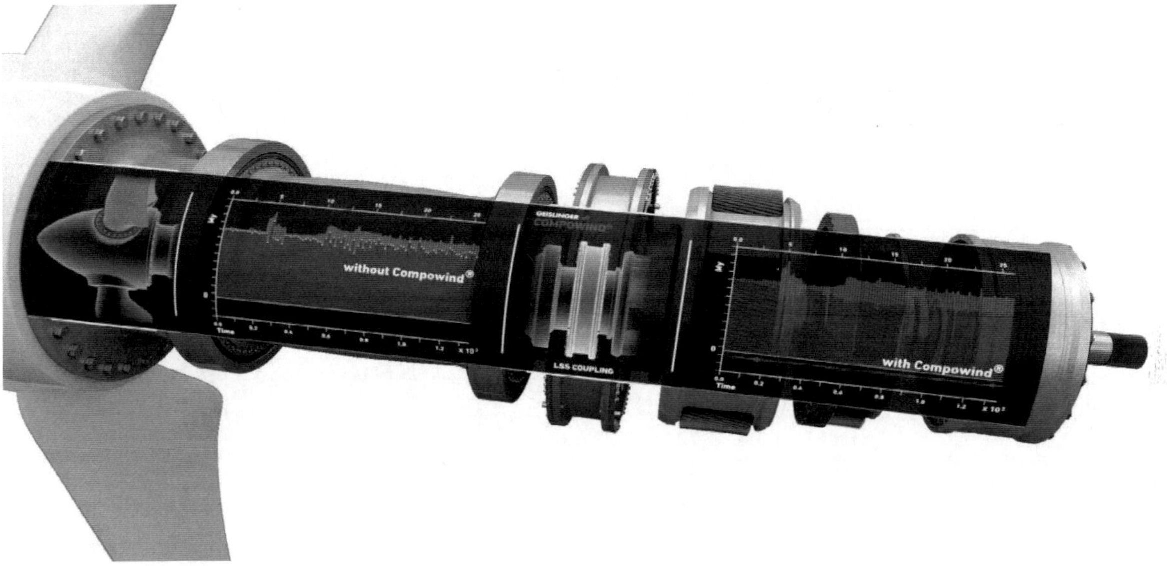

Figure 5. Semi-integrated mid-speed drivetrain with Compowind® LSSC between main shaft and gearbox input shaft. The graph to the right illustrates the bending moment My in the reference model, while the graph to the right describes the effect of the coupling with almost zero bending load and smooth transmission of loads.

First results related to the noise and tonality reducing effect of this LSSC are depicted below (*Figure 6)*. It is a comparison of blade root (left) and blade tip (right) accelerations of the reference model (black graph) against the coupling model (red and mint graphs), giving a good indication to what extent the LSSC takes influence on the dynamic systems behaviour, vibration levels and finally, its effect on noise and tonality.
The mint graph ("CW 220") is representing a stiffness study to learn more about the influence of coupling stiffness on the noise reduction effect. A slight difference can only be observed by the blade tip accelerations caused by the 2nd stage 1st harmonic. This is important since the latest generation of mid-speed powertrains might make a more compact size like a CW 220 (outer diameter 2200 mm) necessary rather than the CW 240 which fits into the design envelope of this generic 6 MW wind turbine model.

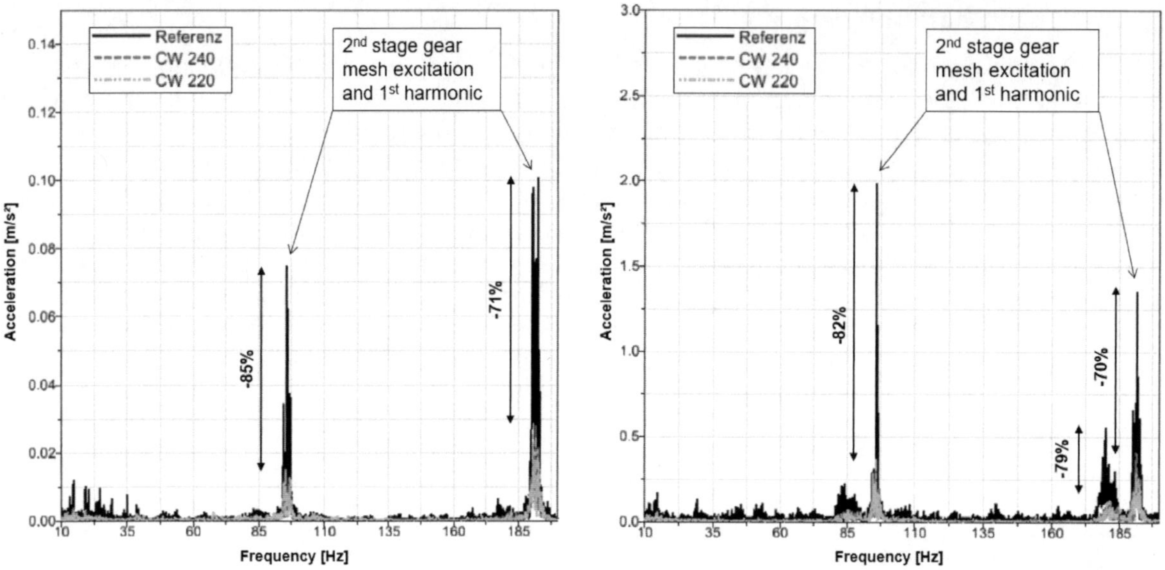

Figure 7. Accelerations in blade root (left) and blade tip) of reference and coupling models.

The effective potential of the noise mitigation capability of the LSSC finally is demonstrated by *Figure 8* by a comparison of the radiated sound power from the rotor blades by up to -16.5 dB.

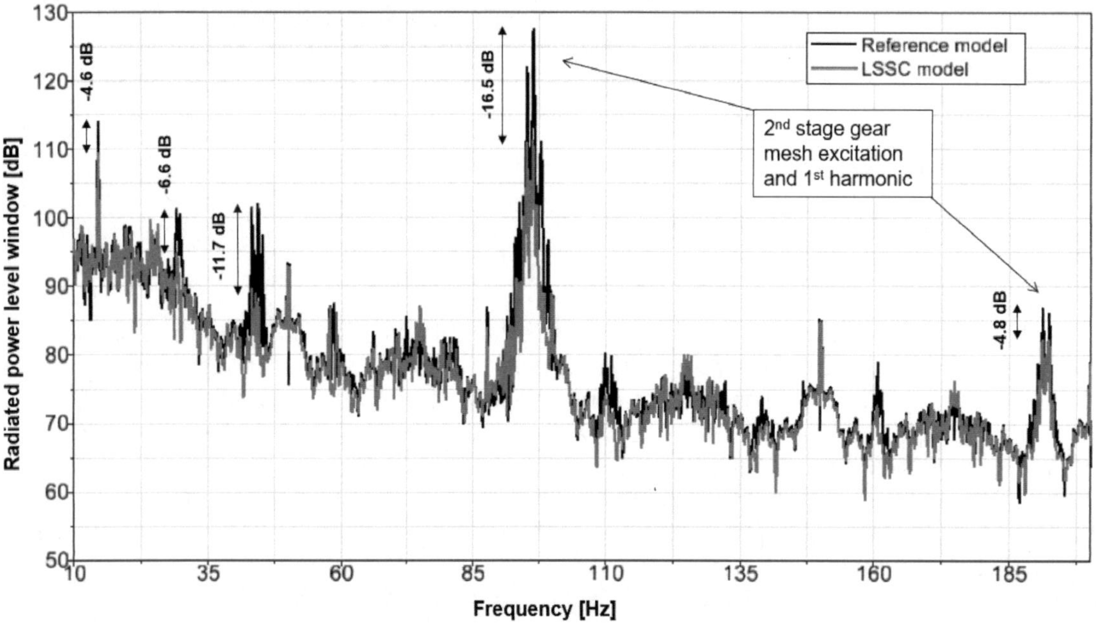

Figure 8. Comparison of radiated power levels from rotor blades.

Accelerations of the tower head and the sound power levels radiated from the tower also improve significantly as demonstrated by *Figure 9*. Same as for the blade investigations, a LSSC stiffness study was carried out with no significant differences in tower accelerations observed (left).

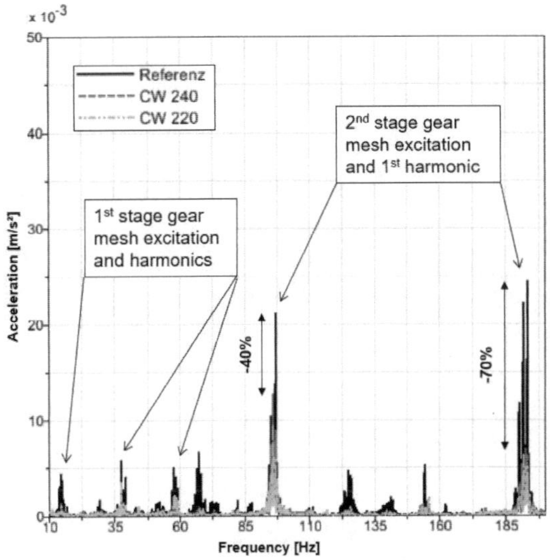

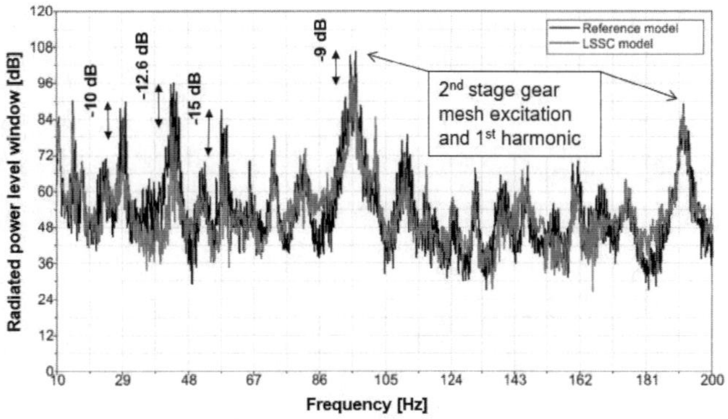

Figure 9. Comparison of tower head accelerations (left) and radiated power levels from tower (right).

3.2 Viscous Damper

A viscous damper consists of an inertia ring coupled to a housing by a special, high-viscous silicone oil (*Figure 10*). Torsional vibrations result in an angular offset between damper housing and inertia ring, applying shear load to the silicone oil, converting the vibration energy into heat, and transferring it to the ambient. Vibration amplitudes are effectively mitigated over a wide frequency spectrum. A torsional damper always is located at the spot of the highest level of energy, which normally is at the origin of torsional excitations, which, in the case of a wind gearbox is the second, respectively the third, planetary stage (*Figure 11).*

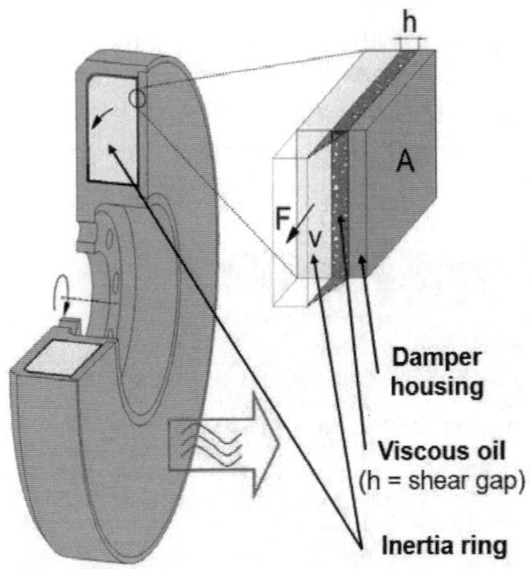

Figure 10. Cross-section of viscous damper.

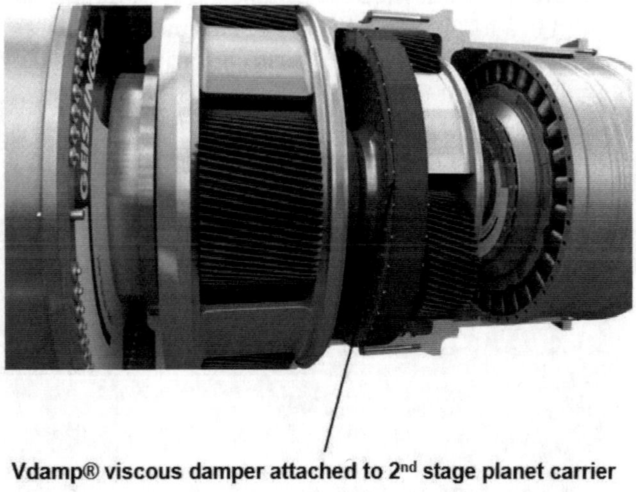

Figure 11. Viscous damper fully integrated to mid-speed drivetrain.

Result of a TVA based on the earlier described 4.X MW wind turbine with a high-speed drivetrain unveil, that the broadband damping effect effectively reduces the amplitude in the frequency band from 140 to 145 Hz by more than 60% (*Figure 12*).

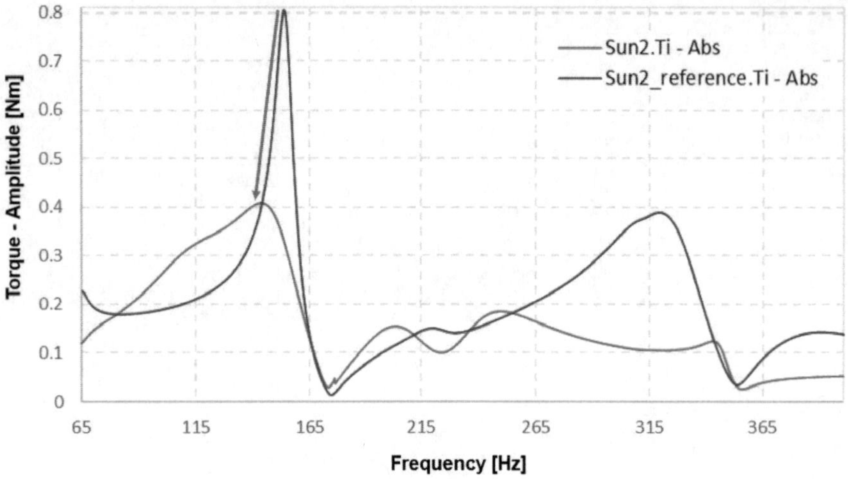

Figure 12. Comparison of amplitudes reference model (blue graph) against damper model with viscous damper (red graph).

Measurements on a gearbox but also on a wind turbine systems level of a mid-speed drivetrain unveiled that, a reduction of over 50% in torsional amplitude of sun 2, respectively ring gear 2, results in a reduction of approximately -6 dB. Anyway, the reduction potential of a viscous damper is dependent on the systems architecture, ratios of excitation energy to mass moments of inertia, eigenfrequencies, the available space for a damper and finally that the dominant eigenmode is torsional.

3.3 Geislinger Coupling

This torsional elastic coupling is a robust all-steel product with fatigue-free spring blades in a radial arrangement, transmitting torque and providing an individual torsional elasticity. In the marine industry is known as the Geislinger coupling. Other than a damper, a coupling is separating the torsional system into two tuned subsystems by introducing elasticity to the torque path. For this reason, vibratory torques can hardly be transmitted. Therefore, vibrations are diminished, and the noise transfer is attenuated significantly. The damping properties of the coupling further reduce the resonance amplitudes, which is an effect of a combination of mechanical and hydrodynamic damping, illustrated by *Figure 13.*

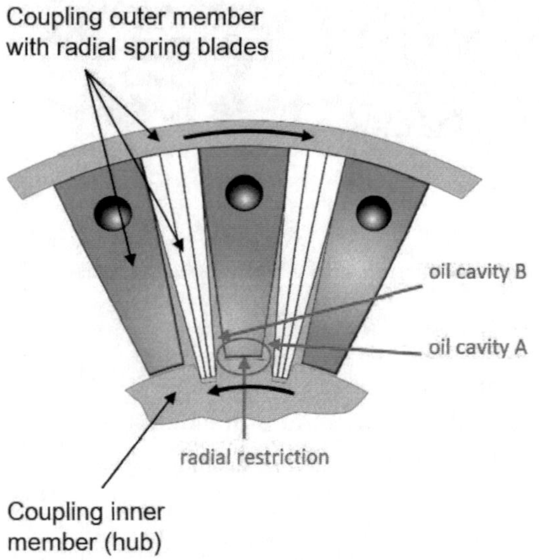

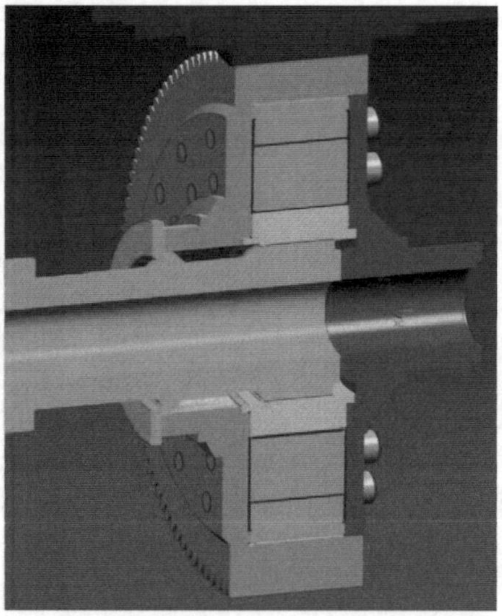

Figure 13. (left) Concept of Geislinger coupling describing mechanical and hydrodynamic damping.

Figure 14. (right) Concept of a Geislinger coupling fully integrated to the gearwheel of the high-speed stage.

Mechanical damping stems from the friction between the spring blades but also from the engagement of the spring blade tips to the grooves of the inner coupling member. The design of the coupling which is fed with oil from the gearbox oil system, allows additional hydrodynamic damping. Torsional vibrations lead to an offset between the outer and the inner coupling member, oil is forced to flow from oil cavity A to cavity B (or vice-versa) through the radial restriction between the coupling members.
For high-speed drivetrains, the coupling can fully be integrated to the gearwheel of the parallel stage (see design concept *Figure 14*) without any requirements on additional designed space. Additionally, this concept is posing a rather cost-effective solution as the torque is rather low at the high-speed side, allowing a compact coupling design. Anyhow, it should be mentioned that a Geislinger coupling technically can be integrated to a med-speed drivetrain also, between the last sun pinion and its bearing set.
The effect of such a coupling on the torsional mode and its ability to shift the resonance frequency to lower levels is illustrated by TVA result in *Figure 15* with an assumption of a comparatively torsional stiff coupling. The dashed arrow indicates the potential of the coupling as the torsional stiffness reduces.

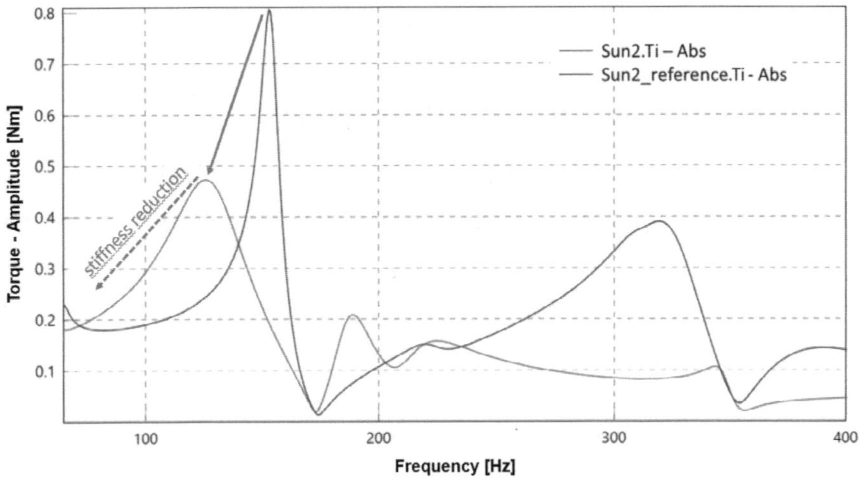

Figure 15. Comparison of amplitudes reference model (blue graph) against coupling model (red graph).

Depending on the system and designed space available related to the torque, the resonance frequency might be shifted to an area beyond of any eigenmodes, fully isolating the system from vibrations.
A Geislinger coupling has not yet been pursued further for such an application since recent priorities were set to the viscous damper introduction described in chapter 3.2 and, particularly to the carbon fibre sun shaft development and validation explained in following chapter 3.4.

3.4 Carbon Fibre Sun Shaft

Powertrain experts in the wind industry are well aware of the effect of reducing the stiffness of the output sun shaft of the last planetary stage of the main gearbox as typical eigenmodes can be dominated by its torsional stiffness. Obviously, a steel sun shaft sets hard limits to desired stiffness reductions. The bandwidth for tuning the torsional stiffness can be extended by introducing a torsional elastic coupling, like described in the previous chapter 3.3 or, by the consideration of a material with different properties than steel, advanced composites. A distinct advantage of a carbon fibre sun shaft is its fully integrability to co-axial gearboxes and the avoidance of a so-called 'add-on' product (*Figure 16).*

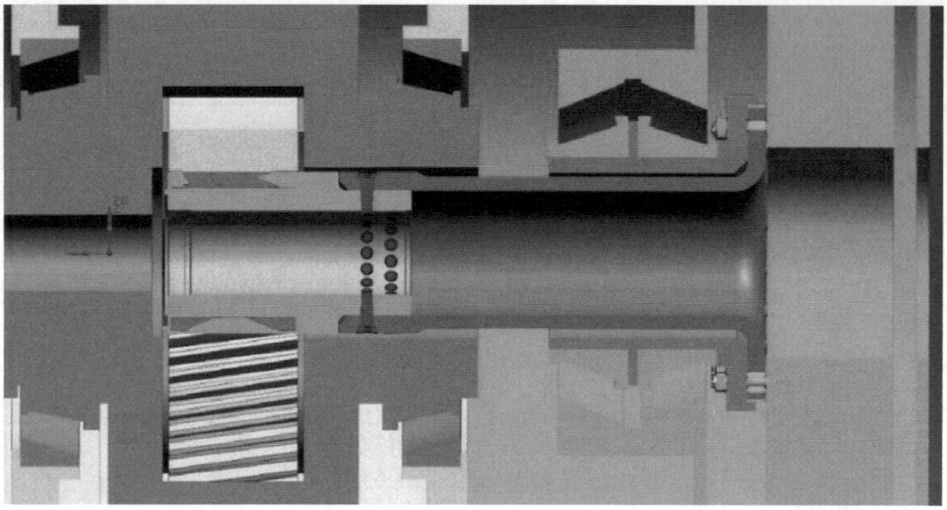

Figure 17. Carbon fibre sun shaft (grey) in a mid-speed gearbox: the carbon shaft is radially bolted to the sun pinion (blue bolts) and features a formed flange to allow direct connection to the generator rotor (green).

TVA results depicted in *Figure 17* delivers a similar effect as seen with the torsional elastic coupling. But, a comparable low torsional stiffness, achieved by extending the carbon shaft towards the generator by means of a formed flange (*Figure 17),* results in a higher shift of the resonance frequency to a lower level and, a reduction of the amplitude by approximately 60%.

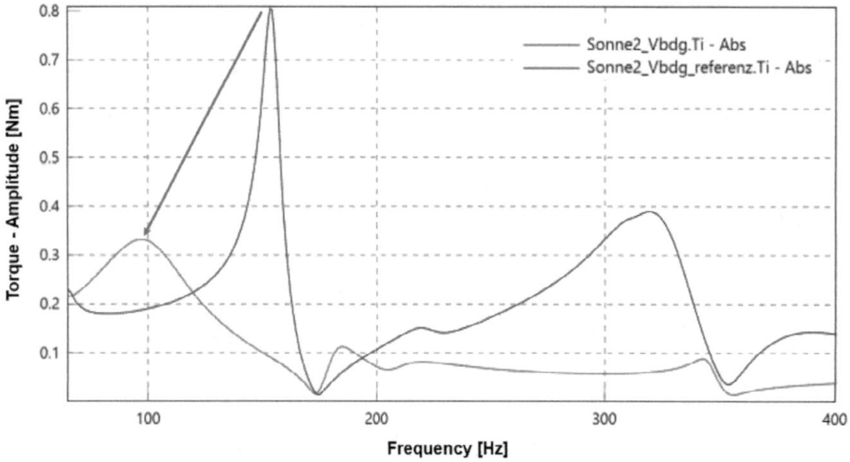

Figure 18. Comparison of amplitudes reference model (blue graph) against coupling model (red graph).

Depending on the volume of control i.e., the available length for a carbon fiber sun shaft or, if it is possible to go for a design with a formed flange with utilized the space up to generator, respectively, if even the space inside the generator is available, the torsional stiffness compared to a steel shaft can reduced by up to 80% and even more. Certainly, depending on the system, a carbon fiber sun shaft opens the opportunity to achieve a tonality-free drivetrain without massive design implications and without an additional component.

A prototype of a carbon fiber sun shaft had already successfully been tested and validated on component and systems level including field testing on tower in a med-speed wind turbine. The validation program also encompassed a component lifetime validation. To allow destructive testing of a certain number of test items, precisely down-scaled specimens used, representing the effective design of the prototypes. A summary of this validation program had been presented recently at a conference (Klönne, M., et al.).

4. Conclusions

The work of this paper is a summary that demonstrates that drivetrain integrated powertrain components bear substantial potential to reduce noise levels and to mitigate or, even eliminate tonality issues. This is of particular importance against the background of ever-increasing power densities of future onshore wind turbines, forcing manufactures in rethinking their present drivetrains architectures and moving towards higher integrated mid-speed systems. The first part of this paper investigated the reduction of structure-borne sound introducing a LSSC made of advanced composites into sound radiating structures of a 6 MW onshore wind turbine. Decoupling the main shaft from the gearbox input shaft by the LSSC results in significant reductions in accelerations at interface points to the large surface structures, and already gives a good indication about the noise attenuating effect of this coupling. The subsequent calculation of the radiated sound power confirmed these results by a reduction of the radiated sound power from the blades of up to -16.5 dB. In combination with the vast reduction of parasitic drivetrain forces, a composite LSSC is posing a valuable technology for future onshore wind turbines.

Although the investigations on the LSSC confirmed the effective sound attenuating properties, additional mitigation might be needed, respectively, in case of a fully integrated mid-speed drivetrain or a traditional high-speed concept, necessary. For this reason, a gearbox integrated torsional damper, a torsional elastic coupling and a carbon fibre sun shaft were investigated. Though, results from dynamic simulations including the calculation of sound radiation are not yet available, results from a TVA are giving a good indication about the potential of every solution. Respectively, field validation of the viscous damper and of the carbon fibre sun shaft already proved its effect. All three options have in common that these are fully integrated to the gearbox and that these provide a rather good ratio of moderate additional cost versus effect and customer value.

As indicated, the gearbox integrated options will be matter of future research i.e., to perform dynamic simulations, a TPA and particularly sound radiation calculations, identical to LSSC investigations. That allows to finally confirm its effect and to enable parameter studies as a sound basis for the design of tonality-free drivetrains.

References

Kari, A., et al. (2017) *Non-Torque Loads in Drivetrains – A Study on the Effective Reduction of Gearbox Loads and the Improvement of the Dynamic System Behaviour*, Conference for Wind Power Drives

Cardaun, M., et al. (2023) *Evaluation of a low-speed stage coupling with regard to structure-borne sound propagation in a wind turbine*, Conference for Wind Power Drives

Windhofer, K., et al. (2021) *Assessment of Noise Attenuating Powertrain Components,* Conference for Wind Power Drives

Cardaun, M., et al. (2022) *Multi-Physical Simulation Toolchain for the Prediction of Acoustic Emissions of Direct Drive Wind Turbines,* Journal of Physics: Conference Series 2265 4, S. 42047

Schünemann, W., et al. (2021) *Identification of relevant acoustic transfer paths for WT drivetrains with an operational transfer path analysis* Forschung im Ingenieurwesen 85, 345–351

Klönne, M., et al. (2023) *Non-Torque Loads in Drivetrains – On Optimizing the NVH Behaviour of WTG Powertrains through Technology Transfer across Industries*, Conference for Wind Power Drives

10th International Conference on Wind Turbine Noise Dublin – 21st to 23rd June 2023

WindTUNE: a new tool for modelling wind farm noise uncertainties

Bill Kayser, UMRAE, Cerema Gustave Eiffel University, France. bill.kayser@univ-eiffel.fr
Simon Bianchetti, UMRAE, Cerema Gustave Eiffel University, France
David Ecotière, UMRAE, Cerema Gustave Eiffel University, France.
Benoit Gauvreau, UMRAE, Gustave Eiffel University, Cerema, France.

Summary

Representative predictions of wind turbine noise require to accurately model the main mechanisms and characteristics of acoustic emission (i.e. extended sound source with aeroacoustic noise generation) and acoustic propagation in outdoor environment (i.e. ground effects and atmospheric properties). As these phenomena fluctuate over time and space, it leads to great uncertainty on Sound Pressure Level (SPL) estimated at local resident buildings/facades. Such uncertainty is not yet properly quantified by engineering noise prediction models. Thus, this paper presents a modeling tool developed in the framework of the French project PIBE, which aims at quantifying the SPL uncertainty involved in wind farm noise predictions. Ultimately, this modeling tool will be freely available online and will help to better understand the risk of noise pollution at each stage of a wind farm's life, in order to guarantee compliance with the regulatory requirements concerning the exposure of local populations.

1. Introduction

Representative predictions of wind farm noise require to accurately model the main mechanisms and characteristics of acoustic emission (i.e. extended sound source with aeroacoustic noise generation) and acoustic propagation in outdoor environment (i.e. ground effects and atmospheric properties). As these phenomena can fluctuate over time and space, it leads to uncertainty on sound pressure level (SPL) at local resident. Such uncertainty is not yet properly quantified by engineering noise prediction models, and this is one of the objective of the PIBE project [1][2] to make progress on this issue.

The aim of this paper is to present a method of uncertainty quantification of SPL due to possible fluctuations or to uncertainty of input environmental parameters. The methodology consists in modeling noise at receiver for many scenarios of influent input parameters thanks to a quasi-Monte Carlo sampling. This allows one to determine the distribution of the SPL induced by the probability distribution of uncertain environmental parameters. In practice, thousands of simulations may be required to conduct such uncertainty analysis, which leads to prohibitive calculation costs. As done in [3], one solution is to replace the initial wind turbine noise model by a metamodel that reproduces the expected SPL with highly reduced computational costs and small errors in the SPL estimation. Thus, the metamodel is built to determine the SPL distribution at receivers locations for a single wind turbine. Then, the same procedure is conducted for each wind turbine of the wind farm, in order to assess the overall SPL spread at receivers locations near the wind turbine farm.

2. The noise modelling

2.1 The source model for a single wind turbine

To model the noise emitted by a single wind turbine, the moving monopoles approach [4] is used. It consists in dividing each blade into segments of varying chord and span using a strip theory to account for non-uniform incidence flow along the blades. The trailing edge noise and the inflow noise are calculated thanks to the Amiet's theory which finally provides the angle-dependent sound power level of each segment [5]. An attenuation term is included to consider the propagative effects between the turbine and a far field receiver. This term is calculated through a propagation model based on the wide-angle parabolic equation (WAPE) without using the effective sound speed approximation [6]. Finally, the noise contributions from all the blade segments are summed at the receivers by assuming that they are uncorrelated [7].
The wind turbine we consider in this paper has a nominal electrical power of 2.3 MW, a rotor diameter of 93 m, a hub height of 80 m, and three blades measuring 45 m in length. The speed of rotation increases linearly from 6 rpm at the cut-in wind speed of 4 m/s (measured at hub height) to 16 rpm at a wind speed of 12 m/s. Further information on the modelling of wind turbine noise can be found in [1][8][9][10].

2.2 Atmospheric effects

The refraction of the acoustic waves is considered through the wind vertical profile $U(z)$ and temperature vertical profile $T(z)$:

$$U(z) = U_{ref}\left(\frac{z}{z_{ref}}\right)^{\alpha},$$

$$T(z) = T_0 + a_T \ln\left(\frac{z}{z_0}\right),$$

where U_{ref} (m/s) is the wind speed at height z_{ref} (m) above ground level (hub height), z (m) is the height above the ground, α is the wind shear factor, T_0 (K) is the air temperature at the ground surface, a_T (K/m) is a refraction coefficient that determine the influence of temperature profile, and $z_0 = 0.13\, h_v$ (m) is the roughness length that depends on vegetation height h_v (m).

The atmospheric absorption is considered in accordance with the standard ISO 9613-1 [11]. It depends on air temperature T_0 (K), atmospheric pressure P_{atm} (Pa) and the relative humidity of air h_r (%) chosen here as 80%.

The WAPE model has the capability to account for turbulence scattering by perturbing the acoustic refractive index [12]. However, a significant number of realizations (typically between 50-100) are required to estimate the sound pressure level (SPL). This computational requirement is prohibitively high for uncertainty analysis purposes. To address this issue, an alternative method proposed in the Harmonoise project is used [13], that consists in correcting the attenuation of SPL in a refracting atmosphere without considering turbulent scattering ($SPL_{noscatter}$). This correction is obtained by including a scattering contribution ($SPL_{scatter}$). The final attenuation term, denoted by ΔL , is calculated as follows:

$$\Delta L = 10\log_{10}\left(10^{\frac{SPL\,noscatter}{10}} + 10^{\frac{SPL\,scatter}{10}}\right),$$

with:

$$SPL_{scatter} = 25 + 10\log_{10}\gamma_T + 3\log_{10}\frac{\omega}{1000} + 10\log_{10}\frac{r}{100},$$

where $\omega = 2\pi f$ with f (Hz) the acoustic frequency, r (m) the source-receiver distance, and γ_T a measure of turbulence strength [14].

2.3 Ground effects

The ground effects are considered using an effective admittance model [15][8]. The sound absorption is modeled through the Miki's impedance model [16] that depends on the airflow resistivity parameter σ:

$$\frac{Z}{Z_0} = 1 + 6.17\left(\frac{\rho_0 f}{\sigma}\right)^{-0.632} + i9.44\left(\frac{\rho_0 f}{\sigma}\right)^{-0.632},$$
$$\frac{k}{k_0} = 1 + 8.73\left(\frac{\rho_0 f}{\sigma}\right)^{-0.618} + i12.76\left(\frac{\rho_0 f}{\sigma}\right)^{-0.618}.$$

The Miki's model should be used in the frequency validity domain: $f > 0.01\,\sigma/\rho_0$ where $\rho_0 = 1.24$ kg·m^{-3} is the density of air.

The scattering by ground roughness is considered through an effective admittance term (see [17]) that depends on 2 parameters: σ_h (m) which is the standard deviation of the ground roughness heights and l_c (m) which is the correlation length of the horizontal variations of the ground.

2.4 The metamodeling

The metamodel aims to replicate the behaviour of the original physics-based model with reduced computation time and a reasonable level of accuracy. The output of the physics-based model is a 2D SPL map with x values between 500 and 3000 meters and z values between 0 and 10 meters, with a resolution of 0.5 meters. To estimate the behaviour of each acoustic receiver using a statistical emulator, the SPL maps must be represented by a limited number of scalars, around ten, to avoid excessive computation time. These scalars will become the emulated quantities. The metamodeling process consists of three steps detailed below.

The first step involves generating a training sample Y, composed of N SPL maps calculated using the physics-based model, that discretize the input parameters space of the model using Latin-Hypercube Sampling (LHS). The centred training sample $\overline{Y}$ is created by removing the mean of the training sample $\overline{y}$ from each SPL map y in the full training sample Y.

In the second step, the dimension of the physics-based model output is reduced using principal component analysis (PCA) of the centred training sample $\overline{Y}$. Each SPL map y can be expressed as a linear combination of principal components Ψ on a reduced subspace such that $y = \sum a \times \Psi$. The principal components Ψ, which are obtained from the reduction of dimensionality using PCA, can be represented in the form of an SPL map with 5001×20 elements. These elements can be seen as "elementary maps". The scalar members of a represent the coefficients that indicate the weight of each principal component Ψ in the final SPL map y.

When calculating a new SPL map, only the coefficients a need to be determined as the principal components Ψ are already known. To predict the relationship between the projection coefficients

a and the inputs X, a fast statistical emulator based on kriging interpolation is used. Kriging interpolation is a linear interpolation method, meaning that predictions at a target point are linear combinations of the training data. It is an unbiased method, ensuring that the predictions at training points match with the data.

2.5 The noise modelling for an entire wind farm

The metamodel is built to model the SPL emitted by a single wind turbine, in a 2D (x, z) domain where x ∈ [500; 3000] m and z ∈ [0; 10] m. In order to assess the SPL uncertainty for an entire wind farm, the methodology is to use the single wind turbine metamodel for each wind turbine using the appropriate distance and propagation angles θi between wind turbines and the receiver (Figure 1). Every noise contributions of all wind turbines are energetically summed to have the wind farm noise at receiver.

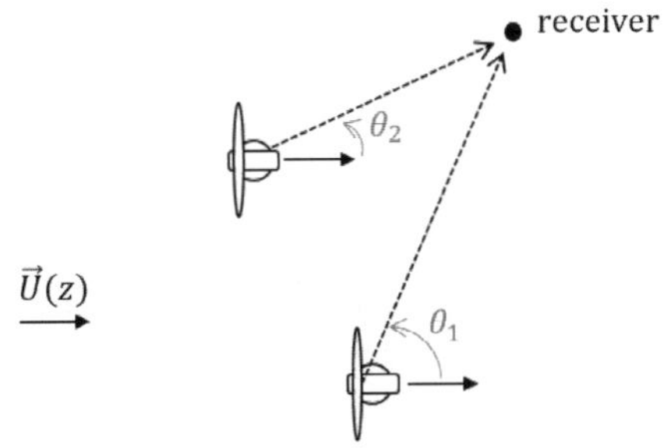

Figure 1: Schematics of the methodology for a 2 wind turbines wind farm

3. Uncertainty analysis

The objective of the uncertainty analysis is to determine the probability distribution P(y) of the output y of the model (i.e. SPL), induced by the probability distributions P(X) of the uncertain parameters X (i.e. environmental parameters). The distribution P(y) is estimated numerically by sampling the distributions of the inputs P(X) to propagate the uncertainty of the inputs X. The process described at §2.5 is repeated for each parameter sampling to finally get the distribution P(y) at receiver.

4. Online tool for uncertainty estimation

The uncertainty modelling has been implemented in a web application: WindTUNE. This app includes the meta-model described above and gives estimate of SPL distribution at receivers located at 500m to 3000m from a wind farm, and at a height from 0.5m to 10m. Although the modelling was designed for wind turbines with a hub height of 80 m and a blade length of 45 m, the application can be used for wind farms if wind turbine dimensions are close to those above.

4.1 User interface

The user interface of WindTUNE is based on a Shiny application [18] and internal calculations are made using R codes [19].

The interface allows to define the wind farm, to specify the input parameter information, to process calculations, and finally to visualize or download results about SPL distribution at receiver.

The wind farm can be implemented manually using coordinates, or through predefined external file containing GPS data.

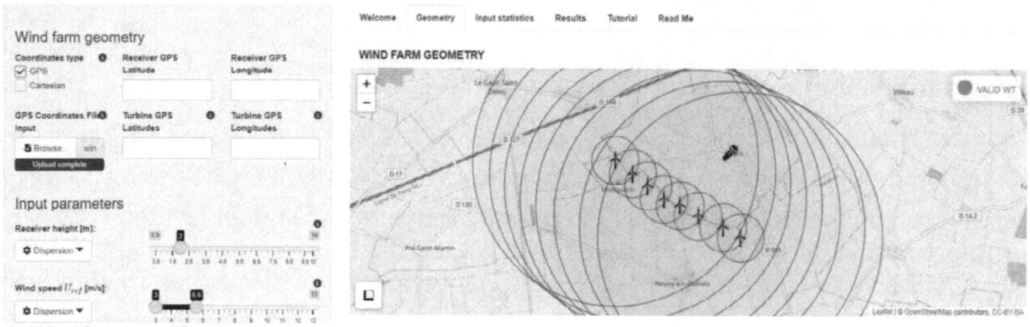

Figure 2: Wind farm specifications

4.2 Input parameters

WindTUNE considers the influence of 7 input parameters (Figure 3) that govern the SPL at receiver (wind speed at hub U_{ref}, wind direction, vertical wind shear factor α, atmospheric turbulence factor γ_T, airflow resistivity of the ground σ, air temperature at ground level T_0, temperature gradient parameter a_T). For each of them, statistic information should be provided: distribution range and type (uniform, normal, Weibull for wind speed, log-normal, user-defined etc). If a distribution is unknown, it is recommended to use a uniform one so as not to favor any particular condition. It should be noted that this choice of uniformity may lead to overestimate uncertainties.

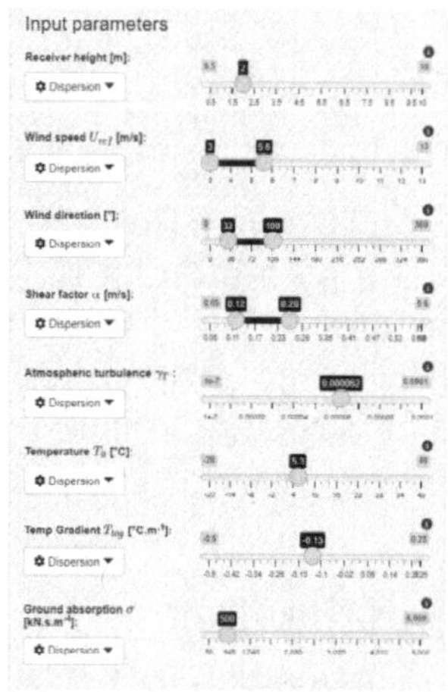

Figure 3: input parameters statistic information

4.3 Outputs

The main output is the SPL distribution at receiver (dB or dBA), normalized by the median value (Figure 4). The distributions of the sound contribution of each wind turbine are also available (Figure 5), as well as some statistics of all distributions (standard deviation, confidence interval etc). A specific tab allows to check the quality of the sampling process of according to each parameter distribution (Figure 6).

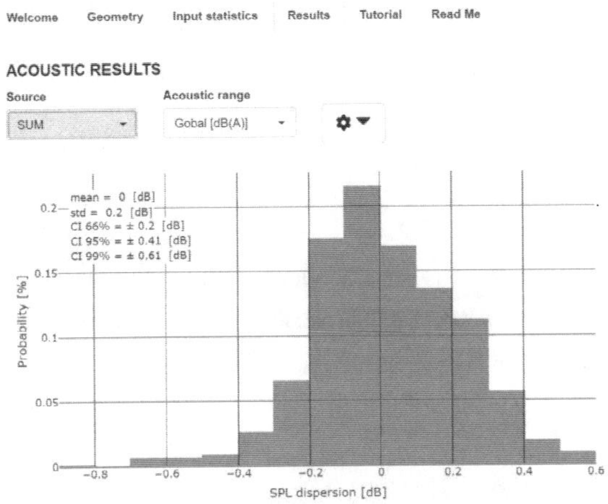

Figure 4: SPL distribution at receiver normalizes by median SPL (dBA)

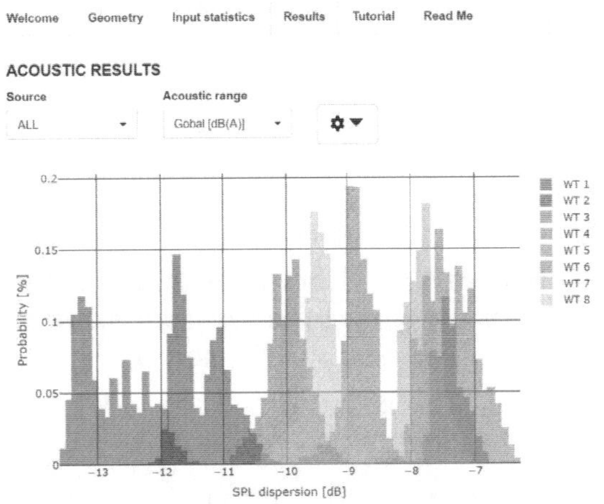

Figure 5: distribution of SPL at receiver contributions of each wind turbine normalizes by median SPL (dBA)

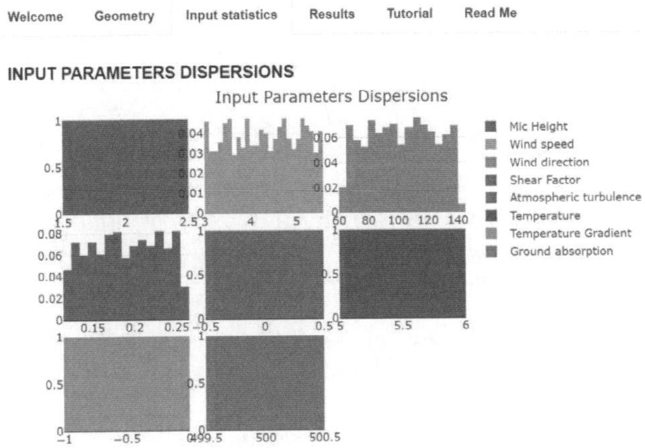

Figure 6: distribution of sampled input parameters

5. Conclusions

This paper presents the open access app WindTUNE that estimates the SPL distribution at a distant receiver from a wind farm, and due to the uncertainty or to the fluctuations of several input parameters that govern sound emission and propagation to the receiver. Calculations are based on a metamodel trained with a wide-angle Parabolic Equation (WAPE) propagation model coupled to a wind turbine sound emission model that uses the Amiet's theory for trailing edge noise and inflow noise generation.

WindTUNE is a useful complementary tool for estimating the variability and uncertainty of the noise level predicted by engineering tools in an impact assessment. It can be used for example to assess the risk of a wind farm exceeding the noise limits. Il will be open-access online at the end of the PIBE project [2].

Acknowledgements

This research is funded by the French National Agency for Research within the convention N°ANR-18-CE04-0011.

References

[1] Ecotière, D., Gauvreau, B., Cotté, B., Roger, M., Schmich, I., Nessi, M.-C., 2019. PIBE: a new French project for predicting the impact of wind turbine noise, in: Proceeding of 8th International Conference on Wind Turbine Noise, Lisbon, Portugal.

[2] https://www.anr-pibe.com/en

[3] Bill Kayser, Benoit Gauvreau, David Écotière, and Vivien Mallet. Wind turbine noise uncertainty quantification for downwind conditions using metamodeling. The Journal of the Acoustical Society of America, 151(1):390–401, January 2022.

[4] B. Cotté. Extended source models for wind turbine noise propagation. *The Journal of the Acoustical Society of America*, 145(3):1363–1371, March 2019.

[5] Yuan Tian and Benjamin Cotté. Wind turbine noise modeling based on Amiet's theory: Effects of wind shear and atmospheric turbulence. *Acta Acustica united with Acustica*, 102(4):626–639, August 2016.

[6] Vladimir E. Ostashev, D. Keith Wilson, and Michael B. Muhlestein. Wave and extra-wideangle parabolic equations for sound propagation in a moving atmosphere. *The Journal of the Acoustical Society of America*, 147(6):3969–3984, June 2020

[7] J. Christophe, J. Anthoine, and S. Moreau. Amiet's Theory in Spanwise-Varying Flow Conditions. *AIAA Journal*, 47(3):788–790, 2009.

[8] Bill Kayser, Benoit Gauvreau, and David Ecotière. Sensitivity analysis of influential parameters for wind turbine noise. In *8th International meeting on Wind Turbine Noise conference*, Lisbon, Portugal, June 2019.

[9] Bill Kayser, Benjamin Cotté, David Ecotière, and Benoit Gauvreau. Environmental parameters sensitivity analysis for the modeling of wind turbine noise in downwind conditions. *The Journal of the Acoustical Society of America*, 148(6):3623–3632, December 2020. Publisher: Acoustical Society of America.

[10] Kayser, B., Ecotière, D., Gauvreau, B., 2022. Modelling the uncertainties of wind farm noise predictions, in: Proc. 51st Internoise Congress, Glasgow.

[11] ISO9613-1:1993. Acoustics — Sound attenuation in free field — Part 1: atmospheric absorption calculation, 1993.

[12] P. Chevret, Ph. Blanc-Benon, and D. Juvé. A numerical model for sound propagation through a turbulent atmosphere near the ground. The Journal of the Acoustical Society of America, 100(6):3587–3599, December 1996.

[13] Dirk van Maercke and Jérôme Defrance. Development of an Analytical Model for Outdoor Sound Propagation Within the Harmonoise Project. Acta Acustica united with Acustica, 93(2):201–212, March 2007.

[14] Erik Salomons, Dirk van Maercke, Jérôme Defrance, and Foort de Roo. The Harmonoise Sound Propagation Model. Acta Acustica united with Acustica, 97(1):62–74, January 2011.

[15] F. G. Bass and I. M. Fuks. Wave Scattering from Statistically Rough Surfaces: International Series in Natural Philosophy. Elsevier, 1979. Google-Books-ID: aLM3BQAAQBAJ.

[16] Yasushi Miki. Acoustical properties of porous materials-Modifications of Delany-Bazley models-. Journal of the Acoustical Society of Japan (E), 11(1):19–24, 1990.

[17] Ray Kirby. On the modification of Delany and Bazley fomulae. *Applied Acoustics*, 86:47–49, December 2014.

[18] https://www.shinyapps.io

[19] https://www.r-project.org/

10th International Conference
on
Wind Turbine Noise
Dublin – 21st to 23rd June 2023

Socio-psychological effects of wind turbine noise: Research activities of IEA Wind TCP Task 39 (Work Packages 4 & 5)

E. A. King, University of Galway, Ireland: eoin.king@universityofgalway.ie
Eugene McKeown, University of Galway, Ireland
Denis O'Hora, University of Galway, Ireland
Lars Sommer Søndergaard, Department of Acoustics, Noise and Vibrations, FORCE Technology, Denmark: lss@forcetechnology.com
Erik Thysell, Department of Acoustics, Noise and Vibrations, FORCE Technology, Denmark: erth@forcetechnology.com
Tom Cronin, Danmarks Tekniske Universitet (DTU), Denmark
Daphne Schössow, Leibniz University Hannover, Germany: schoessow@ikt.uni-hannover.de
Stephan Preihs, Leibniz University Hannover, Germany: preihs@ikt.uni-hannover.de
Jürgen Peissig, Leibniz University Hannover, Germany: peissig@ikt.uni-hannover.de
Gundula Hübner, Martin-Luther-University Halle-Wittenberg + MSH Medical School Hamburg, Germany: gundula.huebner@psych.uni-halle.de
Johannes Pohl, Martin-Luther-University Halle-Wittenberg + MSH Medical School Hamburg, Germany: johannes.pohl@psych.uni-halle.de
Florian Johannes Yanic Müller, MSH Medical School Hamburg, Germany: florian.mueller@medicalschool-hamburg.de
Esther Blumendeller, University of Stuttgart, Germany: blumendeller@ifb.uni-stuttgart.de
Niels Adema, Hanze University of Applied Sciences Groningen, The Netherlands: ni.c.adema@pl.hanze.nl

Summary

This paper presents a report of some of the activities of the International Energy Agency's (IEA) Wind TCP Task 39. By identifying best practices in an international collaboration, Task 39 hopes to provide the scientific evidence to inform improved regulations and standards, increasing the effectiveness of quiet wind turbine technology. Task 39 is divided into five separate work packages, which address the broad wind turbine noise topic in successive steps; from wind turbine noise generation (WP2), to airborne noise propagation over large distances (WP3). The assessment of wind turbine noise and its impact on humans is addressed in WP4, while WP5 is dealing with other aspects of perception and acceptance, which may be related to noise. All WPs contribute to a dedicated Work Package on dissemination (WP1). This paper provides an update of activities primarily associated with the socio-psychological aspects of wind turbine noise (WP4 and WP5). Through the consideration of a wide variety of factors, including measurement technologies, auralisation and psychology, the effects on noise perception, annoyance and its impact on wellbeing and health is being further investigated. This paper presents a discussion of the activities of each member country and highlights some of the key research questions that need to be further considered.

1. Introduction

The IEA Wind Technology Collaboration Programme (TCP) is an international co-operation of 23 countries and sponsor members that share information and research activities to advance wind energy deployment. The goal of IEA Wind TCP Task 39 is to mitigate the generation of negative wellbeing, prevent health effects and consent effects by consolidating the understanding of wind turbine sound emission, propagation and noise perception, in order to accelerate the development and deployment of quiet wind turbine technology.

The integration of wind turbines in the energy system is subject to several environmental, societal and regulatory constraints. An important impact of wind turbines on the community derives from the emission of wind turbine noise. In many jurisdictions, there are concerns about the potential impacts of wind turbine noise on health and wellbeing. Perceptions of wind turbine noise can also negatively affect societal acceptance, a key to the successful adoption of new technologies, both at the local and global levels.

Developing noise mitigation technologies and recommending best practices for regulatory and siting processes is regarded as an important step toward public acceptance. IEA Wind TCP Task 28 (on the Social Science of Wind Energy Acceptance) has advanced the potential for enhanced community engagement to address that particular issue. One goal of Task 39 is to work with Task 28 to align research to reduce the non-acoustic influences on wind turbine acceptance. This combination of effort should eventually facilitate the wider deployment of wind energy. Work to increase the collaboration between these Tasks is ongoing.

The Task 39 work programme is summarised in Figure 1 and includes 5 Work Packages. The overall approach is to address the broad wind turbine noise topic in successive steps, from wind turbine noise generation (WP2), to airborne noise propagation over large distances (WP3). The assessment of wind turbine noise and its impact on humans is addressed in WP4, while WP5 is dealing with other aspects of perception and acceptance which may not be related directly to noise itself. Cross-cutting topics (e.g. amplitude modulation, low-frequency noise, etc) can be used as vectors for interactions between engineering and social/psychological sciences. WP1 is about dissemination and will be also considered in each of the WPs.

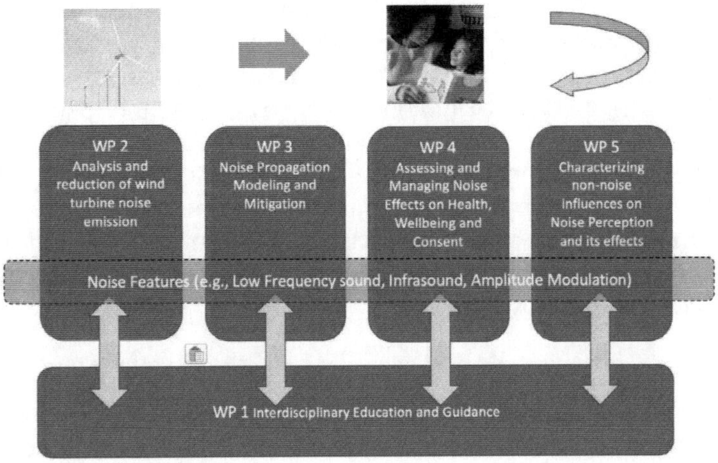

Figure 1: Task 39 Work Programme

Task 39 was initiated in October 2017 and its 1st Phase officially finalised after 3 years at the end of 2020. Phase #2 was approved by the Executive Committee around mid-2021, and a 2nd Phase kick-off meeting was held in September 2021. While the 1st Phase concentrated on engineering models, in the 2nd Phase, the objective is to propose a work programme with a more balanced approach for addressing both engineering and socio-psychological aspects.

1.1 Goals of Task 39 WP4

This WP includes a programme of activities designed to assess the contribution of wind turbine sound to noise perception, annoyance and the effects of these on health, wellbeing and consent. Proposed activities in this task include both lab and field-based psycho-acoustic annoyance testing as well as exploring the possibility of using auralization and stimulus synthesis in annoyance assessments.

1.2 Goals of Task 39 WP5

Social acceptance of wind turbines is driven to some extent by noise produced by wind turbines, but there is evidence of an effect in the reverse direction. That is, sensitivity to wind turbine sound/noise[1] may be driven partly by social acceptance with lower acceptance driving greater sensitivity to such noise. These complex iterative interactions require detailed research to investigate and interactions with Task 28 are likely to be a great benefit to this activity. Regulations in some countries impose a 'penalty' on audible characteristics of wind turbine sound such as tonality. Development of penalty schemes for amplitude modulation is ongoing. Such penalty schemes are predicated on the concept that annoyance is related to a sound level (measured in decibels). A further step assumes that a penalty in decibels can equate the annoyance of a sound with an audible feature with a higher sound level without the audible feature. The annoyance concept is a complex issue, and an investigation is required to validate this principle and estimate penalties if appropriate

2. Activities of Task Participants

WPs 4 and 5 activities have been ongoing since March 2022, with an online meeting gathering experts from different scientific horizons (engineering, noise assessment, psychology). It was

[1] The terms 'noise' and 'sound' are often (incorrectly) used interchangeably. Indeed in preparing this paper we considered the difference between objective 'sound' and perceived 'noise'.

highlighted at an early point that the group might need develop an efficient knowledge exchange program, so experts from different backgrounds could communicate effectively. For example, annoyance is an important concept to both fields, but depending on one's background the discussions on annoyance might deviate into one field. To address this issue, a seminar featuring presentations from experts in both Engineering and Psychology was held, and followed by an open discussion forum. Early meetings also indicated a desire to develop an effective knowledge sharing platform, and it has been proposed that this will be through a shared working documents (hosted on the Open Science Framework) to facilitate the joint definition of technical concepts (from multiple fields). These actives are ongoing and are informed/supported by the collaborative activities of partners, described below.

2.1 Denmark

The department of Acoustics, Noise and Vibration at FORCE Technology focus on some of the unanswered questions from the DecoWind project in the EUDP project "Participation to IEA Wind Task 39" (grant 134-21022). DecoWind was a 3-year Danish research project whose goal was to devise advanced control strategies for wind turbines and farms for minimizing their acoustic impact. It was a collaboration between FORCE Technology, DTU, Siemens-Gamesa Renewable Energy, and EMD International.

In Denmark and several other countries noise is regulated as absolute levels, hence the audibility of the noise source is not directly handled. In the rural areas some of the common sources of noise is vegetation or waves, which masks other environmental sound/noise sources. The effect of masking from vegetation and/or waves has not been studied in much detail in Denmark. The aim of this work package in the EUDP project is to gather data from vegetation and wave sound/noise and use this to form simple models for both vegetation and wave sound/noise. The models are used to estimate the audibility of wind turbines erected in rural areas, considering both temporal effects, spectral effects and effects of wind turbine size, distance and wind shear.

In parallel, the auralization of wind turbine sound/noise is studied, both in the EUDP project with a focus on auralization of offshore wind turbines, and in a Performance Contract with a focus on the auralization of onshore wind turbines.

Separately, the Science, Technology & Innovation (STI) group at DTU Wind and Energy Systems is working on a project called Co-Green, funded by the Independent Research Fund Denmark, using the case of wind turbine noise to explore how different scientific disciplines understand noise, how and to what extent they work together, and how regulations and policies are impacted by their work. Working with Danish wind farm case studies, the project takes lay knowledge of the environment seriously and considers conventional experts as part of the network surrounding the issue of how wind turbine sound is perceived as noise, and what the regulations do about it. The results of the project are envisaged to help towards a more inclusive and co-created approach to wind turbine siting, and to the subject of wind turbine noise, in particular.

2.2 Germany

The Stuttgart Chair of Wind Energy (SWE) at the University of Stuttgart is involved in the interdisciplinary Project Inter-Wind (grants 03EE2023A-D), together with the Karlsruhe Institute of Technology (KIT), Center for Solar Energy and Hydrogen Research Baden-Württemberg (ZSW) and the MSH Medical School Hamburg (MSH). Within the framework of this project, the relation of meteorological, acoustic and ground motion measurements with annoyance reports from residents of a wind farm in southern Germany is being investigated (see Gaßner et al., 2022; Müller et al., 2023).

Since 2020, three measurement campaigns have been carried out at a wind farm on the Swabian Alb in southern Germany. The SWE is has carried out acoustic measurements. One microphone was placed close to a wind turbine (WT) of the wind farm, two other microphones outside and inside residential buildings in the municipality at a distance of 1km from the wind farm. In parallel, the KIT assed ground motions, the ZSW added meteorological and operation data from the turbines. A field experiment on how different operations modes result in annoyance mitigation measures is in progress. Findings from the measurement campaigns and the combined data analyses in relation to reported annoyance will be discussed with the IEA Task.

Additionally, in cooperation the Delft University of Technology and the MSH conducted a first acceptance analyse by residents of the Klixbüll test side in northern Germany. This work of linking sound emission of established and airborne wind energy and social acceptance benefits from intensive exchange with the IEA Tasks 28 and 48.

The Institute of Communications Technology (IKT) at Leibniz University Hannover is working on the publication of sound recordings of wind turbines from our former Project WEA-Akzeptanz (Wind Turbine Acceptance) (see Schössow et al., 2022). The collected extensive data set will be published open-access considering the FAIR-principle. Within the dataset there is meteorological, sound pressure and turbine-specific data. The dataset published initially will hold the recordings of one month of three wind turbines, three microphones and one meteorological mast. Spatial audio and 360° videos will be available on request for some days of the measurement campaign and could be used as stimuli in laboratory studies or as ground truth for auralisation purposes.

In terms of laboratory studies in the course of a student thesis the influence of the presentation format of visual stimuli was investigated (Schössow et al., 2023). For this study the presentation of 360° videos on three projection screens surrounding the participant in the real lab, the same situation but build in VR as well as the pure 360° video were chosen as presentation modes. The results from the study show that the annoyance ratings and distance estimation of the turbine are unaffected by the visual presentation mode. However, the overall immersion and feeling of interaction was significantly better for the 360° video than for the two "flat" presentations.

2.3 Ireland

Since Ireland's first wind farm development in 1992, wind farms have been regulated in terms of noise exposure-response guidelines, in which exposure is measured in terms of a sound pressure level. Since then, the supporting science has improved and the psychoacoustics of wind turbine noise has identified features other than noise that induce annoyance. However, the prevalence and impact of these features on Irish wind farm communities has not been assessed.

In 2022, under the Irish Research Council COALESCE (Collaborative Alliances for Societal Challenges) funding call, the interdisciplinary project 'Wind Sense' was commissioned. The COALESCE call was designed to support the development of interdisciplinary and intersectoral collaboration/capacity in the context of national or global challenges. It is hoped awardees would expand their research activities and build the sustainability of their research agenda through enhanced competitiveness for future success in European or international collaborative funding programmes. The Call supports researchers to form new connections and to consolidate existing national and international knowledge networks as part of a challenge-based approach. A challenge-based approach will bring together resources and knowledge across different fields, technologies, and disciplines, including social sciences and the humanities, and indeed beyond academia, into new sectors. A key component of the COALESCE grant is that it is led by a PI from AHSS, but includes a Co-PI from a STEM field. Given the interdisciplinary goals of the Wind Sense project, concerning the psychoacoustics of wind turbine noise, the project aligns with the strategy of the COALESCE call.

The aim of the 'Wind Sense' project, (led by O'Hora and King at the University of Galway) is to generate WTN annoyance maps for Irish wind farms based on novel sound quality models of the prevalence of WTN features around candidate wind farms and the impact of these features on annoyance. Wind turbine noise annoyance maps will be generated for the candidate wind farms and will allow for the development of a national wind turbine noise annoyance map in Ireland, to inform turbine developers and policy makers. The project sees collaboration with IEA Task 39 Members, and activities are ongoing.

3. Key Research Questions

The following key research questions have been identified by Task 39 participants, and might be considered as research topics in need of further interdisciplinary investigation:

- Health vs Well-Being. These terms are often used interchangeably, but they mean different things. For example, stress due to wind turbine noise may result in an elevated stress experience and thus will impact well-being, but this may not result in health issues in every case.
- Annoyance – what is 'annoyance'? The scientific community needs a robust definition for better understanding of annoyance and associated impacts. If the scientific community does not have an accepted definition, then the general public impacted by developments may become irritated, e.g., different understandings result in different amount of strongly annoyed residents (see Hübner et al., 2019).
- Benchmarks. There is some debate on the developments of benchmarks for annoyance; while some standards are under development/revision as well as other factors (such as a technical specification for 'non-acoustic' factors related to annoyance) being considered by ISO Technical Committees, there appears to be no formally accepted benchmarks for annoyance.
- Mitigation measures. There are a limited number of studies that have performed assessments of (validated) mitigation measures (including, for example, any experiments that quantify how many people are less annoyed following mitigation measures). This could be further complicated by varying planning restrictions across countries; for example, oftentimes once wind turbines are operational it is difficult to assess low-noise operations if such measures were not included in the original planning process. It may be that more flexibility is needed in planning processes in order to allow for the assessment of mitigation measures.
- International cooperation. It would be interesting to examine in detail the different approaches to the management and control of wind turbine noise in different countries. For example, Germany has different noise settings during day/night period, but such an approach is not possible in Denmark, while set-back distances can vary widely from country to country. It would be beneficial to perform cross-country comparisons between planning conditions/restrictions to determine the pros and cons of various approaches in practice.
- Set-back distances. It is unclear if set-back distances have a discernible effect on annoyance; contradictive findings exist. When based on GIS-data, it would seem that distance does not have a significant impact (if emission regulations are applied)
- New sources. Innovative technologies (AWE) will lead to new noise sources. The scientific community will need to adapt to these new sources and assess the potential impact on annoyance. For Task 39, this could lead to potential collaborative opportunities with Task 48 (on Airborne Wind Energy).

4. Conclusions

This paper presents an overview of the activities of IEA Wind TCP Task 39 that are primarily associated with the socio-psychological aspects of wind turbine sound/noise (WP4 and WP5). We would encourage industry to consider these aspects in more detail, as any solutions that address socio-psychological aspects will need industry support. There is a real value on engagement and working with communities, both in terms of engaging the community for solutions, but also preparing them for actual impacts.

Task 39 is working on science-based solutions to these issues, but it is recognised that the issues we are trying to address are very much a transdisciplinary issues. These will require collaborative research transcending individual disciplines to construct knowledge beyond the scope of any single discipline.

References

Gaßner, L., Blumendeller, E., Müller, F. J. Y., Wigger, M., Rettenmeier, A., Cheng, P. W., Hübner, G., Ritter, J., & Pohl, J. (2022). Joint analysis of resident complaints, meteorological, acoustic, and ground motion data to establish a robust annoyance evaluation of wind turbine emissions. Renewable Energy, 188, 1072–1093. https://doi.org/10.1016/j.renene.2022.02.081

Hübner, G., Pohl, J., Hoen, B., Firestone, J., Rand, J., Elliot, D. & Haac, T. R. (2019). Monitoring annoyance and stress effects of wind turbines on nearby residents: A comparison of U.S. and European samples. Environment International, 132. (Doi.org/10.1016/j.envint.2019.105090).

Müller, F.J.Y., Leschinger, V., Hübner, G. & Pohl, J. (2023). Understanding subjective and situational factors of wind turbine noise annoyance. Energy Policy, 173, 113361, https://doi.org/10.1016/j.enpol.2022.113361.

Schössow, D., Bergner, J., Preihs, S. & Peissig, J. (2022). Concept for an open database of acoustical wind turbine measurements. Wind Europe Technology Workshop 2022, Brussels.

Schössow, D., Kawczynski, D., Preihs, S., Peissig, J. (2023). Der Einfluss visueller Präsentationsmodi auf die Wahrnehmung von WEA-Stimuli in Probandenstudien. Fortschritte der Akustik - DAGA 2023, 49. Jahrestagung für Akustik, Hamburg.

Bertagnolio, F et all (2022). DecoWind: Development of low-noise and cost-effective wind farm control technology. Inter-noise 2022. https://az659834.vo.msecnd.net/eventsairwesteuprod/production-inconference-public/3889aefd315848d6bf9656a24c0a820d

Søndergaard, L. S. et all (2021). Wind farm neighbourship investigated by a daily app questionnaire combining weather, noise, and annoyance. 9th International Conference on Wind Turbine Noise

10th International Conference on
Wind Turbine Noise
Dublin – 21st to 23rd June 2023

The many understandings of noise and the opportunities and pitfalls of interdisciplinary research

Julia Kirch Kirkegaard, DTU Wind & Energy Systems, jukk@dtu.dk
Tom Cronin, DTU Wind & Energy Systems, tocr@dtu.dk
Sophie Nyborg, DTU Wind & Energy Systems, sonyb@dtu.dk

Summary

The accelerated deployment of renewable energy generation is seen as the backbone of most energy transition policies around the globe. This continues to put pressure on finding space for wind farm development to fulfil the capacity targets. In turn, these deployments face increasing societal resistance in many countries (Cousse et al. 2020; Lintz and Leibenath 2020), constituting a grand challenge to wind energy (Kirkegaard et al. 2023; Veers et al. 2019). Recently, it has been argued that finding ways for *how* to tackle this challenge, so that wind energy is implemented in a just but timely manner, constitutes a grand challenge in itself and one that can benefit from interdisciplinary research (Kirkegaard et al. 2023)

In this presentation/session, we would like to explore how and whether an interdisciplinary approach to the 'social-technical grand challenge' of wind energy deployment is a viable solution to the entangled challenges of renewable energy deployment and social acceptance. For this, we have been working on using the case of wind turbine noise to explore the potential avenues, but also barriers, to conducting interdisciplinary research in the attempt to tackle this social-technical grand challenge.

Wind turbine noise - the sound produced by wind turbines - has been recognised as one of the most prevalent and tangible sources of public contention in wind farm developments (Borch et al. 2020; Solman et al. 2022), but why is this so? Our contribution to this session is based on research conducted in the Co-Green project funded by Independent Research Fund Denmark that is focused on the issue of how and why wind turbine noise in Denmark is problematised (i.e. how do various groups come to the conclusion that wind turbine noise is a concern?) and politicised (i.e. how are these concerns expressed and have influence), causing controversy.

Based on primary data (interviews) and secondary data (document studies, literature review), we present an analysis of some of the underlying reasons our research has uncovered as to why wind turbine noise is so difficult to solve and 'fix', not only by different disciplines alone (Kirkegaard et al. forthcoming), but also when collaboration across disciplinary boundaries is attempted (Nyborg et al forthcoming).

First, we illustrate how noise is different things to different disciplines (Kirkegaard et al. forthcoming). To engineering science, wind turbine noise is substantially a technically definable issue, that can (and should) be controlled through regulatory governance. The issue is one that can be resolved by 'experts' through reducing the volume of the noise, i.e. the noise level measured in dB(A), by technological innovations in wind turbine design (Wagner 1996), and communicated via calculations, simulations, tests and maps. Here, we find the figure depicting dB(A) vs. types of noise and the wind turbine noise contour maps in EIA reports to be established examples. In turn, the understanding of the health-related discipline (e.g. medical and psychology fields) is based upon the premise that adverse levels or types/qualities of sound can have an impact on people's health and wellbeing due to a somewhat quantifiable effect of the noise on a subject's state of annoyance (e.g., van Kamp and van den Berg (2021); van den Berg 2021). This field has been dominated by an attempt to establish a relation between a measurable 'dose' of noise and a meaningful human response (annoyance). The most evident example of this is the influential dose-response graphs (Pedersen & Waye 2004; based on Schultz 1978) and the comparisons with other forms of noise (e.g. road traffic, trains and planes). This discipline furthermore has increasingly struggled to tackle 'non-acoustic factors' which, in a way, somehow 'pollute' these rather graceful but simple graphs and models. In particular, findings from the social science-based literature coming from the perspective of what is commonly termed "social acceptance", have proved challenging to integrate (e.g. Walker et al. 2015; Wolsink et al. 1993; Haggett 2012). For its part, the social science discipline has detected various sources of resistance and concerns over noise that cannot be related directly to the 'dose', but rather to various others concerns (e.g. visual impact and the manner of public engagement), and which can often not be easily quantified or correlated (also see Taylor and Klenk 2019). Indeed, this discipline does not strive to identify or isolate a public response dedicated to noise, but rather strongly emphasises the complexity and inherent entanglement of various issues that go to make up a particular stance on the subject of wind farm siting.

Second, having outlined this background, it forms the foundation for us to frame a broader discussion of two themes:

- *Do we regulate wind turbine noise in a way that helps to ameliorate people's concerns?* We argue that the different literatures to different degrees have been able to influence noise regulations and standards. In particular, influential graphs such as the dose-response relationship and its underlying logic has been able to travel into noise standards and regulations; however, it is not necessarily able to solve the fundamental issues of different understandings of what noise is, and thus of how it should be solved. This has also raised the significant question as to whether the social acceptance literature – often with relative little quantitative data or visual inscriptions – stands a chance to be translated into policy?

- *Can more interdisciplinary research take place, and at a deeper level, on the issue of noise, and how? What is this aiming to achieve?* In our work we have used empirical data collected from international conferences focused on noise and IEA Tasks concerned with noise to explore the limitations and opportunities of doing interdisciplinary research in this realm. We ask the question: is it at all possible to solve the problem with noise across disciplines if we see noise as fundamentally different things? If reducing the technical volume of noise is not the issue, then how can non-technical disciplines provide guidance on what to do? Or do we need to first agree on what the (noise) issue is, but how, and with what means and ends?

Finally, we would like to end by discussing how to move forward in the wind energy sector if we want interdisciplinarity to take place in the energy transition broadly speaking, and in terms of wind turbine noise specifically. Having shed some new light on the different understandings of what noise is, and the different types of solutions they give rise to, our presentation might give rise to somewhat provocative questions such as 'does noise reduction help reducing wind turbine annoyance?', 'does interdisciplinarity make sense if we look at different things', or 'is it only quantitative data that can be translated into noise regulations and standards'?

References, selected

Borch K., Munk A.K., and Dahlgaard, V. (2020) Mapping wind-power controversies on social media: Facebook as a powerful mobilizer of local resistance, *Energy Policy*, 138(111223). doi: 10.1016/j.enpol.2019.111223

Cousse, J., Wüstenhagen, R., Schneider, N. Mixed feelings on wind energy: Affective imagery and local concern driving social acceptance in Switzerland. Energy Research & Social Science, 70, 101676 (2020).

Haggett, C. (2012) The Social Experience of Noise from from Wind Farms. In Learning from Wind Power, https://doi.org/10.1057/9781137265272.

Kirkegaard, J. K., Rudolph, D. P., Nyborg, S., Solman, H., Cronin, T., Gill, E., and Hallisey, M. (2023, in print): Tackling Grand Challenges in wind energy through a socio-technical perspective. *Nature Energy.*

Kirkegaard, J. K., Nyborg, S., Frantzen, D. N., Cronin, T., and Horst, M. (work-in-progress paper). *Making science relevant for regulation – on visual inscriptions as immutable mobiles and their agency to mobilise regulation around contested wind turbine noise.* To be submitted to *Social Studies of Science, May 2023.*

Lintz, G., Leibenath, M. The politics of energy landscapes: the influence of local anti-wind initiatives on state policies in Saxony, Germany. Energy, Sustainability and Society,10(5), 2020.

Nyborg, S., Kirkegaard, J. K., Frantzen, D. N., Cronin, T., and Horst, M. (forthcoming). Doing interdisciplinarity around the wicked problem of climate change – the case of wind turbine noise To be submitted to *Science, Technology & Human Values, May 2023.*

Pedersen, E., and Persson Waye, K. (2004) Perception and annoyance due to wind turbine noise—a dose–response relationship, *The Journal of the Acoustical Society of America*, 116(6), pp. 3460–3470. doi: 10.1121/1.1815091

Schultz, T. J. (1978) Synthesis of social surveys on noise annoyance. *Journal of Acoustical Society of America*, 64(2), p.. 377-405. doi: 10.1121/1.382013. PMID: 361792.

Solman, H., Kirkegaard, J. K. and Kloppenburg, S.: Governing concerns around wind turbine noise: analysing a noise monitoring app as a digital framing device. *Energy Research & Social Science,* https://www.sciencedirect.com/journal/energy-research-and-social-science/vol/98/suppl/C.

van Kamp, I., and van den Berg, F. (2021) Health effects related to wind turbine sound: A review. In RIVM. doi:10.3397/IN-2021-3088

van den Berg, F. (2021) The hidden meanings in environmental sounds: the case of wind farm projects, *Vibrations in Physical Systems*, 32(1), pp. 1–8. doi:10.21008/j.0860-6897.2021.0.00

Taylor, J. and Klenk, N. (2019). The politics of evidence: Conflicting social commitments and environmental priorities in the debate over wind energy and public health. Energy Research & Social Science, 47, pp. 102-112.

Veers, P., Dykes, K., Lantz, E., Barth, S., Bottasso, C. L., Carlson, O. (2019). Grand challenges in the science of wind energy. *Science* 366 (6464).

Wagner, S., Bareiβ, R., Guidati, G. (1996) Wind Turbine Noise. Springer. ISBN-13: 978-3-642-88712-3 DOI: 10.1007/978-3-642-88710-9

Walker, C., Baxter, J. and Ouellette, D. (2015) Adding insult to injury: The development of psychosocial stress in Ontario wind turbine communities, *Social Science and Medicine*, 133, pp. 358–365. https://doi.org/10.1016/j.socscimed.2014.07.067

Wolsink, M., Sprengers, M., Keuper, A., Pedersen, T. H., and Westra, C. A. (1993) Annoyance from wind turbine noise on sixteen sites in three countries. Proceedings of the European Community Wind Energy Conference, May 2017, pp. 273–276.

WHO (2018), https://www.euro.who.int/__data/assets/pdf_file/0008/383921/noise-guidelines-eng.pdf

10th International Conference
on
Wind Turbine Noise
Dublin – 21st to 23rd June 2023

Noise forecast app: positively impacting non-acoustic factors

Erik Koppen, Arcadis: erik.koppen@arcadis.com
Madelon Ekelschot-Smink, Arcadis: madelon.smink@arcadis.com
Jaap Boon, Arcadis: jaap.boon@arcadis.com

Summary

Literature shows annoyance by wind turbines is related to the sound level. The higher the exposure to wind turbine noise, the more annoyance occurs. However, between communities and between individual persons large differences occur in the annoyance level at a specific sound level. This is caused by a number of factors including non-acoustic factors like situational, personal and contextual factors. The interactive noise forecast app aims to positively impact a number of non-acoustic factors in the operational phase of a wind farm. Also, it aims to create more insight into the conditions in which annoyance occurs, so a more effective noise management and communication strategy is possible. The app has been and is being used in multiple projects comprehending over 100 turbines, over 500 MW wind capacity and over 17,000 addresses. This paper describes the non-acoustic factors the app is targeting, what information is presented and what the interactive part comprehends. General findings, preliminary insights and results generated by application of the app in various projects are presented.

1. Introduction

Wind energy is one of the renewable energy sources we need to reach the climate goals of the Paris Agreement. Nevertheless, residents near wind farms have concerns about the environmental impact of wind turbines. One of their biggest concerns is the impact of wind turbine noise on the living environment and their health. Residents near wind farms under development often have no idea what to expect. Participation in the siting process and clear communication by developers and authorities can give residents a better understanding and take away some of the concerns. However, the day-to-day impact stays unclear to them since this highly depends on variable weather conditions. Another factor that plays an important role is that real noise issues around existing wind turbines are not always addressed properly. Since the level and character of wind turbine noise can change strongly depending on atmospheric conditions, wind farm operators do not always recognize and acknowledge the issue. Lack of understanding can lead to more annoyance and negative publicity which in turn negatively affects residents of wind farms under development.

Arcadis developed an interactive app that targets both acoustic as well as non-acoustic factors that affect how people perceive wind turbine noise. This paper describes the factors the app is targeting and what the app comprehends. Also, insights and results generated by applying the innovative app are presented. At this stage only preliminary and limited results can be presented due to the stages of different projects and the fact that it is currently only allowed to disclose data that cannot be traced back to specific projects (unless it is already public).

2. Reducing annoyance by wind turbines

2.1 Annoyance by wind turbine noise

Annoyance by wind turbines is related to the sound level. The higher the exposure to wind turbine noise, the more annoyance occurs. However, between communities and between individuals large differences occur in the annoyance level at a specific sound level (Janssen, Vos, Eisses, & Pedersen, 2011) (Michaud, Keith, Feder, Voicescu, & et al., 2016). The Community Tolerance Levels derived by Michaud et al. show that certain communities are 8 decibels less tolerant of wind turbine noise than other communities. This is partly caused by the specific character of wind turbine sound like the low-frequency noise content, the degree of amplitude modulation, a possible tonal component, possible masking by ambient noise and the degree of sound insulation of the residences in a specific community. Besides these acoustic factors also non-acoustic factors like situational, personal and contextual factors affect how residents perceive the sound of wind turbines. Situational factors are factors like the visual impact (e.g. impact on the landscape, shadow flicker and aviation obstruction light markings), other sound sources and the attractiveness of the area. Personal factors include aspects like noise sensitivity, expectations, health and safety concerns, economic benefit and attitude towards wind energy, energy companies and authorities. Contextual factors include factors like participation in the planning process, procedural justice and feelings of fairness (van Kamp & van den Berg, 2020). Van Kamp & van den Berg conclude wind turbine sound has a moderate effect on annoyance. Besides sound a range of non-acoustic factors influence the level of annoyance. They therefore imply considering these other factors will help reducing the impact of wind turbine sound. The Knowledge Document Wind Turbine Noise published by the Dutch Knowledge Platform Wind Energy in June 2015 also mentions the for residents unpredictable character of wind turbine noise due to wind variations as a factor influencing the level of nuisance (RIVM, Aarten Wind Solutions, ECN, GGD-GHOR Netherlands & EAE/RUG, 2015).

2.2 Positively impacting non-acoustic factors

In 2017 Arcadis developed an interactive app to positively impact a number of non-acoustic factors in the operational phase of a wind farm that affect how neighbouring residents perceive the wind turbine noise. Also, it aims to create more insight into the conditions in which annoyance occurs, so a more effective noise management and communication strategy is possible. The app called **Noiseforecast.app** (in Dutch: **Geluidsverwachting.nl**) presents as main feature a short-term noise and shadow flicker forecast.

It is expected residents will be less annoyed if they know what level and duration of exposure to expect. The Dutch Knowledge Document Wind Turbine Noise (RIVM, Aarten Wind Solutions, ECN, GGD-GHOR Netherlands & EAE/RUG, 2015) presented the use of a weather forecast at 100 metres height as a potential improvement at operational wind farms. This could give neighbouring residents a better idea of the expected noise in the short term. The noise forecast app offers more than that. It not only presents a general weather forecast and a wind forecast at hub height, but also provides insight into the expected short-term exposure to wind turbine noise and shadow flicker.

The interactive app is an accessible tool for residents to submit, at any moment, feedback on how they personally perceive the noise and possible shadow flicker of nearby wind turbines. It is expected residents will be less irritated when they can easily and anonymously report nuisance. As a resident at a community meeting commented on the use of the app: '*That's nice. If I am annoyed by the wind turbines, I can report it immediately and it will be off my mind*'. Experience learned residents appreciate it if they can provide feedback anonymously. This is not only the case if residents experience nuisance, but also when they are not annoyed. The latter was once pointed out by a resident stating that as a large part of their small community was against the nearby wind turbines, she did not feel comfortable saying she was not.

In a report of the Dutch National Institute for Public Health and the Environment RIVM (Welkers, van Kempen, Helder, Verheijen, & van Poll, 2020) regarding the 2018 WHO Environmental noise guidelines, one of the recommendations for wind turbine noise is that the approach to wind turbines should strongly focus on the non-acoustic factors. As an example, the report mentions a Dutch wind farm where residents can report nuisance through an app. When severe nuisance is experienced, the wind farm operator can decide to shut down the wind turbines. Since the residents feel they have some control they are less annoyed, knowing the wind turbines will be shut down if unacceptable annoyance occurs.

Providing insight into how the sound level relates to weather conditions and ambient noise, will help residents understand better why the sound is (more) audible at specific times. Also, giving residents insight into the energy yield and avoided CO_2 emissions will help them get a better understanding of the benefit of wind turbines and the need for larger turbines. More insight and understanding will enhance trust and social acceptance.

In 2021 Arcadis was involved in a limited survey by students amongst residents regarding what information they would appreciate in the operational phase of a wind project. The survey showed what residents valued most was information on what measures are applied to mitigate noise and how other residents perceive the wind turbine noise. Also, a noise forecast and a monitor registering and presenting historic sound levels were deemed important by most respondents, as was information on the wind speed and direction at hub height. The largest spread in responses occurred for information on the energy production and on when and why certain turbines are not in operation at a specific time. A large share of respondents found this (very) important whilst another large share of respondents thought this information was not important at all.

In 2022 TNO reported about experiences from neighbouring residents of four different wind farms (Peuchen, Kox, Klösters, & Straver, 2022). The report concludes the positive or negative attitude towards a wind farm before it was built is a strong predictor for the perception of the wind turbines once they are operational. For a specific wind farm in the Netherlands an important factor in the mainly negative experience of the wind turbines was the result of lack of communication, participation and transparency by the wind farm developer. This created a lot of concerns and distrust among the residents. Transparent communication will help reducing distrust and concerns. The app can assist in transparent and frequent communication.

2.3 Positively impacting acoustic factors

Besides continuously informing residents, the app also gathers feedback from residents on how they personally perceive the noise and possible shadow flicker of the turbines. The app offers a listening ear to all local residents. The feedback from residents is used to continuously monitor nuisance and to gain more knowledge on the influence of weather and environmental conditions. It also shows how nuisance depends on distance and orientation to the wind farm and how it varies over time. A better understanding of the conditions in which annoyance occurs makes more effective noise management and communication strategy possible. Insight into under what conditions annoyance occurs helps determining a more effective application of noise reducing measures. Also, the app helps monitoring and further optimizing the effect of these mitigating measures. Therefore, the app is not only aimed at impacting non-acoustic factors, but also at impacting acoustic factors.

The Dutch Knowledge Document Wind Turbine Noise (RIVM, Aarten Wind Solutions, ECN, GGD-GHOR Netherlands & EAE/RUG, 2015) stated a regular consultation between operators and residents, and possibly authorities, could improve the situation at operational wind farms. A regular consultation can contribute to the timely identification of issues. This way, operators can propose mitigating measures in above-average nuisance situations. Continuously monitoring the perception of residents via the app assists in a timely identification of issues and provides more specific input for the regular consultations. The wind farm operator will also obtain a better understanding of the position of the local residents because of better insight into how the residents actually experience the wind turbines.

3. Noise forecast app

The interactive app called **Noiseforecast.app (**in Dutch: **Geluidsverwachting.nl**) provides neighbouring residents a hyper-local wind turbine noise forecast. Residents can 24/7 view an up-to-date 48-hour forecast for their home. The noise forecast is based on a combination of high-resolution weather models, wind turbine specifications, the local environment and noise source and propagation calculations. The calculations take into account weather parameters like wind speed and direction at hub height and 10 metres height, temperature, humidity, atmospheric stability, cloud cover etcetera.

The app also presents an estimate per hour for the expected perceptibility of wind turbine noise. This takes into account possible (partial) masking by ambient noise from sources like road traffic, possible industrial facilities, wind-induced noise and wind related rustling of leaves. The perceptibility is estimated based on literature and the characteristics of the local ambient noise sources, taking into account a local weather forecast. The estimated perceptibility is regularly being refined based on the feedback from residents.

The app presents the local residents also a shadow flicker forecast for their home. If shadow flicker is being expected, the app also shows the expected duration of the shadow flicker.

The app not only presents forecast data but is interactive. With the app neighbouring residents can easily and anonymously report how they perceive the wind turbine noise and shadow flicker, 24/7. For wind turbine noise a 7-point scale is used, ranging from 1 (no annoyance) to 7 (highly annoyed). For shadow flicker a 5-point annoyance scale is used. When feedback is submitted, not only the annoyance level but also the exact date and time, and – with consent of the user – the location are registered. With this information the feedback level can be linked to the forecasted sound level, perceptibility, shadow flicker and weather conditions for the specific time and location. Furthermore, the geographical distribution of the feedback responses can be determined. The geographical distribution is only shared with the client or third parties in an anonymous way.

Besides a forecast for wind turbine noise and shadow flicker, the app also presents a local 48-hour weather forecast near ground level, the wind force and direction at hub height, and the expected energy yield in percentage of maximum production. Optionally, the app provides graphs showing the historical energy production per hour, per day, per month and per year.

The app also presents news and info pages to inform neighbouring residents about the wind farm, the app, relevant developments and interesting findings.

Figure 1 presents a selection of screen shots of the app.

Figure 1. A selection of screenshots of the noise forecast app

4. Preliminary results of applying the app

4.1 Experience in projects

The noise forecast app is used in a variety of projects comprehending over 100 turbines, over 500 MW wind capacity and over 17,000 neighbouring residences. Depending on the site the app is made available to the residents by the owner of the wind farm, by municipal authorities and by a provincial authority together with owners of wind farms.

The app was initially tested in a pilot project for a wind farm that was taken in operation the year before. The pilot project ran for a year. All residents within 1.5 kilometre distance from the wind farm were invited to participate. Surprisingly, the residents only reported low nuisance levels. On a nuisance scale of 1 to 7 the reported nuisance levels varied from scores 1 (no nuisance) to 3 (slightly annoyed). In total 88% of all feedback responses indicated score 1, i.e. no annoyance at all. Of the seven nuisance levels to be chosen from, the four highest nuisance levels were never reported. Residents explained they were not or hardly annoyed by wind turbine noise, because the turbine noise was usually masked by the background noise of the nearby motorway.

For another wind farm residents did report considerable annoyance by wind turbine noise. The reported nuisance levels varied from 1 (no nuisance) to 7 (highly annoyed). The feedback from local residents via the app was used to determine under what weather conditions most nuisance occurred. This was crucial in determining the cause of the noise complaints and in investigating possible and efficient mitigating measures. After several improvements were made, measurements demonstrated a significant noise reduction had been achieved. Continuous monitoring via the app showed that as a result the number of negative feedback responses had been reduced considerably.

Other wind farms also show large differences in how residents experience wind turbine noise. This is partly due to acoustic factors like the sound level at residences or the specific character of the turbine sound. However, even if the sound level and characteristics are similar, there are large differences in reported annoyance levels. There are strong indications that non-acoustic factors play an important role, like the level of acceptance or resistance in the development phase. As the pilot project demonstrated masking by ambient noise is also a relevant factor.

Most projects are ongoing and detailed results can or may not be shared in this stage. One case study is presented in the next paragraph, a project in which the involved municipalities recently shared the preliminary results with local residents. The last paragraph of this chapter describes some preliminary general findings based on all wind projects in which the app was or is being applied.

4.2 Case study Wind farm Oude Maas

On behalf of the Dutch municipalities Barendrecht and Hoeksche Waard the noise forecast app is being used for wind farm Oude Maas since March 2022. This wind farm is in operation since July 12th, 2022. The wind farm consists of five 3.6 MW wind turbines with 120 metres hub height.

The turbines are positioned along the river Oude Maas. The wind farm is situated in an agricultural area just south of the river, both to the west and to the east of the motorway A59. This motorway crosses the river through a tunnel. The wind farm is located in the municipality

Hoeksche Waard, but the dwellings north of the turbines are located in the municipality Barendrecht. Residents were concerned about the noise impact of the wind turbines and requested noise measurements before and after construction of the turbines. The municipality Barendrecht did not honour this request, but alternatively decided to apply the Arcadis noise forecast app. The app was selected to inform residents 24/7 about the expected exposure to noise and shadow flicker, and to monitor the perception of the residents and the level of annoyance. The municipality Hoeksche Waard decided to join this initiative.

The closest residences are located at about 550 to 650 metres from the wind farm. In the direct vicinity of the turbines a total of 4196 residences are present within a radius of 2 kilometres from the turbines, of which:

- 86 residences within 1 kilometre distance;
- 789 residences within 1 to 1.5 kilometres distance;
- 3321 residences within 1.5 to 2 kilometres distance.

All these residences were invited to use the app and were included as a forecast location.

The municipalities promised residents to regularly provide feedback on the results of the app. Intermediate results have been published on their websites and in April 2023 the results of the first 8,5 months were shared with the residents in a (digital) public meeting. The main intermediate results of the ongoing project are summarized below.

The number of residents using the app varied from 280 to 650 users per month. The average number of users was 416 per month, about 10% of the number of addresses that were invited to use the app. During the period July 12th, 2022 to April 1st, 2023, residents from in total 63 locations provided feedback on how they perceived the wind turbine noise. This is 15% of the average number of users per month and 1.5% of all residences within 2 kilometres distance. In 8,5 months in total 260 feedback responses have been received with regard to noise, i.e. on average nearly 4 responses per location. Residents from 6 locations provided feedback on how they perceived shadow flicker, 1% of the average number of users per month and 0.1% of all residences within 2 km distance. This means that even though it is easy to submit feedback, most residents and users never submitted feedback.

For shadow flicker in total 17 feedback responses were submitted, but no annoyance was reported. This indicates effective mitigation, since the operators of wind farm Oude Maas promised residents to automatically shut down turbines during periods they (threaten to) cause shadow flicker on neighbouring residences. If the shadow control system had not been correctly initialized this would have shown up quickly in the feedback responses as experiences for another wind farm learned.

Figure 2 shows for noise per distance class the number of feedback responses per nuisance level. Figure 3 shows per distance class the number of unique locations that submitted feedback per nuisance level. The total number of locations that submitted feedback is 13% of all residences within 1000 metres distance, 1.5% of all residences with 1000 to 1500 metres distance and 1.2% of all residences within 1500 to 2000 metres.

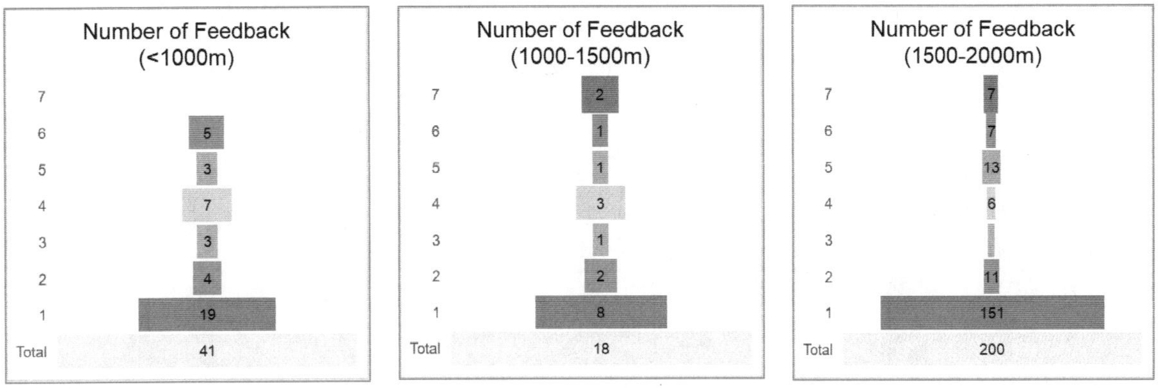

Figure 2. Number of feedback responses per nuisance level as a function of distance to the most nearby turbine.

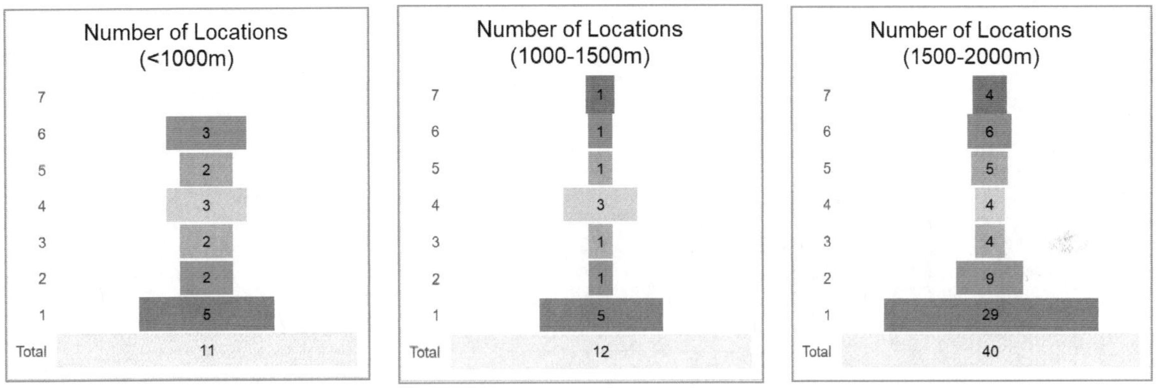

Figure 3. Number of locations submitting feedback per nuisance level as a function of distance to the most nearby turbine. The total number of locations that submitted feedback is lower than the sum of locations per nuisance level, since one location can submit several nuisance levels.

Figure 4 shows per month the number of feedback responses per nuisance level. Figure 5 shows per month the number of unique locations that submitted feedback per nuisance level. In total 178 of all feedback responses (68%) received indicate no annoyance. 39 feedback responses (15%) indicate high annoyance (nuisance level 5-7). Most negative feedback was received during the first five months of operation. The project is still ongoing. It will be interesting to see if the number of (negative) feedback responses will keep decreasing or will increase later in the year due to different weather conditions.

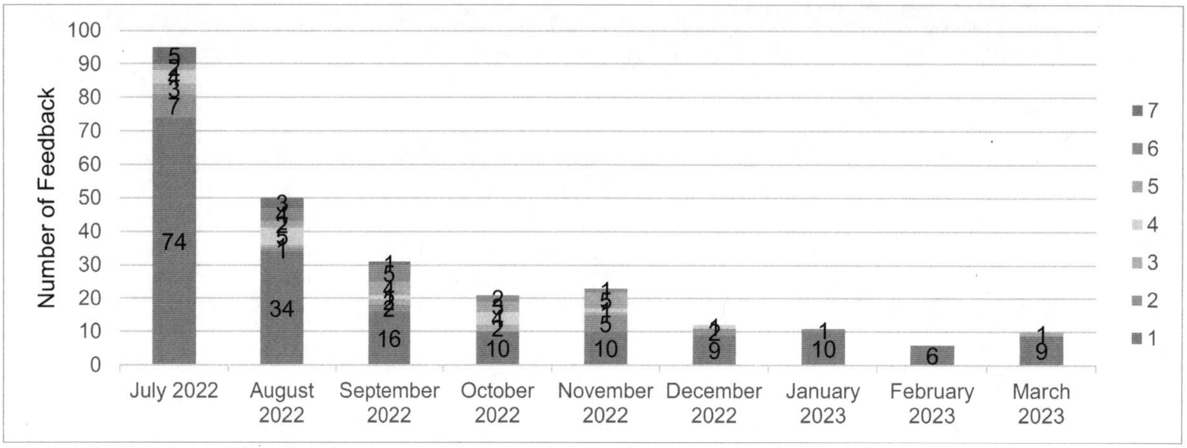

Figure 4. Number of feedback responses per nuisance level, per month

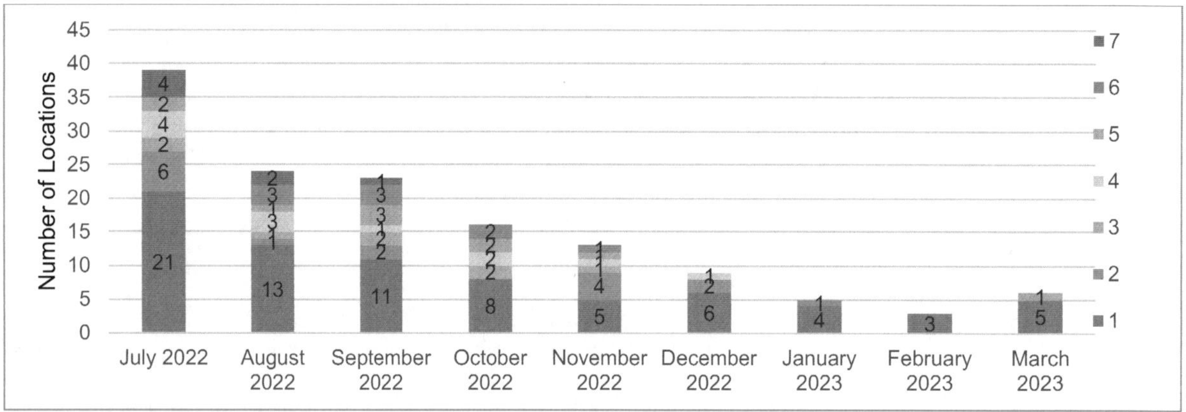

Figure 5. Number of locations submitting feedback per nuisance level, per month

4.3 Preliminary general findings

Up to now about 4600 times feedback has been submitted by residents from about 320 locations near wind farms where the app is made available. The number of feedback responses and number of locations submitting feedback strongly varies per wind farm. The number of residents submitting feedback varies from about 2 to 20% of the population. Of course, the longer the app is active, the more feedback is submitted by more residents. Since most projects are still ongoing this varies per wind farm. Another important factor is the level of nuisance that is experienced. The more nuisance is experienced, the more feedback is submitted. This is to be expected since residents that are annoyed, are more inclined to submit feedback than other residents. A number of residents that is not or slightly annoyed does still provide feedback, but all projects show the number of feedback responses for low nuisance levels declines with time. The way the app is communicated to the residents, what is done with the results, how results are communicated to the public and the method and frequency of communication also affect the number of users and respondents.

Some wind farm operators fear residents will not submit genuine feedback and might abuse the feedback option. When analysing the feedback not only the feedback responses per nuisance levels, but also the number of locations submitting feedback is considered. The analyses show how responses are distributed in time, location and distance to the turbines, and how they relate to forecasted sound levels. In some projects also the relation to turbine Scada data and to sound measurements in the field has been investigated. Though some deviating responses occur, a possible intentional active negative impact on the overall feedback has never had any significance on the overall outcome. Most responses can be explained, and overall residents appear to provide honest feedback. The higher sound levels are forecasted, the more feedback is submitted and, on average, the higher nuisance levels are reported. There is however a difference in how residents respond. Some residents only report nuisance level 7 when they are annoyed, other residents do use the full nuisance scale to indicate the level of annoyance.

Figure 6 gives an idea of the amount of feedback submissions versus the number of locations that submitted feedback in a certain week. It shows the number of locations submitting feedback and the amount of feedback strongly varies per week. The number of feedback responses submitted in a specific week varies from 1 to 4.5 responses per location. On average in a certain week, each location that submitted feedback responded 2.28 times. The feedback responses and the number of locations submitting feedback are strongly correlated. This is an indication of valid feedback.

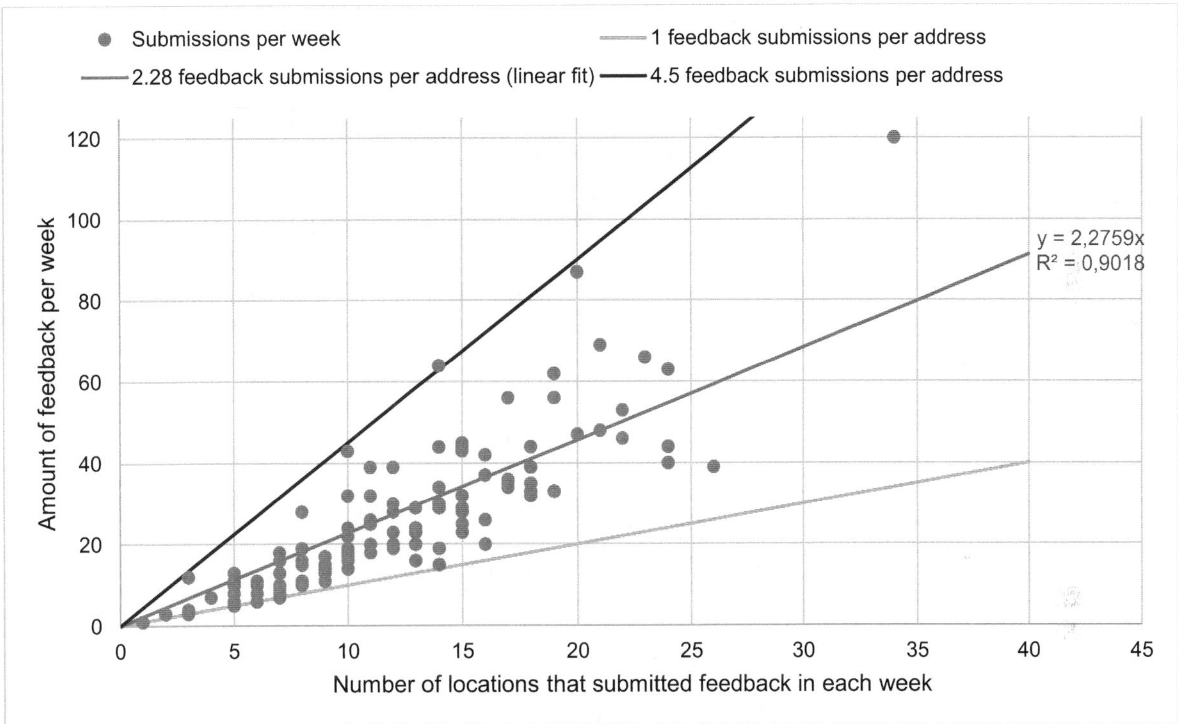

Figure 6. Amount of feedback submissions versus the number of locations that submitted feedback across all wind farms where the app in the specific weeks was implemented. Each dot represents one week of data.

Experience learned residents appreciate the app. Of course, as to be expected not all residents use the app. Not everyone is interested in the wind turbines. In general, after a start-up period about 10 to 30% of the neighbouring residences use the app and about 2 to 20% of the neighbouring residences actually submit feedback. Concerned or annoyed residents will be more inclined to use the app than residents who do not care about the wind turbines and who experience no nuisance. The app has proven an efficient tool to monitor the perception of residents, to identify issues and to identify under what conditions (most) annoyance occurs. The effect of mitigating measures can be monitored without taking actual noise measurements in the field.

An interesting preliminary finding is how reported nuisance levels are distributed over the hours of a day. Figure 7 shows the relative amount of feedback for high nuisance levels (scores 5-7) per hour of the day across all windfarms where the app was implemented. So, these are the hours people were most annoyed by wind farm. This graph was derived by determining the average amount of nuisance levels 5 to 7 that were submitted per hour of a day as a percentage of the average amount of nuisance levels over all 24 hours of a day. This analysis was done for each wind farm separately. To guarantee anonymity with respect to the individual wind farms only the maximum and minimum percentages from the bandwidth of all wind farms are presented in figure 7. Figure 8 presents a similar graph, but then for the three lowest nuisance levels (levels 1 to 3). So, these are the hours the residents submitted feedback through the app and were not or only slightly annoyed by the wind turbines.

Figure 7 shows the relative amount of the three highest nuisance levels is the highest at the end of the evening and the beginning of the night, and around breakfast, lunch and dinner time. When comparing this with the relative amount of the three lowest nuisance levels in Figure 8, it is observed that also feedback regarding low nuisance peaks around mealtimes. It can therefore be concluded that negative feedback around mealtimes mainly increases because the overall feedback increases. This is not the case for the peak in high nuisance levels at the end of the evening and the beginning of the night. This peak occurs at a time the amount of feedback for low nuisance levels is relatively low. Though it is a preliminary result, it is a strong indication that nuisance by wind turbine noise is most prevalent in the late evening and early night. This seems logical since it is the time most people go to bed, and ambient noise levels at this time will be considerably lower than during the day and the beginning of the evening. This means that if noise reducing measures are to be taken, they will likely be most effective in reducing overall annoyance during this period of the day. When more feedback data becomes available, more can be learned about the impact of wind turbine noise on residents. Some learnings will only be relevant and insightful for a specific wind farm, whilst other learnings can be relevant for a large number of wind farms.

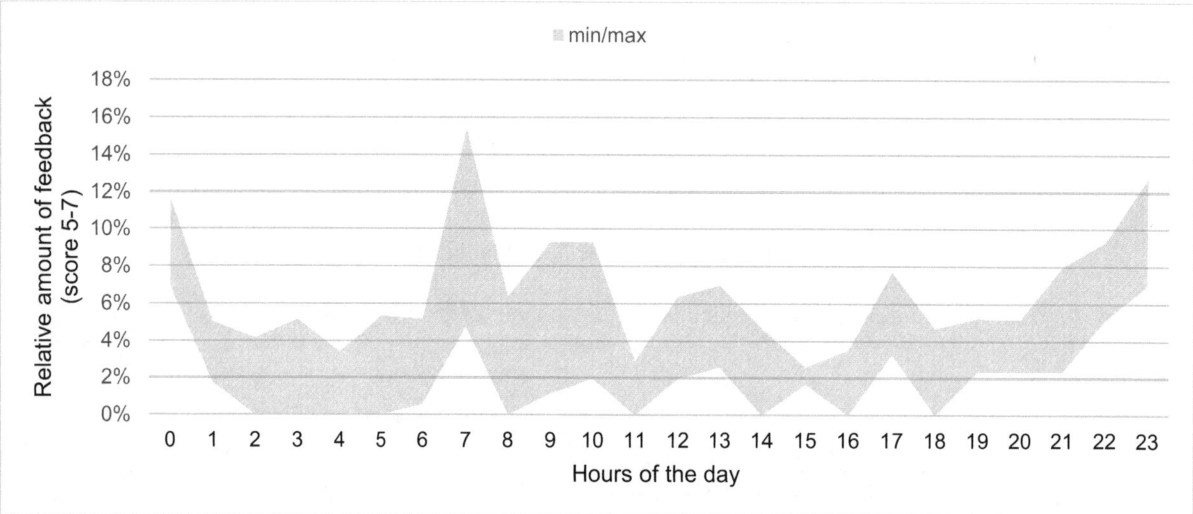

Figure 7. Relative amount of feedback for high nuisance levels (scores 5-7) per hour of the day, shown as a bandwidth ranging from minimum to maximum values across all wind farms where the app was implemented.

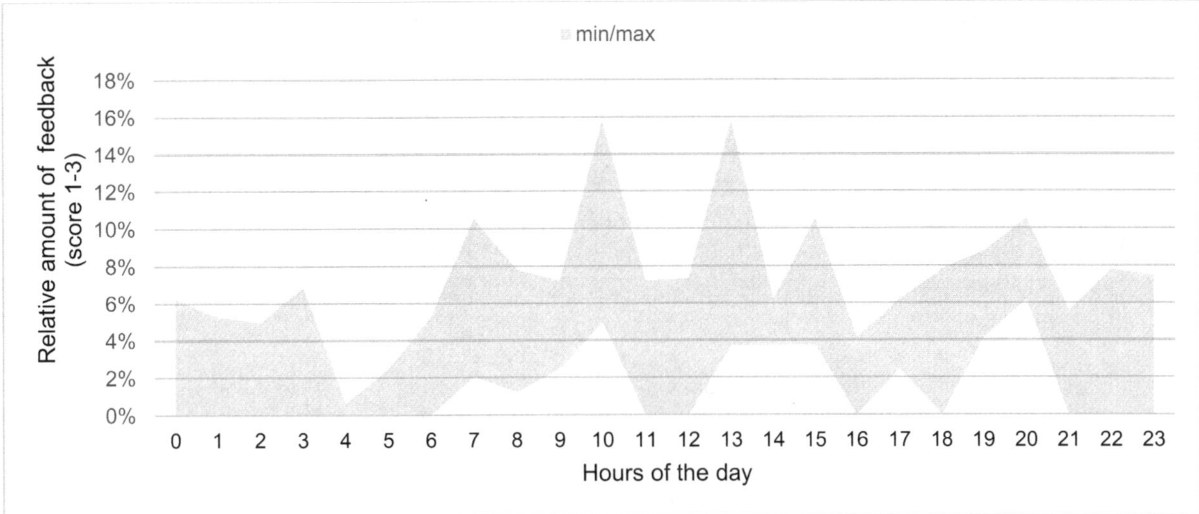

Figure 8. Relative amount of feedback for low nuisance levels (scores 1-3) per hour of the day, shown as a bandwidth ranging from minimum to maximum values across all wind farms where the app was implemented.

5. Conclusions

The noise forecast app is used in a variety of projects comprehending over 100 wind turbines, over 500 MW wind capacity and over 17,000 neighbouring addresses. Most projects are ongoing. The app targets both non-acoustic and acoustic factors that affect the impact of wind turbines on residents. The main feature is that it manages expectations of residents with regard to the exposure to noise and shadow flicker. It is for residents an easy tool to access information and news messages regarding a neighbouring wind farm. The app is interactive and offers a listening ear to all local residents. In the app residents can easily provide feedback on how they personally perceive the wind turbine noise and shadow flicker. This feedback is used to 24/7 monitor nuisance and to gain more knowledge on the influence of weather and environmental conditions on the impact of the wind turbines on residents.

The application of the app to operational wind farms shows concerns from neighbouring residents are not always justified. In situations when they do experience nuisance, the noise forecast app is crucial in determining the conditions under which the nuisance is being experienced. This helps determining the cause and effective mitigating measures. In all situations, the use of the app creates more insight into the actual impact of wind turbines on residents. A better understanding of the weather and environmental conditions in which annoyance occurs makes a more effective noise management and communication strategy possible. More insight and a better mutual understanding between residents and wind farm operators will enhance trust and social acceptance.

References

Janssen, S. A., Vos, H., Eisses, A. R., & Pedersen, E. (2011). A comparison between exposure–response relationships for wind turbine annoyance and annoyance due to other noise sources. *The Journal of the Acoustical Society of America 130(6)*, pp. 3746–53.

Michaud, D. S., Keith, S. E., Feder, K., Voicescu, S. A., & et al. (2016). *Personal and situational variables associated with wind turbine noise annoyance.* The Journal of the Acoustical Society of America.

Peuchen, R., Kox, E., Klösters, M., & Straver, K. (2022). *Beleving windenergie op land; inzichten uit vier windparken.* TNO 2022 P10127.

RIVM, Aarten Wind Solutions, ECN, GGD-GHOR Netherlands & EAE/RUG. (2015). *Kennisbericht Geluid van windturbines (Knowledge document wind turbine noise).* Pilot Kennisplatform Windenergie.

van Kamp, I., & van den Berg, G. P. (2020). *Health effects related to wind turbine sound: an update.* RIVM report 2020-0150.

Welkers, D., van Kempen, E., Helder, R., Verheijen, E., & van Poll, R. (2020). *Motie Schonis en de WHO-richtlijnen voor omgevingsgeluid (2018).* RIVM-report 2019-0227.

INCEEUROPE

10th International Conference
on
Wind Turbine Noise
Dublin – 21st to 23rd June 2023

Standards for regulating environmental impact of wind turbines

Erik Koppen, Arcadis: erik.koppen@arcadis.com
Madelon Ekelschot-Smink, Arcadis: madelon.smink@arcadis.com

Summary

As a result of a motion in the Dutch House of Representatives and the intention in the Dutch coalition agreement 2021-2025 to set clear distance standards for wind turbines, the Dutch Ministry of Economic Affairs and Climate Policy commissioned Arcadis to conduct research into standards for regulating the environmental impact of wind turbines. The goal of the study was to describe effects of different distance standards on nuisance for residents and to go into the advantages and disadvantages of a distance standard compared to specific standards for noise and shadow flicker. As a result of a court verdict in June 2021 the Netherlands currently has no national regulation for wind farms. A literature search has been carried out into the previously applicable standards for wind farms in the Netherlands, how these were established and the underlying considerations. This research has also been carried out for seven other European countries. The research focused on distance standards and standards for noise and shadow flicker. The noise and shadow flicker impact at different distances to a wind farm has been portrayed.

1. Introduction

Commissioned by the Dutch Ministry of Economic Affairs and Climate Policy, research was carried out into standards for wind turbines and the exposure and nuisance related to possible distance standards (Koppen & Ekelschot - Smink, 2022). This research was carried out as a result of a motion in the Dutch House of Representatives and the intention in the Dutch coalition agreement 2021-2025 to set clear distance standards for wind turbines. The goal of this study was to use current literature to identify the effects of different distance standards on nuisance for residents and the advantages and disadvantages of a distance standard compared to specific standards for noise and shadow flicker.

As a result of a court verdict in June 2021 the Netherlands currently has no national regulations for wind farms. As a first step, it was examined how the previously applicable standards for wind farms in the Netherlands were established and what the underlying considerations were. In addition, research was carried out into standards for wind turbines in seven other European countries and the underlying considerations. The study included neighbouring countries Belgium, Germany and (separated by the North Sea) Denmark and the United Kingdom.

Additionally, France, Ireland and Poland were included. The research focused on standards for distance, noise and shadow flicker.

Additionally, also the noise exposure and the shadow flicker duration at different distances to a wind farm were determined, taking into account a number of factors that affect the impact. Also, current insights into the effects of wind turbines on health were discussed. In particular to what extent the effects can be related to the distance to the wind turbines and what a possible distance standard would mean for the percentage highly annoyed persons. The research was used to map out the advantages and disadvantages of a distance standard versus specific standards for noise and shadow flicker.

This paper focusses on the overview of the standards for distance, noise and shadow flicker in the studied countries and on the sound levels and shadow flicker duration that can occur as function of the distance from a wind farm.

2. Distance standards

The distance standards for wind turbines are summarized in Table 1 and visualized in Figure 1. The used references are listed in the bottom of Table 1. Of the eight studied countries, the Netherlands and the United Kingdom do not have a national distance standard. The Flemish Region of Belgium does only require a certain distance to sensitive objects if based on the background noise level a sound level above the target value is permitted. In this case the required minimum distance is 3 times the rotor diameter. Ireland does not have a distance standard, but a target distance of 500 metres. Ireland intends to change this. In the draft for new regulations, a minimum distance of 4 times the tip height with a minimum of 500 metres is assumed to limit the visual impact.

The German state of Bavaria uses the largest distance standards, namely 10 times the turbine tip height (10H). However, the Bavarian state parliament has recently passed a partial relaxation of the 10H rule: for several areas, such as areas near industrial sites, motorways, railways, forests and designated wind priority areas, the distance of wind turbines to residential development is reduced to 1000 metres (Bavarian Government, 2022). The reasoning the Bavarian Minister of Economics and Energy had behind the relaxation was that Bavaria needed to catch up in the production of wind energy (Bavarian Government, 2022). In June 2023 the distance requirement will be relaxed to 800 metres for wind priority areas. Until March 2023 Poland used the distance standard of ten times the turbine height as well. In Poland a new law abolished the '10H' rule (Wardyński & Partners, 2023). Initially the government proposed a new distance standard of 500 metres, but the distance standard was changed to 700 metres in a last-minute amendment (NFP, 2023). The reasoning behind this change, is that it would unblock onshore wind energy, which would enable Poland to meet its 2030 climate targets. This was one of the 'milestones' agreed by Poland with Brussels to receive European funds under the national recovery plan (Euroactiv, 2023) (NFP, 2023).

In the other countries considered, the distance standards vary from 300 metres to 1,100 metres and from 2 times the tip height to 4 times the tip height. Based on the available underlying reasoning for standards, it can be concluded they are mainly set to limit the visual impact of wind turbines. This is also the underlying reason why, in the studied countries, a distance standard is always combined with a noise standard and usually with a shadow flicker standard. The shortest distances seem to be set mainly to limit the visual impact of the wind turbines as objects, in particular to prevent visually oppressive effects caused by the wind turbines. In the German state Lower Saxony, a rule of thumb of 2 times the tip height is used for this, based on case law. For distances less than 3 times the tip height, research into the specific situation is considered necessary to determine whether visually oppressive effects can occur. The larger distance standards also seem to take into account other visual effects.

The distance standard of 1,000 metres in the German state North Rhine-Westphalia was set to protect the nature and landscape and for visually overwhelming effects. It does not apply to individual dwellings and fragmented residential areas. In March 2023, the new coalition agreed to abolish the 1000m standard for repowering projects. For new wind projects in designated wind priority areas the 1000m standard will be abolished in 2025 (Energiezukunft, 2023) (Windindustrie in Deutschland, 2023) (WRD, 2023). The reason for these relaxations is that additional space had to be designated for wind energy in order to meet federal targets. The government presented an interim report on the potential study of wind energy areas, carried out by the State Office for Nature, Environment and Consumer Protection, that stated that 42% more space would be available if a distance standard of 700 metres instead of 1000 metres is implemented (Windindustrie in Deutschland, 2023) (WRD, 2023).

Denmark uses a distance standard of 4 times the tip height to limit the visual nuisance caused by, among other things, light reflections, shadow flicker and obstacle lighting.

Table 1. Summary of distance standards in the studied countries. References are cited at the bottom of the table.

Country	Distance to sensitive objects [m]	Reason/motivation distance standard
Belgium – Flanders	No distance standard, unless a sound level above the target value is permitted based on the background noise level. Then a minimum distance of 3 x rotor diameter is required.	Not known.
Belgium– Wallonia	Minimum distance recommendation: - Power: 100 kW – 1 MW: 350 m - Power >1 MW: 4 x tip height Distance to individual dwellings in case of limitation of visual effects by shielding: 400 m.	Limiting the visual impact, given that a shorter distance is allowed by visual shielding.
Denmark	4 x tip height. This distance does not apply to the home of wind turbine owners.	To prevent visual nuisance due to, among other things, light reflections, shadow flicker and obstacle lighting.
France	500 m for turbines with a hub height > 50 m.	--
Germany – Bavaria	General areas: 10 x tip height. Areas near industrial sites, motor-ways, railways, forests and designa-ted wind priority areas: 1000m. In June 2023 the distance requirement will be relaxed to 800 metres for wind priority areas.	To protect nature and the landscape and to prevent visually overwhelming effects. The 1000m was in November 2022 implemented as a relaxation to be able to catch up in the production of wind energy.
Germany – Lower Saxony	Rule of thumb: 2 x tip height[1)].	Rule of thumb, based on case law, to avoid visually oppressing effects.
Germany – North Rhine-Westphalia	1,000 m, excluding individual homes and fragmented residential areas. Repowering projects and wind priority areas are excluded.	Protecting nature and the landscape and for visually overwhelming effects.
Germany – Other Länder	- Residential areas: 400 to 1,100 m or assessment per case;	Not studied.

Country	Distance to sensitive objects [m]	Reason/motivation distance standard
	- Individual residential buildings and fragmented residential areas: 300 to 1,000 m or assessment per case.	
Ireland (current)	No distance standard, but 500 m is considered acceptable.	Wind turbine noise is normally deemed acceptable at more than 500 m distance.
Ireland (draft new regulation)	4 x tip height with a minimum of 500 m, with exception of applications where shorter distances have been agreed with the relevant owner(s). Exceptions are possible for small-scale wind energy developments for local use. No larger distance is permitted to limit the visual impact.	To limit the visual impact. No longer based on noise pollution, because the WHO determined in 2018 there is no evidence for an acceptable uniform distance between wind turbines and residential areas, since noise propagation depends on more factors than just distance.
Netherlands	No national distance standard, but local or regional standards or guideline values for distances are occasionally used.	In most cases, in the Netherlands, the distance between residences and wind turbines is determined by the noise standard.
Poland	10 x tip height (10H), was changed to 700m in March 2023.	For 10H, external safety: scattering of burning fragments in a fire after a lightning strike during a violent storm. For the change to 700m: to unblock wind energy to be able to achieve 2030 climate targets.
United Kingdom	No distance standard.	--

References per country:
Belgium – Flanders: (Flemish Government, 1995)
Belgium – Wallonia: (Wallonian Government, 2013)
Denmark: (Danish Government, Erhvervsstyrelsen, 2019) (Danish Government, Naturstyrelsen, Miljøministeriet, 2015)
France: (French Government, 2017)
Germany (Bundesministerium der Justiz und für Verbraucherschutz, 2021)
Germany – Bavaria: (Bayerische Staatskanzlei, 2021) (Bavarian Government, 2022)
Germany – Lower Saxony: (State Chancellery of Lower Saxony, 2021)
Germany – North Rhine-Westphalia: (North Rhine-Westphalia, 2021) (Energiezukunft, 2023) (Windindustrie in Deutschland, 2023) (WRD, 2023)
Germany – Other Länder: (Fachagentur Windenergie an Land, 2021)
Ireland (current) (Irish Government, Department of Housing, Local Government and Heritage, 2006)
Ireland (draft new regulation) (Irish Government, Department of Housing, Local Government and Heritage, 2019)
Netherlands: (Province Noord-Holland, 2020) (Province Noord-Holland, 2022) (Municipality Emmen, 2016)
Poland: (Polish Government, 2016) (Wardyński & Partners, 2023) (NFP, 2023) (Euroactiv, 2023)
United Kingdom: N/A

[1)] Deviation is possible if research into the specific situation shows there are no visually oppressive effects. For distances less than three times the tip height, research is required into the specific situation, to determine whether visually oppressive effects can occur.

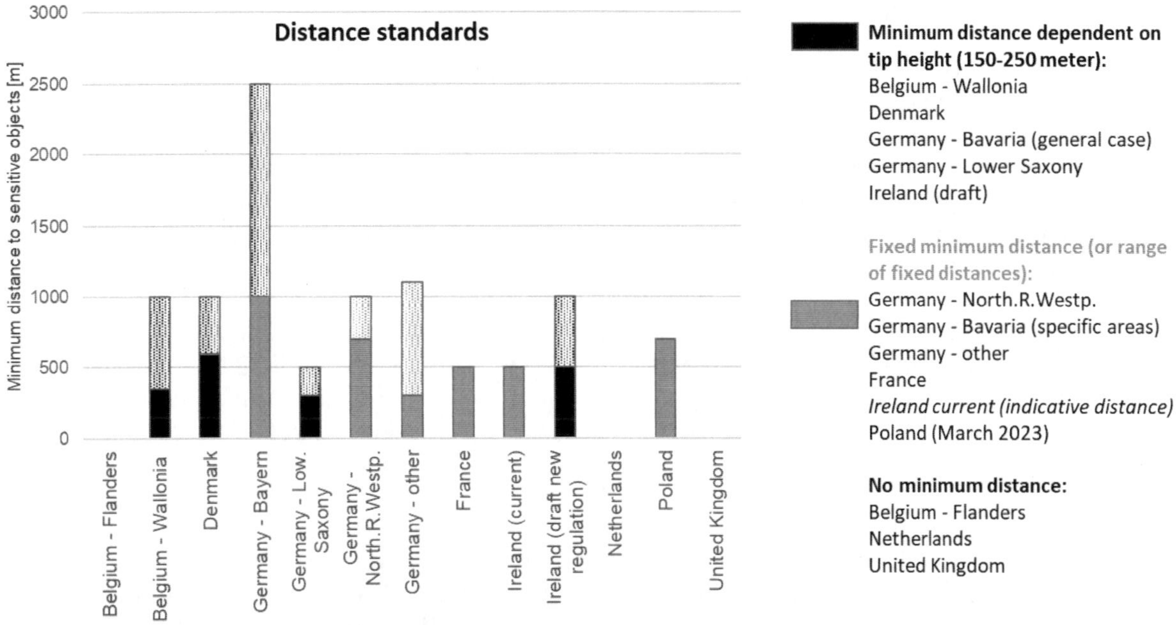

Figure 1. Overview of distance standards for wind turbines. The bandwidths for the black bars – the shaded parts – are due to the dependence on tip height. The image shows bandwidths for tip heights between 150 and 250 metres. The bandwidth for the orange bar – the shaded part – is due to the differences in distance standard between the different federal states. In Flanders, no distance standard applies unless, based on background noise a sound level above the target value is permitted. Then a minimum distance of 3 times rotor diameter applies.

3. Noise standards

More differences than similarities are found between the noise standards of the studied countries:

- Different noise parameters are used for the assessment of wind turbine noise.
- The limit values differ.
- The level of protection for residential areas versus residences in rural areas differs.
- In certain countries fixed limit values apply, in other countries the limit values depend on the land use or background noise level.
- Different calculation methods are used, so even if limit values were the same, the impact of the standards may be different.

The noise standards and their impact are therefore difficult to compare.

In 2010, the Dutch government opted for the use of L_{den} and L_{night} for the assessment of wind turbine noise, in accordance with the definition in the European Environmental Noise Directive. However, this is no obligation since the Directive does not apply to wind turbines. L_{den} and L_{night} are based on annual averaged sound levels. None of the other seven countries studied use annual averaged sound levels to regulate wind turbine noise. These countries use the equivalent sound pressure level L_{Aeq}, a parameter based on the equivalent sound level or a statistical parameter for the assessment of wind turbine noise. In Europe, besides the Netherlands only Norway uses the parameter L_{den} for regulating wind turbine noise (Norwegian Ministry of Climate and Environment, 2021).

As a result of a court verdict in June 2021 the Dutch regulations for wind turbines may no longer be applied to wind farms (Netherlands Council of State, Administrative Jurisdiction

Division, 2021). The new to be implemented regulation for noise could deviate from the previous standard.

The noise standards for wind turbines in the studied countries are summarized in Table 2. The used references are listed in the bottom of this table. Due to different parameters and definitions, limit values in studied countries are difficult to compare. However, a reasonable comparison can be made for the night period. Limit values for the highest equivalent sound level L_{Aeq} in the night period, partly converted from other noise parameters, are summarized in Figure 2. The night period is usually the most critical period for assessing wind turbine noise. For the eight countries studied, the limit value for the equivalent sound level at maximum noise production of the wind turbines (partly converted from other noise parameters) ranges for the night period from 35 dB(A) – at a background noise level of less than 32 dB(A) – to 45 dB(A). In most cases, the limit value for the equivalent sound level is in the range of 39 to 45 dB(A). In other countries the limit value for the night period is usually also within the range of 35 to 45 dB(A), but in the United States some states or districts have limit values of 50 or 55 dB(A) (Koppen & Fowler, 2015).

Table 2 shows most countries have a noise limit which depends on the land use or the background noise level. Of the studied countries the exceptions are the Walloon region of Belgium and the Netherlands. As Koppen & Fowler showed there are more exceptions like Norway, Finland and a number of states and districts in the US. The Flemish region of Belgium, Denmark and Germany apply more strict noise limits to residential areas than to residences in rural areas. In contrary, there are also countries that allow more noise for residential areas than for residences in rural areas, such as Sweden, New Zealand and the Australian state of South Australia (Koppen & Fowler, 2015). Actually, this also applies to France, the UK (mainly in the day period) and to the draft new regulations in Ireland, since in these countries the limit values are related to the background noise level. Some countries choose to keep a quiet rural environment relatively quiet, while other countries choose to allow more noise in rural areas, most likely because of the relatively low population density in rural areas.

In most countries, a penalty is applied to tonal noise from wind turbines, ranging from 1 to 6 dB depending on the strength of the tonal character. The now expired regulation for wind turbine noise in the Netherlands did not include a penalty for tonal noise.

Of the eight studied European countries, only Denmark has a specific standard for low-frequency noise from wind turbines (Danish Government, Miljø- og Fødevareministeriet, 2019). However, according to the draft of the revised wind energy development guidelines, Ireland also intends to introduce a standard for low-frequency noise (Irish Government, Department of Housing, Local Government and Heritage, 2019).

Table 2. Summary of noise standards for wind turbines in the studied countries (in addition to any distance standards as listed in Table 1). References are cited at the bottom of the table.

Country	Noise parameter	Residential areas	Housing in rural areas
Belgium – Flanders	L_{Aeq} at 95% rated power	Day: 44 dB(A)[1)] Evening/night: 39 dB(A)[1)]	Day: 48 dB(A) Evening/night: 43 dB(A)
Belgium – Wallonia	L_{Aeq} [dB(A)]	43 dB(A)	
Denmark	L_r [dB(A)][2)] L_{pALF} [dB][3)]	37 dB(A) L_r at 6 m/s 39 dB(A) L_r at 8 m/s 20 dB L_{pALF} at 6 and 8 m/s	42 dB(A) L_r at 6 m/s 44 dB(A) L_r at 8 m/s 20 dB L_{pALF} at 6 and 8 m/s
France	$L_{50.10min}$ of $L_{Aeq,1s}$ values [dB(A)]	Day: increase of 5 dB(A) with respect to the background noise level, with a lower limit of 35 dB(A) Night: increase of 3 dB(A) with respect to the background noise level, with a lower limit of 35 dB(A)[5)]	
Germany	L_r [dB(A)][5)]	Day: 50/55 dB(A)[6)] Night: 35/40 dB(A)[6)]	Day: 60 dB(A) Night: 45 dB(A)
Ireland (current)	$L_{A90,\ 10min.}$ [dB(A)]	Day: - Background noise level + 5 dB(A) with a lower limit of 45 dB(A) [≈ 47 dB(A) L_{Aeq}][7)] - In quiet environments with a background noise level less than 30 dB(A): 35 to 40 dB(A) [≈ 37 to 42 dB(A) L_{Aeq}] [7)]) Night: 43 dB(A) [≈ 45 dB(A) L_{Aeq}] [7)]	
Ireland (new regulations draft)	$LA_{rated,\ 10min}$[8)]) [dB(A)]	- The background noise level plus 5 dB(A) at an assessment level in ranging from 35 to 43 dB(A) [≈ 37 to 45 dB(A) L_{Aeq}][7)]) - 35 dB(A) at a background noise level of less than 30 dB(A) [≈ 37 dB(A) L_{Aeq}][7)] - 43 dB(A) at a background noise level of 38 dB(A) or more [≈ 45 dB(A) L_{Aeq}][7)]	
Netherlands	L_{den} [dB] L_{night} [dB]	47 dB L_{den} 41 dB L_{night} [≈ 43-46 dB(A) L_{Aeq} at 8 m/s or at (95%) rated power][9)]	
Poland	L_{Aeq} [dB(A)]	Day: 50/55 dB(A)[10)] Night: 40/45 dB(A)[10)]	
United Kingdom	$L_{A90,\ 10min.}$ [dB(A)]	Day: background noise level + 5 dB(A), with a lower limit of 35 to 40 dB(A) [≈ 37 to 42 dB(A) L_{Aeq}][11)] Night: background noise level + 5 dB(A) with a lower limit of 43 dB(A) [≈ 45 dB(A) L_{Aeq}][11)]	

References per country:
Belgium – Flanders: (Flemish Government, 1995) (Flemish Government, n.d.)
Belgium – Wallonia: (Walloon Government, 2021) (Wallonian Government, 2002) (Sertius, Modyva, Pissart AE, 2020)
Denmark: (Danish Government, Miljø- og Fødevareministeriet, 2019)
Germany: (TA Lärm, 1998) (Bauerdorff, 2019)
France: (French Government, 2021) (Dutilleux, 2019)
Ireland (current): (Irish Government, Department of Housing, Local Government and Heritage, 2006)
Ireland (draft new regulation): (Irish Government, Department of Housing, Local Government and Heritage, 2019)
Netherlands: (Dutch Government, 2007) (Dutch government, 2010)
Poland: (Polish Government, 2012)
United Kingdom: (ETSU, Working Group on Wind Turbine Noise, 1996)

1) For residential areas less than 500 m distance from an industrial site, a 4 dB(A) higher limit value applies.
2) The rating level L_r is equal to the equivalent sound pressure level L_{Aeq} corrected for the average meteorological conditions, for times of the day with an increased sensitivity and with a possible penalty for tonal or impulse-like noise.

Country	Noise parameter	Residential areas	Housing in rural areas

3) This requirement for low-frequency noise concerns the level inside the dwelling.444) Assuming a basic background noise level $L_{50.10min}$ of the $L_{Aeq,1s}$ values in the night period of 30 dB(A) for a rural environment and 40 dB(A) for a residential area and the estimate of an additional contribution to the background noise level $L_{50.10min}$ of the $L_{Aeq,1s}$ values of 40 dB(A) due to wind induced and leaf noise at high wind speeds, the $L_{50.10min}$ of the $L_{Aeq,1s}$ noise level at maximum noise production of a wind farm could be approximately 40 and 43 dB(A) respectively.

5) The rating level L_r is equal to the equivalent sound pressure level L_{Aeq} with a possible penalty for clearly audible tones.

6) The lowest limit value applies to a purely residential area, the highest limit value to a general residential area.

7) The equivalent noise level L_{Aeq} is approximately 2 dB(A) higher than the $L_{A90,\ 10min.}$ level. (Irish Government, Department of Housing, Local Government and Heritage, 2019).

8) The rating level $L_{A\ rated,10\ min}$ is equal to the $L_{90.10\ min}$ level plus any penalties for special audible characteristics of the wind turbine noise, such as tonal noise and amplitude modulation with a modulation of 3 dB or more.

9) At a level of 47 dB L_{den}, L_{night} is typically 41 dB. This is an annual average sound level. When the wind is strong, higher levels occur. The difference between the annual average level and the sound level at high wind speeds depends on the type of turbine and the local wind climate. For the Dutch wind climate, the highest equivalent sound pressure level L_{Aeq} is 2 to 5 dB(A) higher than the L_{night} level, but usually 2 to 4 dB(A) higher. The larger the wind turbine hub height and the rotor diameter, the smaller the difference, as the maximum noise is produced a larger proportion of the time.

10) Depending on the function and type of area.

11) The UK scheme states the $L_{A90.10\ min}$ is typically 1.5 to 2.5 dB(A) lower than the $L_{Aeq.10\ min}$. (ETSU, Working Group on Wind Turbine Noise, 1996).

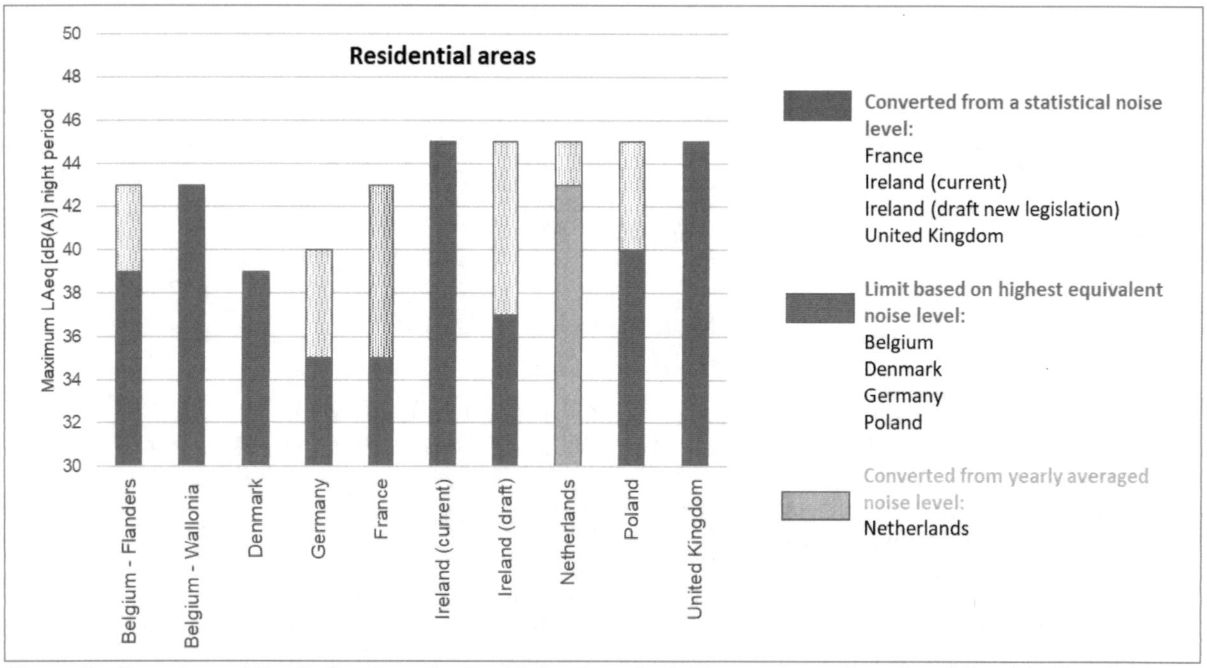

Figure 2. Overview of the limit values for the highest equivalent sound pressure level (L_{Aeq}) in the night period for residential areas, partly converted from other noise parameters. The bandwidths – the shaded parts – are due to the dependence on the background noise level, the zoning of the area, the influence of the local wind climate, etcetera.

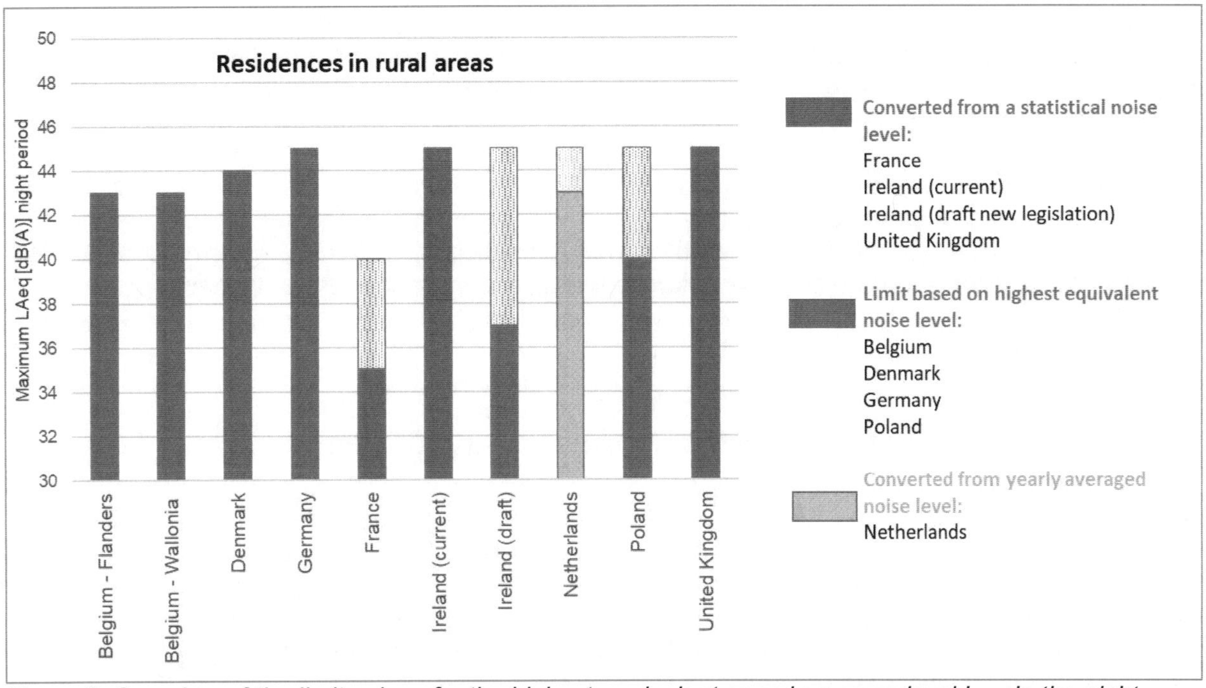

Figure 3. Overview of the limit values for the highest equivalent sound pressure level L_{Aeq} in the night period for residences in rural areas, partly converted from other noise parameters. The bandwidths – the shaded parts – are due to the dependence on the background noise level, the zoning of the area, the influence of the local wind climate, etcetera.

4. Shadow flicker standards

In general, shadow flicker standards are more uniform than noise standards. In 2002, Germany published a guideline for calculating and assessing shadow flicker, based on scientific research (Länderausschuss für Immissionsschutz, 2002). Most countries that have a regulation or a guideline for assessing shadow flicker based this on the German guideline (Koppen, Gunuru, & Chester, 2017).

The German guideline, updated in 2019 (Bund-/Länder-Arbeitsgemeinschaft für Immissionsschutz (LAI), 2020), sets a shadow flicker limit of 30 hours per year and 30 minutes per day for the astronomical maximum possible shadow duration (worst-case scenario). If a shadow flicker control system is used which automatically stalls the wind turbine at times shadow flicker is expected to occur, the actual shadow flicker duration must be limited to 8 hours per year. In Wallonia (Walloon Government, 2021), France (French government, 2021) and Ireland (Irish Government, Department of Housing, Local Government and Heritage, 2006), the aforementioned limit for the astronomically maximum possible shadow flicker is applicable. It is noteworthy that in these countries the shadow flicker only needs to be examined at a relatively short distance and in France only for office buildings. According to the new regulations draft, Ireland intends disallow any shadow flicker on sensitive objects (Irish Government, Department of Housing, Local Government and Heritage, 2019). The operator must take appropriate measures to avoid shadow flicker. Flanders set as limit value of maximum of 8 hours/per year and 30 minutes/day effective shadow flicker (Flemish Government, 1995). This is similar to the German guideline in the situation a shadow flicker control system is used. However, Flanders allows more shadow flicker for sensitive objects - other than residences - in industrial sites. In Denmark, a slightly more shadow flicker is allowed: 10 hours/year effective shadow flicker (Danish government, Naturstyrelsen, Miljøministeriet, 2015). The United Kingdom does not have a shadow flicker standard, but it is understood it is common practice to apply limit values in line with the German guideline. Poland has no shadow flicker standard. Given the distance standard of 10 times tip height which was until recently applicable no shadow flicker for sensitive objects is to be expected. Recently the distance requirement was changed into 700 metres (NFP, 2023). It is not clear if now a shadow flicker regulation is or will be introduced.

The Netherlands has the most deviating shadow flicker standard, because the limit consists of a combination of days per year and minutes per day. Shadow flicker may occur no more than 17 days a year for more than 20 minutes a day (Dutch government, 2007). As a result of a court verdict in June 2021 the shadow flicker regulations may no longer be applied to wind farms (Netherlands Council of State, Administrative Jurisdiction Division, 2021). The new to be implemented regulation is likely to deviate from the previous standard.

5. Sound levels and shadow flicker duration at different distances to a wind farm

Based on calculations for two fictitious wind farms located in the Netherlands – a line arrangement of three wind turbines and a double line arrangement of two times five wind turbines – the sound levels at different distances to a wind farm have been determined. The range in sound levels was determined taking into account differences caused by the lay-out of the wind farm, the number of turbines, the range in sound power levels for 2 to 6 MW turbines, the wind climate, the level of ground absorption and the orientation in relation to the predominant wind direction. The calculations were done according to the Dutch requirements for calculating and measuring wind turbine noise (Dutch government, 2007).

Figure 1 shows the lower and upper limit for the sound levels expressed as L_{den} (as defined in European Environmental Noise Directive) that can occur at a certain distance from a wind farm. The graph shows that the sound level as a function of the distance to a wind farm decreases, but that due to the factors described above a large scatter occurs in sound levels at a specific distance. The larger the distance, the larger the scatter in sound levels. The difference between the lower and upper limit varies from 14 to 18 dB. Under specific conditions like a solitary low-noise turbine or a large wind farm with relatively noisy turbines the sound level could respectively be even lower or higher.

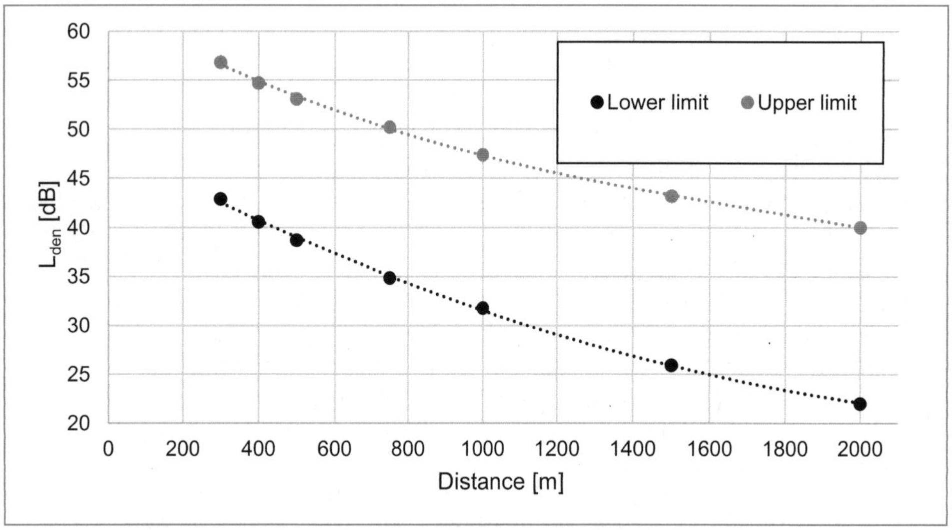

Figure 1. Lower and upper limit of the sound levels expressed as L_{den} [dB] as a function of the distance to a wind farm in the Netherlands

Figure 2 shows the lower and upper limit for the equivalent sound pressure levels at maximum noise production of the wind turbines $L_{Aeq\ max}$ that can occur at a certain distance from a wind farm. This graph too shows the sound level as a function of the distance to a wind farm decreases, but that due to the factors described above a large scatter occurs in sound levels at a specific distance. For the $L_{Aeq,max}$ the difference between the lower and upper limit varies from 11 to 16 dB(A). The difference between the lower and upper limit is smaller than for L_{den}, because $L_{Aeq,max}$ is not affected by the local wind climate.

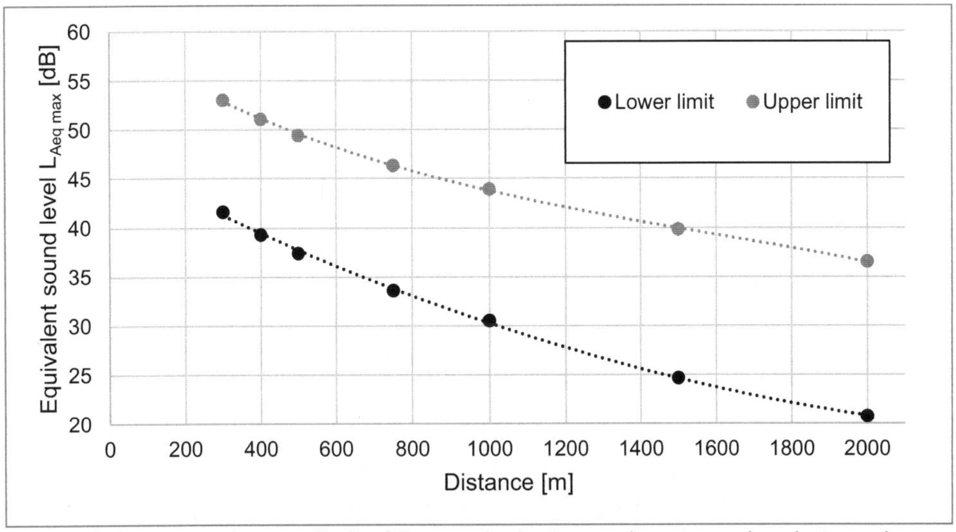

Figure 2. Lower and upper limit of the equivalent sound pressure level at maximum noise production of the wind turbines $L_{Aeq\ max}$ [dB(A)] as a function of the distance to a wind farm

Figure 3 shows the lower and upper limit for the shadow flicker duration that can occur at a certain distance from a wind farm in the Netherlands. This graph shows the shadow flicker duration as a function of the distance to a wind farm also decreases, but that again a large range in shadow flicker duration occurs due to many factors that play a role. For shadow flicker, the specific orientation in relation to the wind farm is a crucial factor. Most shadow flicker occurs southwest and southeast of a wind turbine. This is related to the low position of the sun at sunrise and sunset. Immediately south of a wind turbine, shadow flicker never occurs, because in the Netherlands the sun never shines from the north. If residences can be exposed to shadow flicker, in the Netherlands and many other countries a shadow flicker control system is required to limit the duration of shadow flicker. As a result, it is not the distance to a wind farm that determines the nuisance experienced due to shadow flicker, but the settings of the shadow flicker control system.

The results for noise and shadow flicker demonstrate a distance standard cannot replace standards for noise and shadow flicker. As the previous graphs show, there is a large scatter in protection level at a certain distance. This means that a distance standard does not offer the same protection to all residents. There are too many variables that affect the actual exposure to noise and shadow flicker. Depending on the situation, a certain distance standard can lead to excessive nuisance or offer excessive protection. Excessive protection can limit the possibilities for the development of wind farms. It could be considered to implement a distance standard as an additional requirement. This can offer a certain protection for visual impact and a basic protection for noise and shadow flicker.

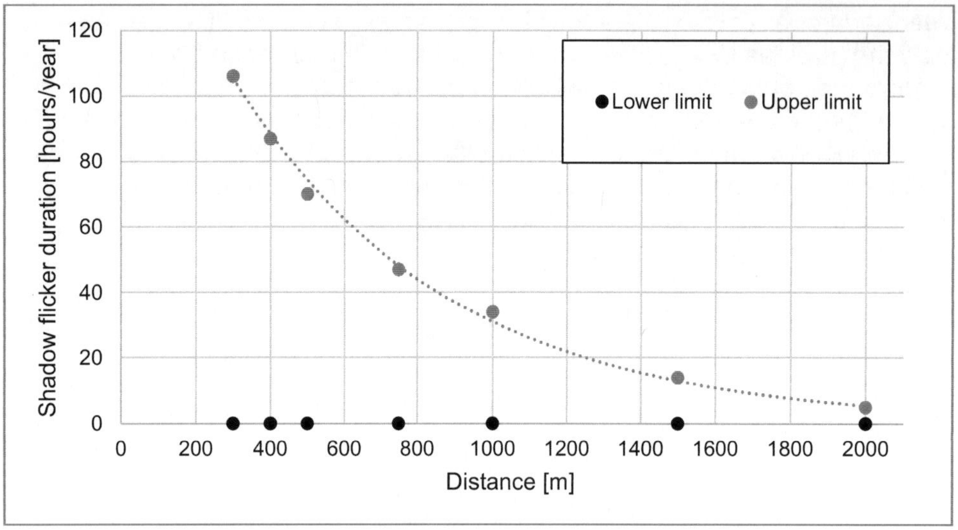

Figure 3. Lower and upper limit of the shadow flicker duration in hours per year as a function of the distance to a wind farm in the Netherlands

6. Conclusions

From the eight studied countries the Netherlands, the Flemish Region of Belgium and the United Kingdom do not have a distance standard for wind farms. In the countries that have a distance standard the required distance to sensitive objects varies from 300 metres to 1,100 metres and from 2 times the tip height to 10 times the tip height. There is no country that regulates nuisance from wind farms solely by a distance standard. For the cases where the underlying considerations were available, the main reason to set a distance standard was to limit visual impact from the wind turbines. This is also the reason why countries combine a distance standard with a noise standard and/or a shadow flicker standard. The shortest distances are mainly set to limit the visual oppressive effect from wind turbines as an object.

All studied European countries have a noise standard for wind turbines. Due to different noise parameters, definitions and calculation methods, limit values in studied countries are difficult to compare. However, a reasonable comparison can be made for the night period. For the eight countries studied, the limit value for the equivalent sound pressure level at maximum noise production of the wind turbines $L_{Aeq,max}$ (partly converted from other noise parameters) ranges for the night period from 35 dB(A) to 45 dB(A). In most cases, the limit value for the equivalent sound level is in the range of 39 to 45 dB(A). In other countries the limit value for the night period is usually also within the range of 35 to 45 dB(A), but in the United States some states or districts have limit values of 50 or 55 dB(A).

The approach for assessing shadow flicker is more uniform. Of the eight studied countries only the UK and Poland have no standard for regulating shadow flicker. Most countries that have a regulation or a guideline for assessing shadow flicker based this on the German guideline for assessing shadow flicker. The Netherlands has the most deviating shadow flicker standard, because the limit consists of a combination of days per year and minutes per day.

The sound levels and shadow flicker duration at different distances show that they decrease with distance. However, due to a number of factors like the lay-out of the wind farm, the number of turbines, the range in sound power levels, the wind climate, the level of ground absorption and the orientation to a wind farm a large scatter occurs in the exposure to noise and shadow flicker at a specific distance.

The results demonstrate a distance standard cannot replace standards for noise and shadow flicker. A distance standard does not offer the same protection to all residents. There are too many variables that affect the actual exposure to noise and shadow flicker. Depending on the situation, a certain distance standard can lead to excessive nuisance or offer excessive protection while limiting the development of wind farms. As an additional requirement a distance standard can offer a certain protection for visual impact and a basic protection for noise and shadow flicker.

References

Bauerdorff, A. (2019). Current regulations for the protection against noise from wind turbines in Germany. *8th International Conference on Wind Turbine Noise 2019.* Lissabon.

Bavarian Government. (2022, November 9). *Bayerisches Staatsministerium für Wirtschaft, Landesentwicklung und Energie.* Retrieved from Aiwanger: "Bayern wird ein Boom-Land für die Windkraft - aktuell sind 59 Windenergieanlagen in der Pipeline - fast 70 Projekte zusätzlich betreuen unsere Windkümmerer": https://www.stmwi.bayern.de/presse/pressemeldungen/pressemeldung/532-2022/

Bayerische Staatskanzlei. (2021, June 1). Art. 82 Windenergie und Nutzungsänderung ehemaliger landwirtschaftlicher Gebäude. *Bayerische Bauordnung (BayBO).*

Bund-/Länder-Arbeitsgemeinschaft für Immissionsschutz (LAI). (2020, januari 23). Hinweise zur Ermittlung und Beurteilung der optischen Immissionen von Windkraftanlagen – Aktualisierung 2019.

Bundesministerium der Justiz und für Verbraucherschutz. (2021). Baugesetzbuch (BauGB) § 249 Sonderregelungen zur Windenergie. In *§ 249 Sonderregelungen zur Windenergie.* Retrieved from https://www.gesetze-im-internet.de/bbaug/__249.html

Danish Government, Erhvervsstyrelsen. (2019, September 6). Bekendtgørelse om planlægning for og tilladelse til opstilling af vindmøller. *BEK nr. 923.*

Danish Government, Miljø- og Fødevareministeriet. (2019, February 07). Bekendtgørelse om støj fra vindmøller. *BEK nr. 135.*

Danish government, Naturstyrelsen, Miljøministeriet. (2015). *Vejledning om planlægning for og tilladelse til opstilling af.*

Danish Government, Naturstyrelsen, Miljøministeriet. (2015). *Vejledning om planlægning for og tilladelse til opstilling af.*

Dutch Government. (2007). *Besluit van 19 oktober 2007, houdende algemene regels voor inrichtingen (Besluit algemene regels voor inrichtingen milieubeheer).* Retrieved november 18, 2021, from https://wetten.overheid.nl/BWBR0022762/2021-07-01

Dutch government. (2007). *Regeling van de Minister van Volkshuisvesting, Ruimtelijke Ordening en Milieubeheer van 9 november 2007, nr. DJZ2007104180, houdende algemene regels voor inrichtingen (Regeling algemene regels voor inrichtingen milieubeheer).* Retrieved november 18, 2021, from https://wetten.overheid.nl/BWBR0022830/2018-04-05

Dutch government. (2010). Besluit van 14 oktober 2010 tot wijziging van het Besluit algemene regels voor inrichtingen milieubeheer en het Besluit omgevingsrecht (wijziging milieuregels windturbines). *Staatsblad van het Koninkrijk der Nederlanden nr. 749.*

Dutilleux, P. (2019). France - Germany: A Comparison of the Acoustic Assessment Procedures. *8th International Conference on Wind Turbine Noise.* Lisbon.

Energiezukunft. (2023, March 10). *Nordrhein-Westfalen - 1.000 Meter Abstandsregel aufgeweicht.* Retrieved from Energiezukunft: https://www.energiezukunft.eu/erneuerbare-energien/wind/1000-meter-abstandsregel-aufgeweicht/

ETSU, Working Group on Wind Turbine Noise. (1996, September). The assessment and rating of noise from wind farms. *ETSU-R-97.*

Euroactiv. (2023, February 6). *Euroactiv.* Retrieved from Poland's controversial wind energy law voted on this week: https://www.euractiv.com/section/energy-environment/news/polands-controversial-wind-energy-law-voted-on-this-week/

Fachagentur Windenergie an Land. (2021, October). Überblick zu den Abstandsempfehlungen zur Ausweisung von Windenergiegebieten in den Bundesländern.

Flemish Government. (1995). *VLAREM II, Besluit van de Vlaamse regering van 1 juni 1995 houdende algemene en sectorale bepalingen inzake milieuhygiëne.* Retrieved januari 6, 2021, from https://navigator.emis.vito.be/mijn-navigator?woId=40985

Flemish Government. (n.d.). *VLAREM II - Bijlagen, Bijlage 5.20.6.1. Richtwaarden voor windturbinegeluid.* Retrieved januari 6, 2022, from https://navigator.emis.vito.be/mijn-navigator?woId=42564

French Government. (2017, March 1). *Code de l'environnement, Section 11: Eoliennes.* Retrieved januari 17, 2022, from https://www.legifrance.gouv.fr/codes/article_lc/LEGIARTI000033933299/

French government. (2021, December 10). *Arrêté du 26 août 2011 relatif aux installations de production d'électricité utilisant l'énergie mécanique du vent au sein d'une installation soumise à autorisation au titre de la rubrique 2980 de la législation des installations classées pour la prot....* Retrieved januari 17, 2022, from https://www.legifrance.gouv.fr/loda/id/JORFTEXT000024507365/

French Government. (2021, December 10). *Arrêté du 26 août 2011 relatif aux installations de production d'électricité utilisant l'énergie mécanique du vent au sein d'une installation soumise à autorisation au titre de la rubrique 2980 de la législation des installations classées pour la prot....* Retrieved januari 17, 2022, from https://www.legifrance.gouv.fr/loda/id/JORFTEXT000024507365/

Irish Government, Department of Housing, Local Government and Heritage. (2006). Wind Energy Development Guidelines. Retrieved 12 3, 2021, from https://www.gov.ie/en/publication/f449e-wind-energy-development-guidelines-2006/

Irish Government, Department of Housing, Local Government and Heritage. (2019, December 12). Draft Revised Wind Energy Development Guidelines December 2019.

Koppen, E., & Ekelschot - Smink, M. (2022, april 19). Onderzoek afstandsnormen windturbines. *Arcadis report D10048616.25.*

Koppen, E., & Fowler, K. (2015). International Legislation for Wind Turbine Noise. *EuroNoise 2015.* Maastricht.

Koppen, E., Gunuru, M., & Chester, A. (2017). International Legislation and Regulations for Wind Turbine Shadow Flicker Impact. *7th International Conference on Wind Turbine Noise.* Rotterdam.

Länderausschuss für Immissionsschutz. (2002). *Hinweise zur Ermittlung und Beurteilung der optischen Immissionen von Windenergieanlagen (WEA-Schattenwurf-Hinweise).*

Municipality Emmen. (2016, June 28). Structuurvisie Emmen, Windenergie NL.IMRO.0114.2015003-S701.

Netherlands Council of State, Administrative Jurisdiction Division. (2021, June 30). Verdict 202003882/1/R3.

NFP. (2023, February 9). *Notes from Poland.* Retrieved from Polish parliament approves law to unblock building of onshore wind farms: https://notesfrompoland.com/2023/02/09/polish-parliament-approves-law-to-unblock-building-of-onshore-wind-farms/

North Rhine-Westphalia. (2021, December 2). *Geltende Gesetze und Verordnungen (SGV. NRW.) mit Stand vom 2.12.2021 § 2 (Fn 2) Mindestabstand für privilegierte Windenergieanlagen.* Retrieved December 3, 2021, from https://recht.nrw.de/lmi/owa/br_bes_detail?sg=0&menu=1&bes_id=29504&anw_nr=2&aufgehoben=N&det_id=517851

Norwegian Ministry of Climate and Environment. (2021, June). Retningslinje for behandling av støy i arealplanleggingen (T-1442/2021).

Polish Government. (2012, October 1). Rozporządzenie Ministra Środowiska z dnia 1 października 2012 r. zmieniające rozporządzenie w sprawie dopuszczalnych poziomów hałasu w środowisku. *Poz. 1109*. Retrieved 12 2, 2021, from http://isap.sejm.gov.pl/isap.nsf/DocDetails.xsp?id=WDU20120001109

Polish Government. (2016, February 19). Ustawy o inwestycjach w zakresie elektrowni. *Druk nr 315*. Retrieved 12 2, 2021, from https://www.sejm.gov.pl/Sejm8.nsf/druk.xsp?nr=315

Province Noord-Holland. (2020, October 22). Omgevingsverordening NH2020.

Province Noord-Holland. (2022). Ontwerp-omgevingsverordening NH2022.

Sertius, Modyva, Pissart AE. (2020, February 3). Evaluation des incidences sur l'environnement de deux projects de plan. *Projet d'Arrêté ministériel relatif aux études acoustiques des parcs éoliens*, pp. 187-193.

State Chancellery of Lower Saxony. (2021). Planung und Genehmigung von Windenergieanlagen an Land in Niedersachsen (Windenergieerlass). In *RdErl.D. MU, ged. ML, ged. MI, ged. MW - MU-52-29211/1/305.*

TA Lärm. (1998). *Sechste Allgemeine Verwaltungsvorschrift zum Bundes-Immissionsschutzgesetz (Technische Anleitung zum Schutz gegen Lärm - TA Lärm) vom 26. August 1998 (GMBl Nr. 26/1998 S. 503).*

Wallonian Government. (2002, July 4). Arrêté du Gouvernement wallon fixant les conditions générales d'exploitation des établissements visés par le décret du 11 mars 1999 relatif au permis d'environnement (M.B. 21.09.2002 - err. 01.10.2002).

Wallonian Government. (2013, July 11). Cadre de référence pour l'implantation d'éoliennes en région Wallonne.

Walloon Government. (2021, February 25). Arrêté du Gouvernement wallon portant conditions sectorielles relatives aux parcs d'éoliennes d'une puissance totale supérieure ou égale à 0,5 MW et modifiant l'arrêté du Gouvernement wallon du 4 juillet 2002 arrêtant la liste des projets soumis à étude d.

Wardyński & Partners. (2023, March 17). *Rules for siting of onshore wind farms liberalised.* Retrieved from Lexology: https://www.lexology.com/library/detail.aspx?g=1bf680dc-6696-4ed5-a818-5e27644c0a8b

Windindustrie in Deutschland. (2023, March). *Ein erster, aber wichtiger Schritt: Abstandsregel für Repowering gestrichen.* Retrieved from Windindustrie in Deutschland: https://www.windindustrie-in-deutschland.de/meldungen/ein-erster-aber-wichtiger-schritt-abstandsregel-fuer-repowering-gestrichen

WRD. (2023, March 9). *Windräder: Land streicht Abstandsregel für Repowering.* Retrieved from WRD: https://www1.wdr.de/nachrichten/windkraft-tausend-meter-regel-repowering-abgeschafft-100.html

10th International Conference
on
Wind Turbine Noise
Dublin – 21st to 23rd June 2023

Experience using the IOA AM method, and how the results may vary with distance and direction

Tom Levet, Hayes McKenzie Partnership Ltd: tom@hayesmckenzie.co.uk

Summary

A method for quantifying the amplitude modulation characteristics of wind turbine noise was proposed by a working group setup by the UK Institute of Acoustics. This paper provides a note on some of the in-field experiences using that method. Several measurement locations at various sites have been analysed. These locations are at a variety of receptor distances, and due to the nature of the surveys occurred in a variety of wind directions relative to the wind turbine(s). The results have been normalised to a relative direction in order to investigate trends within the results. The summaries provided are not intended to be exhaustive of the range of results that may be possible, and there would still seem to be a need to treat results on a site-by-site basis. Nevertheless, it is intended the results may be helpful in furthering the understanding of if, or how, it may be possible to predict where high levels of modulation may occur, for potentially undertaking risk assessments for future projects.

1. Introduction

A working group was setup by the UK Institute of Acoustics in 2014, to propose a method for quantifying the amplitude modulation (AM) characteristics of wind turbine noise, when measured in an outdoor amenity area at a receptor location. Several draft methods were proposed within a discussion document, consultation responses were subsequently considered, and the results of a final reference method was published in a 2016 report [1]. The reference method involves measuring third octave $L_{Aeq,100ms}$ values, dividing the data into 10 s blocks, quantifying the modulation depth for each 10 s block within certain frequency bands, and deriving a modulation depth rating for the 10 minute period by considering the 90th percentile of the most prominent blocks within the 10 minute period.

Noise measurements have been undertaken at a variety of receptor locations, mainly for the purposes of testing compliance with noise related planning conditions. Such planning conditions have generally not included any AM component, however the noise measurements on many occasions have included the logging of third octave band $L_{Aeq,100ms}$ data, for subsequent analysis and investigation.

In each situation, the modulation depth ratings have been calculated for each 10 minute period within the noise survey, with most surveys lasting between 3 to 6 weeks. The wind speed and wind direction have been logged alongside the noise measurements. In each instance this has been derived from the SCADA data, by taking an average of each of the nacelle anemometer readings (and wind direction reading) from each turbine across the site.

The different wind farm sites investigated here ranged in size and turbine model. The various turbine models were in the multi-megawatt range, such that across the 32 receptor locations examined, there was an average of 8.6 turbines to the site, with an average rated power of 2.6 MW.

The results in each situation have been reviewed such that any significant false positives that occurred, due to other things in the environment modulating, have been excluded from the dataset. This generally involved a mixture of tools, such as reviewing modulation frequency spectra or listening to recordings.

The modulation depth ratings have been binned according to wind speed and normalised wind direction. Normalised wind direction has been determined by considering the approximate bearing of the receptor to the closest turbine. This way the results have been anonymised and normalised, such that 0° is a wind direction when the receptor is directly downwind (of the closest turbine), 180° is directly upwind, 270° is directly crosswind (under the downwards going blade), and 90° is directly crosswind on the other side (under the upwards going blade). This is equivalent to imagining the receptor is always situated directly south of the closest turbine and the wind direction is varying.

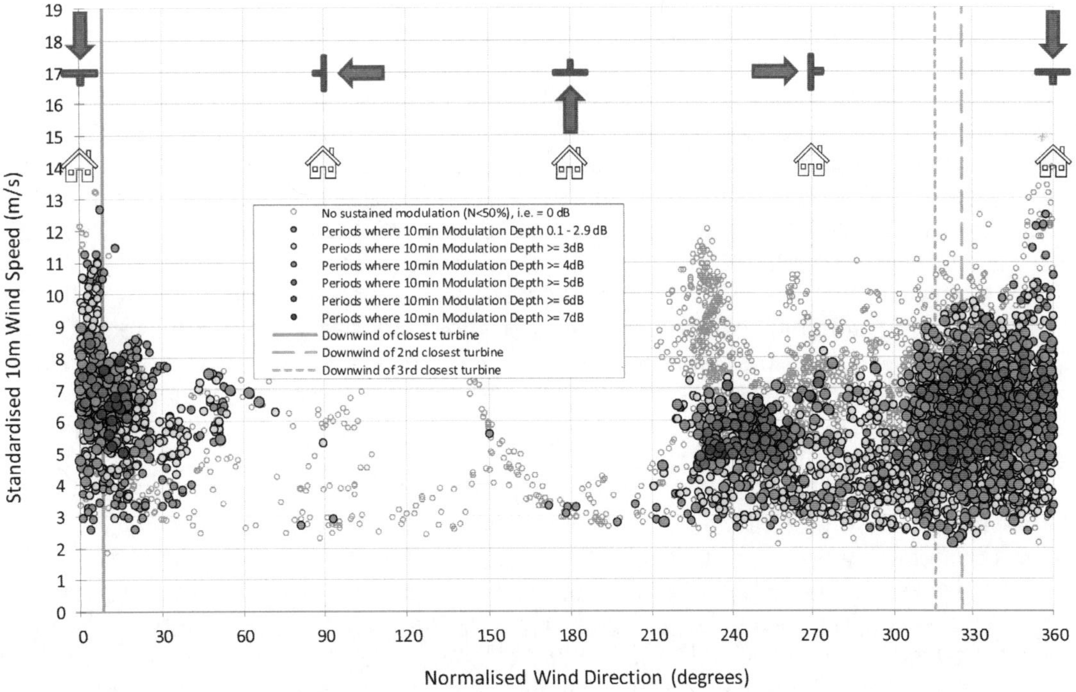

Figure 1: Wind Conditions of Measured Modulation Depth (Band 2: 100 – 400 Hz)

Figure 1 shows an example plot for modulation depth results at a receptor location approximately 600 m away from the nearest turbine. This shows the wind conditions in which various AM results were measured, highlighting periods where the relatively higher modulation depth occurred. It can be seen that for this scenario, the worst modulation generally occurred when directly downwind of the closest turbine, but there were other areas, just off directly downwind and in approximate crosswind conditions, that also exhibited relatively high levels. In this situation it would seem these secondary areas of relatively high modulation depth, occur when downwind and crosswind of the second and third closest turbines.

Table 1: Average Modulation Depth (dB) (Band 2: 100 – 400 Hz)										
Normalised Wind Direction Bin	Standardised 10m Wind Speed Bin (m.s^{-1})									
	3	4	5	6	7	8	9	10	11	12
0° Sector	1.7	2.2	3.6	4.1	4.3	3.8	3.0	2.4	1.8	0.3
30° Sector	0.9	1.0	3.2	3.3	3.1	2.5	-	-	-	-
60° Sector	-	-	0.7	2.0	1.8	-	-	-	-	-
90° Sector	0.2	0.0	-	0.0	-	-	-	-	-	-
120° Sector	0.0	-	-	-	-	-	-	-	-	-
150° Sector	-	0.0	0.0	0.2	0.0	0.0	-	-	-	-
180° Sector	0.2	0.0	-	-	-	-	-	-	-	-
210° Sector	1.3	1.1	0.7	2.7	-	0.0	0.0	0.0	-	-
240° Sector	2.0	2.4	3.8	2.7	0.7	0.0	0.0	0.0	0.0	-
270° Sector	1.6	1.7	2.6	1.2	0.6	0.1	0.0	0.0	-	-
300° Sector	2.3	2.4	2.2	1.6	1.4	0.5	0.4	0.0	-	-
330° Sector	2.2	3.3	4.0	4.4	4.1	2.6	1.2	0.5	0.0	-

Table 1: Average Modulation Depth (Band 2: 100 – 400 Hz), as Function of Wind Speed and Wind Direction

Results have been binned according to 1 m/s integer wind speed bins (standardised 10 m wind speed), and 30° wide wind direction sectors (as for example in Table 1). Results for the frequency ranges of 50-200 Hz, 100-400 Hz, and 200-800 Hz, have been examined separately. An arithmetic average modulation depth has been calculated for each bin, from the 10 minute results, for each frequency band. Where the method returned a result that had less than 50% prominent 10 s blocks, such a 10 minute period was assigned 0 dB modulation depth. Where there were less than 6 data points within a bin, an average result has not been derived. Such a matrix approach is thus similar to that proposed within IEC 61400-11-2 [2] for quantifying AM. By binning the results as a function of wind direction and deriving average modulation depths, this takes account of the frequency of occurrence, and allows for targeted mitigation of the wind conditions where the worst modulation occurs.

Figure 2 shows the measured modulation frequency (where absolute values have been removed for confidentiality reasons) for each 10 minute period (taking the mean of the prominent 10 s blocks), plotted against wind speed, compared to the SCADA derived blade passing frequency of the installed turbine model. The fact that these two parameters follow a similar trend with wind speed, provide indication that the measured modulation (after excluding false positives) does most likely relate to the rotation of the wind turbines, and not to another noise source in the environment.

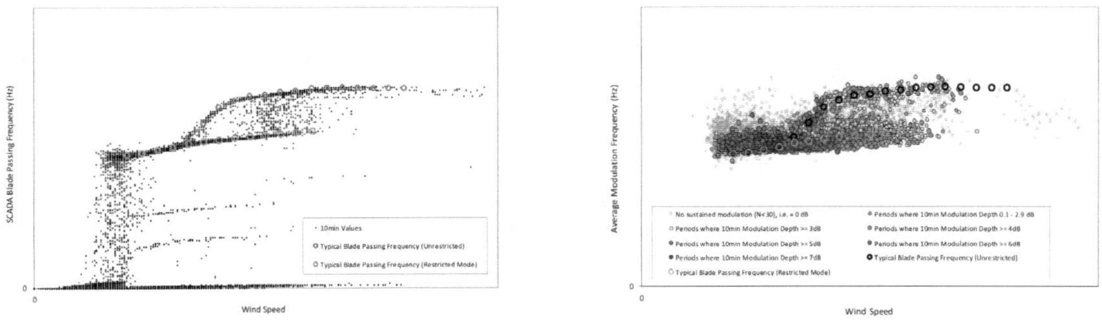

Figure 2: 10 minute Mean Modulation Frequency (Band 1: 50-200 Hz) vs Wind Speed

2. Results

2.1 Receptors at a range of distances

Results have been analysed from receptors at a variety of distances (approximately 250 m to 1500 m) from the closest turbines. Figure 3 below shows results from 37 different receptor surveys, for the directly downwind direction (30° wide sector), detailing the average 10 minute modulation depth (for the three frequency ranges of 50-200Hz (top left), 100-400 Hz (top right), 200-800 Hz (bottom left)) as a function of distance, for the 6 m/s wind speed bin. It was generally the case that 6 m/s standardised 10 m height wind speed was the wind speed where the worst aerodynamic related modulation occurred, and hence most of the analysis shown here has been based on this wind speed.

It can be seen that the modulation depth generally reduces with distance. At the closest residential distances (at approximately 450 – 600 m away), the results for Band 2 (100 – 400 Hz) were in the range of 2.2 – 5.7 dB average modulation depth depending on the receptor/survey. At further distances the average modulation depth reduces, with almost all surveys further than 1 km showing an average modulation depth less than 3 dB. The results from Band 1 tend to show less of a relationship with distance, which in some instances is related to other factors, such as modulating tonal components in the 50-200Hz range. The bottom right panel of Figure 3 below shows an analysis comparing the range of downwind results in the three frequency bands.

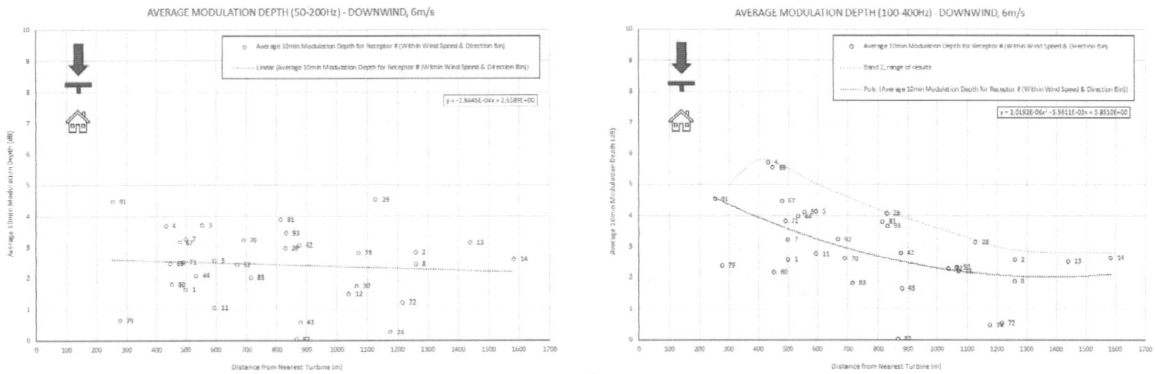

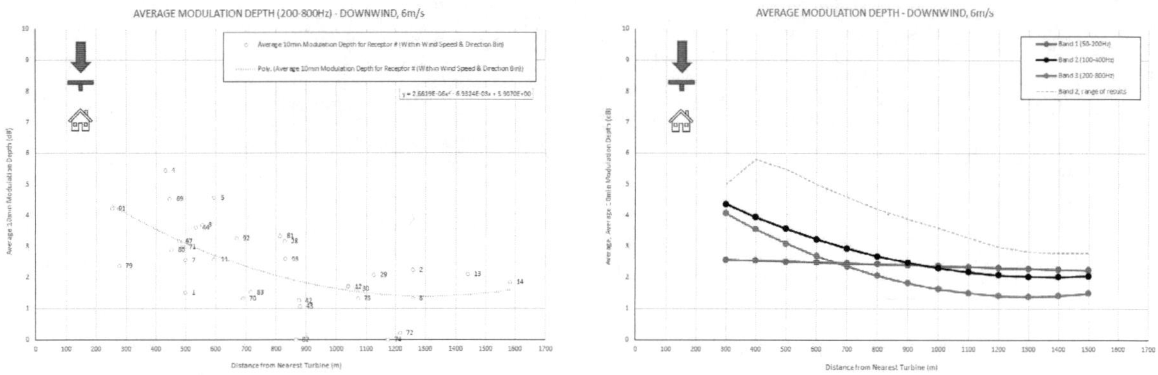

Figure 3: Average Modulation Depth vs Distance, Downwind, 6 m/s

Figure 4 below shows similar analyses, for the upwind results. These show that in upwind directions, the modulation depth generally falls away quicker than downwind directions. For example, when approximately 700 m away, the Band 2 modulation depth was on average approximately 2.9 dB when downwind, but for upwind directions the Band 2 modulation depth was on average 1.4 dB.

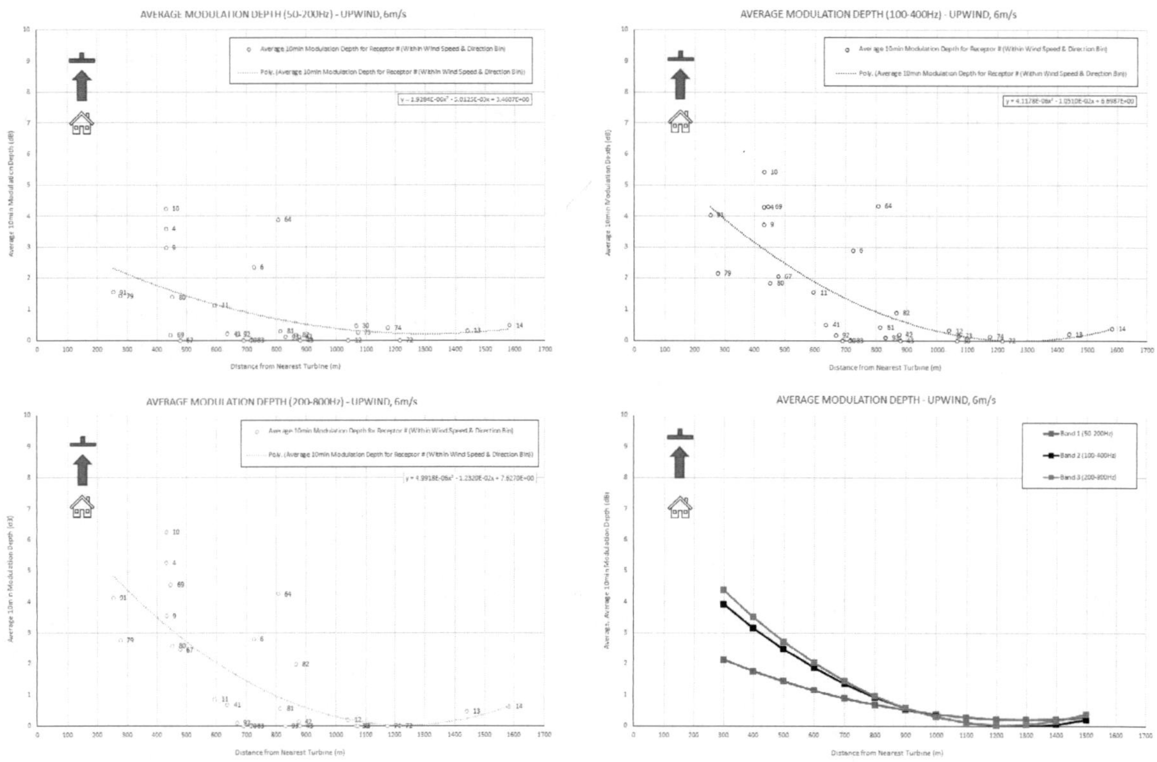

Figure 4: Average Modulation Depth vs Distance, Upwind, 6 m/s

Figures 5 – 6 below show similar analyses, for the two crosswind sectors. Although there were exceptions, most of the surveys showed average modulation depths less than 3 dB, when considering distances greater than 500 m in these crosswind directions.

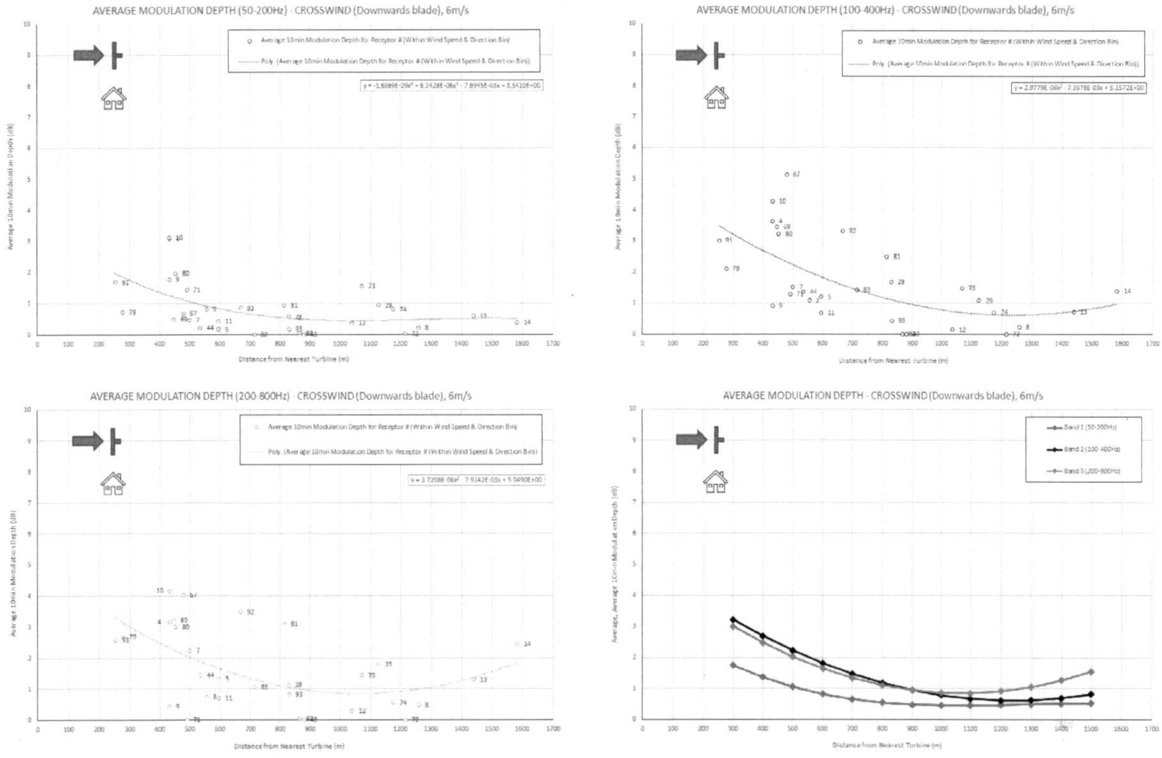

Figure 5: Average Modulation Depth vs Distance, Crosswind (Downwards Going Blade), 6 m/s

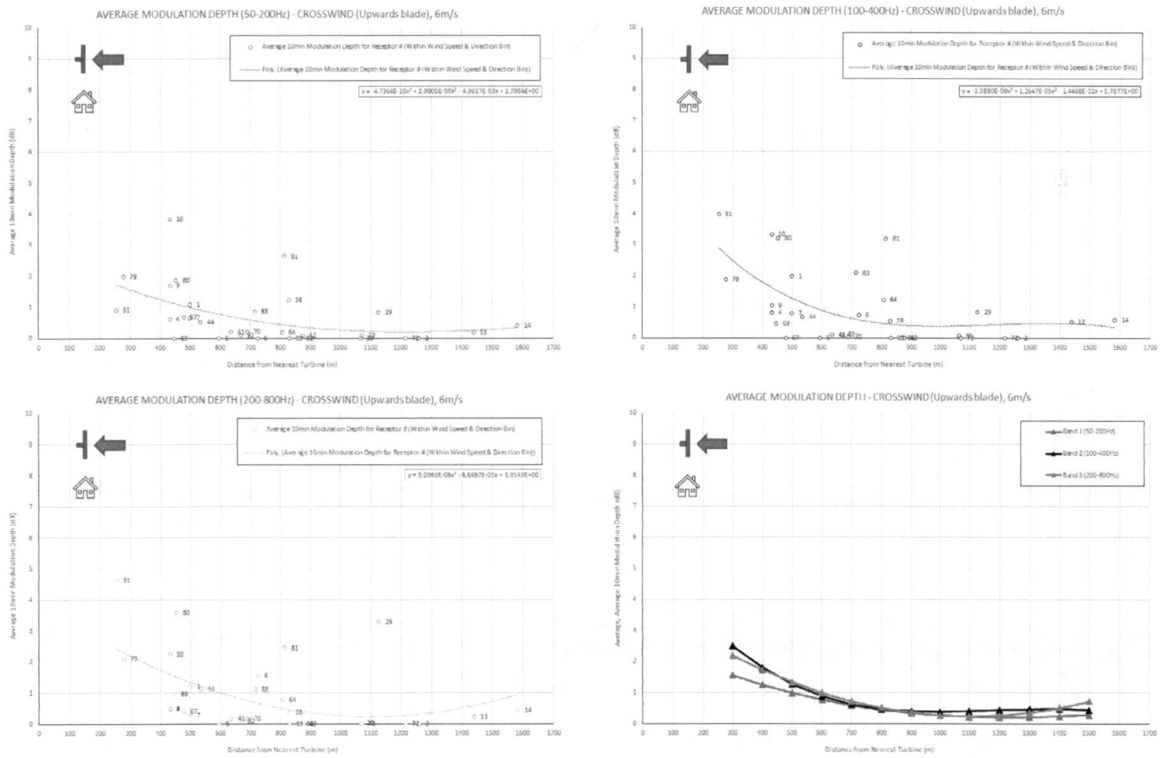

Figure 6: Average Modulation Depth vs Distance, Crosswind (Upwards Going Blade), 6 m/s

Figure 7 summarises the regressions shown for the three frequency bands, showing how the relationships of average modulation depth varied with distance, for the four main wind directions. These curves show the polynomial derived from the average of all receptors, and the error bars represent +/- 1 standard deviation within approximately the same distance (+/- 100 m)

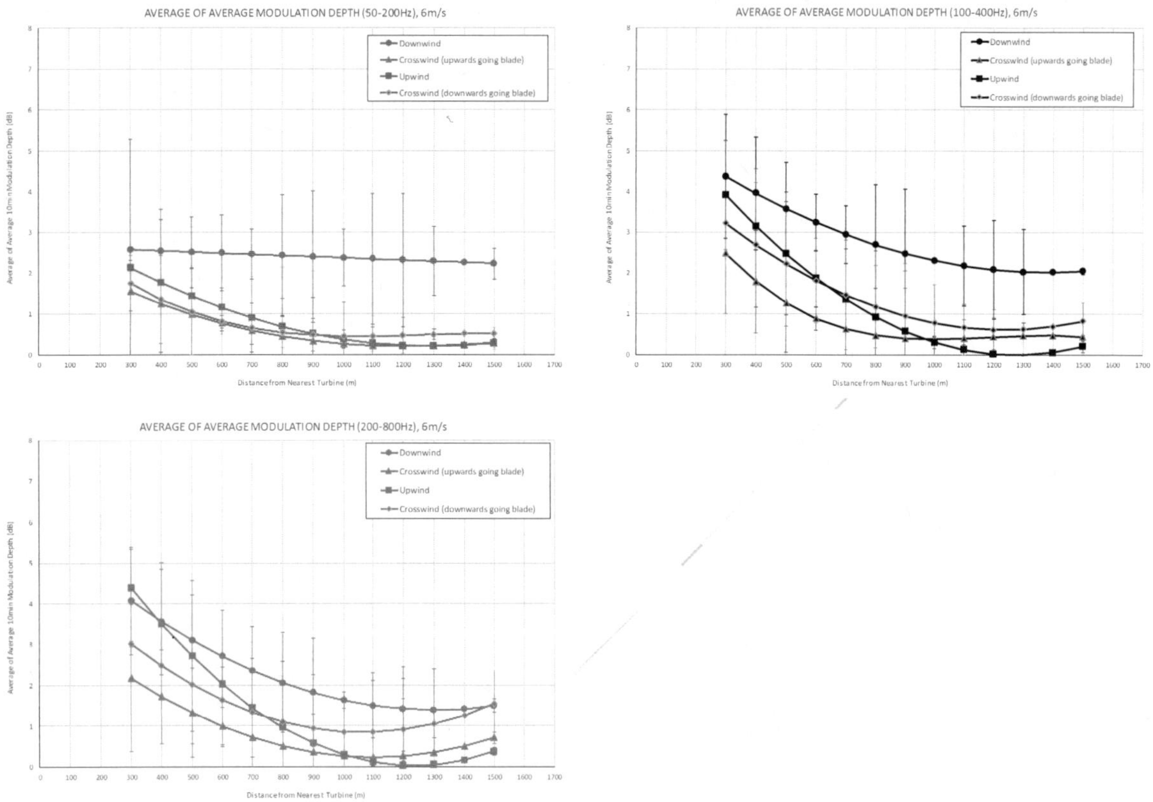

Figure 7: Average of Average Modulation Depth vs Distance, 6 m/s

For Band 1, results in the downwind direction tend to extend further than the other directions. Band 2 results have a similar trend but show more of an uplift at closer distances with generally the higher results of the three bands. Band 3 results tend to have similar levels to Band 2 at the closest distances but tend to fall away quicker with maybe the exception of the crosswind direction when under the downwards going blade. However, it should be noted that the regressions are somewhat tenuous given the inconsistent spread of data points across distance, and more reasonable shapes would be achieved with results from more receptor locations. The shape of the regression curves derived here, are partly due to having fewer results at longer distances.

2.2 Factors that may contribute to the variability

<u>Restricted Modes of Operation</u>

The results here have included some repeat noise surveys, where measurements have been repeated whilst the closest turbine(s) have been running in restricted operational modes for some periods. Figure 8 below highlights some of those instances for the downwind and crosswind (downwards going blade) directions.

Surveys 4, 9, and 10 were measured at the same location (at approximately 430 m). Survey 4 was a series of initial measurements of unrestricted operation, and Surveys 9 and 10 were repeat surveys at a different time of year, that included periods of restricted (Survey 9) and unrestricted operation (Survey 10). Figure 8 shows that for the crosswind condition at 6 m/s, the average modulation depth (of Band 3) for the restricted operation of Survey 9, showed an improvement of over 2 dB compared to that measured in Survey 4.

Surveys 5 and 11 were also measurements from the same location (at approximately 600 m), where Survey 5 was unrestricted operation and Survey 11 was a repeat survey where the closest turbines were in a restricted mode for some wind directions. It can be seen on Figure 8, that for the downwind condition, there was just under a 2 dB reduction in average modulation depth at this location shown by the Survey 11 results.

Surveys 2 and 8 were also measurements from the same location (at approximately 1200 m), where Survey 8 was a repeat survey of restricted operation. It is similar for Surveys 1 and 7 where Survey 7 was a repeat survey of restricted operation at the same location (at approximately 500 m). For these locations, the modulation results for the restricted surveys showed less of an improvement, and in the case of Survey 7 showed somewhat of an increase in the downwind condition. This indicates that noise restricted modes of operation may not always provide an improvement in the measured modulation. There may be confounding factors such as the contribution from other turbines, and the results may be related to the specific turbine type and nature of the restricted mode of operation.

Understanding the nature of the improved modulation results from the restricted operation may be a point of further investigation. For example, it would be useful to understand whether the turbine related 'peak's and 'troughs' of the modulation have converged and less modulation is actually exhibited from the turbines, and/or is it the fact that the lower level of the noise restricted mode in these instances have meant that the same modulation has been masked more amongst the other noise in the environment.

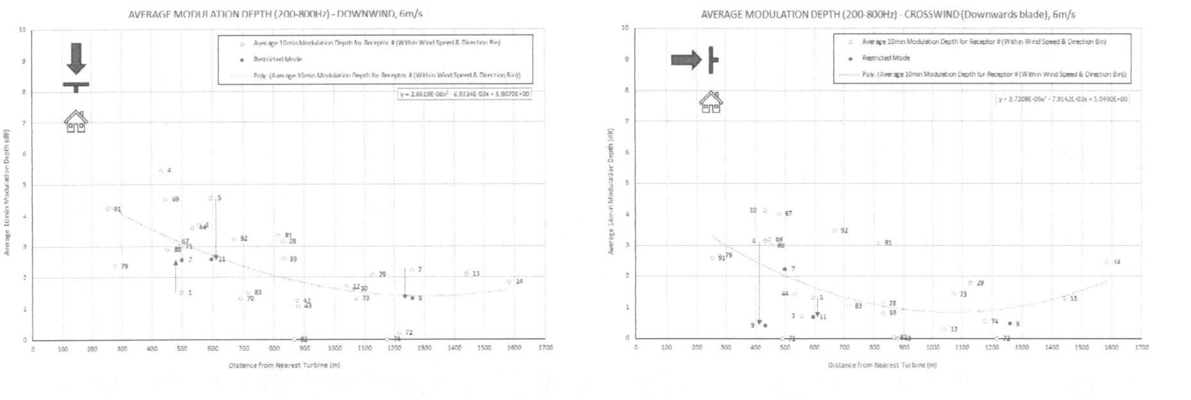

Figure 8: Average Modulation Depth (Band 3: 200-800Hz) vs Distance, Downwind & Crosswind (Downwards Going Blade), 6 m/s, Restricted Modes Of Operation Highlighted

Figure 9 below considers the location at approximately 430 m, and compares the results for various wind directions across the three surveys (although data for the downwind condition was lacking for the repeat surveys).

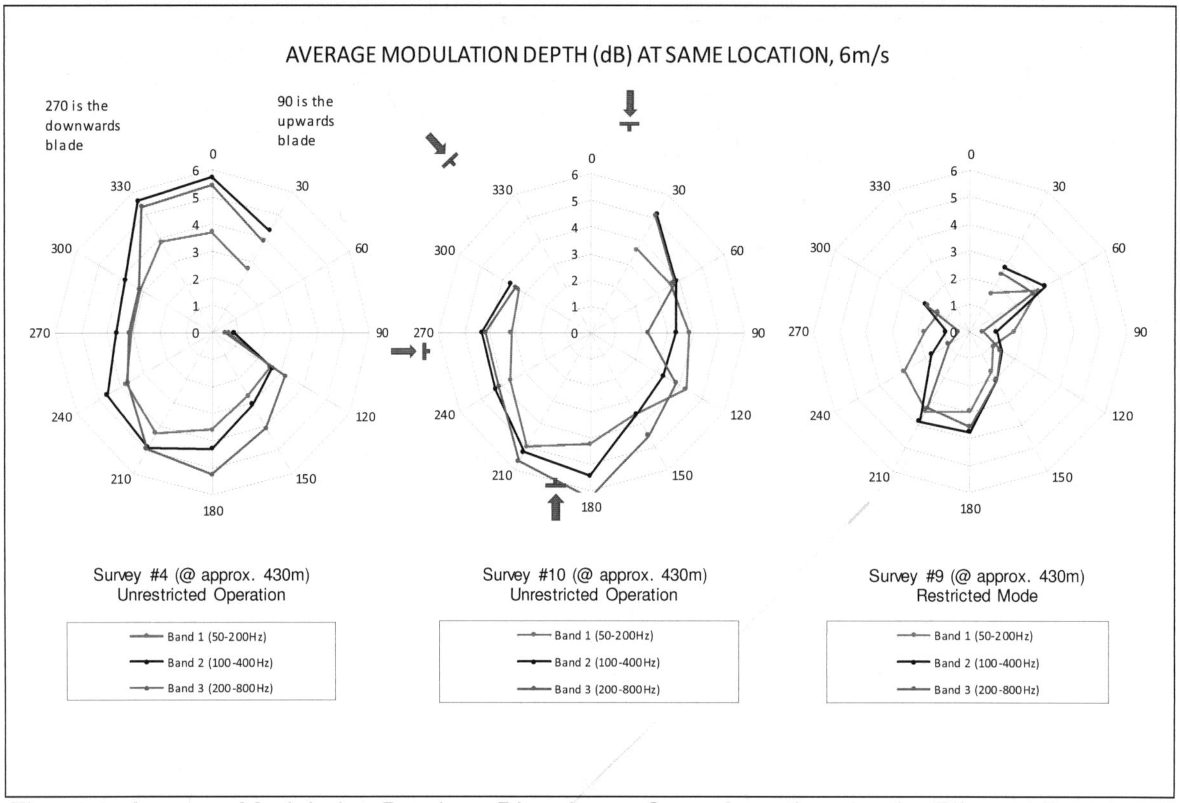

Figure 9: Average Modulation Depth vs Direction at Same Location, 6 m/s, Effect of Restricted Modes Of Operation

It can be seen that the results from the restricted operation shown in the right-hand panel were an improvement upon the initial unrestricted operation shown in the left-hand panel, and were better than the results from periods of unrestricted operation within the repeat Survey 10. It is curious that Survey 10 shows somewhat of an increase in average modulation depth compared to Survey 4, but the difference is mostly within approximately 1 dB for the same wind conditions. The reason for this is not fully understood, however this may be indicative of the uncertainty, or natural variation, that can be expected from using the method under different circumstances at different times of the year Figure 10 below also shows the difference in average modulation depth results between the restricted operation of Survey 9 compared to the original unrestricted operation of Survey 4. This shows the results for Band 2 (100 – 400 Hz) and Band 3 (200 – 800 Hz) for 5 – 7 m/s standardised 10 m height wind speed, as a function of wind direction. It can be seen that for most of the directions where the restriction was applied, there was a notable improvement in the average modulation depth. The 60° and 90° sectors were directions where the restriction was not applied. It can be seen that the results between the two surveys are quite comparable for these directions, although there is some variation in results that may be indicative of the uncertainty that is to be expected from these types of assessments.

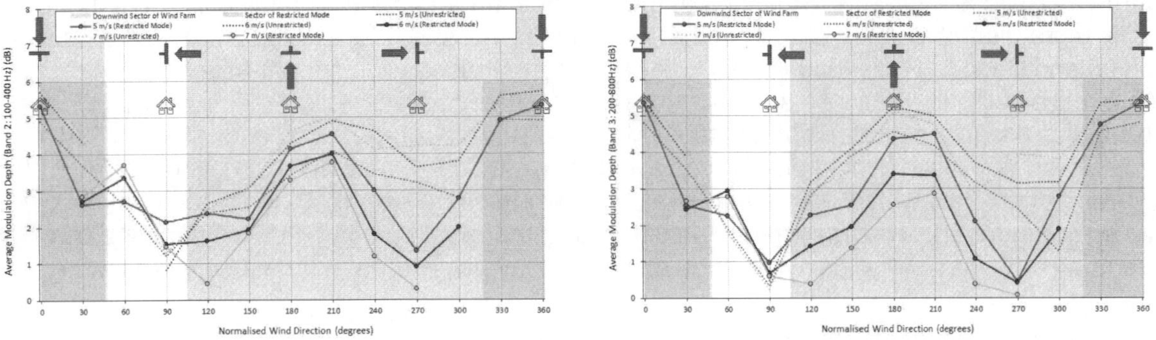

Figure 10: Average Modulation Depth (Band 2: 100-400 Hz & Band 3: 200-800 Hz) vs Wind Direction, 5 – 7 m/s, Effect of Applying Restricted Mode Of Operation, Location ~ 430 m

Figure 11 below considers the slightly more distant location at approximately 600 m, and compares the results between the unrestricted (centre-panel) and restricted (right-hand panel) for different wind directions. For reference, the left-hand panel again shows the unrestricted operation of the closer location previously shown in Survey 4. The receptor locations shown by Survey 4 and Survey 5 are affected by the same turbine model. Thus on Figure 11, the centre panel (compared to the left-hand panel) shows an example of increasing distance, and the right-hand panel (compared to the centre panel) shows an example of applying a restricted mode of operation.

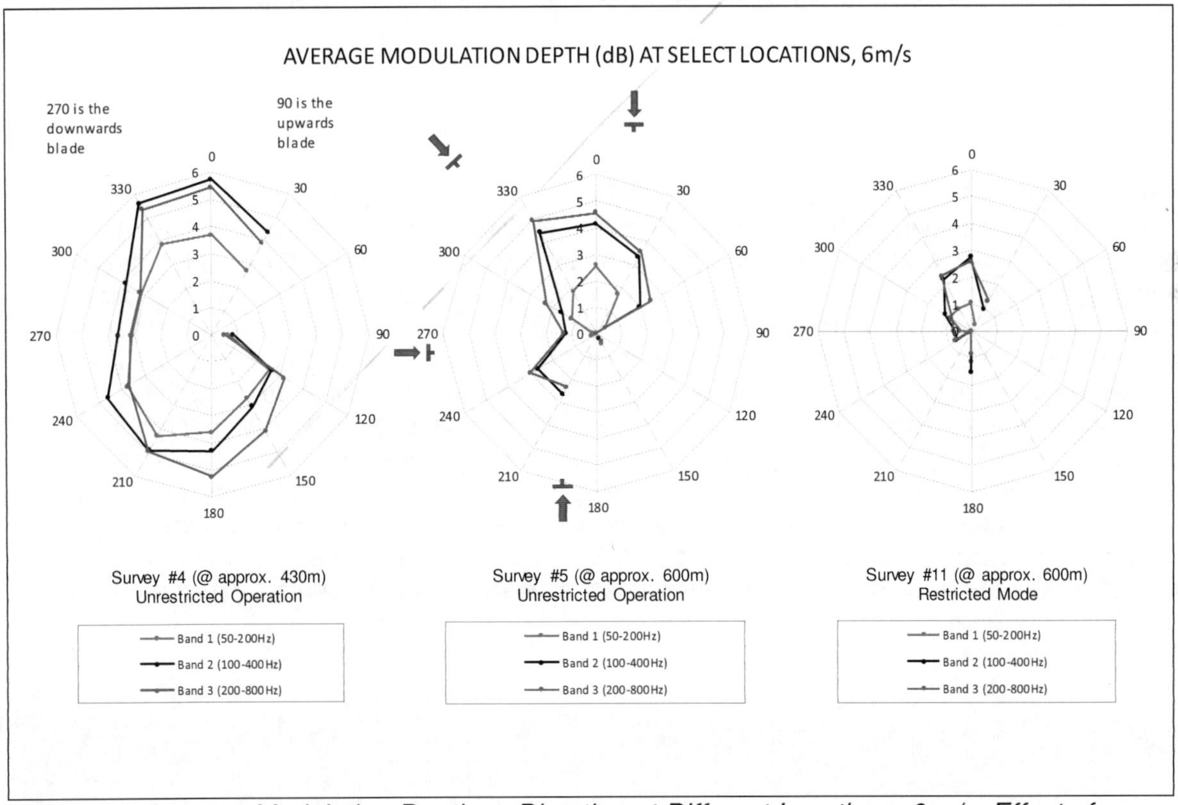

Figure 11: Average Modulation Depth vs Direction at Different Locations, 6 m/s, Effect of Restricted Modes Of Operation

Figure 12 provides another comparison of the unrestricted and restricted noise surveys at the slightly more distant location at approximately 600 m. This shows the results for Band 2 (100 – 400 Hz) and Band 3 (200 – 800 Hz) for 5 – 7 m/s standardised wind speed, as a function of wind direction. Again, it can be seen that for most of the directions where the restriction was applied, there was a notable improvement in the measured average modulation depth. The effect is arguably more noticeable for crosswind (downwards going blade) directions than downwind directions, which was a feature also shown to some extent at the closer location detailed in Figure 10. This may be an indication that the restricted mode in this instance has more of an effect on 'swishing' type modulation that is typically more noticeable in crosswind directions.

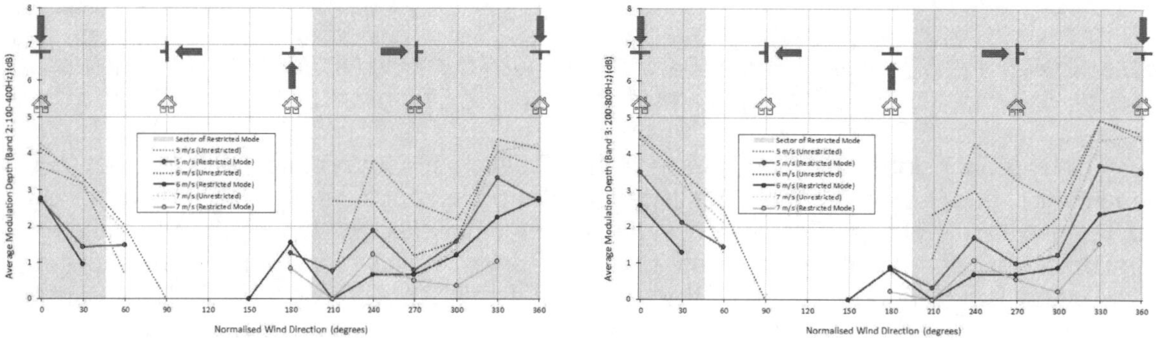

Figure 12: Average Modulation Depth (Band 2: 100-400Hz & Band 3: 200-800Hz) vs Wind Direction, 5 – 7 m/s, Effect of Applying Restricted Mode Of Operation, Location ~ 600m

Turbine Type and Serrations vs Non Serrations

Figure 13 below shows the results from receptors affected by a different turbine type than shown previously in Figures 9 – 12. This compares results from three receptors at increasing distance from the closest turbine. It can be seen that the magnitude of modulation depth shown at approximately 450 m was slightly higher, but was not dissimilar to that shown at approximately 250 m, but there may be peculiarities to measurements taken at close distances. When at approximately 250 m, there were relatively high levels of modulation measured in the crosswind (upwards going blade) direction, that isn't seen at further distances. Further investigation is required to understand the reasons behind this, but this may be related to particular directivity of a noise generation mechanism that is only specific to close distances. The right-hand panel shows results from an even further receptor location (at approximately 810 m), where it can be seen that the results between the three frequency bands are more comparable and Band 1 is as high as the other two bands. This indicates that at further distances, the results become more low frequency biased, and possibly more concentrated on downwind conditions.

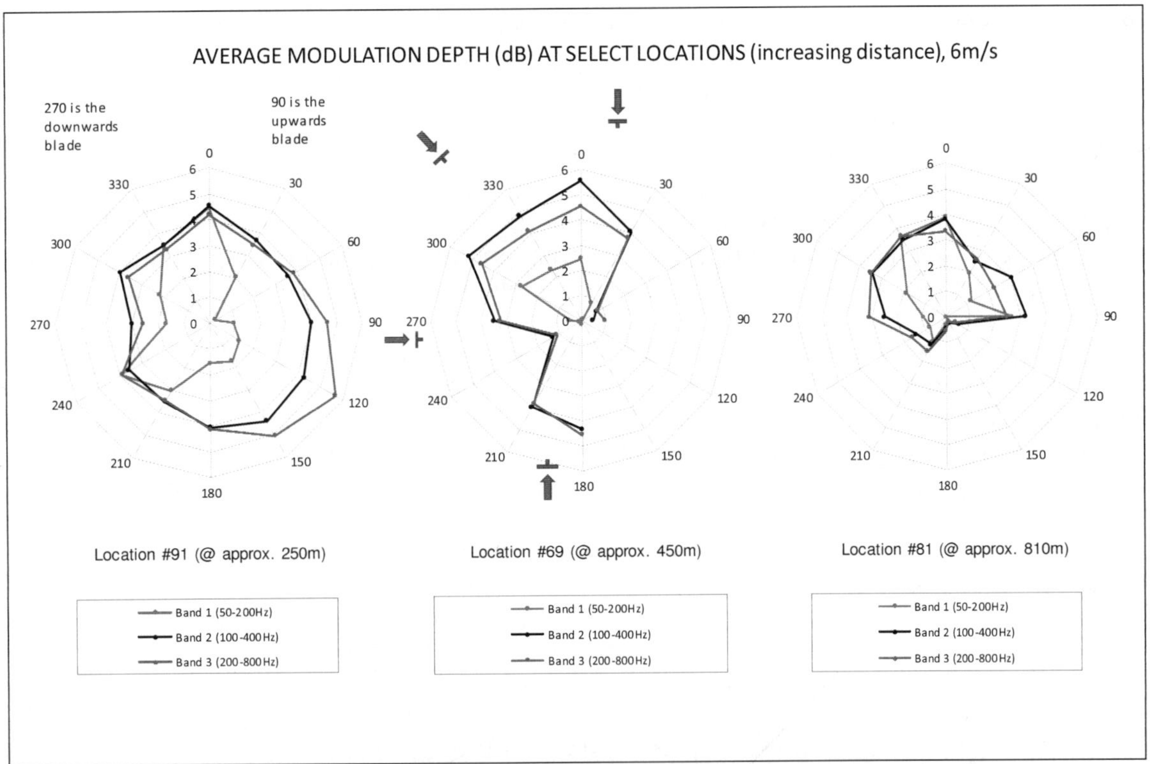

Figure 13: Average Modulation Depth vs Direction at Different Locations, 6 m/s

Amongst the measurements undertaken it has been notable that turbine models with trailing edge serrations have, generally, given some of the lower modulation results. The results are not exhaustive, or necessarily a fair representation, nevertheless Figure 14 below highlights the results from surveys where the closest turbine was a model that had trailing edge serrations fitted. It can be seen that most of the results from turbine models with serrations showed lower results than other surveys at similar distances.

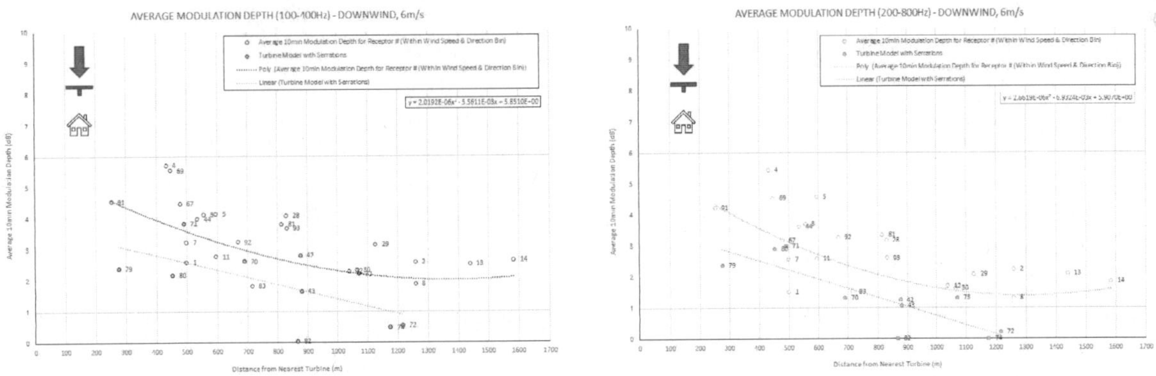

Figure 14: Average Modulation Depth (Band 2: 100-400Hz and Band 3: 200-800Hz) vs Distance, Downwind, 6 m/s, Results from Turbine Models with Serrations Highlighted

Figure 15 below compares the results of three receptors. Surveys 91 and 79 were relatively close (at approximately 250 m and 280 m) to turbines with and without trailing edge serrations. It should be noted that these are not the exact same turbine type that has had serrations added, but simply illustrative examples of one turbine type that doesn't have serrations on the trailing edge, and another that does. Although these are not directly comparable, the results from the turbine with serrations are notably lower than the receptor at a similar distance to a turbine without serrations.

The right-hand panel of Figure 15 shows results from a more distant receptor (at approximately 450 m), that happens to be affected by a mixture of the turbine types measured in the left-hand and centre panels. It is interesting that the results shown for Survey 80 (right-hand panel) would not be dissimilar to the average of the two other panels. It is the case that for the Survey 80 location, 0° in this analysis is when it is downwind of one of the serrated turbines shown in the centre panel, and the relatively high areas of modulation shown at 90° and 240° is when it is downwind of one of the non-serrated turbines shown in the left-hand panel. The limited results here suggest that, if one measures at say 250-300 m away from a particular turbine model, one may be fairly close to covering the range of results that might be expected at more distant residential distances, bar some exceptions. In terms of a risk assessment and predicting what levels of modulation may occur at receptor distances, there may be benefit in having a mid-distance measurement for the turbine model under analysis. In a similar way that tonality is measured at source for each turbine model (as part of the IEC 61400-11 declaration), there may be an argument for doing a similar classification, potentially at a medium distance, for levels of amplitude modulation.

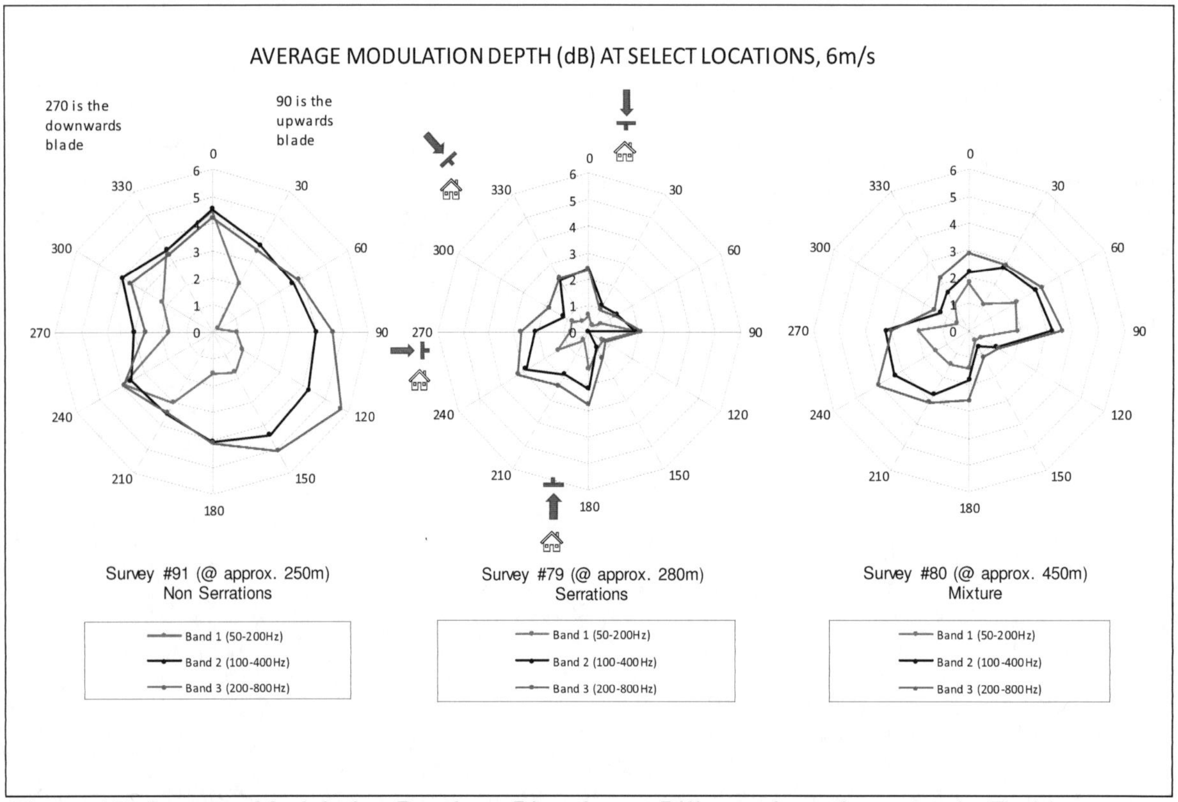

Figure 15: Average Modulation Depth vs Direction at Different Locations, 6 m/s, Turbine Models Non Serrations / Serrations

Other Factors and Collaboration

It would appear that the modulation results are not fully described by distance and wind direction and that other factors also have an influence. There may be merit as part of future works, to investigate other aspects such as: turbine model, or background noise level (in the relevant frequency ranges), and if these factors correlate with the range in measured modulation results.

The dataset shown is only a certain number of receptors that have happened to be measured. The dataset has not been designed to give a representative range of possibilities equally spread across a range of distances and situations. To this end, there would be benefit in a collaboration exercise with other parties to build upon such a database. The results of which could be anonymised. It would be expected that the addition of results at more receptor locations may help to provide clearer indication of systematic trends.

3. Conclusions

Amplitude modulation depth results from a variety of receptor distances were examined and results were analysed as a function of normalised wind direction to investigate trends within the results.

The measured modulation depth generally tended to decrease with increasing distance from the nearest turbine, with levels in the downwind direction extending further when compared to other directions.

For the dataset considered, the highest results generally occurred for 6 m/s standardised 10 m height wind speed, with results for Band 2 (100 – 400 Hz) mostly exhibiting higher levels than the other two frequency ranges. The results for Band 1 (50 – 200 Hz) can be complicated by the presence of modulating tonal components in that frequency range.

Noise surveys that have included periods where the closest turbines have been operating in a restricted mode of operation have, more often than not, exhibited lower average modulation depths than comparable periods of unrestricted operation, however there have been exceptions to this rule.

Some noise surveys of turbines with trailing edge serrations have shown relatively low modulation depth results, indicating this may be a relevant feature of consideration.

References

[1] Institute of Acoustics, IOA Noise Working Group (Wind Turbine Noise), Amplitude Modulation Working Group (2016). *Final Report. A Method for Rating Amplitude Modulation in Wind Turbine Noise.* Version 1, dated 09/08/16.

[2] IEC 61400-11-2. Wind Energy Generation Systems. *Part 11-2: Measurement of wind turbine noise characteristics in receptor position.*

10th International Conference
on
Wind Turbine Noise
Dublin – 21st to 23rd June 2023

Experimental investigation of noise from a wall-mounted swept tip blade in a wind tunnel

Oliver Lylloff, Department of Wind and Energy Systems, Technical University of Denmark (DTU), Frederiksborgvej 399, DK-4000 Roskilde: ollyl@dtu.dk
Andreas Fischer, Department of Wind and Energy Systems, Technical University of Denmark (DTU), Frederiksborgvej 399, DK-4000 Roskilde.

Summary

A wall-mounted swept tip segment of a wind turbine blade (tip model) is tested in an acoustic Kevlar-walled wind tunnel at free stream velocities ranging from 20 to 80 m/s (corresponding to chord-based Reynolds numbers $4.9 \cdot 10^5$ to $2.0 \cdot 10^6$). The tip model used is the result of a design optimization focused on tip extensions for wind turbine blade upscaling. The trailing edge and tip vortex noise spectra are determined by integration of acoustic images generated with a microphone array using beamforming techniques. Aerodynamic lift and drag coefficients are determined from 128 surface pressure tabs on the model and related to the acoustic results. The results indicate, that tip vortex noise is dominant at high angles of attack (corresponding to high lift coefficients) and low flow-speeds. At higher flow speeds, trailing edge noise is the dominant source of acoustic output. This suggests, that tip vortex noise is important to take into account when wind turbines are operating in low wind speeds or noise curtailment. Additionally, the acoustic spectra dependence on velocity is estimated for trailing edge and tip vortex noise. The results indicate, that trailing edge noise scales with a power between 5 and 6, and similar for tip vortex noise, but only at high lift coefficients. Despite the special model used, the presented methodology clearly shows the benefit of using acoustic imaging techniques to distinguish noise sources in a wind tunnel, and can pave way for improved tip vortex noise models in the future.

1 Introduction

Wind turbine noise is comprised of different noise generation mechanisms, but the main contributing source is generally considered to be aerodynamic noise from the trailing edge of the blades. The dominating sources are located at about 80% to 95% of the blade radius, where flow velocities are highest [1]. Therefore, much of the research conducted on wind turbine noise has focused on two-dimensional trailing edge noise, where the noise is considered to scale with the flow speed to a power of 5. Trailing edge serrations are nowadays widely used in the industry to decrease the overall sound emission of a wind turbine [2].

Tip vortex noise (tip noise) was not considered as a dominant noise mechanism on modern wind turbines. It was already in the 1990s demonstrated that it could be reduced by gradually reducing the chord length towards the tip [3], [4] as shown in Figure 1. However, since trailing edge noise is mitigated more and more successfully, tip noise might become relevant again for modern wind turbine designs. As it has not been relevant in the last 20 years there are not many engineering models to predict tip noise available (one example of such a model is the one by Brooks, Pope, and Marcolini [5]) and relevant data sets to develop such a model for modern wind turbines are rare. Hence, the goal of this study is to provide an experimental methodology based on wind tunnel tests, that can lead to tip noise model validation.

Early experimental work on airfoil tip noise and models was done in refs. [6], [7] on a NACA0012 airfoil at low Reynolds numbers. The studied flow conditions are insufficient for modelling modern day wind turbines [8], but the experimental methodology has proven very useful. In the study by Brooks and Marcolini [7], tip noise was obtained by 'subtracting' 2D and 3D airfoil noise spectra (both should produce similar TE noise). At zero-lift ($\alpha = 0$), tip noise was assumed to be negligible, hence 2D and 3D noise spectra should be similar, and a fair agreement was observed, when corrected for different span-lengths. An interesting outcome of the study was, that the noise power scaling law, known for trailing edge noise, was not observed for tip noise [7]. These early studies was later included in the so-called BPM model [5], which is well-know for predicting trailing edge noise. In a more recent study [9], also on a NACA0012 airfoil, the BPM model predictions for tip noise are found to agree with measurements at higher Reynolds numbers, but only at low frequencies. The study proposed an empirical model extension to BPM that agrees better with measurements. A tip noise scaling law of flow speed to power 7.5 was found but a physical interpretation was not evident [9]. An extensive study on different airfoil tip models are given in refs. [10], [11], which couples noise measurements with flow visualizations (PIV). A summary of studies on tip noise in wind tunnels are given in Table 1.

This paper is structured as follows. In Section 2, the wind tunnel, experimental setup, post-processing techniques, and the tip model are described. In Section 3, the aerodynamic and acoustic results are shown. Aerodynamic lift and drag coefficients are determined, and the acoustic spectra, images and scaling properties are studied. The results and methodology are discussed in Section 4 and conclusions are given in Section 5.

2 Methods

In this section, a description of the wind tunnel and measurement methods are given.

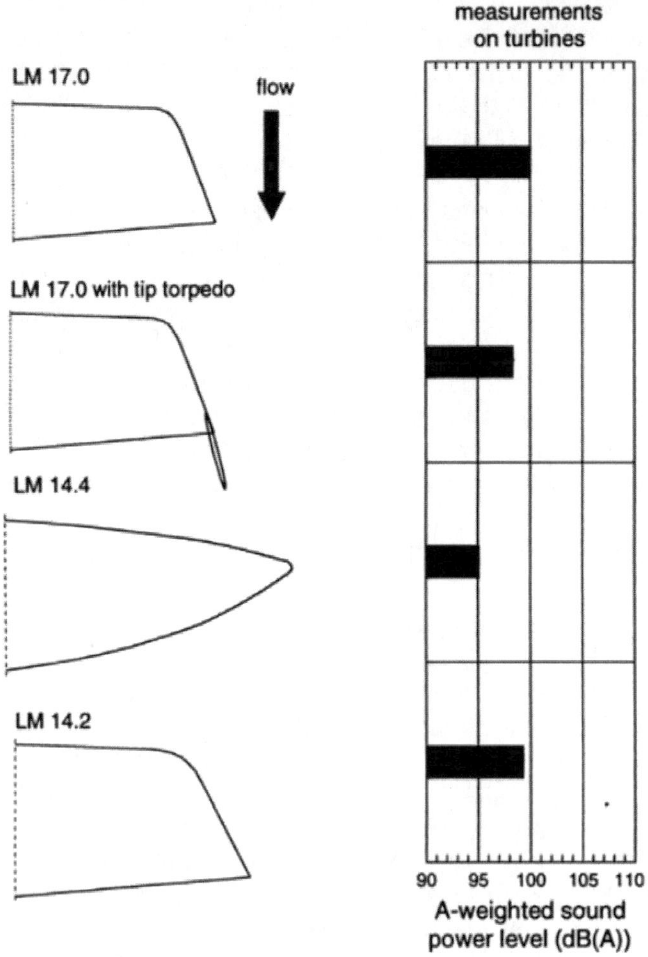

Figure 1: Illustration from [3].

2.1 Poul la Cour Tunnel

The Poul la Cour Tunnel (PLCT, Technical University of Denmark (DTU), Risø Campus, Roskilde, Denmark), is a university-owned wind tunnel dedicated to wind energy research. It was commissioned in 2018 and is capable of both aerodynamic and aeroacoustic measurements. The wind tunnel is comprised of a closed-loop airline with acoustic absorbent treatment and a fan with a nominal power of 2.4 MW. A maximum flow speed of 105 m/s can be achieved in the test section. The test section dimensions are $2 \times 3 \times 9$ m (H×W×L) and it has interchangeable side walls: Hard walls in aerodynamic configuration, and tensioned Kevlar walls in acoustic configuration. The later is utilized in this study. The design of the acoustic setup was inspired by the Virginia Stability wind tunnel [14]. The benefit of the Kevlar-walled configuration is that sound from the test item can transmit almost unhindered through the Kevlar wall and be captured by acoustic equipment, while the flow

Airfoil(s)	Tip shape	Re [-]	U_0 [m/s]	Ref.
NACA0012	Flat & round		$40, 70$	[7]
NACA0012	Flat & round		?	[5]
NACA0012	Flat	$8.0 \cdot 10^5 - 1.6 \cdot 10^6$	$30 - 60$	[9]
NACA0012 & NACA0018	Flat	$2.3 \cdot 10^5 - 3.3 \cdot 10^5$	$35{,}50$	[12]
Unknown type	Flat	$3.0 \cdot 10^5 - 1.1 \cdot 10^6$	$30 - 100$	[13]
8 different NACA	Flat and round	$1.0 \cdot 10^5 - 2.3 \cdot 10^5$	$5 - 50$	[11]
Custom design	Round	$4.9 \cdot 10^5 - 2.0 \cdot 10^6$	$20 - 80$	This study

Table 1: Previous work on tip noise in wind tunnels.

is retained inside the test section, and only small corrections are needed compared to an open-jet configuration [15]. Surrounding the test section is an anechoic room with a free-field condition that was tested according to ISO 3745 [16]. It is close to an ideal free field above frequencies of 125 Hz. However, in the frequency range between 200 Hz and 3150 Hz the deviation from ideal free field conditions is $\pm$ 2 dB which is slightly higher than allowed according to ISO 3745.

2.2 Microphone array methods

Acoustic measurements are conducted with an 84-channel microphone array (1/4" B&K Type 4985) situated in the anechoic room, outside the test section at a distance of 2.3 m from the tip model. Acoustic images are computed with conventional frequency-domain beamforming [17], denoted Delay-and-sum (DAS), and Clean-SC [18]. Source integration is used to extract acoustic spectra from three different spatial regions (see Fig. 2). The trailing edge and tip integration regions are 0.5 m wide (chord-wise) and 0.8 m high (span-wise). The airfoil integration region is 1.5 m by 1.6 m. The integration regions are positioned 0.2 m from ceiling and floor to reduce the influence of junction noise and reflections from the floor. The resulting integrated spectra from each of the regions are normalized to 1 m span.

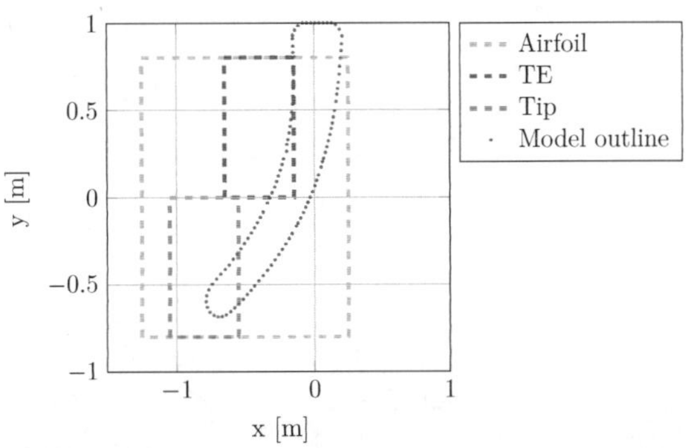

Figure 2: Integration regions used in the study. Flow direction is right to left.

The two different post-processing techniques, DAS and Clean-SC, have different use-cases. To illustrate this, an example of the spectra produced by the two methods (using the trailing edge integration region shown in Fig. 2) are shown in Fig. 3. In a broad frequency range, between 800 Hz and 3000 Hz, there is good agreement. At lower frequencies, DAS produces higher levels, due to a poor resolution of the acoustic images. The spectral shape, however, is very smooth compared to Clean-SC. At higher frequencies (above 4000 Hz), Clean-SC shows a fluctuating behavior, that is likely due to background noise. In the following, Clean-SC is used for computing integrated spectra, while DAS is used to show acoustic images.

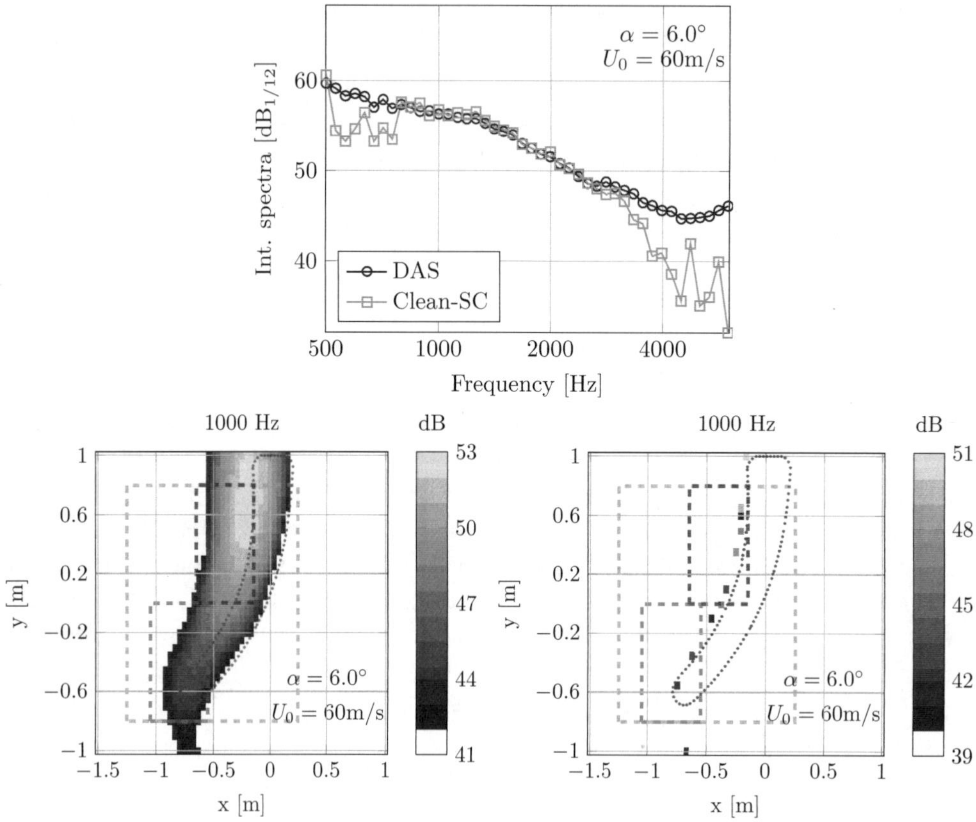

Figure 3: Comparison of post-processing techniques. Top: Integrated spectra. Bottom left: DAS acoustic image. Bottom right: Clean-SC acoustic image.

2.3 Blade tip model

The swept blade tip model [19] (See Fig. 4), was chosen because of its availability. The swept design was the result of an aeroelastic optimisation within load constraints. It is not representative of the blade tip of modern wind turbines, but the tip noise mechanism is the same as for a more traditional design.

The root section of the model was covered by a fairing (see Fig. 4b) to reduce the noise source at the junction of the model and the wind tunnel ceiling, because such a noise source would not be present on a wind turbine blade. Additionally, if the junction noise source was not reduced, it might influence the tip noise levels at low frequencies when the resolution of the microphone array is poor.

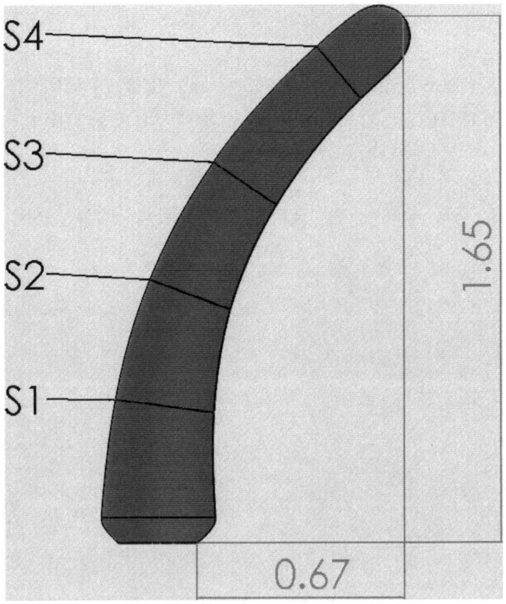

(a) Model sketch of airfoil. The model was mounted in the ceiling of the wind tunnel, hence it appears upside-down in subsequent figures. Flow direction is left to right.

(b) Picture of airfoil model in tripped configuration mounted in wind tunnel test section with fairing at root.

Figure 4: Swept blade tip model.

3 Results

In this section, the aerodynamic and acoustic results of the tested configurations are presented. In Table 2 an overview of the experimental data collected is given. The angle-of-

Configuration	AoA [deg.]	Re [-]	U_0 [m/s]
Tripped	$-20:2:20$	$4.9 \cdot 10^5$	20
Tripped	$-20:2:20$	$9.7 \cdot 10^5$	40
Tripped	$-14:2:20$	$1.4 \cdot 10^6$	60
Tripped	$-14:2:18$	$2.0 \cdot 10^6$	80

Table 2: Tested configurations in this study.

attack and aerodynamic coefficients presented in the following are uncorrected and computed only for the root-segment of the blade (the section below the ′S1′ line in Fig. 4a). A standard 2D aerodynamic Kevlar-wall correction [20] is insufficient for the analysis the swept tip blade used in this study. For future work, a 3D correction as the one described in ref. [21] would be relevant for this use case.

3.1 Aerodynamic results

The lift coefficient as function of angle-of-attack (AoA) and lift over drag are shown in Figure 5. In the following results, three particular cases are chosen to couple the acoustic

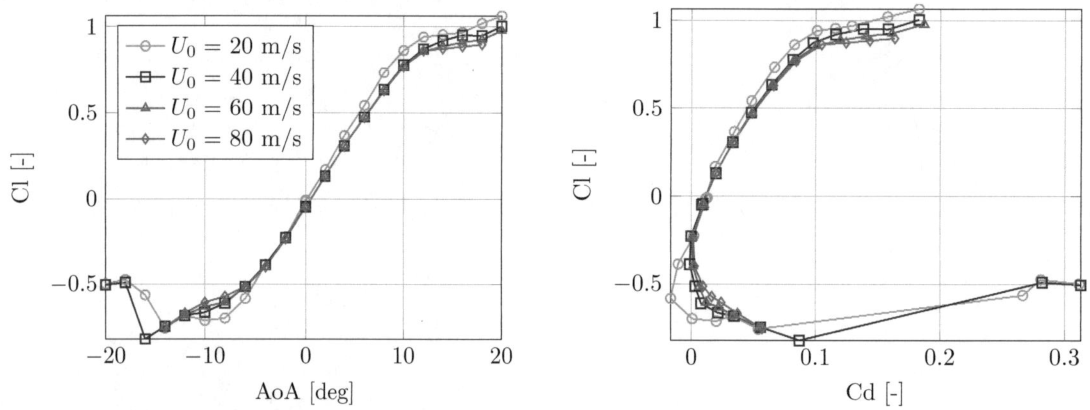

(a) Lift coefficient C_l as function of angle-of-attack (AoA).

(b) Lift over drag.

Figure 5: Aerodynamic results.

and aerodynamic results: A case with attached flow ($C_l = 0.5$, AoA $= 6°$), a case with partly stall behavior ($C_l = 0.8$, AoA $= 10°$), and a case in deep stall ($C_l = 1.0$, AoA $= 14°$).

3.2 Acoustic results

Acoustic images are shown in Figs. 6 to 8 and noise spectra captured from spatial integration of the acoustic images are shown in Figs. 9 to 11.

The three cases considered are: Attached flow $C_l = 0.5$ (shown in Fig. 6), partly stall $C_l = 0.8$ (shown in Fig. 7), and deep stall $C_l = 1.0$ (shown in Fig. 8). Comparing the different acoustic images, it is clear, that in the attached flow case, trailing edge noise is becoming increasingly dominant for increasing flow speed. However, one exception is at $U_0 = 20$ m/s, where the tip noise source is pronounced at higher frequencies. This is also observed in the integrated spectrum in Fig. 9. At increasing angle-of-attack, going into partly stall, the tip noise source is more prominent in the acoustic images (Fig. 7), particularly towards higher frequencies. Although at 1000 Hz, trailing edge noise is still dominant at $U_0 = 80$ m/s. Moving to deep stall (Fig. 8), the noise sources are more evenly distributed along the airfoils trailing edge, except at $U_0 = 20$ m/s, where a tip noise source is still present. The same trends are observed in the integrated spectra in Fig. 10 and

Fig. 11. Interestingly, in the case for partly stall, the cross-over frequency where tip noise becomes more dominant than trailing edge noise is moving as function of flow speed. At $U_0 = 20\,\mathrm{m/s}$ it is 700 Hz, at $U_0 = 40\,\mathrm{m/s}$: 1200 Hz, $U_0 = 60\,\mathrm{m/s}$: 2000 Hz, and at $U_0 = 80\,\mathrm{m/s}$ it is 3000 Hz (see Fig. 10).

3.2.1 Overall Integrated Spectrum Level

The general trends observed in the acoustic images and integrated spectra, in the previous section, can be summarized by computing overall integrated spectrum levels (OAISL) by a summation of integrated spectra (from Clean-SC) of the three difference spatial regions (airfoil, trailing edge, and tip) in the frequency range 0.5 kHz-5 kHz. Results are shown in Fig. 12 as function of C_l. This metric allows for a direct coupling between aerodynamic and acoustic observations, but lacks the frequency dependence that was described in the previous section. In general, tip noise is only dominant at $U_0 = 20\,\mathrm{m/s}$ and $C_l > 0.7$. However, when going into deep stall, trailing edge noise again dominates. At $U_0 = 40\,\mathrm{m/s}$ this tendency is vaguely observed, but tip noise is generally $3 - 5$dB lower. At higher flow speeds, tip noise is more than 10 dB lower than trailing edge noise.

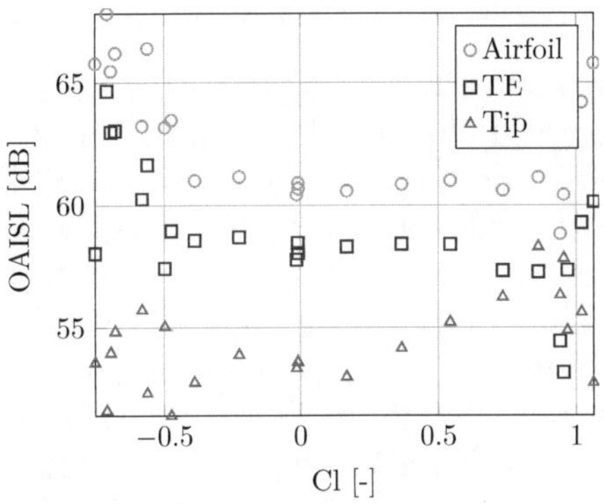

(a) $U_0 = 20\,\mathrm{m/s}$

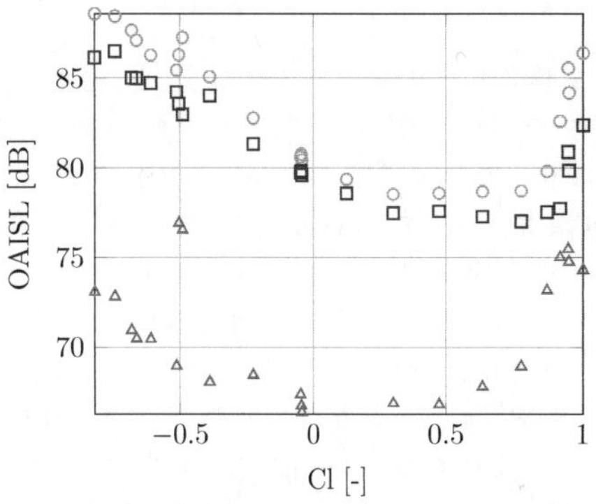

(b) $U_0 = 40\,\mathrm{m/s}$

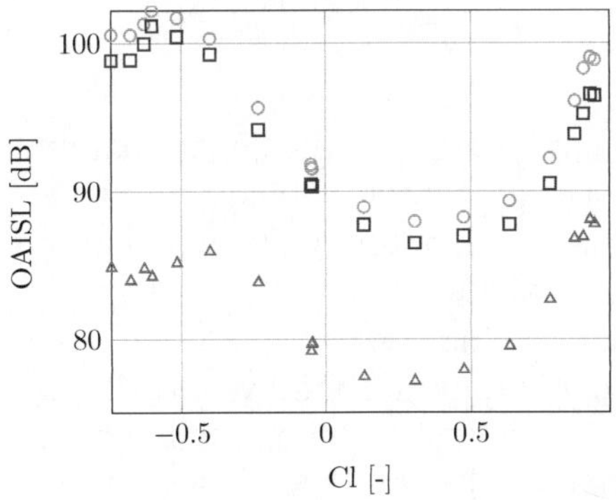

(c) $U_0 = 60\,\mathrm{m/s}$

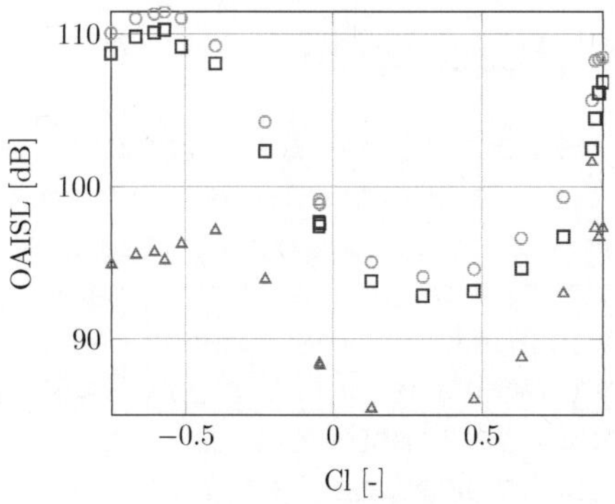

(d) $U_0 = 80\,\mathrm{m/s}$

Figure 12: Overall Integrated Spectrum Level (OAISL) computed from integrated spectra using Clean-SC in frequency range 0.5 kHz-5 kHz.

3.2.2 Velocity scaling

Using the overall integrated spectrum level (OAISL), computed above, the velocity scaling is assessed. Results and regression lines for the three flow cases considered, are shown in Fig. 13.

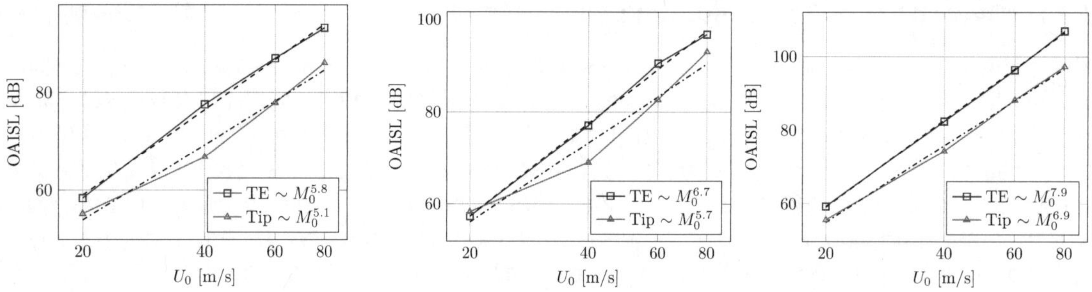

(a) Scaling at $C_l = 0.5$. (b) Scaling at $C_l = 0.8$. (c) Scaling at $C_l = 1.0$.

Figure 13: Velocity scaling at three different C_l values and regression lines as function of free-stream Mach number M_0.

With a limited set of data points, regression lines are subject to large uncertainty, but the observed tendency is an increase in power coefficient as function of C_l. The trailing edge noise of a extruded aerofoil section scales with the Mach number to the power of 5 under the assumption that the boundary layer turbulence scales with the flow speed to a power of 2 [22]. Brooks and Marcolini [7] empirical found a scaling of the mach number to the power of 5 for tip noise which is in line with our results for attached flow ($C_l = 0.5$). Three dimensional flow effects along the swept trailing edge might cause the difference in scaling of the trailing edge noise compared to classical literature. At $C_l = 0.8$ and $C_l = 1.0$ the flow is partially or fully detached from the model. The high values of the scaling exponent might be caused by this flow condition.

4 Discussion

The acoustic imaging technique has proven to be a useful tool for identifying different noise generation mechanisms on a blade tip model. Developments in wind tunnel design and post-processing methods, over the last couple of decades, have improved the acoustic image resolution greatly, to an extent where small details can be studied with high precision. These developments also put extra weight on the choice of acoustic imaging technique. In this study, DAS and Clean-SC was chosen, which are two well-known and established methods within the acoustic imaging community, and their mutual benefits and disadvantages were briefly described. But there are other methods available, that might be relevant for this particular use case, e.g., [13]. One issue is the poor resolution at lower frequencies, which is somewhat solved by using Clean-SC. In the acoustic images, e.g., Fig. 6 at 1000 Hz, reflections from the floor is seen to extend into the tip integration region. This effect is even stronger at lower frequencies, which can lead to overestimated levels.

The choice of integration regions is another point that could be relevant to study in future work. In this study, equally-sized rectangular regions were placed approximately over the trailing edge and tip sections, but smaller regions shaped to the curvature of the model is another possibility. The effects of such regions have not been studied in the literature, and it is unclear if it has a benefit over the conventional rectangular regions. One possible future development could be a more direct coupling between the acoustic integration regions and aerodynamics of individual sections of the model, e.g., using the segments shown in Fig. 4a.

The aerodynamic properties of the root segment of the tip model was used throughout this study. It is quite certain, that the flow properties on the segment near the tip is different than the segment at the root, and therefore the coupling between the aerodynamic and acoustic results are subject to some degree of uncertainty. To eliminate this uncertainty, future wind tunnel work on blade tip noise should use a more representative wind turbine blade tip, such as the design "LM 14.4" shown in Fig. 1.

In the section about velocity scaling, a simple power law regression line was used to estimate the power coefficient, under the assumption that the Mach number is the only dependent variable. This might be true for trailing edge noise, but it is not evident that this is also the case for tip noise. For instance, the empirical model developed in ref. [9] has additional dependent variables.

5 Conclusion

An experimental wind tunnel methodology for investigation of blade tip noise was presented. Acoustic images produced with a microphone array and state-of-the-art post-processing techniques were used to extract noise from the trailing edge and tip regions of the blade. The acoustic spectra were compared at different flow velocities and angles of attack and related to the aerodynamic flow properties. It was found, that tip noise is dominant at low flow speeds, and at high angles of attack, corresponding to lift coefficients above 0.7. At higher flow speeds, tip noise is only dominant in the high frequency range. The coupling between acoustic results and the aerodynamic properties could be further improved in future work by implementing a 3D Kevlar-wall corrections, as described in ref. [21], and computing flow properties on individual segments of the blade tip model. Knowledge of the boundary layer properties could shed further light on the complex coupling between aerodynamics and acoustics, and be investigated with, e.g., a hot-wire probe.

With the increasingly successful mitigation of trailing edge noise, tip noise might be the next dominant noise source to tackle in future wind turbine design, and with the presented methodology in this study, the development of new and more precise tip noise models are within reach.

Acknowledgements

The present work is conducted as a sub-task of the IEA Wind TCP research Task 39 (Quiet Wind Turbine Technology). DTU participation was supported by the EUDP project

Jr. nr. 134-21022 funded by the Danish Energy Agency (Energistyrelsen).

References

[1] S. Oerlemans, P. Sijtsma, and B. Méndez López, "Location and quantification of noise sources on a wind turbine," *Journal of Sound and Vibration*, vol. 299, no. 4, pp. 869–883, 2007. DOI: `https://doi.org/10.1016/j.jsv.2006.07.032`.

[2] S. Oerlemans, "Wind Turbine Noise Mitigation," in *Handbook of Wind Energy Aerodynamics*, B. Stoevesandt, G. Schepers, P. Fuglsang, and Y. Sun, Eds., Cham: Springer International Publishing, 2022, pp. 1447–1462. DOI: `10.1007/978-3-030-31307-4{\_}73`.

[3] H. Klug, T. Oston, M. V. Lowson, *et al.*, "Aerodynamic noise from wind turbines and rotor blade modification," English, in *International European Union wind energy conference, Gothenburg, Sweden*, 1996, pp. 782–784.

[4] S. Wagner, R. Bareiß, and G. Guidati, *Wind Turbine Noise*. Berlin, Heidelberg: Springer Berlin Heidelberg, 1996. DOI: `10.1007/978-3-642-88710-9`.

[5] T. F. Brooks, D. Pope, and M. A. Marcolini, "Airfoil self-noise and prediction," *Nasa Reference Publication*, no. 1218, 1989.

[6] A. R. George and S. T. Chou, "BROADBAND ROTOR NOISE ANALYSES.," *NASA Contractor Reports*, 1984.

[7] T. F. Brooks and M. A. Marcolini, "Airfoil tip vortex formation noise," *AIAA journal*, vol. 24, no. 2, pp. 246–252, 1986.

[8] P. Fuglsang and H. Aagaard Madsen, "Application of aeroaccoustic models to design of wind turbine rotors," English, in *Aero-acoustic noise of wind turbines noise prediction models*, B. Maribo Pedersen, Ed., Technical University of Denmark. Department of Fluid Mechanics, 1997, pp. 39–47.

[9] D. J. Moreau, C. J. Doolan, W. N. Alexander, T. W. Meyers, and W. J. Devenport, "Wall-Mounted Finite Airfoil-Noise Production and Prediction," *AIAA Journal*, vol. 54, no. 5, pp. 1637–1651, May 2016. DOI: `10.2514/1.J054493`.

[10] T. Zhang, T. Geyer, C. de Silva, J. Fischer, C. Doolan, and D. Moreau, "Experimental Investigation of Tip Vortex Formation Noise Produced by Wall-Mounted Finite Airfoils," *Journal of Aerospace Engineering*, vol. 34, no. 6, Nov. 2021. DOI: `10.1061/(asce)as.1943-5525.0001315`.

[11] T. Zhang, D. Moreau, T. Geyer, J. Fischer, and C. Doolan, "Dataset on tip vortex formation noise produced by wall-mounted finite airfoils with flat and rounded tip geometries," *Data in Brief*, vol. 28, p. 105 058, Feb. 2020. DOI: `10.1016/j.dib.2019.105058`.

[12] T. F. Geyer, D. J. Moreau, J. Giesler, P. M. Hall, E. Sarradj, and C. J. Doolan, "Measurement of the noise generated by wall-mounted airfoils of different thickness," in *2018 AIAA/CEAS Aeroacoustics Conference*, American Institute of Aeronautics and Astronautics Inc, AIAA, 2018. DOI: `10.2514/6.2018-3796`.

[13] G. R. Yakhina, M. Roger, A. Finez, *et al.*, "Localization of swept free-tip airfoil noise sources by microphone array processing," *AIAA Journal*, vol. 58, no. 8, pp. 3414–3425, 2020. DOI: 10.2514/1.J058231.

[14] W. J. Devenport, R. A. Burdisso, A. Borgoltz, *et al.*, "The Kevlar-walled anechoic wind tunnel," *Journal of Sound and Vibration*, vol. 332, no. 17, pp. 3971–3991, Aug. 2013. DOI: 10.1016/j.jsv.2013.02.043.

[15] S. Glegg and W. Devenport, *Aeroacoustics of low mach number flows: Fundamentals, analysis, and measurement.* 2017, pp. 1–537.

[16] "Acoustics – Determination of sound power levels and sound energy levels of noise sources using sound pressure – Precision methods for anechoic rooms and hemi-anechoic rooms," International Organization for Standardization, Geneva, CH, Tech. Rep., 2012.

[17] D. H. D. H. Johnson and D. E. Dudgeon, *Array signal processing: concepts and techniques.* PTR Prentice Hall, 1993, pp. 1–533.

[18] P. Sijtsma, "CLEAN based on spatial source coherence," *International journal of aeroacoustics*, vol. 6, no. 4, pp. 357–374, 2007.

[19] T. Barlas, G. R. Pirrung, N. Ramos-García, *et al.*, "Wind tunnel testing of a swept tip shape and comparison with multi-fidelity aerodynamic simulations," English, *Wind Energy Science*, vol. 6, no. 5, pp. 1311–1324, Oct. 2021. DOI: 10.5194/wes-6-1311-2021.

[20] A. Fischer, C. Bak, O. Lylloff, *et al.*, "Cross validation of the aerodynamic and acoustic measurements in two Kevlar-walled wind tunnels," *Journal of Physics: Conference Series*, vol. 2265, no. 2, p. 022 103, May 2022. DOI: 10.1088/1742-6596/2265/2/022103.

[21] H. Ura, M. Shigemi, T. Hirotani, and T. Homma, "Wall Interference Correction Method for Kevlar Wall Test Section," *Journal of Aircraft*, vol. 57, no. 5, pp. 889–900, Sep. 2020. DOI: 10.2514/1.C035864.

[22] J. E. F. Williams and L. H. Hall, "Aerodynamic sound generation by turbulent flow in the vicinity of a scattering half plane," *Journal of Fluid Mechanics*, vol. 40, no. 04, p. 657, Mar. 1970. DOI: 10.1017/S0022112070000368.

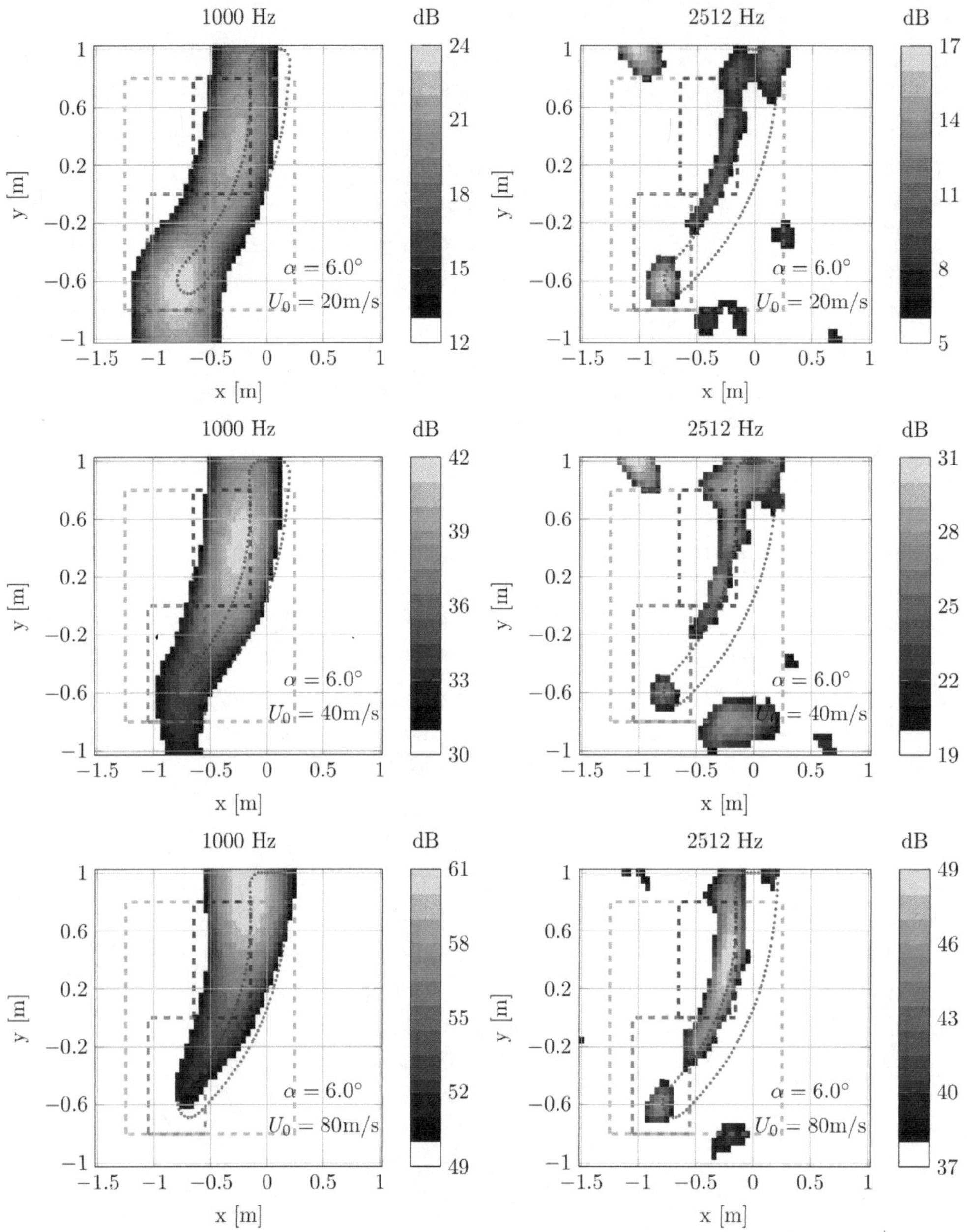

Figure 6: Acoustic images computed with conventional beamforming (DAS). Flow direction is right to left. The angle-of-attack α corresponds approximately to a $Cl = 0.5$, i.e., attached flow.

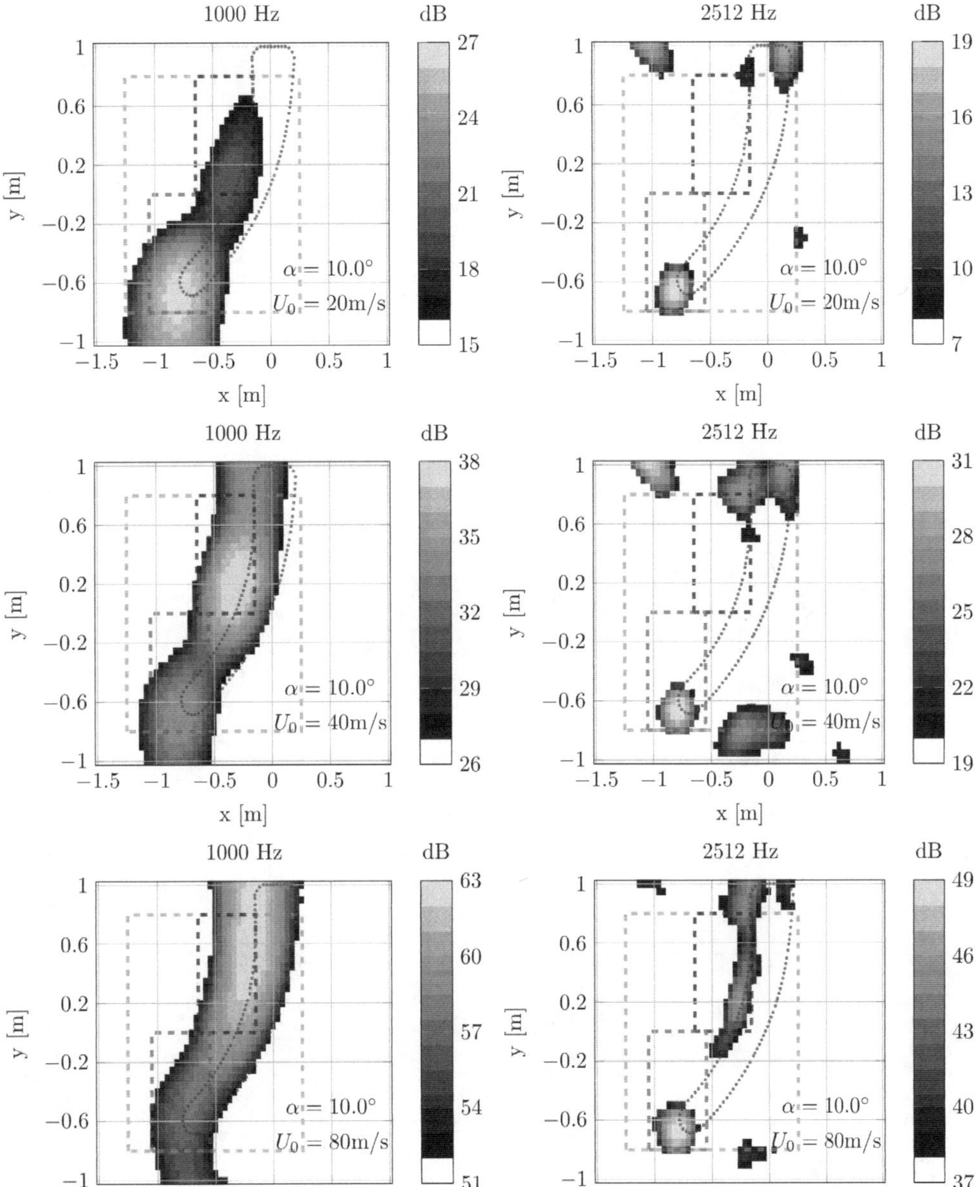

Figure 7: Acoustic images computed with conventional beamforming (DAS). Flow direction is right to left. The angle-of-attack α corresponds approximately to a $Cl = 0.8$, i.e., partially separated flow.

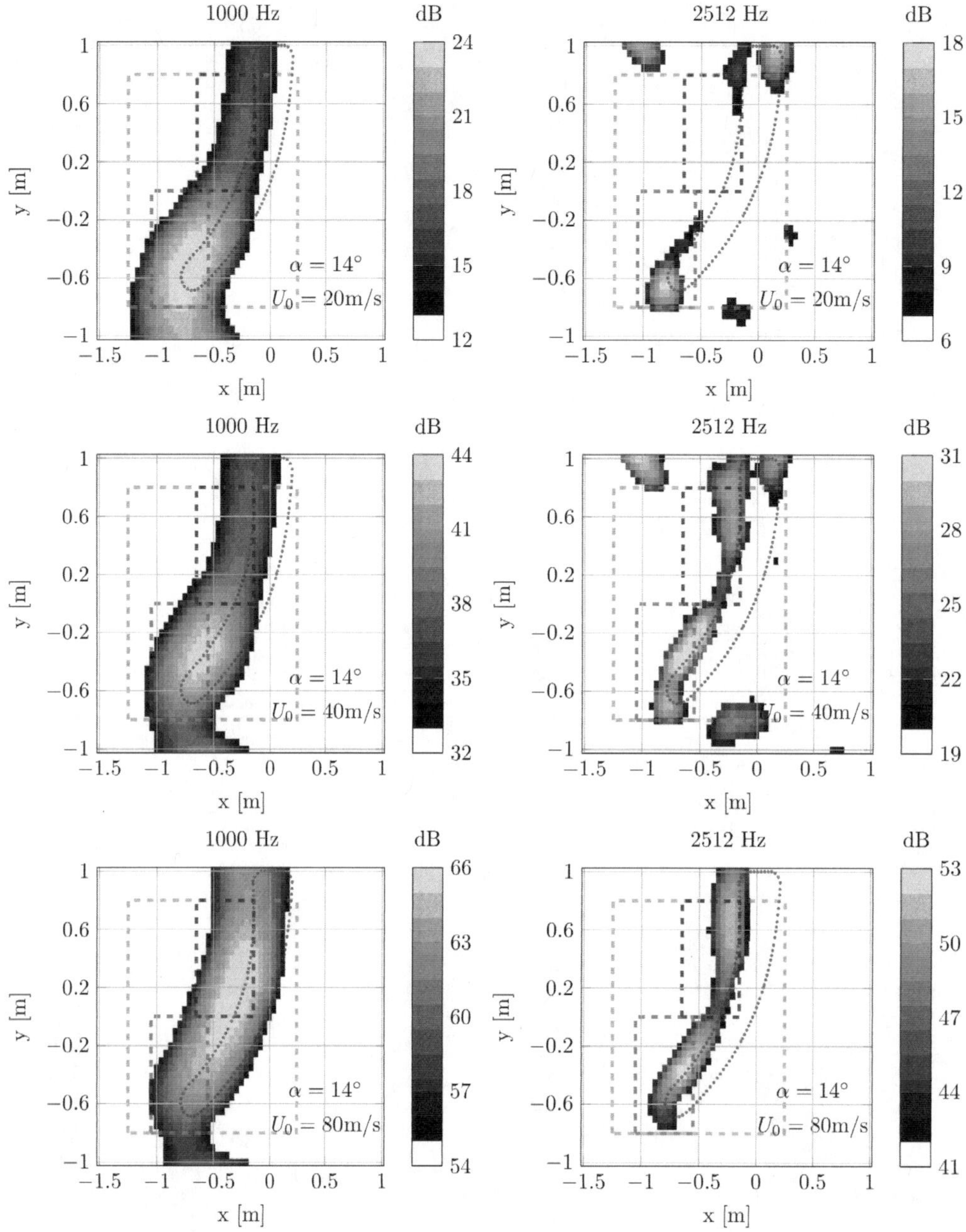

Figure 8: Acoustic images computed with conventional beamforming (DAS). Flow direction is right to left. The angle-of-attack α corresponds approximately to a $Cl = 1.0$, i.e., deep stall.

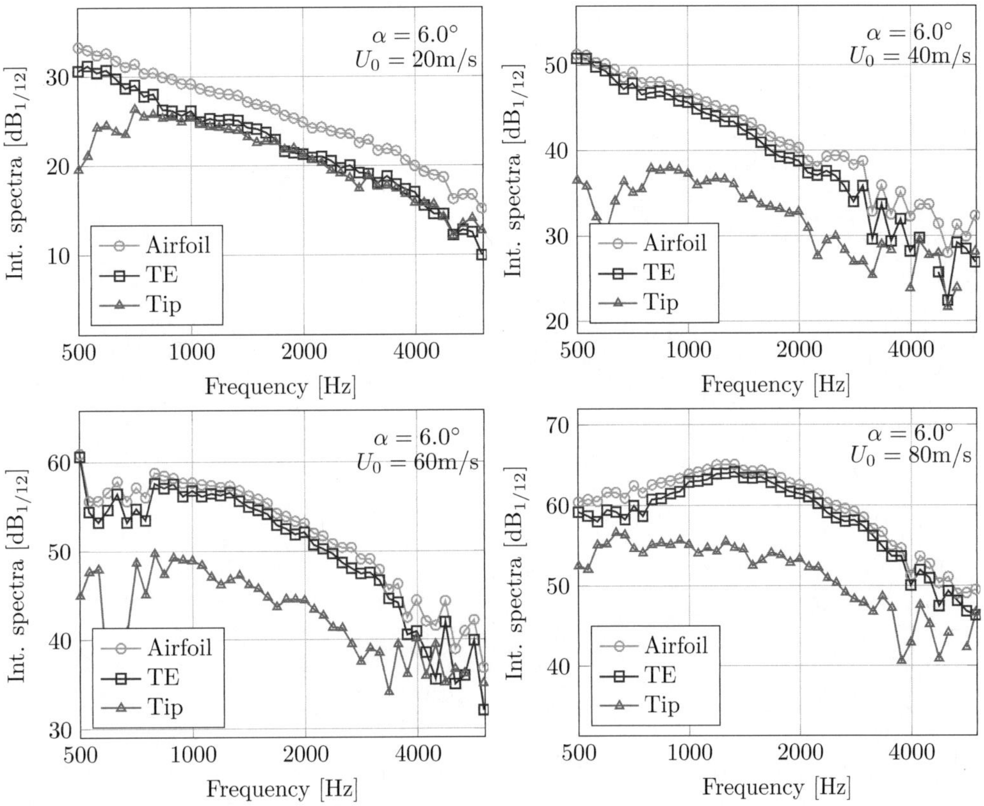

$\alpha = 6.0°$
$U_0 = 20\text{m/s}$
Int. spectra [$\text{dB}_{1/12}$]
Frequency [Hz]
Airfoil
TE
Tip
$\alpha = 6.0°$
$U_0 = 40\text{m/s}$
$\alpha = 6.0°$
$U_0 = 60\text{m/s}$
$\alpha = 6.0°$
$U_0 = 80\text{m/s}$

Figure 9

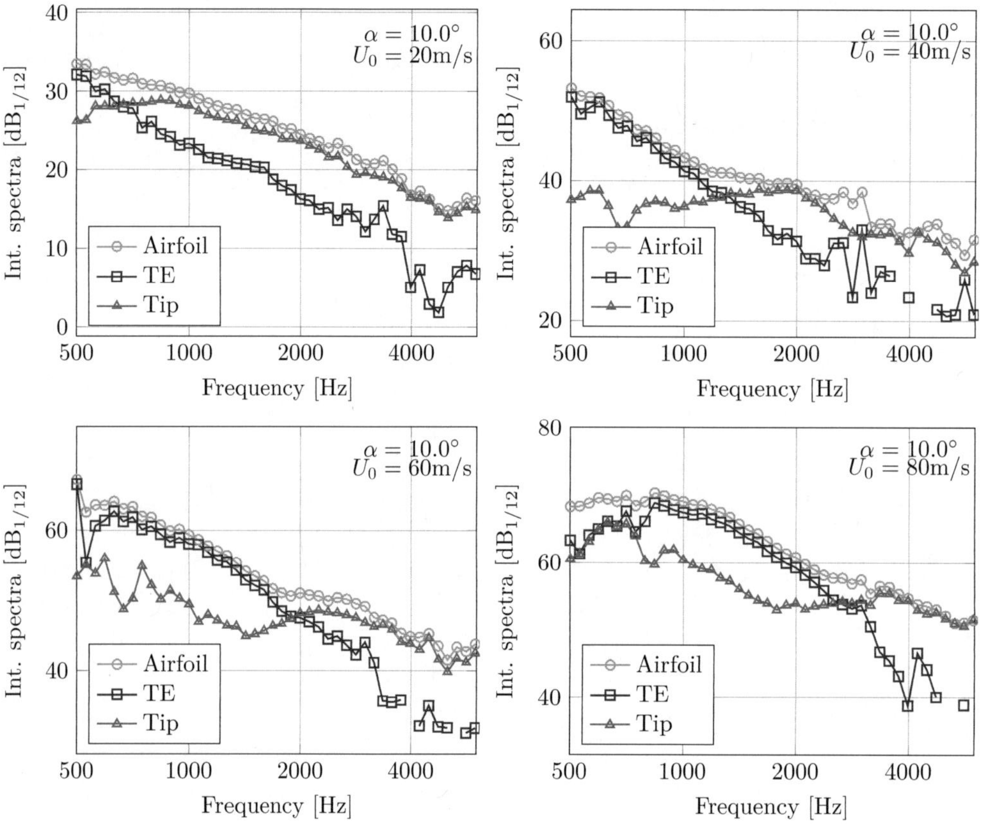

$\alpha = 10.0°$
$U_0 = 20\text{m/s}$
$\alpha = 10.0°$
$U_0 = 40\text{m/s}$
$\alpha = 10.0°$
$U_0 = 60\text{m/s}$
$\alpha = 10.0°$
$U_0 = 80\text{m/s}$
Airfoil
TE
Tip
Int. spectra [$\text{dB}_{1/12}$]
Frequency [Hz]
500
1000
2000
4000

Figure 10

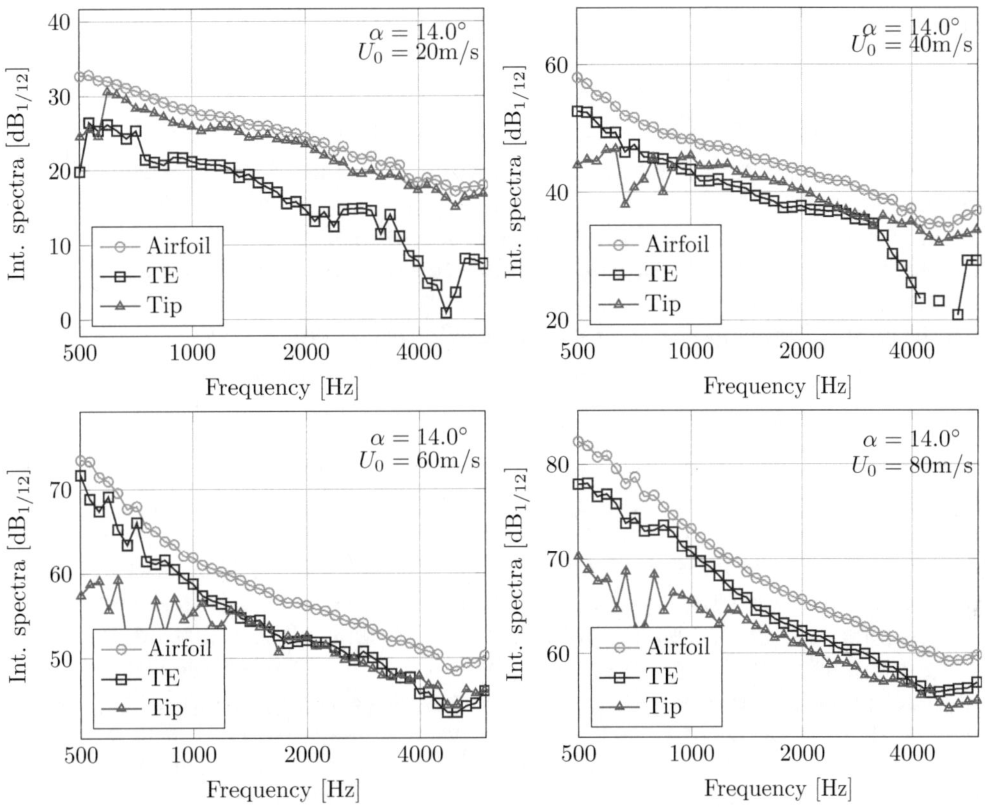

α = 14.0°
U0 = 20m/s
α = 14.0°
U0 = 40m/s
α = 14.0°
U0 = 60m/s
α = 14.0°
U0 = 80m/s
Airfoil
TE
Tip
Int. spectra [dB1/12]
Frequency [Hz]

Figure 11

10th International Conference
on
Wind Turbine Noise
Dublin – 21st to 23rd June 2023

Influence of atmospheric boundary layer characteristics and source height on sound propagation from a 5 MW wind turbine

Neeraj Paul Manelil[1], Keerthana Mohan[1,2], Johannes Nicolaas Theron[1], Leo Höning[1], and Bernhard Stoevesandt[1]
Corresponding author: neeraj.paul.manelil@iwes.fraunhofer.de
[1]Fraunhofer Institute for Wind Energy Systems, Department Aerodynamics, CFD and stochastic Dynamics, Küpkersweg 70, 26129 Oldenburg, Germany.
[2] CSIR-Structural Engineering Research Centre, Chennai, India.

Summary

In response to the growing concern over noise pollution generated by wind turbines and its adverse impact on living organisms, the current research endeavours to contribute to the wind energy industry by improving our understanding of sound propagation. The investigation focuses on the transmission of monotonic sound waves of 100 Hz and 300 Hz through a neutral turbulent atmospheric boundary layer (ABL). Variability of the ABL has been investigated by describing two different surface roughness values that effectively influence ABL characteristics. This study examines the influence of sound source height, viz. 25m, 88m and 151m, on the spatial distribution of sound pressure level (SPL) in the vicinity of a 5MW wind turbine (1km upwind and 2km downwind). These source heights correspond to hub height and extreme blade tip heights of a typical 5MW wind turbine. The velocity fields have been generated using an actuator-line model embedded in a large-eddy simulation (LES) domain A two-dimensional Finite Difference Time Domain (FDTD) acoustic propagation solver, developed in-house, is used for the prediction of sound pressure levels. The spatial derivatives in the governing linearized Euler equations are discretized using an 11-point central differencing scheme while explicit time stepping is performed using an optimized low-dispersion and low-dissipation 4th order Runge-Kutta method. The ground surface is set as a perfectly reflecting boundary and a porous absorbing layer with a width of 20m is implemented on the remaining boundaries to minimize spurious reflections. Contours describing the distribution of relative sound pressure levels (RSPL) and SPL over the two-dimensional domain of size 3000 m x 1000 m is reported in the study. The transmission Loss at different locations along the ground for the test cases are also presented. The study indicates that the characteristics of the atmospheric velocity field, such as turbulence intensity and inflection point, have the potential to change sound pressure levels at far field regions up to 2 km in the downwind direction.

1. Introduction

Onshore wind energy is a highly cost-effective source of renewable energy, with a Levelized Cost of Electricity ranging from 3.94 to 8.29 €Cent/kWh, making it second only to photovoltaics, which range from 3.12 to 11.01 €Cent/kWh (Kost et al., 2021). This has led to an increase in the number of onshore wind turbines, but with it come concerns surrounding the environmental and safety impacts of this technology. In particular, noise has become an increasingly significant health problem in recent years, as exposure to excess noise can lead to stress, sleep deprivation, cognitive impairment, hypertension, and cardiovascular disease (Basner et al., 2014). As a result, there is growing demand for accurate noise prediction tools that can take into account the effects of complex atmospheric boundary layer characteristics.

The topic of wind turbine noise can be divided into two broad categories: noise generation and noise propagation. The primary source of noise from wind turbines is the aero-acoustic source generated by the interaction between the turbine blades and the incoming airflow. The mechanism of noise generation is complex and depends on various factors, including blade geometry and turbulence characteristics. Although detailed models exist to predict aerodynamic noise generation (Cotte, 2019), noise propagation studies commonly represent it as a point source (Lee et al., 2016; Prospathopoulos and Voutsinas, 2007).

The sound pressure level in the far field of a wind turbine is strongly influenced by atmospheric conditions and the turbine wake (Colas et al., 2022). The wind shear profile in the atmospheric boundary layer (ABL) can cause acoustic waves to refract, resulting in higher overall sound pressure levels in the far field than in a stationary atmosphere (Barlas et al., 2018). Additionally, the velocity deficit region behind a wind turbine creates a focusing effect on sound wave propagation, leading to localized zones in the far field with significant noise levels (Barlas et al., 2017). The formation of these focalization zones is highly dependent on ABL characteristics.

This work utilizes an Actuator Line/Large Eddy Simulation (LES-AL) method to produce the atmospheric boundary layer (ABL) fields necessary for investigating the propagation of acoustic waves. When it comes to simulating wind turbine flow, blade-resolved CFD simulations are computationally expensive, especially for MW-class turbines. Therefore, many studies in the literature have utilized Actuator Line (AL) based CFD simulations to evaluate the aerodynamics of wind turbines, such as assessing loads and studying wakes (Sorensen & Shen, 2002; Mikkelsen et al., 2007; Sarlak et al., 2014; Apsley and Stansby, 2020). This approach involves a lifting line representation of the turbine blades, where forces are evaluated using 2D aerfoil polars. These forces are then applied as source terms in the filtered Navier-Stokes equation. Since the blade's geometry is not modelled, the computational effort is significantly reduced.

This study aims to examine the impact of atmospheric boundary layer (ABL) characteristics, sound source heights, and source frequency on the perceived sound of a 5 MW wind turbine located 1 km upwind and 2 km downwind. To achieve this, we coupled the flow-field results of an LES-AL solver with a Finite Difference Time Domain (FDTD) acoustic propagation solver. We investigated two different ABL shear profiles by varying the surface roughness length. The study also examined the influence of source height by placing the monopole sources at the lower blade tip, hub, and upper blade tip locations. Furthermore, we evaluated the impact of source frequency by comparing the distribution of relative sound pressure levels from monopole sources of different frequencies.

2. Problem Definition

The schematic diagram of the LES-AL computational domain is presented in Fig, 1a. The turbine is placed 1000 m from the inlet of the computational domain and is subject to a turbulent ABL field generated from precursor simulations assuming neutral stratification. The domain extends 2000 m in the downwind direction and has a width and height of 1000 m in the lateral and vertical

direction. For the present study, a simplified variant of the NREL 5 MW reference turbine (Jonkman et al, 2007) has been considered, which has a diameter of 126 m and does not include shaft tilt and blade pre-cone. The variability in ABL characteristics is studied by conducting two different simulations with the roughness length z0 = 0.0015 and 0.015.
The vertical plane passing through the middle of the turbine is chosen for acoustic propagation, as shown in Fig. 1b. To bring out the effects of sound source height, the point sources are placed at the wind turbine location corresponding to hub height and extreme vertical tip location of the wind turbine, specifically at 25 m, 88 m, and 151 m. Additionally, propagation simulations are carried out at source frequencies of 100 Hz and 300 Hz at hub height to compare the frequency effects. Propagation of sound in a stationary field is also performed for comparison of results with turbulent ABL.

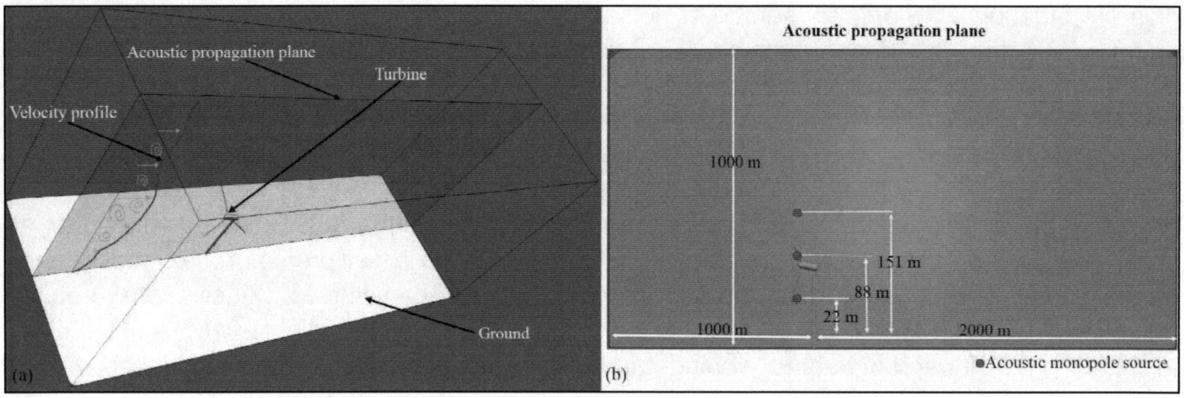

Figure 1: Schematic diagram of a) CFD domain b) acoustic propagation plane.

3. Numerical Methodology

3.1 Flow solver

Boundary Layer wind field generation
The generation of boundary layer wind characteristics and AL based CFD simulations have been carried out using Simulator for Wind Farm Applications (SOWFA) built based on OpenFOAM. The solver for generating boundary layer wind characteristics is based on buoyantBoussinesqPimpleFoam of OpenFOAM.

$$\frac{\partial \bar{u}_i}{\partial t} + \frac{\partial}{\partial x_j}(\bar{u}_j \bar{u}_i) = -2\varepsilon_{i3k}\Omega_3 \bar{u}_k - \frac{\partial \tilde{p}}{\partial x_i} - \frac{\partial}{\partial x_i}\left(\frac{\bar{p}_0(x,y)}{\rho_0}\right) - \frac{\partial}{\partial x_i}(\tau_{ij}^D) - g_3 z \frac{\partial}{\partial x_i}\left(\frac{\rho_k}{\rho_0}\right) + \frac{1}{\rho_0} f_i^T$$

Where $2\varepsilon_{i3k}\Omega_3\bar{u}_k$ is the Coriolis force from planetary rotation $\frac{\partial \tilde{p}}{\partial x_i}$ is the density normalised pressure gradient $\frac{\partial}{\partial x_i}\left(\frac{\bar{p}_0(x,y)}{\rho_0}\right)$ is horizontal mean driving pressure gradient τ_{ij}^D is shear stresses $g_3 z \frac{\partial}{\partial x_i}\left(\frac{\rho_k}{\rho_0}\right)$ is buoyancy term and f_i^T is other density normalised forces.
The framework can generate horizontally homogeneous atmospheric boundary layer. The size of the computational domain is 3000 km (streamwise) x 3000 km (spanwise) x 1000 km (vertical) for neutral atmospheric stability condition. The rotational speed of the turbine is 9.156 rpm, and a time step size of 0.025 s has been adopted. The computational domain was represented with a structured grid with grid cells of 5m in all directions. The total number of grid cells in the domain is 73.4 million. The lower surface of the domain has been considered with roughness, modelled with wall shear stress model and temperature flux. Periodic boundary conditions were imposed on the sides of the domain. The upper surface of the domain has a slip boundary condition. The

precursor simulations of ABL have been carried out for sufficiently long period of time of about 20,000 s, to obtain fully developed turbulent structures.

Actuator line method

Sorensen and Shen (2002) introduced the concept of actuator line model of wind turbine blades. Instead of resolving the rotor blades, they are represented by actuator lines with aerodynamic forces radially distributed along the blade span. At every discretised point along the actuator line, the velocity and local angle of attack are derived from the CFD simulation of flow field. With this information, the aerodynamic forces are evaluated using a lookup table of 2D aerofoil polars, based on blade element theory. The evaluated aerodynamic force is distributed across the grid cells surrounding the turbine in the CFD domain as a body force using a Gaussian projection. The body force per unit volume f_i^T (as appearing in Eq (1)) from the actuator line model of the turbine is described by

$$f_i^T = \frac{F_i^A}{\varepsilon^3 \pi^{3/2}} exp\left[-\left(\frac{r}{\varepsilon}\right)^2\right] \quad (1)$$

where f_i^T is the projected body force per unit volume (N/m^3), F_i^A is the force calculated at the point on the actuator line (N), *r* is the distance between point on the actuator line and the point where force is applied (m) and ε is the Gaussian kernel defining the projection width (m). The governing equations have been solved using the pressure implicit splitting operation (PISO) algorithm, in conjunction with Rhie-Chow interpolation to avoid pressure-velocity decoupling.

3.2 Acoustic propagation solver

In order to investigate the phenomena of refraction, scattering and dissipation of acoustic waves due to propagation through a non-homogenous atmospheric boundary layer, a two-dimensional propagation solver based on Finite Difference Time Domain Method is developed.

The approach explicitly solves the linearized Eulers equation (Eq. 2) that governs the propagation of acoustic wave on a well resolved finite difference grid. This equation is given by:

$$\begin{aligned} &\frac{dp}{dt} + \rho_0 c_0^2 \nabla . v = kQ \\ &\frac{dv}{dt} + (v.\nabla)V_0 + \frac{\nabla p}{\rho_0} = 0 \end{aligned} \quad (2)$$

Where, *p, v* and V_0, *t* is acoustic pressure, acoustic velocity, flow velocity vector and time respectively. ρ_0 is the mean density, c_0 is the mean sound speed and *k* is the adiabatic bulk modulus of the media. Q represent the mass source and is equal to zero at every grid location except in the vicinity of source location. Here source is described as a tapered harmonic signal given by Eq. 3.

$$Q\begin{cases} A/2(1 - cos(\pi t/T_1))cos(2\pi f) \\ Acos(2\pi f) \end{cases} \quad (3)$$

Where, A is the source amplitude, T_1 is the duration of taper and *f* is the frequency of the source. The solution is marched explicitly in time ensuring a courant number smaller than 0.8. The time marching is performed using low-dispersion and dissipation fourth-order Runge-Kutta algorithm described by Berland et al., 2006. Discretization of the spatial derivatives is achieved using a 11-point central difference scheme reported by Bogey, 2004. The difference scheme is switched to high order backward scheme at the boundaries to preserve accuracy. In the present scheme the grid spacing is chosen in such a way that there are 4 grid points within the wavelength of the source signal. Thus, propagation of high frequency signal requires much higher computational resources. Additionally, the fields are filtered after every iteration using an explicit damping function (Bogey, 2004) to remove spurious oscillations. The ground surface is considered perfectly reflecting and an absorbing porous media of 20 m width is prescribed in all the other boundaries.

4. Results and Discussion

4.1 Validation

The results of AL based CFD simulations from the present study have been validated against with blade-resolved CFD simulations available in literature (see Fig. 2 a). The axial force per unit length of the blade of 5MW RWT at 9 m/s from AL-CFD simulations compares well with blade-resolved CFD simulations. Figure 2b compares the transmission loss reported by Dallois et al. 2001 with that from the present solver. This figure describes the transmission loss estimated from a receiver at a height of 10 m along the down-wind direction of a flow field with constant wind velocity of 20 m/s. The source has a frequency of 680 Hz and is located 5m above the ground.

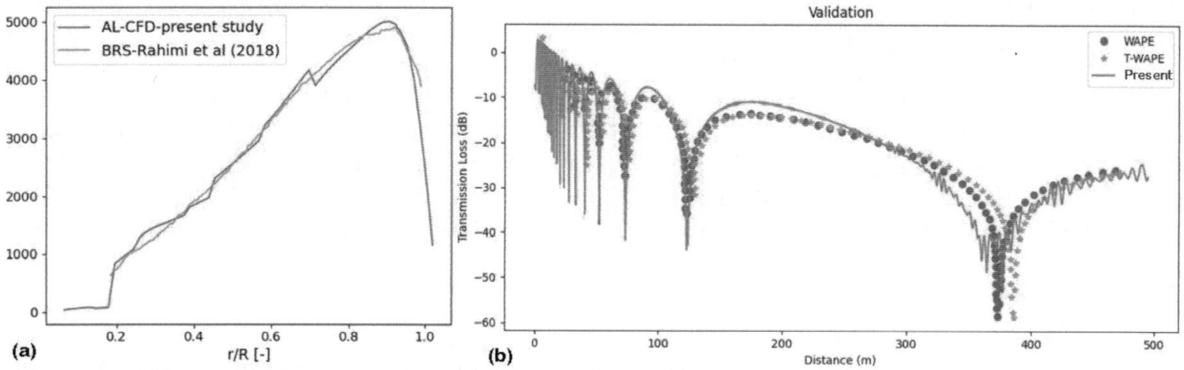

Figure 2: a) validation of LES-AL simulation b) validation of acoustic results with Dallois et al. 2001

4.2 Flow field

The velocity field at a plane in streamwise direction (xz plane) for two ABL characteristics are shown in Fig. 3a They are representative of shear generated, low-speed turbulent structures, with more turbulent structures in the lower part of domain close to ground for case (b). The velocity profile incident on the turbine is presented in Fig 3b. They are characterised by power law coefficients of 0.07 and 0.17, respectively as shown in Fig. 3b.

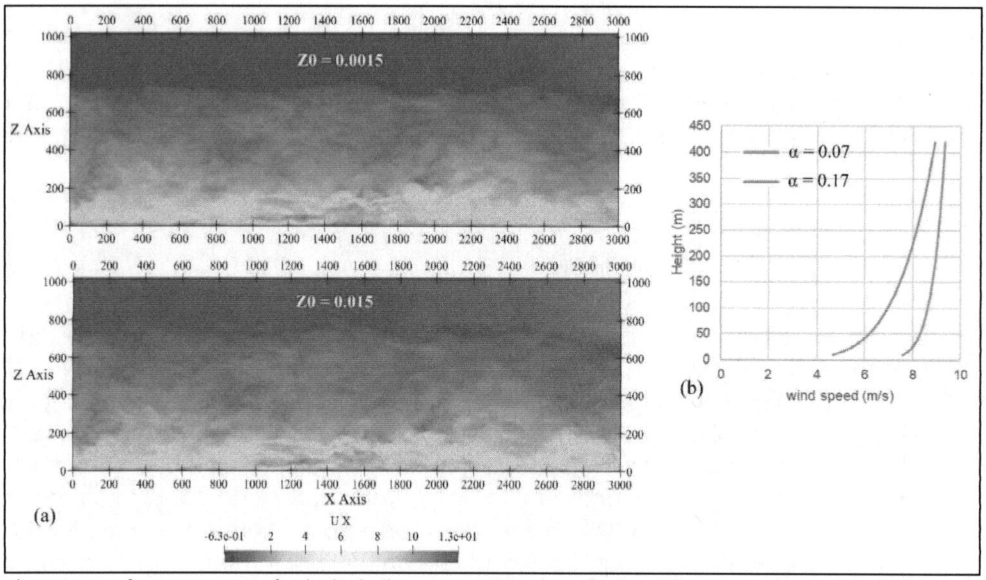

Figure 3: a) contours of x component of velocity in the propagation plane for the different roughness length, b) vertical velocity shear profile at a location before the wind turbine fir to power law.

4.3 Effect of ABL characteristics

The distribution of relative sound pressure level in the propagation plane for a stationary field (homogeneous), turbulent ABL fields with z0=0.0015 and z0=0.015 for a source frequency of 100 Hz and source height of 88 m (hub height) is presented in Figure 4. The contours of the stationary field serve as a baseline case for comparison. It is evident that the contours for z0=0.0015 and z0=0.015 demonstrate the influence of atmospheric conditions on the propagation of acoustic waves, while the baseline case does not. The presence of a shadow zone beyond 500 m upwind of the wind turbine for the turbulent fields is apparent upon closer examination of the upwind region. The RSPL distribution for z0=0.015 shows a greater number of refracted rays in the downwind region compared to that for z0=0.0015, which is in line with the greater shear for z0=0.015 (see Fig. 3b). The contours at the ground level beyond 2000 m in the downwind direction clearly distinguish between the two roughness lengths, with a larger number of refracted rays for z0=0.015, while there is a shadow like zone for z0=0.0015.

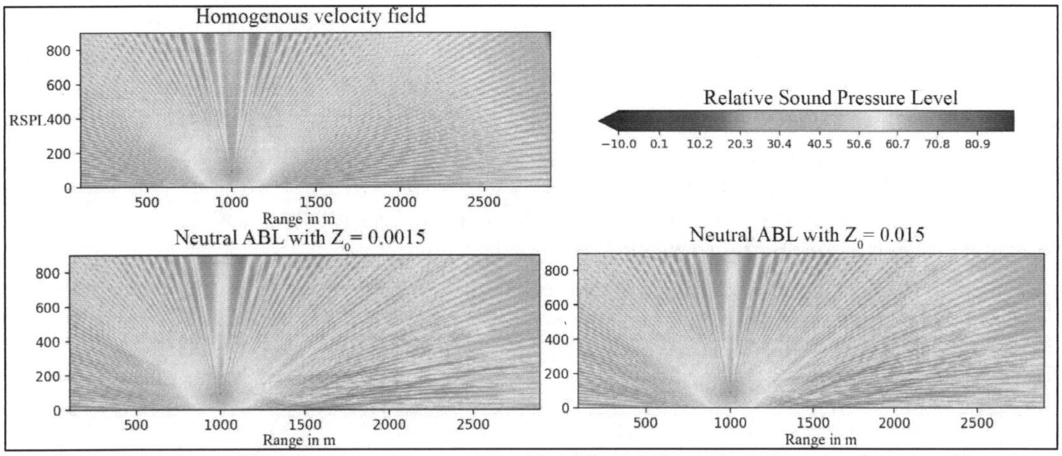

Figure 4:Contours of Relative Sound Pressure level for different atmospheric boundary layer conditions.

The transmission loss on the ground surface for the cases discussed above are presented in Fig. 5. The transmission loss for the stationary field is indicated using the orange line plot, while that from the turbulent fields is presented using the blue line. This graph clearly indicates the presence of a shadow zone in the upwind direction. The presence of the shadow zone is identified by the lower position of the blue line with respect to the orange line. In the downwind direction, fluctuations of TL are observed for both cases of turbulent fields while TL variations indicate a smooth decay for the stationary field. These fluctuations are a result of atmospheric refraction as well as focusing phenomenon by the turbine wake. On comparing the results from the two roughness lengths, it is noted that higher sound levels are observed in the far-field for a higher wind shear profile.

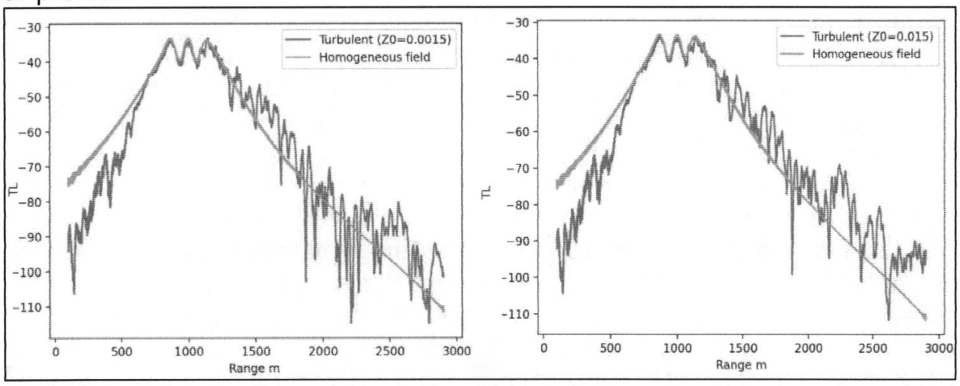

Figure 5: Comparison of transmission loss at ground surface for roughness length of 0.0015 and 0.015 with that of stationary field

4.4 Effect of acoustic source height

The height of the sound source has a significant impact on the acoustic propagation characteristics around a wind turbine. As the source height increases, the directivity of the source changes and affects the sound pressure level (SPL) at various locations. The sound propagation from the rotor is most significant in the vertical plane, where the noise is radiated upwards and downwards. The upward propagation is influenced by atmospheric conditions, whereas the downward propagation is influenced by ground reflection, turbine wake and refraction. In addition, the sound field distribution also varies with the height of the receiver. Therefore, it is important to study the impact of source height on acoustic propagation around a wind turbine for accurate prediction of sound levels at different locations.

The RSPL distribution for different source heights (Hs) is illustrated in Figure 6. The blue region in the figure indicates the significant presence of a shadow zone in the upwind region for Hs = 25 m, which becomes less prominent for Hs = 88 m and almost negligible for Hs = 151 m. For Hs = 25 m, a consistent red region (indicating higher sound levels) is observed near the ground in the downwind direction, indicating the presence of refracted and re-refracted rays. This phenomenon of atmospheric refraction is more pronounced at lower source heights since the sound rays undergo multiple refractions. However, for source heights of 88m and 151m, the dominant effect is the focusing phenomenon of the turbine wake, resulting in higher sound pressure levels in the downwind direction. This effect can be clearly observed for Hs = 151 m, indicated by the dark red region at ground level at a distance of 2000 m from the wind turbine. Comparing the focalizing zones for Hs values of 88 m and 151 m, it is noted that such location extends further downstream with increasing source heights.

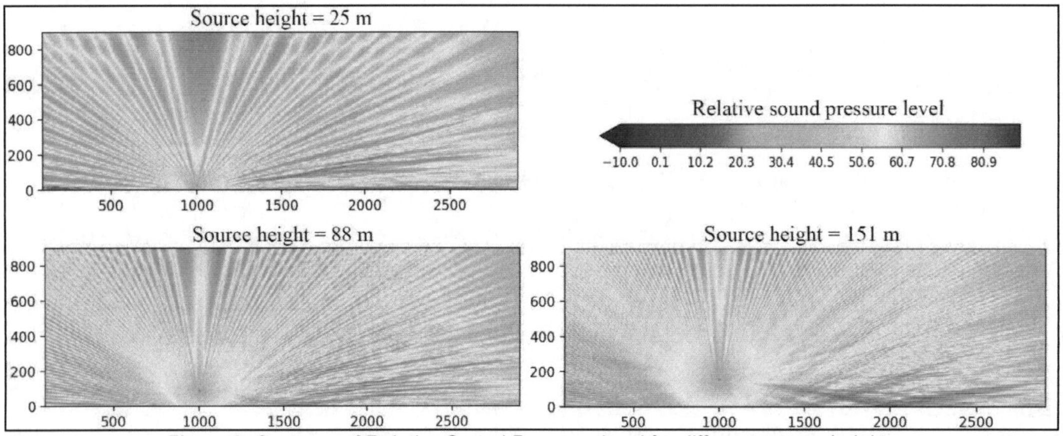

Figure 6 :Contours of Relative Sound Pressure level for different source heights

In Figure 7, the transmission loss on the ground surface is shown for three different cases of source heights. The TL curve for Hs = 151 m deviates the least from the TL of the stationary field, indicating that the shadow zone is minimal for this height. The influence of refracting acoustic rays is clearly evident in the TL plot for Hs = 25 m, where the sound levels in the far field in the downwind direction are significantly higher than those observed for the stationary field. The TL plot for Hs = 151 m indicates the presence of a wide acoustic focussing zone in the range of 900 – 1500 m downwind from the wind turbine. This phenomenon is also visible in the corresponding contour plots in Figure 6. These regions are created due to the velocity gradient and turbulence induced by the turbine. Such locations are referred to as bursting zones, as reported by Barlas et al. in 2017. The focusing effect is greater at the tip of the turbine because of the strong vertical velocity gradient in this region, leading to strong downwind refractions towards the busting zones. The TL plots also reveal the presence of acoustic shadow zones between the focalization locations (see Hs = 151 m in Figure 7). These low acoustic zones are a counter effect of strong focusing phenomenon because the acoustic waves are guided away from these valley regions.

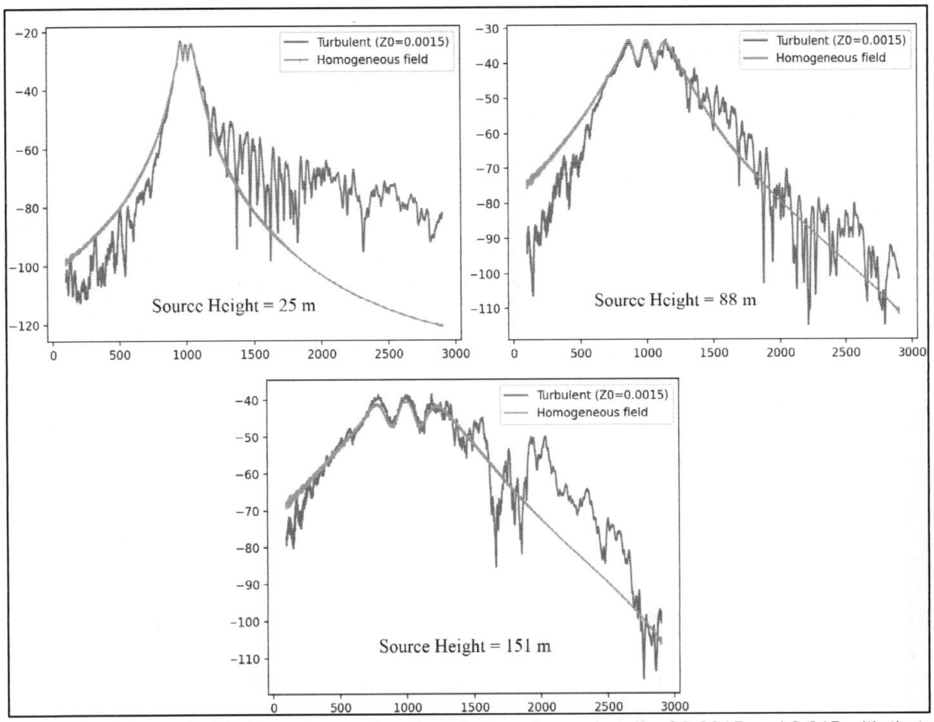

Figure 7:Comparison of transmission loss at ground surface for roughness length of 0.0015 and 0.015 with that of stationary field

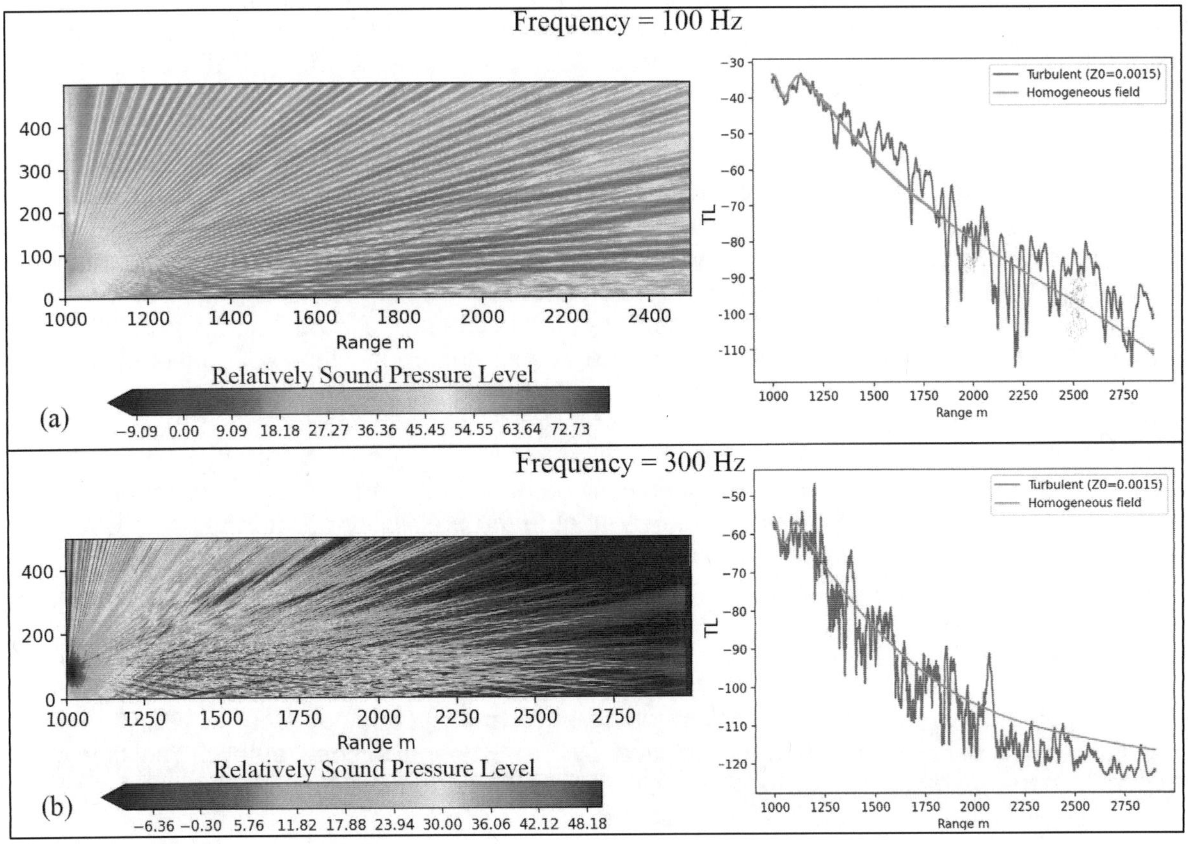

4.5 Effect of acoustic frequency

The frequency of a sound source is a critical factor in determining the characteristics of acoustic propagation in the far field. Sound waves interact with the surrounding environment differently based on their frequency. High frequency waves are more susceptible to atmospheric absorption and refract more easily. Figure 8a illustrates the spatial distribution of RSPL and transmission loss downwind to the wind turbine with a source frequency of 100 Hz at a height of 88m. The influence of atmospheric refraction is evident in the form of peaks and valleys in the transmission loss plot. Figure 8b shows the spatial distribution of RSPL and transmission loss downwind to the wind turbine with a source frequency of 300 Hz at the same height. The TL curve for 300 Hz shows more fluctuations compared to 100 Hz, indicating a greater number of refracted rays. Both RSPL distribution and the TL curve reveal that acoustic signals decay more rapidly at 300 Hz than at 100 Hz. As the distance from the turbine increases, the sound pressure level (SPL) in the far field for 100 Hz tends to increase in a turbulent atmospheric boundary layer (ABL), while SPL for 300 Hz tends to decrease compared to the stationary field.

5. Conclusions

It is known that atmospheric conditions and wind turbine wake deficit has the capacity to alter the direction of acoustic wavefront. In the present study such effects are investigated by coupling the flow field generated from a Large Eddy Simulation/Actuator Line (LES-AL) flow solver with a Finite Difference Time Domain (FDTD) based acoustic solver. The acoustic source is represented as a simplified monotonic point source. Representative scenarios have been considered to investigate the influence of atmospheric boundary layer characteristics, acoustic source height and frequency of acoustic source. The spatial distribution of relative sound pressure level in a representative two-dimensional plane covering region 1 km in the upwind direction and 2 km in the downwind direction of the turbine. These contours clearly establish interference patterns, atmospheric refraction and focussing phenomenon of the wind turbine wake. Acoustic propagation studies that account for such large far field region is limited in literature. The transmission loss at the ground surface along the length of the computational domain is also reported here, the results from acoustic propagation through two differently sheared atmospheric boundary layer indicate that the acoustic refraction and focussing zones are present further away from the wind turbine in the downwind direction for atmospheric boundary layers with greater shear. Investigation on acoustic source location show that the focussing phenomenon is severe when the noise source is placed at the upper tip of the wind turbine. The strong vertical gradient between the turbine wake and wind velocity facilitates the downward refraction of the acoustic waves. The results comparing the propagation from two different acoustic source frequency show that higher frequency signals get decay faster in the downwind direction. The results from this study clearly establish the complexities in noise propagation from wind turbine. Although, the noise levels reported in this study in the far field region are low, the sound levels are significantly higher than that predicted by simple engineering decay models. This study does not completely negate the possibility of harmful noise levels in the far field. Further investigation considering terrain effects and wide spectra of atmospheric boundary layer conditions needs to be performed to answer this problem.

Acknowledgements

The research work is funded by Indo-German Science and Technology Centre (IGSTC) Post-Doctoral Industrial Fellowship (PDIF) received by the first and second author. The first and second authors gratefully acknowledge their hosting and the computational resources provided by Fraunhofer Institute for Wind Energy Systems.

References

Apsley, D. D., and Stansby, P. K., "Unsteady thrust on an oscillating wind turbine: Comparison of blade-element momentum theory with actuator-line CFD," Journal of Fluids and Structures, 98, 103141, 2022.

Barlas, E., Wu, K. L., Zhu, W. J., Porté-Agel, F., & Shen, W. Z. (2018). Variability of wind turbine noise over a diurnal cycle. Renewable Energy, 126, 791-800.

Barlas, E., Zhu, W. J., Shen, W. Z., Kelly, M., & Andersen, S. J. (2017). Effects of wind turbine wake on atmospheric sound propagation. Applied Acoustics, 122, 51-61.

Basner, M., Babisch, W., Davis, A., Brink, M., Clark, C., Janssen, S., & Stansfeld, S. (2014). Auditory and non-auditory effects of noise on health. The lancet, 383(9925), 1325-1332.

Berland, J., Bogey, C., & Bailly, C. (2006). Low-dissipation and low-dispersion fourth-order Runge–Kutta algorithm. Computers & Fluids, 35(10), 1459-1463.

Bogey, C., & Bailly, C. (2004). A family of low dispersive and low dissipative explicit schemes for flow and noise computations. Journal of Computational physics, 194(1), 194-214.

Colas, J., Emmanuelli, A., Dragna, D., Stevens, R., & Blanc-Benon, P. (2022). Effect of a 2D Hill on the Propagation of Wind Turbine Noise. In 28th AIAA/CEAS Aeroacoustics 2022 Conference (p. 2923).

Cotté, B. (2019). Extended source models for wind turbine noise propagation. The Journal of the Acoustical Society of America, 145(3), 1363-1371.

Dallois, L., Blanc-Benon, P., & Juvé, D. (2001). A wide-angle parabolic equation for acoustic waves in inhomogeneous moving media: Applications to atmospheric sound propagation. Journal of Computational Acoustics, 9(02), 477-494.

Jonkman, J., Butterfield, S., Musial, W., & Scott, G., "Definition of a 5-MW reference wind turbine for offshore system development (No. NREL/TP-500-38060)," National Renewable Energy Lab (NREL), Golden, CO, United States, 2009.Kost, C., SHAMMUGAM, S., FLURI, V., PEPER, D., MEMAR, A. D., SCHLEGL, T. (2021) *Levelized cost of electricity-renewable energy technologies.*" Report, Fraunhofer institute for solar energy systems ISE.

Lee, S., Lee, D., and Honhoff, S. (2016). "Prediction of far-field wind turbine noise propagation with parabolic equation," J. Acoust. Soc. Am. 140(2), 767–778.https://doi.org/10.1121/1.4958996.

Mikkelsen, R., Sørensen, J. N., Øye, S., and Troldborg, N., "Analysis of Power Enhancement for a Row of Wind Turbines Using the Actuator Line Technique," Journal of Physics: Conference Series – The Science of Making Torque from Wind, Vol. 75, No. 012044, 2007.

OpenCFD Ltd. (ESI Group), "OPENFOAM – The Open Source CFD Toolbox," available online at http://www.openfoam.com.

Prospathopoulos, J., and Voutsinas, S. (2007). "Application of a ray theory model to the prediction of noise emissions from isolated wind turbines and wind parks," Wind Energy 10, 103–119.https://doi.org/10.1002/we.211

Sørensen, J. N. and Shen, W. Z., "Numerical Modeling of Wind Turbine Wakes," Journal of Fluids Engineering, Vol. 124, 2002, pp. 393–399.

Sarlak, H., Pierella, F., Mikkelsen, R., and Sørensen, J. N., "Comparison of two LES codes for wind turbine wake studies," In Journal of Physics: Conference Series, Vol. 524, No. 1, p. 012145, IOP Publishing, 2014.

Troldborg, N., Actuator Line Modeling of Wind Turbine Wakes, Ph.D. thesis, Technical University of Denmark, Lyngby, Denmark, 2008.

10th International Conference
on
Wind Turbine Noise
Dublin – 21st to 23rd June 2023

Wind turbine directional noise reduced operations – description and measurement

Isaac Old, RSG, isaac.old@rsginc.com
Timothy Carson, RWE Clean Energy, LLC

Summary

Wind turbines can be placed into NRO (noise reduced operation) modes to reduce sound emissions. These modes can be implemented for all meteorological conditions and time periods, or for specific conditions (wind speed and direction) or time periods. The current paper describes how they work and provides measurement results from a project where they are applied.

1. Introduction

There are a number of methods that can be employed to reduce wind turbine sound emissions. For example, during the siting process, setback distances can be increased. After construction, sometimes serrated trailing edge technology can be installed, noise reduced operations (NROs) can be programmed, and various tools such as free layer damping and tuned mass dampers can be used to reduce tonal sound and vibration. The most flexible of these options is NROs, as they rely on turbine operational programming instead of installing a physical tool.

This paper includes:

- A description of how NROs work;
- When NROs can be implemented; and
- Measurement results from a project with NROs implemented.

2. Description

NROs are achieved by predominantly reducing trailing edge noise produced by wind turbines. This is done by slowing turbine rotational speed by pitching the blades out of the wind [1], [2]. This reduces the angle of attack of the blade and reduces trailing edge noise [2]. Unlike use of aerodynamic aids such as serrated trailing edges, it also simultaneously reduces power output due to reduced lift. The method works since trailing edge noise scales to the fifth power of air flow velocity [3] allowing for sound emissions reductions of several decibels. NROs can be implemented all the time or for specific conditions. These conditions can include time-of-day, wind direction, and wind speed. Condition-specific NRO allows for more efficient project operation while meeting applied sound level limits [4].

3. Validation Measurements

3.1 Procedures

To test the effectiveness of directional NROs (NRO applied to specific wind directions), sound level meters were set up at two locations in a wind power project that has directional NROs implemented. The project has two different types of turbines, the first type (near site B) is a turbine with a shorter hub height and the second (near site A) has a taller hub height, with a difference between the two hub heights of about 25 meters. The area was generally hilly with scattered clumps of trees. In all cases, the closest turbine was visible from the sound level meter.

Sound level meters used were all ANSI/IEC Class 1 units set to log 1/3 octave band sound levels over a minimum frequency range of 20 Hz to 10 kHz (1/3 octave bands) at one second intervals for the duration of the monitoring period. Microphones were covered with 180 mm hydrophobic windscreens and were mounted on stakes at an approximate height of 1.4 meters above ground level. Anemometers were co-located with the sound level meters to measure microphone height wind speeds. Depending on location, sound level meters ran for a period of two to three weeks.

Turbines within 2.4 km of each monitor location were shut down for ten minutes multiple times per night to allow for background sound level measurement. This was used to calculate turbine-only sound levels for the one-hour periods before and after the shutdown. Data including transient sound events (cars, animals, aircraft, etc.) were removed from the results entirely.

The turbine near Monitor B is set to go into directional NRO from 203 to 23 degrees for wind speeds at and above 7.9 m/s and the closest turbine to Monitor A is set to go into directional NRO from 90 to 225 degrees for wind speeds at and above 8 m/s.

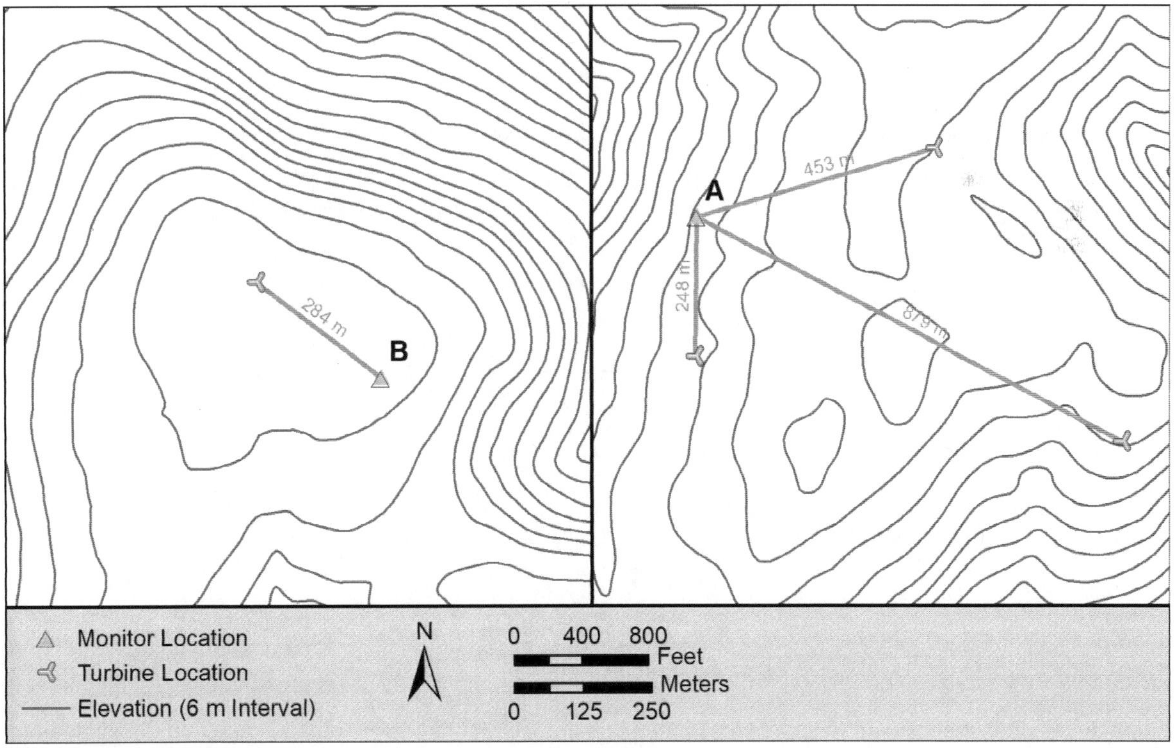

Figure 1: Sound monitor locations

3.2 Results

Results shown for Monitor A focus on periods where the closest turbine shifted into and/or out of NRO instead of general sound level trends. General sound level trends were less clear due to the proximity of other turbines. Monitor B will focus more on general trends.

3.2.1 Monitor A

Figure 2 and Figure 3 show one hour before to one hour after shutdown periods at Monitor A. Along with the sound pressure levels, the charts show the hub height wind speed (yellow dotted line), the wind direction (blue dots), direction range when the microphone was with 45 degrees of downwind from the turbine (green box) and the direction range that the turbine was in NRO (blue box). This means that if the wind direction is within the "NRO Range", the NRO is active and if it is outside that range it is inactive. If the wind direction is within the "Monitor Downwind" range it means that the sound monitor was within 45 degrees of downwind of the turbine. In Figure 2 there are two times when the wind direction goes just outside of the NRO range (at around 22:38 and 23:52) resulting in a clear increase in sound levels. Because the wind direction changes are relatively small the change in sound level is not attributed to directivity pattern. Wind speeds are essentially constant for the second increase, meaning sound emissions would not increase. For the first increase, wind speeds do increase slightly however, this turbine type has already reached its maximum sound emissions at 9 m/s. The same is seen in Figure 3 at 02:35 when the wind direction passes out of the NRO range, resulting in a clear sound level increase. At this location the turbine NRO is specified to produce an almost 3 dB reduction in A-weighted sound level, though this is unlikely to be realized due to influence from nearby turbines that are not in NRO for the same direction range. Changes in sound level shown here are approximately 1.5 dB.

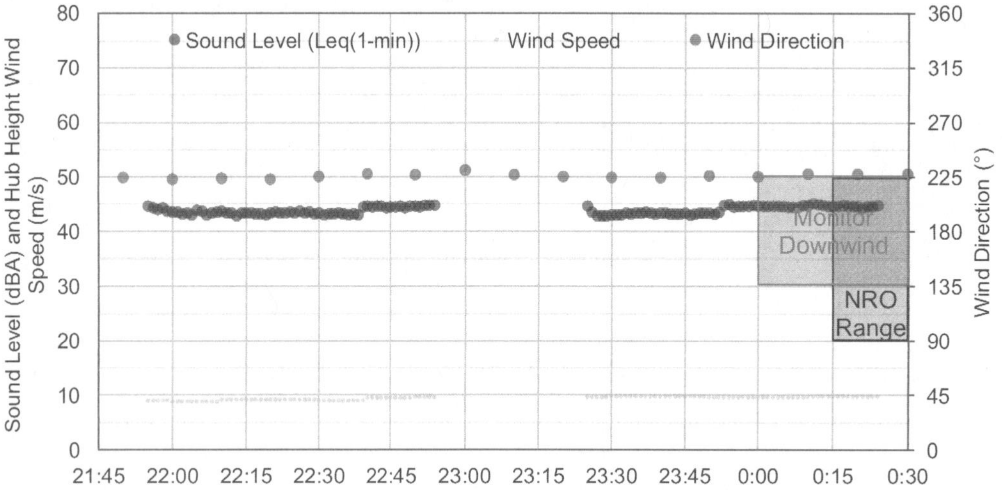

Figure 2: Shutdown period 1 at Monitor A – The NRO range (blue shading) overlaps with the downwind range (green shading)

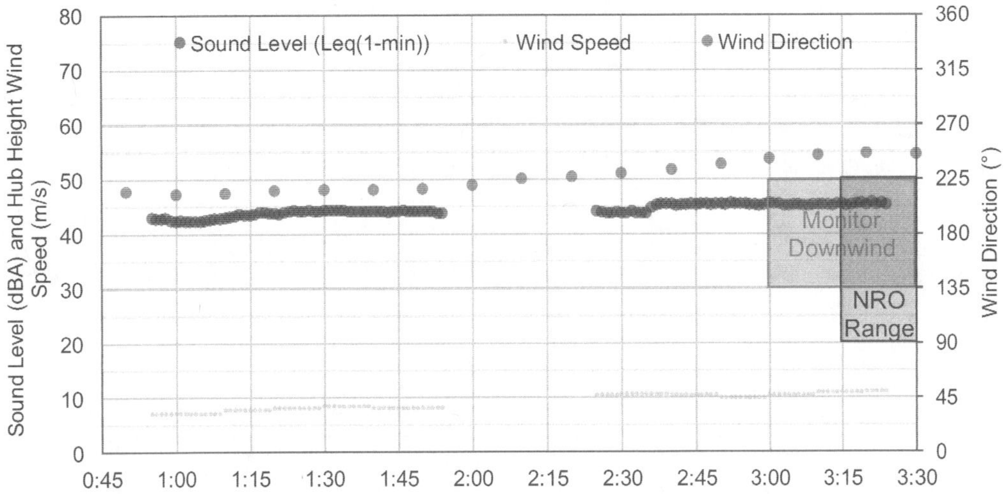

Figure 3: Shutdown period 2 at Monitor A - The NRO range (blue shading) overlaps with the downwind range (green shading)

3.2.2 Monitor B

Figure 4 shows sound levels by hub height wind speed for periods that were in and out of the directional NRO range. This shows that above 7 m/s (when NROs start), the typical sound level is higher when turbine is not in directional NRO. The NRO setting used for this turbine is specified to provide just under 2 dB of attenuation. Data in Figure 4 shows around 3 dB of attenuation, depending on wind speed. This is also shown in Figure 5, where sound levels are broken up by 22.5-degree wind speed bins. For hub height wind speeds of 8 and 9 m/s, wind directions within the NRO range show lower overall sound levels. For 7 m/s, where NROs are not active, this trend is gone. Figure 6 shows the individual 1-minute L_{eq}s for the same data. This shows the same basic trends in more detail. One additional apparent trend is the overall decrease in levels around 225 degrees. This is likely the crosswind dipole source directivity decrease.

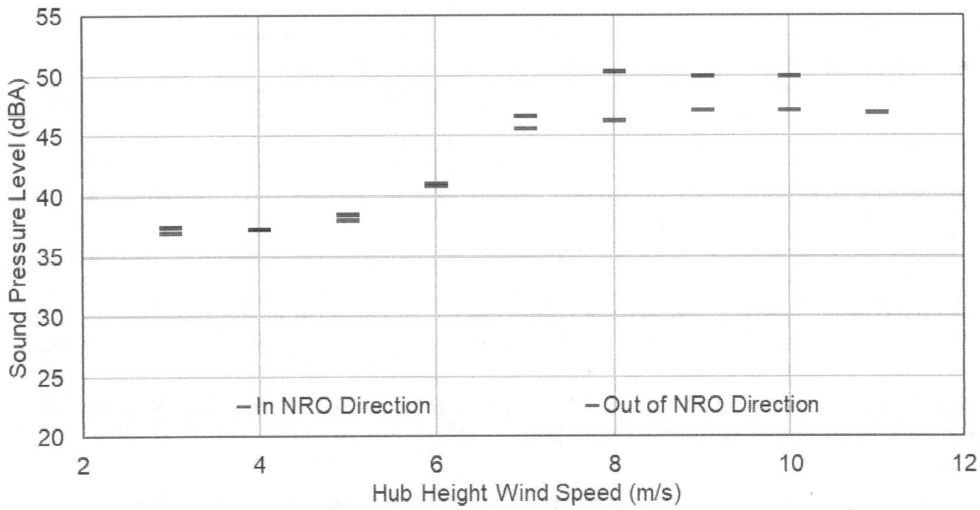

Figure 4: Sound pressure levels by hub height wind speed wind speed for wind directions both in and out of the directional NRO range. Data points are the median of 1-minutes L_{eq}s for each wind speed bin.

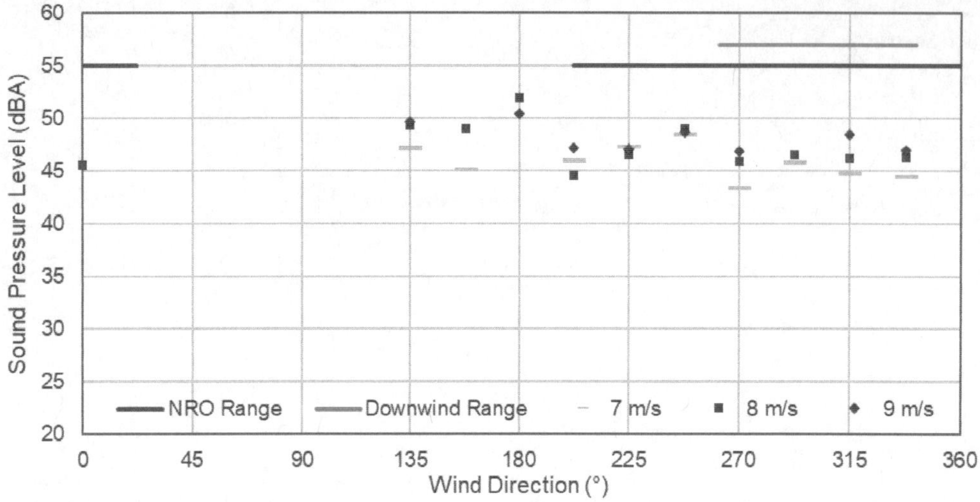

Figure 5: Sound pressure levels by wind direction and hub height wind speed. Data points are the median of 1-minute L_{eq}s for each 22.5-degree bin.

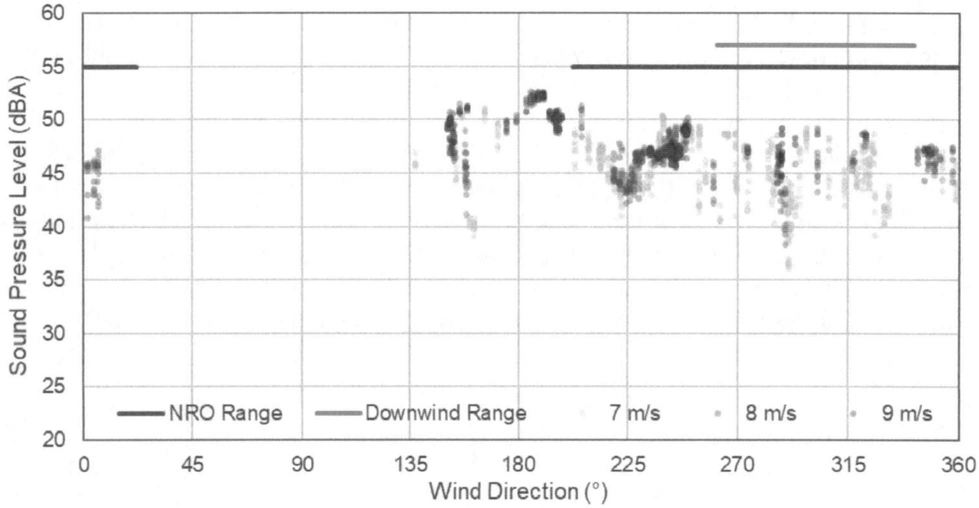

Figure 6: Sound pressure levels by wind direction and hub height wind speed. Data points are 1-minute L_{eq}s.

4. Discussion

The purpose of this study was to give background information on NRO and to evaluate the effectiveness of directional NROs in reducing sound levels from wind turbines. Data presented above show reduced sound levels at monitor locations with wind speeds and directions in the specified NRO ranges, indicating that NRO is an effective tool to reduce sound levels around wind projects during certain wind speeds and directions.

Shortcomings of the data are that the study design does not allow disambiguation of source and propagation directivity from effects of the directional NROs. This does not invalidate the results since monitor locations were downwind of the turbines within the NRO range. Since the worst case propagation directivity and source directivity direction is downwind [5], [6], it means that reduced sound levels in the downwind direction still indicate that directional NROs are effective, it just confuses the extent to which they are effective. The extent to which they are effective is also specific to the turbine and the particular mode used. Due to how close the monitors are to the turbines (less than 300 meters), propagation directivity will be relatively small [5]. Further study should be done, focusing on differentiating the effects of directional

NROs from source and propagation directivity, particularly at distances both closer and further away from the turbines than what was monitored here.

5. Conclusions

This paper describes how directional noise reduced operations (NROs) work and presents measurement results from a wind power project where they were implemented. Conclusions are as follows:

- NROs work by pitching wind turbine blades out of the wind, reducing tip speed and aeroacoustic sound generation. NROs can be implemented on a turbine for all conditions (wind speed, wind direction, time of day, etc.) or for specific conditions and combinations of conditions.
- Measurement results show the expected sound level reductions from the turbines at wind directions where they were placed in NRO.
- Further measurements need to be done to reaffirm these results, isolate the effects of directional NRO from source and propagation directivity, and quantify relative effectiveness at distances closer and further from the turbines than the current measurements.

References

[1] S. Oerlemans, "Wind Turbine Noise Mitigation," in *Handbook of Wind Energy Aerodynamics*, Springer Cham, 2022, pp. 1447-1462.

[2] D. Bowdler and G. Leventhall, Eds., Wind Turbine Noise, Brentwood, Essex: Multi-Science Publishing Co. LTD, 2011.

[3] M. Asheim, P. Moriarty and D. Munoz, "Aeroacoustic Noise Mitigation Investigation for Wind Turbine Blades," in *5th International Conference on Wind Turbine Noise*, Denver, 2013.

[4] B. Petitjean, A. Ambekar, R. Drobietz and K. Kinzie, "Noise Optimized Wind Park Operation," in *6th International Meeting on Wind Turbine Noise*, Glasgow, 2015.

[5] M. Coulon, "Directivity Noise Attenuation Values for Large Wind Turbines - Research based on long term measurements," in *6th International Meeting on Wind Turbine Noise*, Glasgow, Scotland, 2015.

[6] A. Fukushima, K. Yamamoto, H. Uchida, S. Sueoka, T. Kobayashi and H. Tachibana, "Sudy on the Amplitude Modulation of Wind Turbine Noise: Part 1 - Physical investigation," in *Internoise*, Innsbruck, Austria, 2013.

10th International Conference on Wind Turbine Noise Dublin – 21st to 23rd June 2023

Closing in on the Wind Turbine "Sasquatch" – Whose Name is "Annoyance"

William (Bill) K.G. Palmer P. Eng. trileaem@bmts.com

Summary

A research project under way is described, the objective of which is to determine an objective measure to predict annoyance from wind turbines. Some would state categorically that there is nothing specific in the noise profile of wind turbines to cause annoyance. Claims have declared that wind turbine annoyance is the result of stress, and that stress is the result of misinformation about adverse impacts. Annoyance from wind turbines is perceived like the "Sasquatch", a mythical being, for which there is no actual evidence. Yet, a fraction of credible individuals attest that when near operating wind turbines they are irritated, or annoyed, and suffer adverse impacts. When they separate themselves from wind turbines, or when the wind turbines shut down, the individuals find the adverse conditions diminish. However, when they are again exposed, the conditions reoccur. Over time a sensitivity seems to develop, so that the annoyance and adverse conditions occur with reduced exposure. This research project examines the acoustic, environmental, and wind turbine operational conditions existing when impacted individuals report annoyance. Factors such as wind turbine visibility, wind speed and direction, as well as the noise resulting from ambient winds are also considered. The project seeks to determine if the annoyance could be arising independent of the wind turbine noise profile, or from misinformation. Insights arising from the research are discussed, as the project circles closer, and closer, to substantiate a verifiable measure of the character of the wind turbine "Sasquatch."

1. Introduction

The proceedings of the Wind Turbine Noise Conferences indicate that the subject of annoyance from wind turbine noise has not gone away. While the number of papers has gone up and then down, the number of mentions of the word "annoyance" continued to rise.

Conference	# of Papers	# of Mentions "annoyance"
WTN 2005	29	78
WTN 2013	72	406
WTN 2021	40	438

Table 1 – Mentions of "Annoyance" at International Wind Turbine Noise Conferences

The word commonly found used with "annoyance" is "subjective." The challenge this presents is that "subjectivity" is in the mind of the beholder. We are told, the only way to make a fair "subjective" assessment is to assemble an impartial panel of observers, commonly called a "jury." Yet, even in a court of law, this presents challenges. Lawyers make their arguments as to whether potential jurors are representative "peers" for their client or the community, or if they should be rejected as having an ingrained bias. Anyone who has been called as part of the pool of potential jurors for a court case can speak to the mystery of determining "a panel of your peers." There are many tests reported in the wind turbine noise literature, where panels of observers are assembled, to listen to recordings of wind turbine noise and make an assessment of "annoyance." However, any subjective test is open to challenge. Were the test subjects really peers? Were the conditions or setting the same as experienced by residents reporting impact? Were the tests sustained for days on end? In reports of community observations, questions might be asked such as were the turbine parameters (output, size, height, number, and separation distance) representative? Subjective assessments such as, "I am annoyed by the noise from wind turbines," are often countered by, "You do not like wind turbines because you are jealous that your neighbours profit, while you don't." There must be a better way to demonstrate respect for each other. Sitting down with, and really listening to those reporting concerns is a beginning place.

Clearly, an "objective" measure of annoyance, that is not dependent on a representative jury exposed in a similar setting, sustained for a similar duration would be preferred. There are objective measures for measuring noise, most commonly perhaps by A-weighted sound level. However, even when these measures are modified in various ways such as Leq, or Lden, they are still challenged by other sound in the environment from wind, human activity, wildlife, or the special qualities of sound to be able to measure annoyance. This paper gives an overview of am approach to find an objective measure for the annoyance from wind turbine sound.

2. Listening to those Impacted

Listening to real people pour out their heart raises concerns about the ethical principles of doing so. Will their confidentiality be assured? Will they remain anonymous? (Unless they specifically gave permission to share their specific details, as some did.) Can they be assured that the listener will not substitute the interviewer's opinions or biases, in place of those of the one being listened to? University studies, the Code of Ethics of Professional Engineers Ontario, the Institute of Noise Control Engineers, and the Acoustical Society of America, have specific requirements regarding ethical practices, with regard to conducting surveys of the public. For clarity, the information reported in this paper are not the result of a formal survey, in which informants are asked to give formal informed consent, to whom specific questions were asked. Instead, they were the result of informal conversations, in which the participants volunteered information casually, or of their presentations before public forums. Often, this information was predicated by a statement such as, "you are involved in wind turbine noise, isn't there something you can do?" At no point were the participants offered benefit, or harm from providing or withholding information, and ethical principles of regarding the duty to public welfare were held paramount in including comments in this paper.

Those who volunteered information were not considered to be expressing an attitude of disdain for care of the environment. A tutorial presented by George A. Luz, PhD titled, "Some People are More Noise Sensitive than Others" presented at the 161th Meeting of the Acoustical Society of America in Seattle, WA, in May 2011 comes to mind. Luz noted that, "The most outstanding impression of those people who were noise sensitive was that they were typically friendly, generous and sociable and very much aware of their environment." If a common perception might be summarized, it was that the informants expressed hurt. They had honestly reported their concerns and impacts to those considered to be responsible to act, but their reports had been dismissed or not acted on.

Many of that volunteered information did so publicly in deputations before the members of the Multi Municipal Wind Turbine Working Group. (MMWTWG). This working group is formally constituted under the regulations of the Ontario Municipal Act, with public meetings, and public record of meeting minutes. The Working Group is composed of elected representatives from Municipal Councils and a citizen appointee from member councils, from a number of Municipalities in the Province of Ontario concerned about the impact of wind turbines on citizens.

2.1 Summarizing Issues Raised (not necessarily attributed to wind turbines, but by those living in the environment of multiple wind turbines within less than 1 km)

- Some reported change in behaviour of domesticated animals (such as horses, goats, dogs or cattle) after the commencement of operation of wind turbines in their environment. Presumably these animals had no attitude of jealousy, or miss-information.
    - One man reported on a specific change in behaviour of ponies, trained to draw a cart. The man showed me the stable previously housing his cart ponies. After the wind turbines had started up, the ponies which had been stabled fine before, had kicked holes in the walls in the stable. He noted that after the wind turbine start-up he would sometimes visit the stable to find the ponies "all lathered up" as if they had been out for a run, even though they were only standing in the stable. On another occasion, the ponies, while harnessed to their cart had suddenly bolted, and run through a wire fence, cutting themselves up. He noted that after this event, he had given the ponies away to relocate them away from the wind turbines, and they had reverted to their previously docile behaviour.
    - The same man reported on changes in behaviour of the family dog, to not want out, as it had previously. Others reported in change in behaviour of their family pet dogs as well.
    - Another family reported changes in behaviour of goats, and another in changes in behaviour of a dairy herd, requiring the family to relocate.
- The same man who had reported the change in behaviour of his ponies reported changes in his personal health, including a bleed (a stroke was how he described it) in an eye. He reported that his wife, who was away from the house most of the day, at work, experienced no adverse effects. Anecdotally he reported adverse impacts occurring in several neighbours, which were not followed up on. They left the home.
- Another gentleman reported difficulty in sleeping after the wind turbine started up. His family physician had prescribed sleeping tablets. He noted that when away from home on vacation, the "slept like a baby" but on return home, again his sleep deprivation recurred. He also reported balance instability. His wife was not impacted. They moved from the environment, and the condition disappeared, although the gentleman passed away shortly after. Sleep deprivation was reported by a number of others, again, a condition which disappeared when away from home, but returning when back at home. As before, not all family members appeared to be impacted.
- Digestive issues, or nausea, were reported by some.
- Headaches were a common report, for the one reporting, or for other family members.
- Some reported changes in control of diabetes, or changes in control of blood pressure, or other cardio-vascular issues, with some requiring relocation to address the issue.
- Tinnitus or sensations of vibration transmitted into homes were reported by some.
- Some addressed the necessity to change work schedules, to relocate residence, or to retire prematurely from work due to difficulty in sleeping, due to concerns of work errors, or due to health deterioration.
- Some identified specific issue with tonality of the sound, reporting a rising and falling "wooing".
- Specific changes in sound during conditions of freezing rain or hot, still summer nights were reported by a number of people, using terms such as, "pounding" intensity.

- Some reported being able to perceive if nearby wind turbines were operating or not on awakening, even without viewing the turbines, or hearing specific sounds.

2.2 Investigation of Issues Raised

- While it was not possible by the author to do a detailed investigation of each issue raised, for a period of over 15 years, the author has conducted investigations and collected acoustic data at over 20 sites in over 8 different wind power developments, with at least 4 different wind turbine types (Vestas V80, Vestas V82, Enercon E82, and Siemens SWT 2.3 – 101) and at a number of sites at least 5 km distant from wind turbines. The level of detail collected in each investigation has increased over the 15-year period.
    - Initially the information collected was a simple record of 1 minute duration readings at the sites using calibrated A-weighted and C-weighted sound level meter readings and wind speed monitoring at 1.5 metres above ground level, along with the associated wind power development output level and the nearest Environment Canada weather station information.
    - By 2010, the data collected progressed to 1 to 2 minute recordings of the sound pressure level from a calibrated Knowles BL-21994 microphone with a 60-mm primary and 300 mm secondary wind screen. All those recordings are on file.
    - By 2013, data collection progressed to making recordings of the sound pressure level using a calibrated Earthworks M30-BX microphone with a flat frequency response from 9 Hz to 30 kHz (although measured to be flat lower then 9 Hz) using a 90-mm primary and 450 mm secondary wind screen.
    - From 2017, data collection progressed from intermittent records to a continuous record collected at first one, and then several sites using a "2 channel SAM Scribe" monitoring system that collects and records a continuous string of 10-minute sound samples. The SAM-Scribe was purchased by an Ontario resident to collect data at their home, with assistance in setup and monitoring by the author. Since 2020, the resident has loaned the SAM Scribe system to the author for monitoring at the homes of other impacted residents. Roughly a 5-year continuous record of data is now available from the SAM Scribe system, principally at two different wind power developments, with Siemens SWT-101 and Vestas V82 wind turbines, as well as some recordings distant from the wind turbines.
    - Additional data has been collected from time to time to verify the data collected by the SAM Scribe using an ACO Pacific system. This system uses an IEC 61094-4 (Measurement Microphone) compliant 7046 free-field microphone and a 4012 pre-amplifier. The pair have a rated frequency response ±2 dB from 2 Hz to 20 kHz. Additionally, data has been collected using the Earthworks M30BX microphone, and using a pair of Superlux ECM-999 measurement microphones.
    - A further source of data has been recordings performed at sites using an external MOVO omnidirectional Measurement Microphone (rated as flat from 35 Hz to 18 kHz) protected with a primary "muff" type windscreen, used in a protected location away from direct wind exposure, as an external microphone on an iPhone. While not initially thought of as an acoustical monitoring device, performance of the pair give remarkable results. They permit recording a calibration signal from a 94-dB calibrator, and provide a simple method for recording a simultaneous video and calibrated audio file that can be easily transmitted for later analysis.
- Analysis of the collected acoustical data from the various methods has been conducted using the Faber Acoustics application Electroacoustics Toolbox version 3.9.10 on a 3.6 GHz intel Core i5 iMac computer system running macOS 10.13.6.

3. Progression to determine an objective measure for annoyance

Listening to those impacted suggested that annoyance might arise from a number of different pathways. Initially, to determine if a common parameter might be identified, analysis focused on the times identified by the residents at the monitoring sites as annoying, or irritating, to the regulator, the Ontario Ministry of the Environment. While residents do not identify every situation considered as annoying, they do log sample times they consider as typical examples. Recording is by phone to a Ministry "Spills Line" and generally includes a brief description of the condition, the local environmental conditions, and a "rating" of annoyance from 1-10 as requested by the Ministry contact person, although there are no specific criteria for this rating. Progression from analysis of these "annoyance" conditions to a full analysis near and far from wind turbines is described in this section.

3.1 Results of the initial analysis

The key to the analysis technique used on this paper arose from a comment made in discussion at the Wind Turbine Noise Conference in 2021 by Andy McKenzie PhD BSc FIOA, of Hayes McKenzie in the UK. Andy noted that in the UK it was common to use LA90, the A-Weighted sound pressure level exceeded 90% of the time, effectively as the background sound pressure level. This suggested a clue to determine an annoyance measure of the classical signature "swish / or / swoosh" sound variation of a wind turbine.

A simplified display of the cyclical nature of the wind turbine sound might be considered as a sine wave. In reality the situation is considerably more complex. Impacted residents are often impacted by more than one wind turbine. Thus, the composite sound level, while varying cyclically, will be more complex than a simple sine wave.

The difference between the L90 value (the quiet times) and the L10 value (the loud times) gives an assessment of the change in sound level from quiet to loud. While not an exact measure of the value of the "swoosh" it is a simply determined parameter. The parameter gives a readily available measure of cyclical change in sound pressure level near wind turbines. The difference was calculated for both LZ10-LZ90, and for LA10-LA90. These values can be found from modern sound level meters or assessment applications such as the electroacoustics toolbox. The analysis results consistently showed that in the situations identified by the residents as annoying, LZ10 exceeded LZ90 by a value in the order of 6 dB or more, while LA10 was not more than 3 dB higher than LA90. Hence, an initial assessment of an objective measure to signify annoyance was LZ10-LZ90 > 6 dB, while LZ10-LA90 was less than 3 dB.

Figure 1 shows a display of the electroacoustic toolbox sound level meters for LZ10, LZ90, LA10 and LA90. These are for a 2-minute recording sample at a site with 4-Vestas V82 wind turbines within 1000 metres. The 181.5 MW array of 110 wind turbines generated 129 MWh for the hour of the sound sample, The Environment Canada average wind speed for the same hour at the nearest monitoring site was 6.9 metres per second. The display shows the difference between LZ10 and LZ90 to be greater than 10 dB, while the difference between LA10 and LA90 was less than 3 dB. A factor not seen in the static figures, but will be shown in the conference presentation, is how the lower frequency 1/3 octaves "dance" up and down, while the higher frequency 1/3 octaves change little.

Listening to such examples, as will be demonstrated in the conference presentation, shows that such a case clearly portrays the "swoosh." However, in cases where the LA10 exceeds LA90 significantly, (as for example if there is a lot of traffic noise, or bird cries) then the wind turbine "swoosh" is less apparent, and is less likely to be identified as annoying. Similarly, it was found that if the turbines were shutdown then the LZ10-LZ90 was reduced, and again the situation was perceived as less annoying. Thus, the 6dB minimum for LZ10-LZ90, and a 3dB maximum for LA10-LA90 seemed to be reasonable criteria for further analysis.

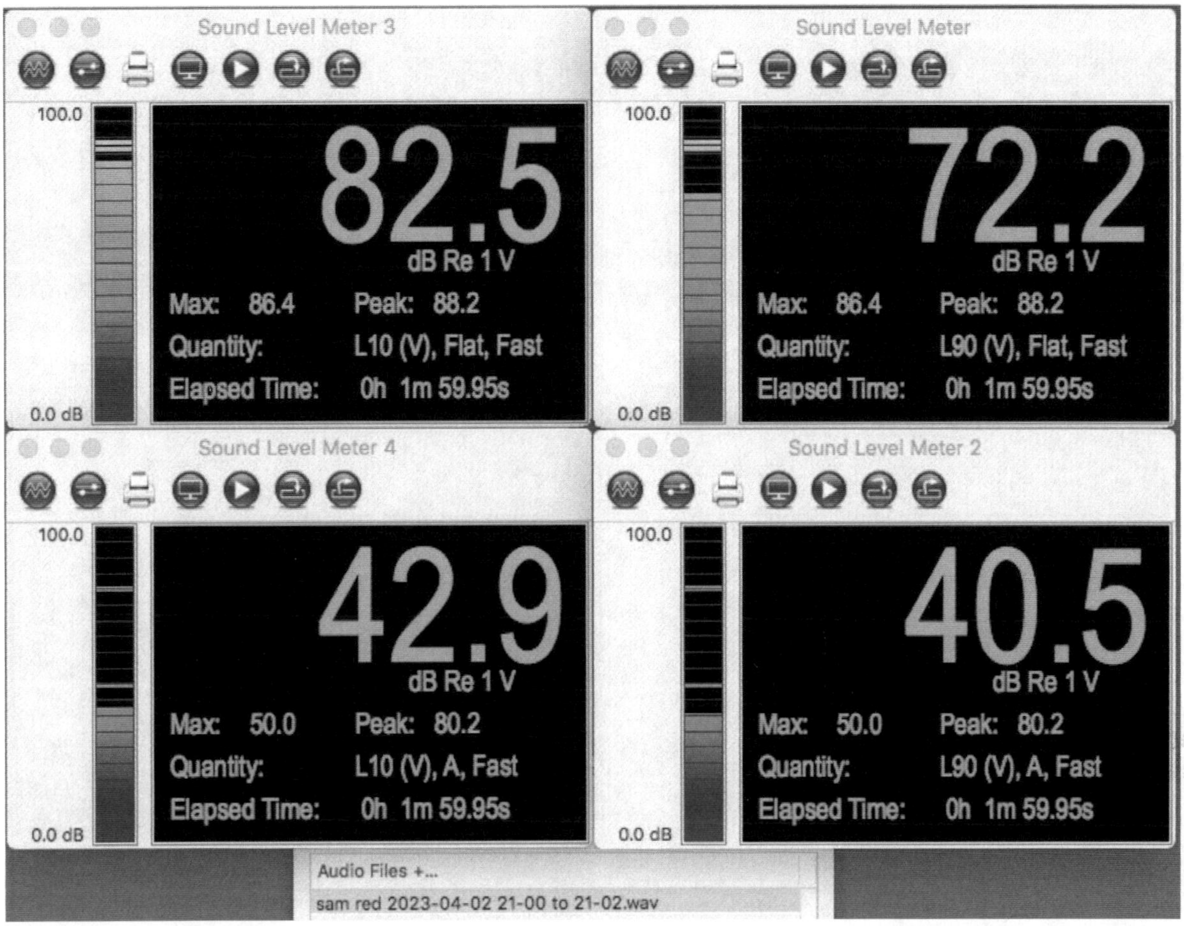

Figure 1 – A typical case perceived as annoying (electroacoustic toolbox LZx and LAx results)

3.2 Verification of the initial analysis as a measure of a wind turbine parameter

To verify that the measurement was not simply a measurement of wind noise, a simultaneous set of recordings were taken at a site > 6 km distant from the nearest wind turbine. This is considered as the "remote site" if further discussion. Turbines at the monitoring site are within view from remote site. The remote site also has on site wind speed and direction monitoring which show close correlation to the nearest Environment Canada monitoring location. The results at the remote site are shown in Figure 2. At this remote site, LZ10 was the same as LZ90 close to the wind turbines, and LA10 was within 1 dB of LZ90 at the site close to the turbines. LZ10-LZ90 was somewhat higher at 14.8 dB, and LA10-LA90 was also higher at 7.3 dB. Five 2-minute data samples in the 10 minutes prior to and after the presented data for both the wind turbine site and the remote site were calculated. The five samples were similar, although particularly intense gusts in the last sample near the wind turbines would have placed it outside the criteria for being considered as annoying.

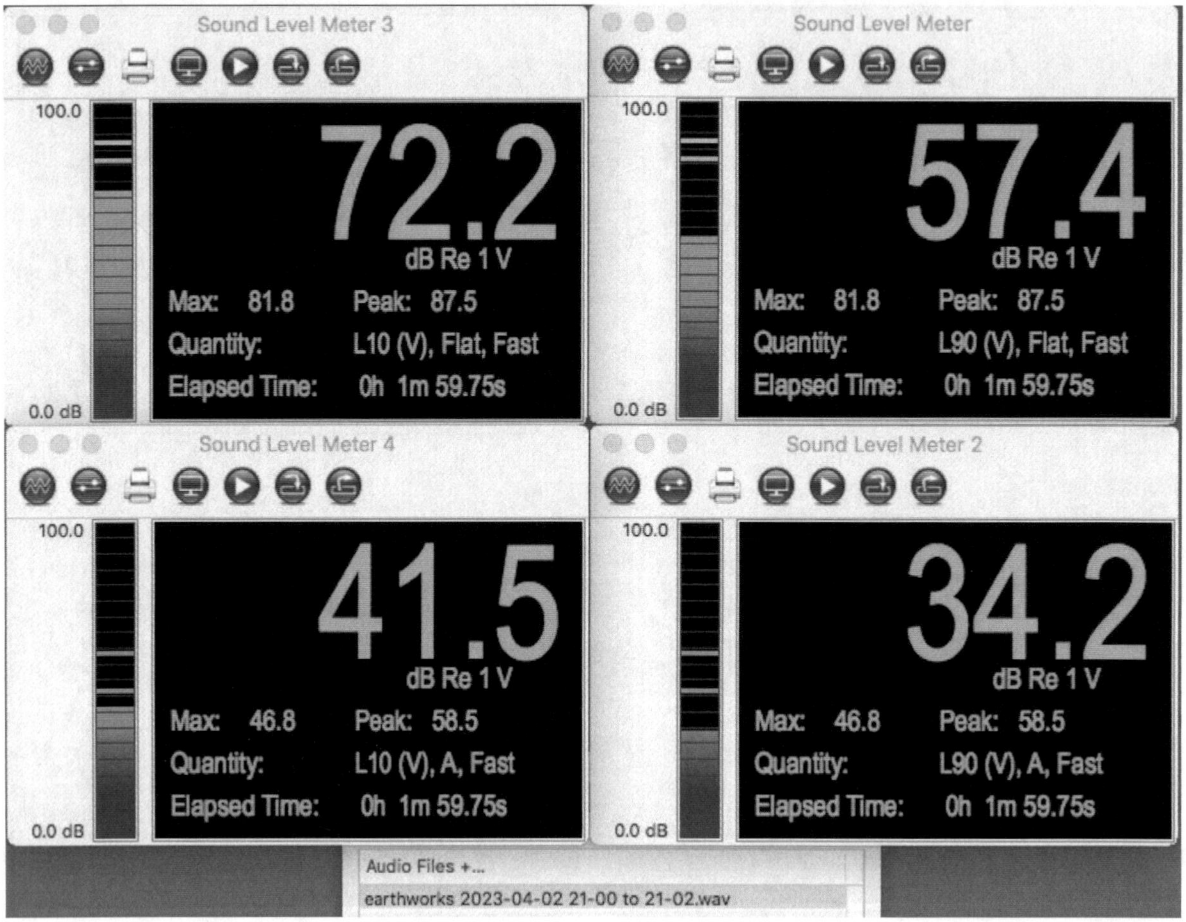

Figure 2 – Analysis of data recorded at same time as in Figure 1, remote from wind turbines

Date 2023-04-02 Time as Shown	Near Wind Turbines		Remote from Wind Turbines	
	LZ10-LZ90 (dB)	LA10-LA90 (dB)	LZ10-LZ90 (dB)	LA10-LA90 (dB)
20-50 to 20-52	7.6	2.1	13.1	6.4
20-55 to 20-57	8.5	2.1	9.4	3.2
21-00 to 21-02	10.3	2.4	14.8	7.3
21-05 to 21-07	9.7	2.9	14.0	4.3
21-10 to 21-12	14.7	5.4	10.5	4.5

Table 2 – Five samples near and remote from wind turbines in period of Figures 1 and 2

3.3 Further analysis underway to verify annoyance criteria

Ongoing analysis continues to verify the criteria indicating conditions consistent with a judgement by residents of annoying conditions. Simultaneous data collection at a site near wind turbines and remote from wind turbines continues. Analysis of over 100 hours of data continues to confirm that the criteria of LZ10-LZ90 > 6 dB and LA10-LA90 < 3 dB only present themselves remote from wind turbines rarely (at a frequency of about 7 times per 100 cases), This has been detected only during conditions of heavy rain, particularly when water droplets are falling from the secondary windscreen to hit the protection at the top of the primary windscreen. The microphone records this similar to a “drum thump”, and are not representative of actual conditions. Ontario regulations as an example do not permit collection of wind turbine noise samples during precipitation, and these are only a subset of those conditions.

Near the wind turbines, conditions meeting the criteria of LZ10-LZ90 > 6 dB and LA10-LA90 < 3 dB occur quite frequently. The frequency of this condition being met had been approximately 20 times per 100 cases. This criteria has been tested against previous cases identified to the Ministry of the Environment by residents as annoying with high correlation.

Data collected in the past at various wind turbine locations is being tested against the criteria. The criteria are showing that it has good potential for use as a screening technique. The technique provides a measurable assessment criteria, independent of subjective assessment.

4. Conclusions

Work underway is getting closer to presenting a formal paper demonstrating a measurable criteria to match subjective assessments of annoyance. The criteria shown to be effective is:

LZ10-LZ90 > 6 dB and LA10-LA90 < 3 dB

This is important as it reduces the need to assemble, and expose, a representative panel of "peers" of noise sensitive persons to assess annoyance. It also demonstrates respect for complaints filed by individuals of adverse impacts when exposed to wind turbines for sustained periods. A criteria to assess, and thus enable prevention of adverse impacts is particularly important due to planned expansion of wind turbine to meet rising electricity needs. Work has shown that the criteria responds well to the conditions near wind turbines, while being largely independent of wind noise. Work to date has shown that the criteria can provide a useful screening tool. Further development is ongoing to help remove the necessity for a listening test to address outside influences. To date a listening test is needed to differentiate influences such as road traffic, aircraft, and spurious noise arising from rain droplets penetrating the microphone windscreen, or windscreen "bumping" during gusts.

References

Ontario Professional Engineers, Code of Ethics. https://www.peo.on.ca/licence-holders/code-ethics

Institute of Noise Control Engineering USA, Canons of Ethics. https://www.inceusa.org/pub/

Acoustical Society of America, Ethical Principles. https://acousticalsociety.org/ethical-principles/

Proceedings of International Conferences on Wind Turbine Noise, 2005, 2013, and 2021.

Luz, G.A. (2011) *Some People are More Noise Sensitive than Others* Tutorial Presented at Acoustical Society of America, 161th Meeting, Seattle, WA. (Cites Moreira & Bryan, 1972)

Moreira, N. M., **Bryan**, M. E. (**1972**). *Noise annoyance susceptibility*. Journal of Sound and Vibration, **1972**, 21, 449-462.

INCEEUROPE

10th International Conference
on
Wind Turbine Noise
Dublin – 21st to 23rd June 2023

Continuous monitoring of wind turbine noise

Vitor Rosão: vitorrosao@schiu.org
Vinícius Duarte: viniciusduarte@schiu.org
Alexandra Reis: alexandrareis@schiu.org
João Almeida: joaoalmeida@schiu.org
Daniel Fernandes: danielfernandes@schiu.org
Alice Ramos: aliceramos@schiu.org
SCHIU, Vibration and Noise Engineering. Portugal. www.schiu.org

Summary

This paper presents the specificities and results of a cost-effective continuous monitoring system that was designed specifically for wind turbine noise analysis. The system utilizes Class 1 sound level meters, which are synchronized with other data sources such as audio/video recordings and meteorological data from local stations. The results are published on a web platform in near real-time, allowing for the identification of specific noise events and the automatic control of relevant parameters. The system can detect tonal characteristics using the ISO/TS 20065:2022 method and amplitude modulation, both of which are factors typically associated with wind turbine noise discomfort.

1. Introduction

This article presents the considered constraints, and their basis, and associated specific features for the development of the continuous noise monitoring system for wind farms, based essentially on sound levels, partly already developed and partly under development, by SCHIU in Portugal.

The development of this system aimed at wind farms (wind turbine noise) is associated with the growing interest of the technical acoustics community, and others, regarding this topic (see, for example, Vitor Rosão 2021).

It is clarified that the essential objective of this monitoring system is to characterise the so-called incident noise [non-consideration of possible sound reflections at the receiving point; see Directive 2002/49/EC and ISO 1996 (part 1: 2016; Part 2: 2017)] coming from a wind farm (wind turbine noise), perceived, outside, next to noise-sensitive receivers (in Portugal it is usual to define a noise-sensitive receiver by: "*the building house, school, hospital or similar or leisure space, with human use*").

The following conditions and the following associated aspects are highlighted and listed, which will serve as the basis for naming the subsequent chapters:

- A continuous noise monitoring system, based on the recording of sound levels, and which is intended to operate autonomously, without human presence, needs to have some form of control of which noise sources are at the origin of the sound levels that are being recorded, otherwise higher sound levels may be detected that do not originate from the noise source of interest. In the development of this system, 3 complementary ways were considered that help to detect, without human presence in real time, which sound levels are and are not associated with the source of noise in question:

    1. Audio/video recording in the measurement area, which allows a posteriori human or other analysis to try to distinguish "audibly" and/or "visually", the type of noise source that is generating the recorded sound levels.

    2. Synchronised recording of parameters other than sound levels, but which may have an influence on sound levels (from source noise or background noise), namely wind speed and direction (ideally through systems that do not generate "extra" sound levels"; an ultrasound-based system was used, with no moving parts) and rainfall, but also other typically relevant meteorological data, namely air temperature and humidity.

    3. Synchronised recording of sound levels, audio and video, in a location other than near to the receivers under analysis, in particular close to the noise source (in this case close to the wind turbine, or wind turbines) with greater influence on the sound levels perceived at the receiving point, to associate any variations in the sound levels of the noise source in question with the variation in sound levels at the receiver point (if the receiver is very close to the source, only one measurement point may be sufficient).

- One of the objectives of the continuous monitoring system developed, is to obtain results that can be used to verify compliance or non-compliance with certain technical acoustic requirements. The developed system is considering, for now, the following requirements and possibilities:

    4. Legal requirements (Portugal), in particular the following aspects:

        i. Limit values, parameters used and reference periods.

        ii. Noise Rating Level.

        iii. Tonal characteristics.

        iv. Impulsive characteristics.

        v. Long term average (see chapter "5.1 Limit values, parameters used and reference periods").

    5. Possibility of other requirements, in particular:

        i. Other frequency weightings.

        ii. Other tonal characteristics methods.

iii. Other impulsive characteristics methods.

iv. Other Amplitude Modulation methods.

Figure 1 shows the general aspect of the main page of the continuous monitoring system platform developed so far.

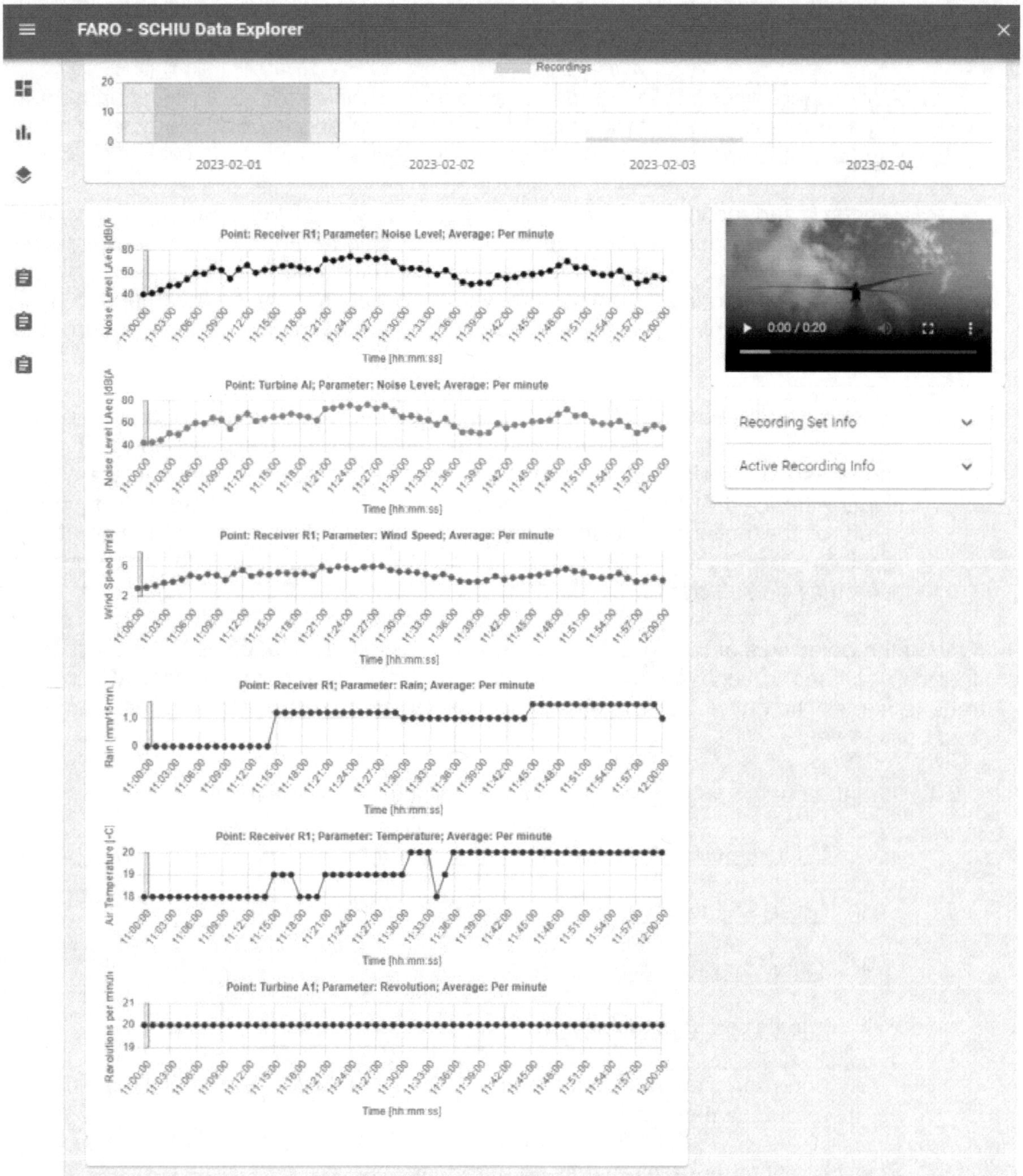

Figure 1: General aspect of the main page of the continuous monitoring system platform developed

2. **Audio/video recording**

In terms of audio/video recording, the developed continuous monitoring system has as main interests and specific features:

i. To have the best possible quality (in constant evolution) so that the audio/video that is possible to hear/see a posteriori, in the “office”, is as close as possible to what a human being, with normal hearing/vision, would be able to perceive if he/she was hearing/seeing the sound/sources on site in real time.

ii. Maintenance of human privacy. An algorithm (in constant evolution) has been developed that allows, in terms of audio, to distinguish the type of noise source in presence, but, in the case of human conversation, it does not allow to distinguish what is being said. In visual terms a filter was used that allows you to see the shapes but does not allow you to identify the person, or other private information, such as vehicle number plates.

2.1 **Current quality**

The current quality of the audio and video of the developed monitoring system, as shown in the example available at the following link, and which already has the audio and visual “filters” applied, allows distinguishing the sources of noise that typically occur:

https://drive.google.com/file/d/1RRn3xuxasZVdHVGFAcZu3VsW5WFuITz-/view?usp=sharing

2.2 **Current audio privacy algorithm**

The developed audio privacy algorithm sought, essentially, to maintain the possibility of human detection of the type of noise source present, but not to allow determining what was being said, in the case of recording any human conversation.

An example speech is presented in the following links, in the original version and in the version in which SCHIU's audio privacy algorithm was applied.

- Original: https://drive.google.com/file/d/1Y9LdkFrloy2qODh-QxGTWdJ8TYr-jy_W/view?usp=sharing.
- With application of audio privacy algorithm: https://drive.google.com/file/d/1yuVgJ3IDZhFjpEPAaqoJXINBvNqbQWo9/view?usp=sharing.

3. **Other parameters**

The developed system registers, in addition to the raw data of the sound levels (see next subchapter “6.1Raw data”), the following parameters:

- Air temperature and humidity.
- Wind speed and wind direction.
- Rainfall.

The previous Figure 1 shows, e.g., the possibility of join analysis of noise levels, wind speed, rain and temperature at Receiver R1.

4. **Close to the noise source**

The system developed allows you to obtain, from a particular noise source of interest (or other location deemed relevant) in this particular case, from one or more wind turbines of interest, the same parameters / data that can be obtained from a Sensitive Receiver, i.e.:

- Raw sound level data.
- Audio/video recording.
- Air temperature and humidity.
- Wind speed and wind direction.
- Rain.

Close to the wind turbine is possible also obtain the rotation speed of the turbine (see the next subchapter "4.1 Wind turbine rotation speed calculation algorithm")

The previous Figure 1 shows, e.g., the possibility of join analysis of:

- at Receiver R1: noise levels, wind speed, rain and temperature.
- at Turbine A1: noise levels, revolutions per minute

4.1 **Wind turbine rotation speed calculation algorithm**

Given the importance of the rotational speed of the wind turbines (see, for example, Luca 2012), an algorithm was developed which, based on the sound levels and video recorded with the wind turbines, allows counting the number of rotations per minute, as exemplified in the available video at the following link:

https://drive.google.com/file/d/16Dxux6YC1uKiyjVHrfWAetjvRL1kA8NU/view?usp=sharing

5. **Legal requirements (Portugal)**

In Portugal there are no specific requirements for Wind Turbine Noise, only requirements for the so-called Permanent Noisy Activities, contained in Article 13 of DL 9/2007 (Mainland Portugal and Autonomous Region of Madeira) and Article 25 of Regional Legislative Decree 23/2010/A (Autonomous Region of the Azores).

5.1 **Limit values, parameters used and reference periods**

In short, the requirements (limit values, parameters and reference periods) are as follows:

- Maximum Exposure Criterion:
    - Essentially dependent on the acoustic classification assigned by the municipality to the site:
        - Mixed Zone: $L_{den} \leq 65$ dB(A); $L_n \leq 55$ dB(A).
        - Sensitive Zone: $L_{den} \leq 55$ dB(A); $L_n \leq 45$ dB(A).

- Unclassified zones until classified or equated: $L_{den} \leq 63$ dB(A); $L_n \leq 53$ dB(A).

  - The parameters, with A weighting in frequency, must be representative of the annual energy average and of the following periods, in Portugal:

    - Mainland and Autonomous Region of Madeira:

      - L_d (Day Level): 7h-20h; L_e (Evening Level): 20h-23h; L_n (Night Level): 23h-7h.

    - Autonomous Region of the Azores:
      - L_d (Day Level): 7h-21h; L_e (Evening Level): 21h-23h; L_n (Night Level): 23h-7h.

    - The L_{den} parameter is determined based on the following equations:

      - Continental and Autonomous Region of Madeira:

$$L_{den}=10\times log\frac{1}{24}\left[13\times10^{\frac{L_d}{10}}+3\times10^{\frac{L_e+5}{10}}+8\times10^{\frac{L_n+10}{10}}\right]$$

      - Autonomous Region of the Azores:

$$\mathrm{L_{den}} = 10\times \log\left[\frac{14\times10^{\frac{L_d}{10}}+2\times10^{\frac{L_e+5}{10}}+8\times10^{\frac{L_n+10}{10}}}{24}\right]$$

- Discomfort Criterion.

  - Essentially corresponds to the difference between the Noise Rating Level, A weighted, L_{Ar} , of the total noise (particular noise of the noise source in question plus background noise) and the Continuous Equivalent Sound Level, A weighted, L_{Aeq} , of the background noise (total noise without noise from the noise source in question).

  - Parameters should be representative of the most critical month:

    - Day time (in Portugal: 7h-20h or 7h-21h): $L_{Ar} - L_{Aeq} \leq 5$ dB.

    - Evening time (in Portugal: 20h-23h or 21h-23h): $L_{Ar} - L_{Aeq} \leq 4$ dB.

    - Night time (in Portugal: 23h-7h): $L_{Ar} - L_{Aeq} \leq 3$ dB.

The developed continuous monitoring system thus allows, for each monitoring point:

- Define day, evening and night time.

- The constants that add to L_e (C_e; typically 5 dB) and L_n (C_n; typically 10 dB) in the L_{den} determination equation.

- Obtain the average daily energy values of L_d, L_e, L_n e L_{den}, depending on the previous definitions.

- Obtain the average monthly energy values of L_d, L_e, L_n e L_{den}, according to the previous definitions and thus determine the most critical month.

- Obtain the average annual energy values of L_d, L_e, L_n e L_{den}, according to the previous definitions.

- Possibility of determining times when, in a reasoned and justified manner, the particular noise of the noise source of interest is negligible, and the associated values may be considered characteristic of the background noise.

5.2 **Noise Rating Level**

The Noise Rating Level, L_{Ar}, in the Portuguese legislation (DL 9/2007) is given by:

$$L_{Ar} = L_{Aeq} + k_1 + k_2 \quad (1)$$

Where k_1 = 3 dB, if tonal characteristics are detected in the recorded noise, and k_2 = 3 dB, if impulsive characteristics are detected in the recorded noise.

5.2.1 **Tonal characteristics**

The method included in Annex I of DL 9/2007, for detecting the tonal characteristics of noise within the evaluation time interval, consists of verifying, in the frequency domain and in third octave bands , whether the sound level of one band exceeds the adjacent ones by 5 dB(A) or more. In these cases, the noise shall be considered tonal.

5.2.2 **Impulsive characteristics**

The method to detect impulsive noise characteristics within the evaluation time interval consists of determining the difference between the equivalent continuous sound level, L_{Aeq}, measured simultaneously with impulsive and fast time weighting filters. If this difference is greater than 6 dB(A), the noise shall be considered impulsive.

6. **Possibility of other requirements**

6.1 **Raw data**

To allow the greatest possible versatility, it was decided that the system would register raw information, duly calibrated, with a sampling frequency of 48 kHz.

This option thus allows to carry out further types of analysis at a later time, namely different frequency weightings or different time weightings.

6.2 **Current requirements**

For now, the system allows automatic verification of the following requirements:

- Portuguese legislation: DL 9/2007.
- Criteria for the assessment of low frequency noise disturbance from the University of Salford 2011 document.
- Criteria for L_{den} < 45 dB(A), according to WHO 2018 document.

6.3 **Current frequency weightings**

For now, the system allows the use of the following frequency weightings:

- IEC 61672-1: 2013:
  - A, B and C weighting.
- ISO 7196:1995:
  - G weighting.

6.4 **Current tonal characteristics methods**

For now, the system allows the use of the following tonal characteristics detection methods:

- Portuguese legislation (Annex I of DL 9/2007).
- ISO/TS 20065:2022.

6.5 **Current impulsive characteristics methods**

For now, the system allows the use of the following impulsive feature detection methods:

- Portuguese legislation (Annex I of DL 9/2007).
- ISO/PAS 1996-3:2022.

6.6 **Current Amplitude Modulation**

Given the relevance of the Amplitude Modulation characteristics, for Wind Turbine Noise the system uses the method of the IOA 2016 document.

7. **Conclusions**

We have tried so far, and will continue to try in subsequent developments, that the continuous noise monitoring system, aimed at Wind Turbine Noise, is developed with the best-known state of the art, as explained in the previous chapters.

We also tried to make the system as versatile as possible so that, if necessary, justifiable and relevant, new types of analysis can be easily introduced.

It is therefore expected that the system can be effectively useful in controlling noise, perceived by Human Receivers (mainly at houses, but also schools and health facilities) that often exist in the vicinity of Wind Farms.

References

Gazette of the Portuguese Republic (2007) Decree-Law 9/2007, from January 17th.

Gazette of the Portuguese Republic (2010) Regional Legislative Decree 23/2010/A, from June 30th.

Institute of Acoustics (IOA) Noise Working Group (Wind Turbine Noise) (2016) *Final Report A Method for Rating Amplitude Modulation in Wind Turbine Noise. Version 1.*

International Electrotechnical Commission (2013) *IEC 61672-1: Electroacoustics: Sound level meters - Part 1: Specifications.*

International Organization for Standardization (1995) *ISO 7196: Frequency-weighting characteristic for infrasound measurements.*

International Organization for Standardization (2016) *ISO 1996-1: Acoustics: Description, measurement and assessment of environmental noise: Part 1: Basic quantities and assessment procedures relating to the assessment and management of environmental noise.*

International Organization for Standardization (2017) *ISO 1996-2: Acoustics: Description, measurement and assessment of environmental noise: Part 2: Determination of sound pressure levels.*

International Organization for Standardization (2022) *ISO/PAS 1996-3: Acoustics: Description, measurement and assessment of environmental noise: Part 3: Objective method for the measurement of prominence of impulsive sounds and for adjustment of* L_{Aeq}.

International Organization for Standardization (2022) *ISO/TS 20065: Acoustics: Objective method for assessing the audibility of tones in noise: Engineering method.*

Luca Fredianelli, et al. (2012) *The suitable parameters to assess noise impact of a wind farm in a complex terrain: a case-study in Tuscan hills*. Euronoise 2012, Prague, Czech Republic.

Official Journal of the European Communities (2002) *Directive 2002/49/EC of the european parliament and of the council of 25 June 2002.*

University of Salford (2011) Andy Moorhouse et. al. *Proposed criteria for the assessment of low frequency noise disturbance.*

Vitor Rosão, et al. (2021) *Necessary adjustments in ISO 9613-2 and CNOSSOS (industries) methods for noise forecasting in Wind Farms*. Euronoise 2021, Madeira, Portugal.

World Health Organization (WHO), Regional Office for Europe (2018) *Environmental Noise Guidelines for the European Region*.

INCE EUROPE

10th International Conference on Wind Turbine Noise Dublin – 21st to 23rd June 2023

Schall_KoGe - measurements, simulations, and validation of wind turbine noise in complex terrain

Arthur Schady DLR Institute of Atmospheric Physics (IPA) Oberpfaffenhofen
arthur.schady@dlr.de
Katharina Elsen DLR Institute of Atmospheric Physics (IPA) Oberpfaffenhofen
katharina.elsen@dlr.de

Summary

The project that was carried out from 2018-2022 was aimed at the realistic forecast of wind turbine noise and tested methods based on comprehensive long-term measurements of meteorology and sound as well as operational data from wind turbines at two sites. Precise flow models were used, as were various sound forecast and propagation models, and all results were compared with the measurement data.

This article presents the main results and problems of the project. The propagation conditions were classified, which allows the comparison of measurement results and model calculations. The source parameterization of the wind turbine was extensively examined so that a simplified yet realistic noise forecast is possible.

Finally, a scientific model was compared with conventional calculation methods and the differences are presented.

1. Introduction

In the joint project Schall_KoGe (sound in complex terrain), sound propagation models were further developed for the efficient investigation and evaluation of the sound of wind energy plants. They have demonstrated their practicality through successful application in the interpretation of measurement data from one campaign in simple terrain in Harsewinkel (Germany, 2020) and the Perdigão-2017 campaign in Portugal. At both sites the wind turbines were 'switched off' during defined time, so that the influence of noise emission of the turbine on the environment could be clarified. With the sub-system of flow simulation, the underlying synoptic situation, local dynamics at the double ridge of Perdigão, and the wake behind the wind turbine on the southwestern ridge were calculated and interpreted. With the sound propagation calculations, the influence of orography in Perdigão on the propagation of sound emitted by the WEA was interpreted. Simulation results for dynamics and sound propagation were verified through measurements on case studies. Sensitivities and dependencies between inflow (weather), topography (forest, hills), and properties of the turbine wake and sound propagation were identified through systematic simulations of idealized scenarios.

2. Measurement Data Analysis

The validation of model results with measurements requires wind turbine sound immission data that are not compromised by background noise of other sources. Sound assessment of wind turbine noise can be difficult in long-term noise monitoring, which includes all different situations of noise emission from the turbine itself and the sound transmission to the microphone, but also from all surrounding noise sources. Therefore, in the following the location and the distances of the used microphones with respect to the wind turbine is described as well as the important analysis steps within the project of Perdigão. In Harsewinkel the terrain is totally flat and similar distance for microphones were kept to the wind turbine as in Perdigão. This region is treated as a rather simple situation, not described here in detail.

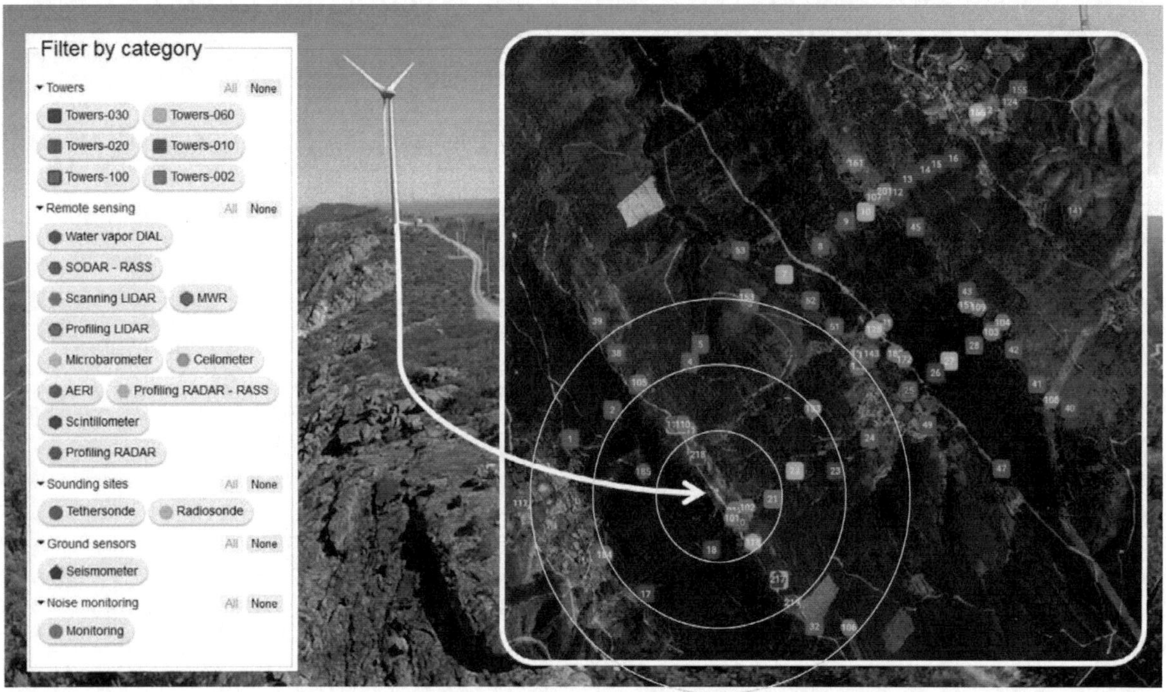

Figure 1: The extensive measuring instrumentation in Perdigão (Portugal) consisting of many meteorology measuring masts, LIDAR and microphones. The circles indicate the distance to the wind turbine in 250m increments.

2.1 Data source

Long-term high-resolution atmospheric and sound data were collected in Perdigão (Portugal) in a large measurement campaign (Fernando, 2019). As a part of the Schall_KoGe project the acoustic data and a part of the meteorological data have been analysed in order to study the sound propagation in vicinity of a wind turbine over complex terrain (see Data Repository).

2.2 Data Classification

We assumed that comparable meteorological situations lead to comparable sound pressure levels. As a large number of data was collected and a sufficient variety of meteorological situations was covered within the measurement period, the data can be classified according to defined meteorological situations and afterwards be analysed and compared with each other.

As a first step, the most important meteorological parameters were identified and discretized building so called meteorological classes. Hereby, the number of parameters and classes strongly depend on the amount of data points available for the analysis.
Therefore, depending on the distribution of measured data and time averaging we classified the parameters according to table 1:

Table 1: meteorological parameter used for classification of specific situations, based on measurements of the tower closest to the wind turbine.

Parameter	Class 1	Class 2	Class 3	Class 4
WD, Wind direction [°]	∈ [25°, 85°] (upwind)	∈ [205°, 265°] (downwind)	-	-
TI, Turbulence intensity	< 0.08	∈ [0.08, 0.16[	≥ 0.16	-
dT/dz, Temp.grad. [K/100m]	< −0.2 (unstable)	∈ [−0.2, 0.2] (neutral)	> 0.2 (stable)	-
du/dz, Wind shear [1/s]	< 0.0	∈ [0.0, 0.02[	∈ [0.02, 0.04[	≥ 0.04
LF, Relative humidity [%]	< 60	∈ [60, 80[	≥ 80	

This results in a total of 2 x 3 x 3 x 4 x 3 = 216 Situations. Additionally, it was found that many times appeared a so-called Low-Level Jet (see Figure 2), what encouraged to build two additional classes one with and another without LLJ.
Provided 10-minute mean values of meteorological and sound data, using the classification given in table 1, the most frequent situations were extracted (compare table 2) and then used to analyse comparable sound propagation conditions.

Table 2: the 5 most common situations

frequency	du/dz	dT/dz	LF	TI	WD	Sit.#
147	> 0.04	stable	< 60	< 0.08	upwind	79
98	∈ [0.02, 0.04]	unstable	< 60	> 0.16	upwind	53
98	∈ [0.0, 0.02]	unstable	< 60	> 0.16	upwind	18
92	∈ [0.0, 0.02]	unstable	< 60	> 0.16	downwind	19
84	> 0.04	unstable	< 60	∈ [0.08, 0.16[	upwind	74

Having identified the most common situation, the next step was to reconstruct as good as possible the associated wind field simulation.

3. Wind field Simulations

A sound propagation model is desired that includes all specific situations influencing the sound immission to accurately reproduce measured values. Nevertheless, input data often is not available to predict the sound level for given circumstances. It is primarily the wind field that drives the wind turbine and creates sound, but the same wind field also changes the sound propagation, such that different levels arrive at the receiver. Therefore, an important work package in the project was dedicated to a reconstruction of the wind field that is as realistic as possible. For this purpose, many meteorological sensors were adapted at the measurement site and used in the project for the validation of the wind field simulation results.

3.1 Model Setup

To reconstruct the wind field different models have been used. First the Weather Research and Forecasting (WRF) model (Skamarock et al., 2008) shall be mentioned. The large number of meteorological instruments involved during the measurement campaign was used to verify a WRF long-term simulation over 49 days with a horizontal grid resolution of 200m (Wagner et al., 2018). It was found that the wind distribution was simulated surprisingly well in spite of the long simulation period and that nocturnal low-level jets (LLJ) were mainly driven by thermally induced pressure gradients. As an example, the simulation of a LLJ over the double ridge on 8 May 2017 is shown in Fig. 2.

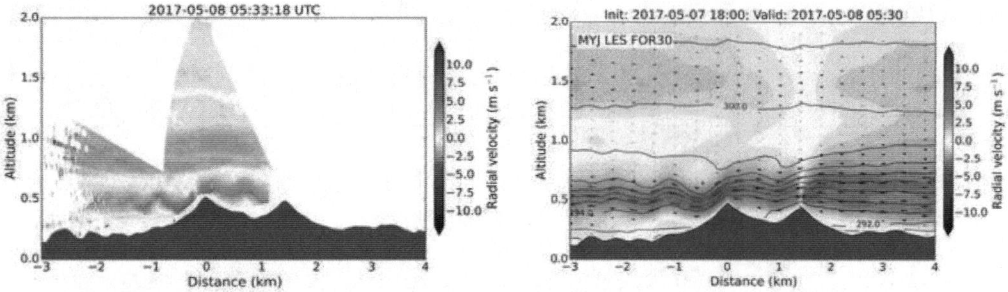

Figure 2: Radial velocity measured by the DLR lidar on 8 May 2017 at 05:33 UTC (left). Simulated radial velocities for a WRF run with forest parameterization (right)

In this project the numerical set-up consisted of up to 4 nested model domains with horizontal resolutions of 5, 1, 0.2 and 0.04 km. Vertical level distances were kept constant at 40 m below 1000 m altitude and stretched above to a model top at 200 hPa (about 12 km height). The number of horizontal grid points of the domains were 300x300, 251x251, 251x251 and 251x251 for the respective domains. In the vertical direction about 90 grid points were used. The initial boundary conditions were taken from ECMWF operational analyses.

This rather complex and expensive calculation was necessary to understand single effects of the flow over the double ridge. At the same time we adapted a rather simple method – a diagnostic wind field model - to reconstruct the wind field in Perdigão as well as in Harsewinkel based on measurement data from the towers or LIDAR.

4. Sound Propagation Models and Results

In the project the high-fidelity sound propagation model AKU_KoGe was used and developed. This model accounts for three-dimensional inhomogeneities in the air, the effects of the uneven ground and a variable definition of source setup. It is based on the Lagrange solution of particle motion (Heimann and Gross; 1999). The model was previously also applied to wind turbine noise by Heimann et al. (Heimann, Käsler, & Gross, 2011).
In the model a high number of sound particles are released at the position of one or several source points from where they propagate into the surrounding airspace. Each particle carries a certain amount of sound energy depending on the strength of the source and the starting direction to account for the source directivity. The particles travel with the effective speed of sound along rays which can be curved according to the local gradients of wind speed and temperature. The ray coordinates are determined in terms of position and direction of the travelling particles (Pierce, 1981).
For the sound propagation the vertical temperature and wind gradients can be summarized by the effective speed of sound c_{eff}, which results vectorially from the sum of the speed of sound and the influence of temperature and wind.

4.1 Wind Turbine Noise - Sound Assessment

As already mentioned, switching the wind turbine (WT) on and off was a great help in interpreting the noise data. Figure 3 shows the sound spectrum during a 3-hour measurement phase where the operational time of the WT can easily be detected. We developed some method to detect the wind turbines noise under different meteorological conditions (Schady, 2020). The result of this method is, that the detectability depends on daytime and rotational speed of the wind turbine. The difference in broad band absolute sound pressure level between wind turbine operation and switched off was rather low, mostly below 10dB, many times below 5dB.

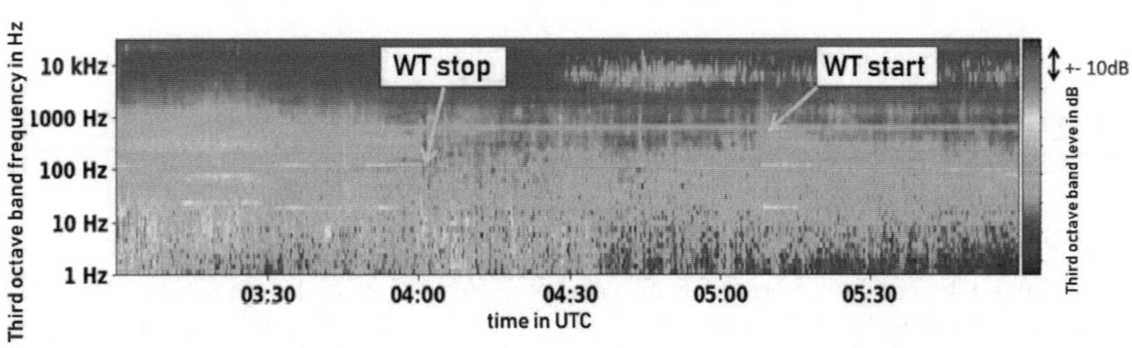

Figure 3: The spectrogram of one microphone in 125m distance on 22.05.2017 in the period between 3:00-6:00 UTC. The colour scale shows the sound pressure level of the individual one-third octave bands. WT stop and start is indicated and visible by the interruption of some one third bands.

4.2 Wind Turbine Noise - Sound Prediction

Model calculations were carried out using the previously described Lagrangian (particle) model (Heimann and Gross; 1999) that was coupled to the results of the flow solver. Adjustments to the model setup were made with regard to mesh size, time steps, as well as frequency spectrum and meteorology. A large number of test cases were calculated in order to analyse and document the strengths and weaknesses of the respective model setup. Additional comparison was provided with standard method of sound level prediction. The standard method is applicable only for simple geometry and topography. Therefore, in this case only the results for Harsewinkel are compared.
Only on behalf of sound propagation simulation one can distinguish different influences on the resulting sound immission. Here, artificial boundary conditions e.g. special source configuration or different wind fields, are used in the model and their effect on the sound field is examined. Figure 4 shows the result of an already complex situation with topography and meteorological input, but with simplified point source at hub height.

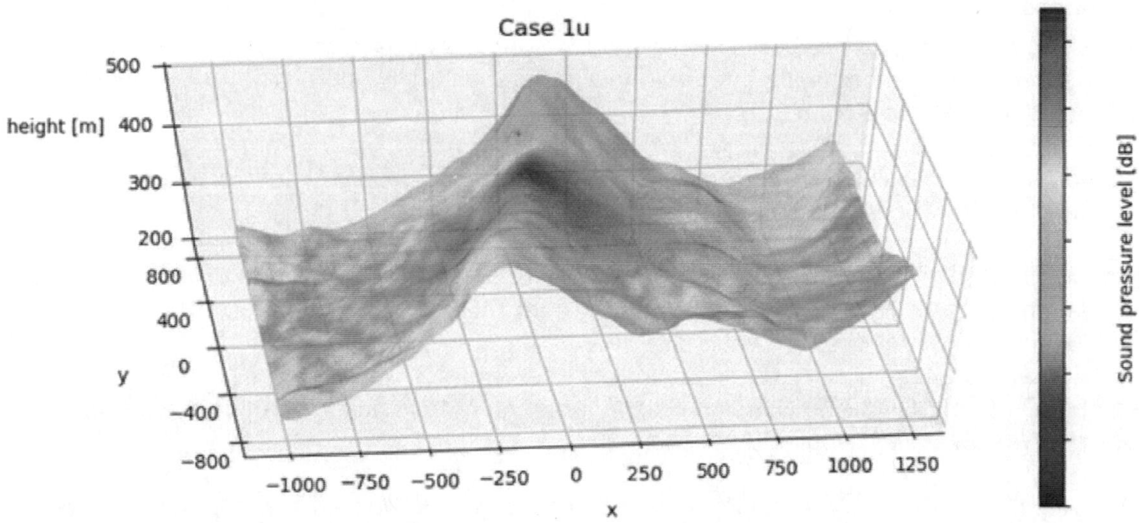

Figure 4: Topography of Perdigão with simulated surface sound pressure level in sound propagation with simulated meteorology and a single sound source

Table 3 shows the total sound level of the measurement and the Aku_KoGe simulation as well as the results of the calculations according to the standard for comparison. The simulation was carried out with settings that were as realistic as possible and evaluated at the measurement locations NMT500 and NMT1000 in Harsewinkel.

Table 3: Difference in sound level between calculated and measured values (i.e. calculation - measurement) at immission locations NMT_500 and NMT_1000. Positive level: simulation louder than measurement, negative level: simulation quieter than measurement

Messpunkt	Aku_KoGe	ISO 9613-2	Interim
NMT_500	+2.7	-2.6	-0.6
NMT_1000	+0.3	-4.4	-6.6

Open remained the comparison of measured and predicted situations for Perdigão, because no emission data were available for validation.

5. Conclusions

In the Schall_KoGe project, models were improved to investigate and evaluate wind energy plant sound efficiently. They were successfully applied to interpret data from the Perdigão-2017 campaign in Portugal as well as for simple terrain in Harsewinkel (Germany). Simulations of idealized scenarios helped to identify relationships between inflow, topography, turbine wake, and sound propagation. Verification through measurements was performed on case studies. The work continues to analyse all data from both measurement campaigns.

6. Acknowledgements

The studies were supported by the Federal Ministry of Economics and Energy within the framework of the Schall_KoGe project with the funding code: 0324337B. Furthermore, we would like to thank the project partners namely Tanja Stuettchen from WRD GmbH for information on the operation of the wind turbine and IAG for the good co-operation within the project.

References

K. M. Elsen and A. Schady (2021) *Different sound source setups in the simulation of wind turbine sound propagation*. In 9th International Conference on Wind Turbine Noise, number May, pages 1–6, Remote from Europe

Fernando, H. J. et.al. (2019). *The Perdigão: Peering into Microscale Details of Mountain Winds*. Bull. Amer. Meteor. Soc., 100, 799–819.

Heimann, D., Käsler, Y., Groß, G. (2011) *The wake of a wind turbine and its influence on sound propagation*. Meteorologische Zeitschrift, Borntraeger Science Publishers. doi: 10.1127/0941-2948/2011/0273

Heimann, D., Groß, G. (1999) *Coupled Simulation of Meteorological Parameters and Sound Level in a Narrow Valley*. Applied Acoustics, 56

Pierce AD. (1981) *Acoustics—an introduction to its physical principles and applications*. New York: McGraw–Hill

Skamarock, W. C. et al., (2008) *A description of the Advanced Research WRF Version 3.*, NCAR technical note, Mesoscale and Microscale Meteorology Division, National Center for Atmospheric Research, Boulder, Colorado, USA.

J. S. Wagner, T. Gerz, N. Wildmann, K. Gramitzky, (2018) *Long-term simulation of the boundary layer flow over the double-ridge site during the Perdigão 2017 field campaign*, Atmos. Chem. Phys. Disc., doi: 10.5194/acp-2018-997

Data Repository (2018): https://perdigao.fe.up.pt/datasets/thredds/catalog/ancillary/catalog (last seen 14.04.2023)

10th International Conference
on
Wind Turbine Noise
Dublin – 21st to 23rd June 2023

A comparison of methods for identifying wind turbine sounds in big data sets

Daphne Schössow, Stephan Preihs, and Jürgen Peissig
Institute of Communications Technology (IKT), Leibniz University Hannover, Germany
Correspondence: schoessow@ikt.uni-hannover.de

Susanne Könecke, Tobias Bohne, and Raimund Rolfes
Institute of Structural Analysis (ISD), Leibniz University Hannover, Germany
Correspondence: s.koenecke@isd.uni-hannover.de

Summary

For the investigation of long-term acoustic measurements in wind farms, methods for the selection of wind turbine noise are needed. In this paper, methods for identifying dominant wind turbine noise and for classifying sound scenes in a wind farm are presented, applied to a measured one-month data set and validated with manually labelled data.

For identifying dominant wind turbine noise, four methods given in the literature are applied. They are all based on statistical, acoustic criteria and differ in their complexity. Dominant wind turbine noise is correctly identified by all methods. In the case of low or no wind turbine noise, statistical criteria are not sufficient. Here, methods that also consider rotor speed show better results.

For the classification of sound scenes, two methods are used - a simple method based on acoustic and turbine-related criteria, and a more complex method using machine learning in the form of a convolutional neural network (CNN). In the examples of this work, the classes *Wind Turbine*, *Wind* and *Silence* are predicted well using simple criteria such as limiting the rotor speed. Bird sounds and other disturbing sounds are classified less good. The CNN-based method uses high-resolution time signals for classification, but coarse windowing was applied to the predictions for an easier comparison of the results to the other method. With normalized audio, the classes *Wind Turbine* and *Aircraft* are classified well. Broadband sounds, such as wind noise, are predicted less good. The classification of the class *Bird* did not yield good results for the given data, but it is suspected that it would work better with lower time scales than one minute. The prediction of the class *Silence* only works without normalization, such that further work needs to be done in this area.

1. Introduction

Performing and evaluating measurements of noise emission and immission from wind turbines is a complex topic. When analysing those, it is especially important to separate the noise of wind turbines from extraneous noise, such as overflights or animal sounds. For large data sets listening to the signals is not effective. Thus, having long time data sets, methods for identifying wind turbine noise (WTN) and other sound sources are required.

In this paper, methods for both, for identifying dominant wind turbine noise and for classifying sound scenes in a wind farm are applied to a measured one-month data set. The measurements and the data are described in Sec 2. Subsequently, the methods used for the identification and classification are presented in relation to the state of the art. For identifying dominant wind turbine noise, four methods from the literature are described in Sec 3.1. The first method for classification applies statistical and turbine-related criteria to identify disturbance noise and WTN (Sec. 3.2). The second method uses machine learning in the form of a CNN to classify audio data (Sec. 3.3). The evaluation and validation of all methods is performed by a comparison with hand labelled data (Sec. 4). In the conclusion, the results from the validation are summarized as well as limitations and possible applications of the methods are explained.

2. Measurements

The one-month data set was recorded as part of a measurement campaign in the project "WEA-Acceptance" [8]. The measurements were conducted at a wind farm in northern Germany. The landscape of the measurement site was characterized by flat grassland. The data set recorded contains acoustic, meteorological and turbine-specific data. For the acoustic measurements, a sound level meter was equipped with a cone and two wind screens. At a height of 1.70 m, sound pressure levels (1 Hz), one-third octave spectra (1 Hz) and high-resolution time signals (51 kHz) were recorded at a distance of 178 m to the nearest wind turbine. In addition, the SCADA data from the nearest (WT1) and two neighbouring turbines (WT2, WT3) were acquired. This data includes, for example, rotational speed, output power and wind speed. The data set also contains meteorological measurement data from a 100 m-mast, although this is not used in this work. An overview of the measurement site, the acoustic measurement setup and information about the nearest wind turbine are illustrated in Fig. 1. Further information about the measurements are given in Könecke et al. (2022) [7].

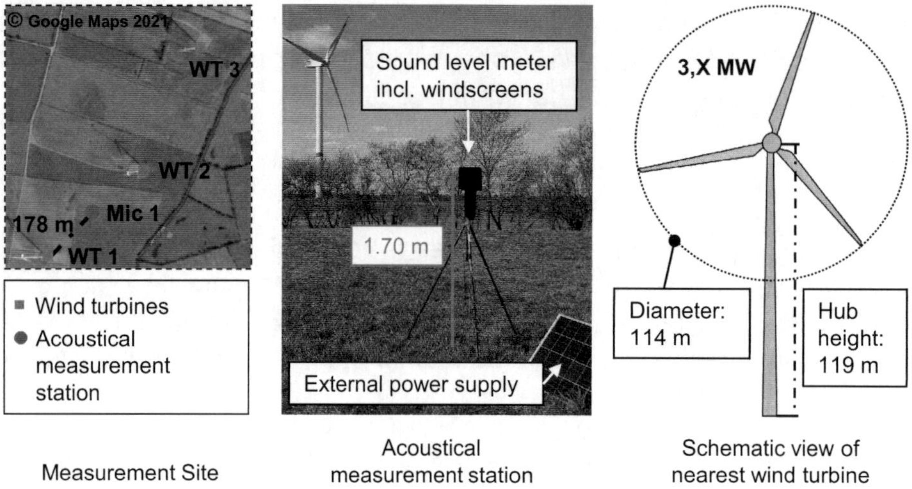

Fig. 1 Overview of the measurements.

3. Methods

In this paper, methods identifying dominant WTN and methods classifying sound scenes in a wind farm are examined and presented in the following sections. While the identification process aims to select dominant WTN without interfering sounds, the classification process considers audible WTN and other sound sources.

3.1. Identifying dominant wind turbine noise

In large data sets, dominant WTN can be identified and selected based on frequency-dependent and statistical criteria, as well as by considering environmental conditions.
To select data with dominant WTN, van den Berg (2004) [15] introduced the criterion:
(1) $L_5 - L_{95} \leq 4$ dB(A).
Here, the temporal fluctuation of the sound levels is examined using the statistical A-weighted percentile sound levels. While WTN results in rather constant values, the levels of disturbance noise scatter strongly. Öhlund and Larsson (2015) [9] additionally formulated the criteria:
(2) 'The A-weighted one-third octave band sound level from 800 Hz and above contribute less than 1.5 dB to the total A-weighted sound level if the total sound level is above 25 dB(A).'
(3) 'Calculation of free field spreading from every turbine (treated as point sources) contribute to a total SPL of 30 dB(A)or above at the immission point.'
Criterion (2) refers to the frequency spectrum of the WTN at the point of immission. Over distance, high-frequency components are absorbed and, at the same time, vegetation noise, for example, has more energy at high frequencies [1, 9]. The criterion guarantees that the A-weighted sound levels, calculated between 800 and 20,000 Hz, contribute less than 1.5 dB(A) to the total sound pressure level. The criterion (3) ensures that the sound power level produced by the wind turbine is at a sufficient level.
The selection criteria (1-3) have been applied, adapted and extended by other authors [4, 8]. In addition to acoustic criteria, Conrady (2020) [3] and Martens (2020) [8] also use turbine-related criteria. The detection of WTN is only possible if the rotor speed, given in revolutions per minute

(rpm), is within a certain (turbine-specific) range. At low rotational speeds, the sound emission is low, so that masking with disturbing noise is likely. To avoid the masking caused by wind-induced noise, the range of wind speed can additionally be limited [2].
The spectral resemblance method is presented in Conrady (2019)[2]. In this method, disturbance noise is identified by comparing measured (L_{mea}) and estimated immission spectra (L_{im}). Measured data are assumed to be WTN if the normalized curves of the immission spectra are similar between 400 Hz and 1600 Hz. Martens (2020) [8] identifies the spectrum of the WTN immission at rated speed and compares it with measured spectra. The procedure is similar to Conrady (2019) [2] and criterion (3), but the spectra are not normalized and equivalent sound pressure levels are used. In addition, background noise measured (L_{BGN}) is included in Martens (2020) [8]. A more comprehensive selection method that combines the criteria of Öhlund and Larsson (2015) [9], turbine-related criteria and spectral resemblance with further analysis techniques is also presented in Conrady (2019) [2].

In this paper, different approaches of identifying dominant WTN are compared with manually labelled data. The approaches and their selection criteria are summarized in Tab. 1. Note, that the method from Conrady (2019) [2] includes the criteria (1) - (3), rotational speed, wind speed and the spectral resemblance method. The more comprehensive selection method considering further analysis techniques is not part of the comparison. Moreover, the approach from Martens (2020) [8] has been further developed and adapted since 2020. Instead of equivalent sound pressure levels, single frequencies are compared. However, the approach is still referred as Martens (2020) [8]. In the evaluations, the site- and turbine-specific limiting values, such as the range of rotor speed and wind speed, were adjusted (see Tab. 1).

Table 1 Overview of approaches and selection criteria for the identification of dominant WTN.

Approach	Selection criteria
Van den Berg (2014) [15]	see (1)
Öhlund and Larsson (2015) [9]	see (1) - (3)
Conrady (2019) [2]	see (1) - (3) mean(abs($L_{\mathrm{im,norm}}(f) - L_{\mathrm{mea,norm}}(f)$)) < 5 dB for f = 125 to 2500 Hz rpm > 9 u < 9 m/s
Martens (2020) [8]	see (1) - (3) abs($L_{\mathrm{im}} - L_{\mathrm{mea}}$) < 5 dB for f = 125 to 2500 Hz rpm > 9 u < 9 m/s $L_{\mathrm{mea}} > L_{\mathrm{BGN}}$+3 dB

3.2. Classifying sound scenes using acoustical and turbine-related criteria

A rather simple approach to classify sound scenes in a wind farm is based on acoustical and turbine-related criteria, which are also presented in Sec. 3.1. The classes used and their selection criteria are summarized in Tab. 2. For the classification of audible WTN, the rotational speed is used as an indicator. Wind speed is chosen as an indicator for wind noise, as it clearly increases with higher wind speed. The limiting values for the rotor speed and for the wind speed are specific for the wind farm and its environment. Other wind farms may have different wind conditions, so that the values might have to be adjusted. The classification of bird sounds uses the measured frequency spectrum and the fact that those sounds are mostly in the high frequency range. Accordingly, birds are classified if the sound level measured increase between 3150 and 20000 Hz. The method of spectral resemblance introduced in Conrady (2019) [2] is used to classify non-defined disturbance noises in the frequency band of the wind turbine. Silence is identified, when the wind turbine is off and the measured equivalent sound pressure level is lower than 50 dB.

Table 2 Overview of approaches and selection criteria for the classification of sound scenes.

Classes	Selection criteria
Wind Turbine	Rotor speed between 7 and 13 rpm
Wind	Wind speed higher than 9 m/s
Bird	Sound level measured increase between 3150 and 20000 Hz
Silence	Rotor speed is lower than 7 rpm
	Measured equivalent sound pressure level is lower than 50 dB
Unidentifiable	mean(abs($L_{\text{im,norm}}(f) - L_{\text{mea,norm}}(f)$)) > 5 dB for f = 125 to 2500 Hz

3.3. Classifying sound scenes using a convolutional neural network

Utilizing deep learning based approaches such as convolutional neural networks (CNN) for sound scene classification gained popularity over the last years (see e.g. [5], [14]). The initial step, the providing of suitable training data, is the most time-consuming and crucial part of this procedure, while the prediction usually has high computational costs.

For the classification of different sounds within the soundscape, a multi-label CNN from the work of Poschadel et al. [10], [11] is used. As output, the neural network can discriminate between 12 classes as listed below:

- *Aircraft* - aeroplanes as well as helicopters
- *Bird* - all kind of bird calls, flapping of wings
- *Cricket* - chirr sounds from insects
- *Frog* - frog calls
- *Hiss* - high- or low-pitched noise from an unknown source (could be from wind turbine)
- *Machine* - periodic artificial tonal sounds (approx. 4500 Hz, generated by sonar)
- *Rain* - rainfall of different intensities
- *Silence* - absence of all sounds
- *Speech* - human speech

- *Traffic* - any land vehicles, such as tractors and agricultural machinery
- *Wind* - gusts of wind, howling, rustling of leaves, noise due to insufficient windscreens (high wind speeds)
- *Wind Turbine* - sound of operating wind turbine(s)

The CNN was trained with audio files from a different measurement campaign, which was conducted in a different windfarm in a similar environment. Additionally, for underrepresented classes, which are all but *Bird* and *Wind Turbine*, augmentation was applied to clips, that solely represented that class and no other, to enlarge the number of samples. For this, either pitch shifting (up or down by 1 or 2 semitones) or time stretching (speed between 0.9 and 1.1) was used, as these are common augmentation techniques for sound classification according to [13]. For more details on the architecture and design decisions, please refer to the work of Poschadel et al. in [10] and [11]. The neural network itself is not publicly available and was not developed for generalized use. With the current setup, the CNN is fed with Mel spectrograms of 0.5 s fragments of RMS-normalized audio and gives probability values in the range [0,1] for each class as output. A value of 0.5 was chosen as the prediction threshold, which means all probabilities equal to or above 0.5 will be seen as the class being *present* and any value below will be seen as the class being *absent*.
Post-processing of the prediction probabilities was applied in the form of rolling means, since all sound sources last longer than the temporal resolution given. This also smoothes the prediction probabilities and thus reduces outliers. A hamming window with a length of 5 s is used as a compromise between long-term sounds like *Wind Turbine* and short-term sounds like *Bird*.

4. Evaluation of Methods

To compare the prediction quality of the different methods, six 30 min audio recordings with varied acoustical characteristics according to Tab. 3 were chosen for evaluation. These recordings were labelled manually by one listener and cross-checked by two others to get a ground truth to compare against. Yet this does not mean that these labels are perfectly correct, especially since no fixed windows were labelled but rather the start and end of a section in which a certain sound could be heard. For example, one person might label a far away aeroplane while another would not as it is barely noticeable. For the evaluation, the start and end points of the labelled sections were rounded to full minutes, as the data used for the classification method described in Sec. 3.2 only had a resolution of 1 min.

4.1. Evaluation Metrics

To evaluate the goodness of the predictions, each recording is analysed separately. First, the evaluation of identifying dominant WTN is presented in Sec. 4.2, comparing the four different methods presented in Sec. 3.1. Then the classification of sound scenes using acoustical and turbine-related data is analysed in Sec. 4.3. Finally, the results of the CNN are presented in Sec. 4.4. Here, each class c, that can be predicted by the network, is evaluated on its own.
For each class TP (true positives - an occurrence was predicted and is present), TN (true negatives - no occurrence was predicted and is not present), FP (false positives - an occurrence was predicted but is not present), and FN (false negatives - no occurrence was predicted but is present) are counted in 60 s intervals for each recording individually.
Accuracy (Acc) describes the ratio of correctly predicted labels (both positive and negative) to all

Table 3 Selected audio recordings for evaluation. All times are in UTC.

ID	Date & Time	Wind speed	Active wind turbine	mean SPL	Sound scene description
1 - low rpm	4.4. 17:30	4-5 m/s	WT1, WT2, WT3	30-40 dB(A)	WT noise, bird calls
2 - pure WTN	5.4. 01:00	5-6 m/s	WT1, WT2, WT3	> 40 dB(A)	WT noise
3 - noisy birds	5.4. 06:30	7-8 m/s	WT1	> 40 dB(A)	WT noise, lots of bird calls
4 - calm	7.4. 19:30	2-3 m/s	none	< 30 dB(A)	calm, only background noise
5 - aeroplane	14.4. 13:30	8-11 m/s	WT1 (with shutdown)	> 40 dB(A)	WT noise, bird calls, aeroplane, noticeable - high wind noise
6 - high wind	24.4. 12:00	10-11 m/s	WT1	> 40 dB(A)	high wind noise

labels. However, the significance of accuracy is limited in case of the CNN as the number of (non-)occurrences per class is imbalanced, e.g. the class *Wind Turbine* being present most of the time in five of six examples, but the class *Aircraft* is only present briefly in the end of one example.

$$Acc_c := \frac{\text{TP} + \text{TN}}{\text{TP} + \text{TN} + \text{FP} + \text{FN}}$$

Precision (Prec) describes the relation of correct positive predictions to all positive predictions, so the ratio of how much of the predicted is correct.

$$Prec_c := \frac{\text{TP}}{\text{TP} + \text{FP}}$$

Recall (Rec) describes the relation of correct positive predictions to the sum of positive labels, so the ratio of how many of the actual positive labels were predicted.

$$Rec_c := \frac{\text{TP}}{\text{TP} + \text{FN}}$$

The F1-score reports the harmonic mean of precision and recall, which both ideally should be high.

$$F1_c := \frac{2 \cdot Prec_c \cdot Rec_c}{Prec_c + Rec_c}$$

In our evaluation, the definition of Röder et al. [12] is used, such that precision and recall get a defined value even if the class is not labelled as ground truth at all for one recording. If TP, FP, and FN are all 0 for a class, then $Prec_c = Rec_c = F1_c = 1$,. If it is only valid that the class is not labelled as ground truth (TP $= 0$), but it was detected nevertheless (FP > 0), then $Prec_c = Rec_c = F1_c = 0$.

4.2. Evaluation of identifying dominant WTN

In order to evaluate the methods for selecting dominant WTN, their results are compared with self-labelled data. Three classes were used for labelling:

- *dominant WTN,*
- *dominant WTN with nearly inaudible disturbance noise*, and
- *dominant WTN with audible disturbance noise.*

The classification of whether WTN is dominant or not is highly subjective. It is clearly pointed out that the results can only provide general findings. The accuracy and F1-score are given per method and for each recording in Tab. 4. Herein, the calculations are related to the class *dominant WTN*. Since precision and recall do not provide any additional information in this comparison, they are not given in this section. In Fig. 2, the comparison of the predicted and labelled data for examples 1 to 4 is shown. Examples 5 and 6 are shown in the appendix A (Fig. 6), since they provide no specific information. These examples are predominantly characterized by high wind noise, which is detected as disturbance noise by all methods. The accuracy and F1-score for examples 5 and 6 is therefore 100 % (Tab. 4).

Example 2 is characterized by dominant WTN. The rotor speed is high, while no disturbing noises are audible in the recordings. All methods classified the data correctly. Due to small rotor speeds, the WTN in example 1 is lower. For two sections, dominant WTN with audible disturbance noise (in this case birds) was labelled. Using the methods given by van den Berg (2014) [15] and Öhlund and Larsson (2015) [9], dominant WTN is classified more often. The accuracy with respect to dominant WTN is 53.33 and 56.67 %, and the F1-score is 0 %. Since the rotor speed is taken into account in Conrady (2019) [2] and Martens (2020) [8], no dominant WTN is classified with these methods. Wind turbine and especially bird sounds characterize example 3. The two classes *dominant WTN with nearly inaudible disturbance noise* and *dominant WTN with audible disturbance noise* were labelled. The methods presented by van den Berg (2014) [15] and Öhlund and Larsson (2015) [9] classify dominant WTN. For both methods, the accuracy with respect to dominant WTN is 46.67 % and F1-score is 0 %. The methods according to Conrady (2019) [2] and Martens (2020) [8] are stricter and have stronger selection criteria, so that very few periods are classified as dominant noise. Example 4 is described by a very calm sound scene with a few bird sounds, so that no dominant WTN was labelled. With the methods of Conrady (2019) [2] and Martens (2020) [8], no dominant WTN is identified. The 'Van den Berg'-method, on the other hand, classifies many periods with dominant WTN (Acc=20 %. F1=0 %). In case of no or only small WTN, it is difficult to select only on the basis of statistical levels. It has already been mentioned in van den Berg that the criterion may meet small sound levels, even though the sound levels do not only emit from the wind turbine but also from the environment. In this work, the statistical levels were determined for 1min-intervals and used for the selection. It might be better to choose a longer period, such as van den Berg (2014) [15] (5 min period) and Öhlund and Larsson (2015) [9] (10 min period).

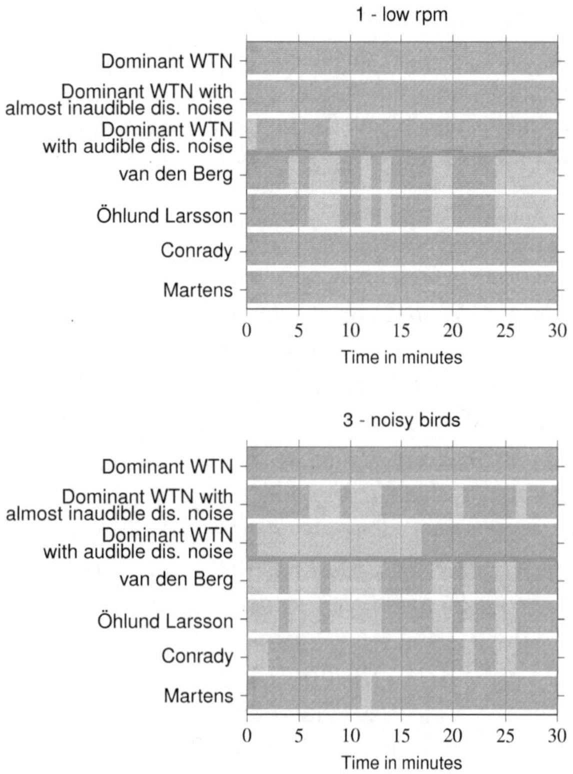

Fig. 2 Manual and methodical labelling results for selecting dominant WTN. Top row: recording 1, 2; bottom row: recording 3, 4. An orange section stands for the sound being present, and a blue section for it not being present (for the majority of the minute.). "Dis. noise" is the abbreviation of disturbance noise.

Table 4 Accuracy (F1-score) of recording 1 to 6 in percent.

Recording No. / Approach	1	2	3	4	5	6
Van den Berg (2014)	53.33 (0)	100 (100)	46.67 (0)	20 (0)	100 (100)	100 (100)
Öhlund & Larsson (2015)	56.67 (0)	100 (100)	46.67 (0)	93.33 (0)	100 (100)	100 (100)
Conrady (2019)	100 (100)	100 (100)	83.33 (0)	100 (100)	100 (100)	100 (100)
Martens (2020)	100 (100)	100 (100)	96.67 (0)	100 (100)	100 (100)	100 (100)

4.3. Evaluation of classifying sound scenes using acoustical and turbine-related criteria

Using acoustical and turbine-related criteria, the classes *Wind Turbine*, *Wind*, *Silence*, *Bird* and *Unidentifiable* are addressed. For the reference of the class *Unidentifiable*, all self-labelled classes except *Wind Turbine*, *Wind*, *Silence* and *Bird* were combined. The reference accordingly includes the classes *Aircraft*, *Frog*, *Machine*, *Rain* and *Traffic*. The results of the comparison are shown in Fig. 3. The evaluation measures accuracy and F1-score are listed in Tab. 5. The measures precision and recall are given in appendix B (Tab. 9).

The class *Wind Turbine* is predicted very well for all examples. Only in example 6, the F1-score is with 96.55 below 100 %. Here, very high wind noise masks the WTN. This is not detected with the single indicator 'rotor speed'.

At high (example 6) and low wind speeds (example 1-3), the class *Wind* is also predicted well (F1 > 98.31 %). At moderate wind speeds (example 5), which are below the defined limit of 9 m/s, the F1-score is 0 %. With optimized limits, better results might be achieved.

If WTN is present, the class *Silence* is predicted correctly. Only in example 4, the class *Silence* was predicted by the manual and methodical labelling. Here, the F1-score is 89.80 %.

The class *Bird* is predicted for all examples. The F1-scores of examples 1 and 3 to 6 are between 86.27 and 100 %. In example 2, no bird sounds are present, but they were predicted to 100 % (F1 = 0 %). The criterion for classifying bird sounds must be optimized. In a first step, the frequency range can be adjusted. Possibly, the criterion should also be changed. There are several possibilities here. A peak in the high frequency range could be an indicator for bird sounds. Furthermore, the energy components in the high frequency range could be considered. However, there is a risk of confusion with wind noise.

The F1-scores of the class *Unidentifiable* are between 0 and 63.64 %. The comparison between the immission spectrum at rated rotor speed and the measured spectra leads to errors at low rotor speeds. If the wind turbine doesn't operate at rated speed, the measured noise is classified as *Unidentifiable*. In addition, the comparison of the spectra takes place between 125 and 2500 Hz. Accordingly, wind noise or individual birds could also be classified as *Unidentifiable*.

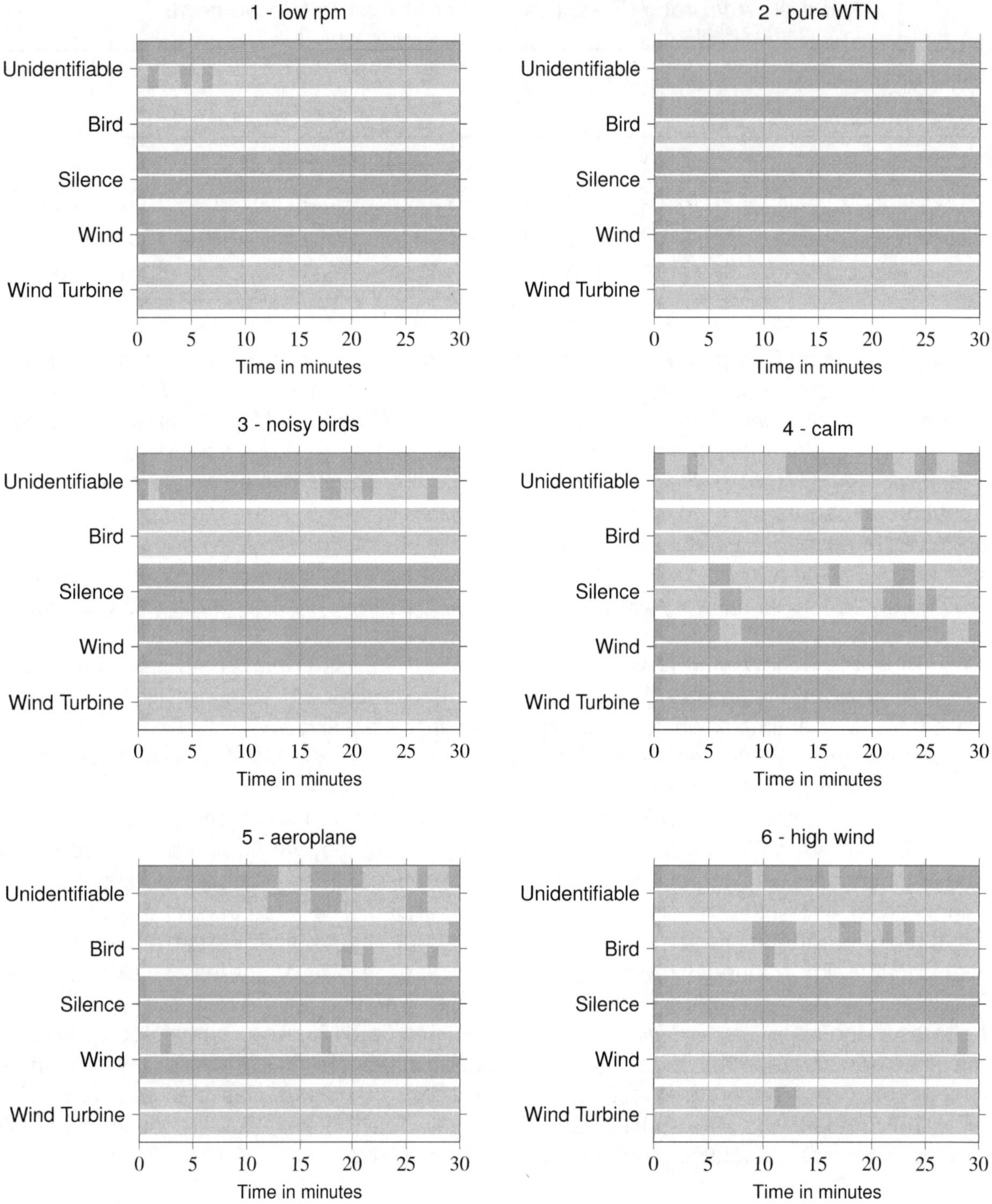

Fig. 3 Manual and methodical labelling results for all examples and selected classes. Top row: recording 1, 2; middle row: recording 3, 4; bottom row: recording 5, 6. The top row of a class shows the result of hand labelled and the bottom row the methodical labelled data. An orange section stands for the sound being present, and a blue section for it not being present (for the majority of the minute.)

Table 5 Accuracy (F1-score) of recording 1 to 6 in percent.

Recording No. / Class	1	2	3	4	5	6
Unidentifiable	10 (0)	96.67 (0)	60 (0)	46.67 (63.64)	40 (43.75)	10 (18.18)
Bird	100 (100)	0 (0)	100 (100)	96.67 (98.31)	86.67 (92.86)	76.67 (86.27)
Silence	100 (100)	100 (100)	100 (100)	83.33 (89.8)	100 (100)	100 (100)
Wind	100 (100)	100 (100)	100 (100)	86.67 (0)	6.67 (0)	96.67 (98.31)
Wind Turbine	100 (100)	100 (100)	100 (100)	100 (100)	100 (100)	93.33 (96.55)

4.4. Evaluation of the CNN

As the prediction results have a resolution of 0.5 s, but the evaluation is done in 1 min segments as a compromise to compare the methods, it is checked whether the median of one such segment is greater than the prediction threshold of 0.5. If that is the case, the class is deemed to be predicted for that segment. The results for accuracy, precision, recall and F1-measure per class per recording can be studied in Tab.s 6, 7, and 8. A visualization of the prediction intervals can be seen on the left side of in Fig. 5 for examples 1, 2, and 4. The figures of the other three examples can be found in the appendix C (Fig. 7).

Values for the sound sources *Cricket*, *Hiss*, and *Speech* are omitted in all tables as these sounds do not appear in any of the examples and are in all cases correctly not predicted either. The values for *Machine* are also ignored, as this class was very specific for the training data and does not appear in that form in our data. In our data, a different kind of high-pitched noise, most likely the movement of the nacelle, was labelled *Machine*, but as the net was trained on a different tone, it is never predicted.

Table 6 Evaluation measures of recording 1 (left) and 2 (right) in percent.

Class	Acc	Prec	Rec	F1
Aircraft	100	100	100	100
Bird	73.33	100	73.33	84.62
Frog	66.67	0	0	0
Rain	100	100	100	100
Silence	100	100	100	100
Traffic	100	100	100	100
Wind	100	100	100	100
Wind Turbine	100	100	100	100

Class	Acc	Prec	Rec	F1
Aircraft	53.33	0	0	0
Bird	100	100	100	100
Frog	100	100	100	100
Rain	100	100	100	100
Silence	100	100	100	100
Traffic	100	100	100	100
Wind	100	100	100	100
Wind Turbine	3.33	100	3.33	6.45

There are some false positives of the class *Frog* for examples 1 and 3 while its non-occurrence is otherwise correctly predicted. When the data with finer granularity is consulted, it seems like these

Table 7 Evaluation measures of recording 3 (left) and 4 (right) in percent.

Class	Acc	Prec	Rec	F1
Aircraft	100	100	100	100
Bird	86.67	100	86.67	92.86
Frog	86.67	0	0	0
Rain	100	100	100	100
Silence	100	100	100	100
Traffic	100	100	100	100
Wind	100	100	100	100
Wind Turbine	100	100	100	100

Class	Acc	Prec	Rec	F1
Aircraft	100	100	100	100
Bird	3.33	0	0	0
Frog	100	100	100	100
Rain	60	0	0	0
Silence	16.67	0	0	0
Traffic	93.33	0	0	0
Wind	86.67	0	0	0
Wind Turbine	40	0	0	0

Table 8 Evaluation measures of recording 5 (left) and 6 (right) in percent.

Class	Acc	Prec	Rec	F1
Aircraft	80	100	33.33	50
Bird	33.33	100	31.03	47.37
Frog	100	100	100	100
Rain	100	100	100	100
Silence	100	100	100	100
Traffic	100	100	100	100
Wind	16.67	100	10.71	19.35
Wind Turbine	83.33	100	83.33	90.91

Class	Acc	Prec	Rec	F1
Aircraft	100	100	100	100
Bird	56.67	100	40.91	58.06
Frog	100	100	100	100
Rain	100	100	100	100
Silence	100	100	100	100
Traffic	100	100	100	100
Wind	36.67	100	34.48	51.28
Wind Turbine	80	100	78.57	88

FPs are misclassifications of *Bird* as depicted in Fig. 4.
Predicting *Wind* seems to be challenging for the CNN with F1 values between 0 and 51.28 % for example 4-6, where noticeable wind is present. However, the spectrum of wind noise in these examples is rather broad, with faint rustling of grass (4) up to such high noise that the audio is clipping (6). Probably the training data is not fitted for this purpose. In [11] *Wind* showed a rather low precision in the test set and the assumption was made that a confusion of *Wind* and *Wind Turbine* is present. This does not show in our examples in the given resolution.
Detecting *Wind Turbine* sound in the soundscapes works well for most examples when it is actually present. The precision values for all 5 cases are at 100 %, the recall lies between 78.57 % and 100 %. Only in example 2 with a lack of other background noise, the recall is significantly lower, with only 3.33 %. This example instead shows a high FP rate of the class *Aircraft*. A confusion of these two classes was also already observed in the previous publication [11]. This could originate from human misclassification in the initial training data, as the sounds can be hard to distinguish.

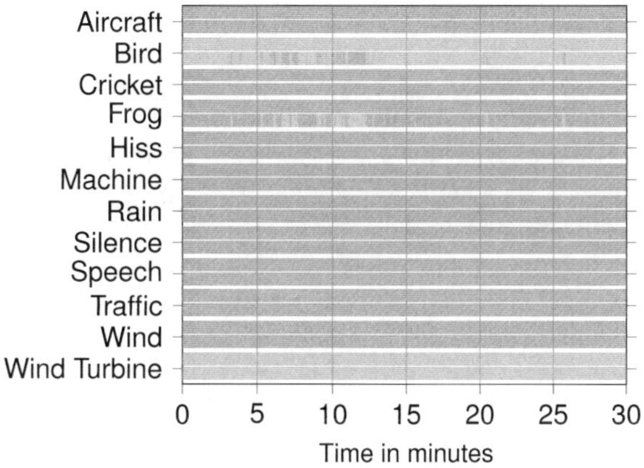

Fig. 4 Prediction results of example 3 in the original prediction resolution of 0.5 s. It shows that *Frog* is likely to be predicted when the classification of *Bird* drops below 0.5. The top row per class shows the hand labelled ground truth data, the bottom row the result of the CNN. An orange section stands for the sound being present, and a blue section for it not being present.

The predicting capabilities of *Bird* range between F1 values of 0 % and up to 92.86 % for all examples (except 2, where no bird calls can be heard). The precision is 100 % in all cases except example 4, while the recall lies significantly lower. While example 1 and 3 likely show a confusion with *Frog*, this is not the case for example 4-6. However, the labelling of bird calls in the ground truth data was taken very liberally in terms of not marking each individual call but rather longer time periods where birds were active. Thus, also a significant amount of times without birds might be marked as such. A higher resolution may show better classification results.

Rain is only present in the example 4, but not predicted. The false negatives might, once again, be due to the training data. In it, comparatively heavy rain was used, while in the present example only single drops can be heard.

The class *Traffic* is also only present in example 4, but does not get predicted at all. However, the sound is only faintly audible and may have not been labelled by a different person.

Predicting actual occurrences of *Aircraft* showed F1 values of 100 % respectively 50 % for examples 4 and 5. The segments of TPs have in common, that other noise sources were less present or the *Aircraft* noise was very prominent. The more distant sounds in example 5 were not detected.

While the non-occurrences are predicted fine, *Silence* is not predicted at all in the case of the calm example (4). This instead shows high misclassifications of *Wind Turbine*.

Further investigation of example 4 - calm sound scene misclassifications

The previously stated bad prediction results for example 4 were further investigated. This example is representative of calm periods during night, when not much sound can be heard. Thus, the overall sound pressure level is rather low. If normalization is applied to such sound, even the faintest background noise becomes very loud. To investigate the influence of normalization, the spectrograms were fed into the net without applying normalization to the audio data first. This produced significantly better results for the classification of *Silence* and removes all false positives of *Wind Turbine* in case of example 4. For example 2, the F1 value of *Wind Turbine* also increased significantly. The FP of

Aircraft disappeared, but even more FP of *Frog* are introduced instead. In case of example 1, which has the second lowest mean SPL value, the prediction results of *Aircraft*, *Bird* and *Wind Turbine* became worse than before. A confusion of *Aircraft* and *Wind Turbine* happened here most likely.
A comparison of the prediction results with and without using normalization of the audio snippets is shown in Fig. 5 for the examples 1, 2, and 4. For the other three examples, the overall prediction rates stay about the same for both cases.

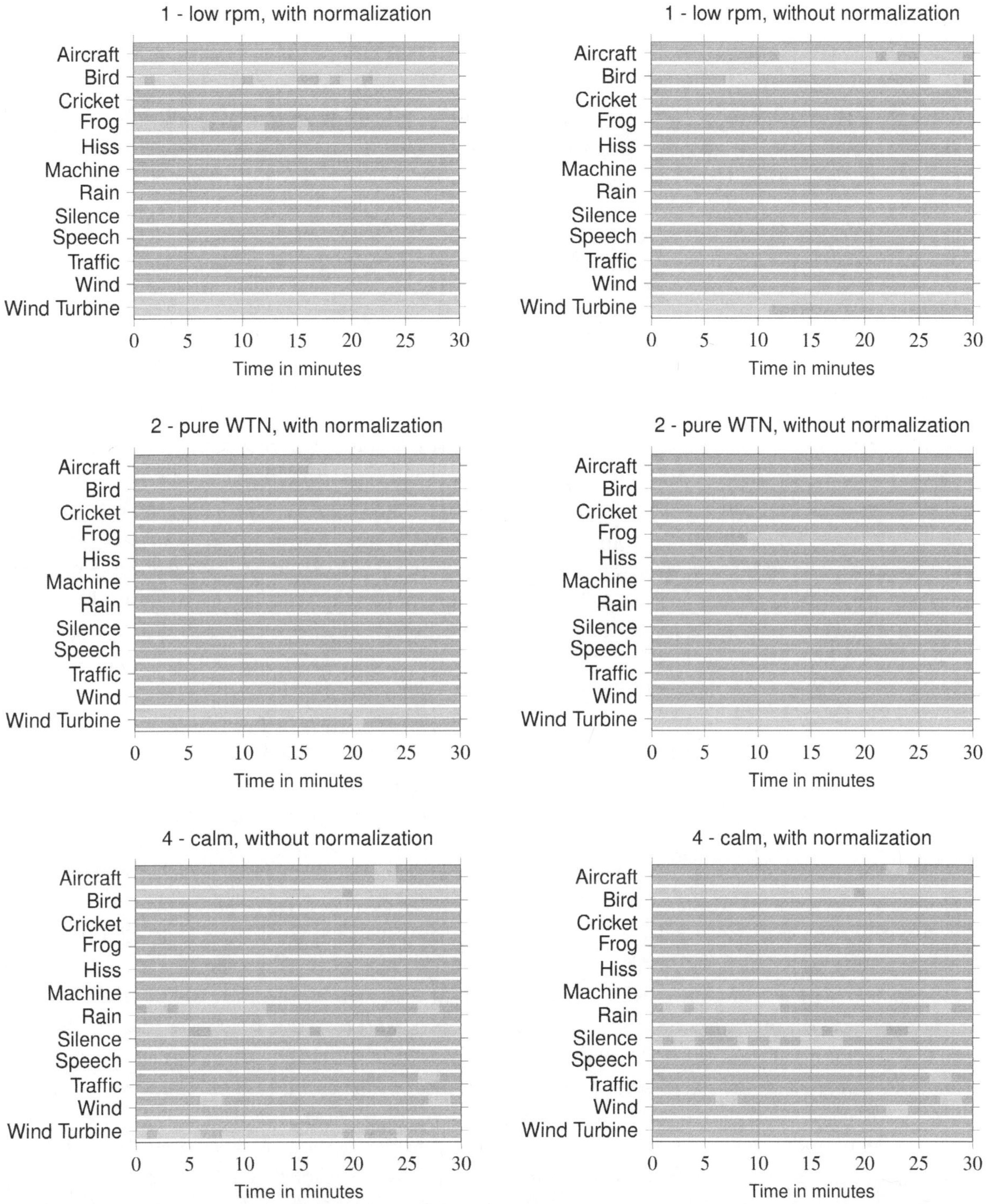

Fig. 5 Predictions of (from top to bottom) example 1, 2, and 4. On the left normalization was applied to the audio, on the right this step was skipped. The top row per class shows the hand labelled ground truth data, the bottom row the result of the CNN. An orange section stands for the sound being present, and a blue section for it not being present (for the majority of the minute.)

5. Conclusion and Discussion

For the identification of dominant WTN, methods from van den Berg (2014) [15], Öhlund and Larsson (2015) [9], Conrady (2019) [2] and Martens (2020) [8] were tested and validated. The methods from Conrady (2019) [2] and Martens (2020) [8] show a good agreement with the labelled data. If the WTN is barely audible or non-existent, predictions with statistical levels, such as in van den Berg (2014) [15], are inconsistent. Nevertheless, all methods show good results regarding wind noise and recognize it as disturbing noise. The choice of method identifying dominant WTN depends on the data situation, the objective of the study and the required strictness of the selection criteria.

To classify sound scenes in a wind farm, two methods are tested that differ in complexity. The first method uses simple approaches such as limiting the rotor speed and the wind speed. With those criteria, the classes *Wind Turbine*, *Wind* and *Silence* are well predicted in the sample data set. An adjustment of the criteria identifying bird noise and other disturbing noises (*Unidentifiable*) is necessary. In the authors' opinion, this simple approach will give less good results with data recorded at a greater distance to the turbine.

The second method uses a CNN, which was pretrained on acoustical data from a similar setting, to classify sound scenes. The predictions showed discrepancies between the different detectable classes, but also between the chosen examples. The prediction of wind turbine noise works well with F1 scores of over 88 %, if wind turbine noise and other background noise is actually present. A suggestion for usage would be to take information from the SCADA data of the nearest wind turbine into account to pre-select time windows when the turbine is actually running. Additionally, the SPL value could be taken into account for detecting likely periods of *Silence*. Even though rerunning the prediction without normalization showed improvements in the predictions of some classes, the CNN was not trained with non-normalized data, and it is thus not guaranteed that this will work reliably. In terms of other sound sources, the results of the neural network showed that dominant *Aircraft* noise is detected with a F1 score of 100 % and in case of a mix of near and distant *Aircraft* sound with more background noise with a F1 score of 50 %. The prediction of bird calls, which is a very common sound in the given soundscape, show big variances in the F1 scores of the different examples, ranging from 0 % to 92.86 %. Anyhow, this might be due to the comparably coarse time windowing of both the hand-labelling and the prediction. The investigations of Poschadel et al. [11], which were conducted with smaller time windows, showed with 80 % on their test set a better overall F1 score for *Bird*. Thus, better results might be achieved with a finer timescale on our data. A further evaluation measure could also be to check the results against detections from BirdNet [6]. Finally, not all of the original classes of the CNN proved to be useful in another scenario. For example, the class *Machine* is unlikely to appear at different locations. Other likely more common sounds, like the movement of the nacelle, can not be predicted at all with the current setup. Further, an uninformed user might misinterpret the given class labels. Some classes like *Wind* or *Rain* might have a too broad spectrum of noises and could improve by having different sub-classed based on their intensity. However, it was not planned by Poschadel et al. to apply the CNN to a different scenario at all and for this the reusing for this measurement campaign showed overall comparable results for most of the present classes.

Summing up, the investigated methods show differences in both, results and complexity. The simpler method (see Sec. 3.2), which uses averaged and statistical levels as well as SCADA data to classify sounds, needs little computing time and capacity and thus can be easily applied even to big amounts of data. The criteria used can be easily changed if other environmental conditions or different types of wind turbines are given. This method is useful if a fast and brief description of a sound scene is sufficient. A limited use of this method is also possible even if the time signal is not given. The CNN (see Sec. 3.3) takes significantly more time and computational resources for predicting the sounds in a given audio file. In return, the net provides data in a much higher resolution and can distinguish between more and different kinds of sounds. Nevertheless, without putting in the time and effort to retrain the neural network, the predictions are limited to the trained classes, which might not be suitable for all surroundings.

Possible uses for the classification of different sounds are to investigate the occurrence of sounds statistically and also the dependency on specific environmental conditions, e.g. bird calls happening at specific daytimes. It can also be used to get a brief description of the contents of a sound scene either to get hints on which time windows to avoid or to use, depending on the requirements. This could be used to preselect audio data for listening studies. Combining the classification with the identification of dominant wind turbine noise, the emission and immission of wind turbines can be investigated easily. In the end, it is a matter of the data at hand and the goal of the investigation, which method to choose.

Acknowledgement

Within in the project "WEA-Acceptance" (project ref. no. 0324134A) and "WEA-Acceptance Data - Database for benchmarking & validation" (project ref. no. 03EE3062), the research at Leibniz University of Hannover was funded by the Federal Ministry for Economic Affairs and Climate Action by an act of the German Parliament. The Institute of Communications Technology and the Institute of Structural Analysis are part of the Center for Wind Energy Research For-Wind.

References

[1] Bolin, K. (2009). Prediction method for wind-induced vegetation noise. *Acta Acustica united with Acustica*, 95(4):607–619.

[2] Conrady, K. (2019). *Wind Turbine Sound in Cold Climates*. Dissertation, Uppsala: Acta Universitatis Upsaliensis.

[3] Conrady, K., Bolin, K., Sjöblom, A., and Rutgersson, A. (2020). Amplitude modulation of wind turbine sound in cold climates. *Applied Acoustics*, 158.

[4] Conrady, K., Sjöblom, A., and Larsson, C. (2018). Impact of snow on sound propagating from wind turbines. *Wind Energy*, 21(12):1282–1295.

[5] Hershey, S., Chaudhuri, S., Ellis, D. P., Gemmeke, J. F., Jansen, A., Moore, R. C., Plakal, M., Platt, D., Saurous, R. A., Seybold, B., et al. (2017). CNN architectures for large-scale audio classification. In *2017 ieee international conference on acoustics, speech and signal processing (icassp)*, pages 131–135. IEEE.

[6] Kahl, S., Wood, C. M., Eibl, M., and Klinck, H. (2021). Birdnet: A deep learning solution for avian diversity monitoring. *Ecological Informatics*, 61:101236.

[7] Könecke, S., Hörmeyer, J., Bohne, T., and Rolfes, R. (2022). A new base of wind turbine noise measurement data and its application for a systematic validation of sound propagation models. *Wind Energy Science [preprint]*.

[8] Martens, S., Bohne, T., and Rolfes, R. (2020). An evaluation method for extensive wind turbine sound measurement data and its application. *Proceedings of Meetings on Acoustics, Acoustical Society of America*, 41.

[9] Öhlund, O. and Larsson, C. (2015). Meteorological effects on wind turbine sound propagation. *Applied Acoustics*, 89:34–41.

[10] Poschadel, N., Gill, C., Preihs, S., Martens, S., Bergner, J., Rolfes, R., and Peissig, J. (2020). Machine Learning basierte Klassifikation von Außenschallszenen für Lärmmessungen an Windenergieanlagen. In *Fortschritte der Akustik - DAGA 2020, 46. Jahrestagung für Akustik*, Hannover.

[11] Poschadel, N., Gill, C., Preihs, S., and Peissig, J. (2021). CNN-based multi-class multi-label classification of sound scenes in the context of wind turbine sound emission measurements. In *Proceeding of the INTERNOISE 2021*.

[12] Röder, M. and et al. (2015-2017). Precision, Recall and F1 measure - Dividing by 0. URL: https://github.com/dice-group/gerbil/wiki/Precision,-Recall-and-F1-measure#dividing-by-0. Online; last access 03-04-2023.

[13] Salamon, J. and Bello, J. P. (2017). Deep convolutional neural networks and data augmentation for environmental sound classification. *IEEE Signal processing letters*, 24(3):279–283.

[14] Shrivastava, H., Yin, Y., Shah, R. R., and Zimmermann, R. (2020). Mt-gcn for multi-label audio-tagging with noisy labels. In *ICASSP 2020-2020 IEEE International Conference on Acoustics, Speech and Signal Processing (ICASSP)*, pages 136–140. IEEE.

[15] van den Berg, G. P. (2004). Effects of the wind profile at night on wind turbine sound. *Journal of Sound and Vibration*, 277(4-5):955–970.

Appendix A - Evaluation of identifying dominant WTN

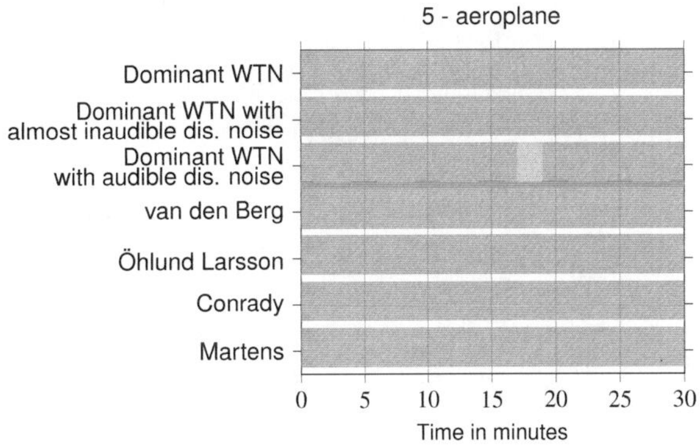

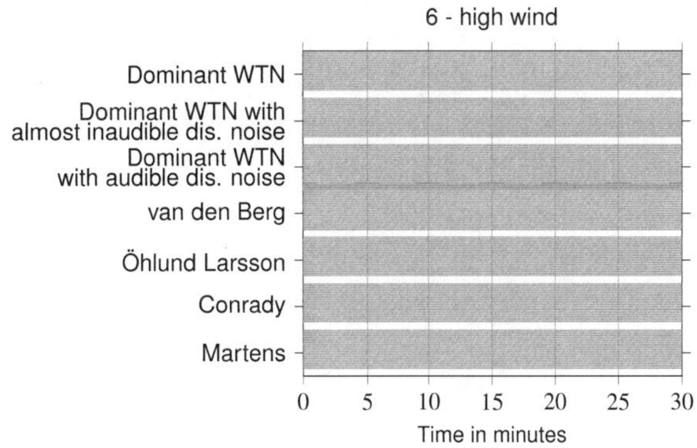

Fig. 6 Manual and methodical labelling results for selecting dominant WTN - Examples 5 and 6. An orange section stands for the sound being present, and a blue section for it not being present (for the majority of the minute.). ***Dis. noise*** **is the abbreviation of disturbance noise.**

Appendix B - Evaluation of classifying sound scenes using acoustical and turbine-related criteria

Table 9 Precision (Recall) of recording 1 to 6 in percent.

Recording No. / Class	1	2	3	4	5	6
Unidentifiable	0 (0)	0 (0)	0 (0)	46.67 (100)	31.82 (70)	10 (100)
Bird	100 (100)	0 (0)	100 (100)	96.67 (100)	96.30 (89.66)	75.86 (100)
Silence	100 (100)	100 (100)	100 (100)	91.67 (88)	100 (100)	100 (100)
Wind	100 (100)	100 (100)	100 (100)	0 (0)	0 (0)	96.67 (100)
Wind Turbine	100 (100)	100 (100)	100 (100)	100 (100)	100 (100)	93.33 (100)

Appendix C - Evaluation of the CNN

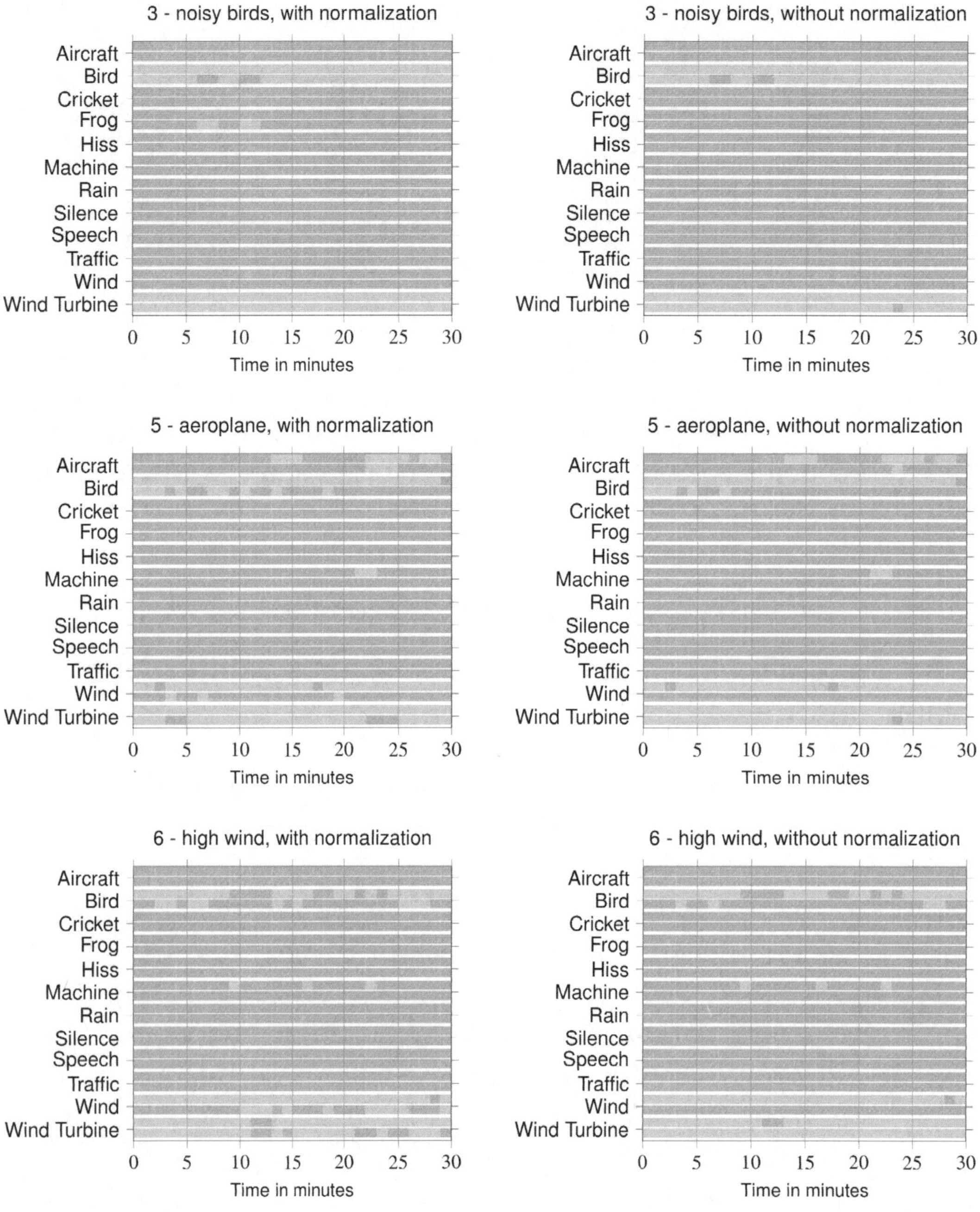

Fig. 7 Predictions of (from top to bottom) example 3, 5, and 6. On the left normalization was applied to the audio, on the right this step was skipped. The top row per class shows the hand labelled ground truth data, the bottom row the result of the CNN. An orange section stands for the sound being present, and a blue section for it not being present (for the majority of the minute.)

INCEEUROPE

10th International Conference
on
Wind Turbine Noise
Dublin – 21st to 23rd June 2023

Numerical Study on the Reduction of Blunt Trailing Edge Noise by the Use of Vortex Generators

F. Seel, L. Blazejewski, T. Lutz, E. Krämer
University of Stuttgart, Institute of Aerodynamics and Gas Dynamics, Germany

Summary

The size of horizontal axis wind turbines will continuously increase leading to new challenges in aerodynamics and aeroacoustics. One of these challenges is the design of the blade root section: For structural reasons, a very high blade thickness in the root area is necessary which results in strong adverse pressure gradients and potentially flow separation. In order to delay these separations, so-called flatback airfoil shapes can be used. The flat back shape decreases the strong adverse pressure gradients. In the wake behind the blunt trailing edge, as the airfoils are bluff bodies, a von Kármán vortex street forms and induces strong pressure fluctuations on the surface of the blunt trailing edge. These periodic fluctuations are the reason for blunt trailing edge noise. The following work shows that very cheap and easily retrofittable vortex generators are able to reduce these noise emissions. To investigate this effect, Delayed Detached Eddy Simulations on an extruded DU97-W-300 flatback airfoil are conducted. From an aerodynamic point of view, it is shown that vortex generators placed at a relative chordwise position of 20% or 45% have a positive effect on the boundary layer velocity distribution through a momentum transfer from the upper parts of the boundary layer towards the lower parts. In contrast, vortex generators placed at a chordwise position of 70% reduce slightly the velocity inside the boundary layer compared to the case without vortex generators. By means of Proper Orthogonal Decompositions of the flow fields, it is demonstrated that the very dominant first mode, arising from the von Kármán vortex street, is reduced by up to 15%, for the chordwise vortex generator position at 20%, and redistributed into higher modes by the effect of the vortex generators. Lastly, with regards to aeroacoustics, the use of a Ffowcs-Williams and Hawkings Code showed that the reduction of the tonal hump resulting from blunt trailing edge noise is reduced by up to 7 dB when using vortex generators.

1. Introduction

In the context of growing size of modern horizonal axis wind turbines (HAWT), new multi-disciplinary design challenges emerge. Regarding the large blade size, high relative airfoil thicknesses are required in the root section for structural reasons. In order to increase the aerodynamic performance, the airfoil shape of those sections has to be designed with the aim to maintain attached flow conditions. This is only achievable through large chord lengths because a certain amount of curvature (i.e. adverse pressure gradient) cannot be exceeded. For onshore HAWT, the overland transport sets limits to the chord length and for this reason, so-called flatback airfoils, designed with a Blunt Trailing Edge (BTE), are often used. Through this shape an increase of the lift curve slope and maximum lift coefficient compared to the closed airfoil with the same relative thickness is obtained [1]. As a further consequence, the large displacement leads to an increase in drag compared to the closed airfoil and as a part of the pressure recovery takes place behind the airfoil a von Kármán vortex street forms. A detailed review on the formation mechanism of the vortex street for flatback airfoils can be found in [2].

Beside of the drag increase, the alternating vortices of the vortex street in the near wake produce pressure fluctuations on the BTE leading to tonal noise emissions of dipol character known as Blunt Trailing Edge Noise (BTEN) [3]. The noise character of BTEN is dependent on the height of the blunt trailing edge h_{BTE} and the displacement thicknesses δ_1 on both sides of the airfoil. As soon as h exceeds $\delta_1/4$, a distinct hump appears in the spectrum and, as h/δ_1 increases further, the bandwidth of the peak narrows [4]. The BTEN tonality appears at the frequency of the vortex shedding (VS) f_{VS}, and can be characterised by the Strouhal number of the flatback. In contrast to the well-established Strouhal number for cylinders, for lifting surfaces the viscous boundary layer regions must also be considered in the height term [4]. Consequently, the Strouhal number for flatback airfoils is defined as:

$$St = \frac{f_{VS} \cdot (h_{BTE} + \delta_{1,SS} + \delta_{1,PS})}{U_\infty} \tag{1}$$

where h_{BTE} corresponds to the height of the BTE, δ_1 to the displacement thickness on the suction side and the pressure side at the trailing edge and U_∞ to the inflow velocity. The usual range of St is at 0.23 to 0.24 [4].

Regarding BTEN mitigation, different approaches have been investigated, all targeting the reduction in amplitude of the vortex street. Blake [4] summarized a multitude of studies and brought together results for different BTE shapes in order to reduce BTEN. He concluded that wedge-like extents of the BTE deliver the largest noise reduction. Yang et al. [5] investigated wavy trailing edges and concluded that the reduction in tonal noise is closely connected to the reduction in aerodynamic drag. Another mitigation technique, based on a splitter plate fixed at the middle of the BTE, was investigated experimentally [1] and numerically [6, 7] and shows significant tonal noise reduction by undermining the VS.

In the present work, the flatback version of the DU97-W-300 airfoil, proposed by Barone et al. [6], is evaluated by means of computational fluid dynamics (CFD) regarding the impact of vortex generators (VGs) on the BTEN. VGs are vane-like passive devices placed on the surface of the blade within a certain inclination angle to the local inflow. Hence, they create streamwise vortices which allow a momentum transfer inside the boundary layer (BL). Traditionally, VGs are employed to delay stall onset on airfoils with closed trailing edge, as they allow the flow to overcome higher adverse pressure gradients than without VGs. According to the experimental investigations of Baldacchino et al. [8] on the closed trailing edge version (DU97-W-300) of the airfoil considered in this work, the VGs should be placed in chordwise direction at about 20% of the chord length which corresponds to the beginning of the main pressure recovery on the airfoil suction side. In this work, the VGs are not aimed to reduce separation but to decrease the BTEN. Hence, the impact of the VGs will be evaluated at three different chordwise positions (20%, 45% and 70%). To the best knowledge of the authors, VGs have not yet been investigated regarding their impact on BTEN, or even more generally, regarding their impact on the von Kármán vortex street of a flatback airfoil. Thus, the following objectives are stated for this work:

- Provide insight into the interaction of streamwise convecting VG vortices and the von Kármán type VS of the flatback airfoil
- Quantify the influence of the VG placement in chordwise direction on lift, drag, von Kármán VS and blunt trailing edge noise

In the second chapter, the numerical methods are presented including the numerical setup and a short description of the aerodynamic and acoustic solvers. The third chapter presents a validation against experimental data. In the fourth chapter the results of the studies of the case without VGs and with the three different VG positions are presented, first in terms of aerodynamics, under various aspects such as lift and drag coefficients, BL velocity profiles, Reynolds stresses and a modal analysis through a snapshot Proper Orthogonal Decomposition (POD). After the aerodynamic results, the sound pressure level spectrum for all cases, and thus the influence of the VGs on the BTEN, is presented.

Table 1: Summary of the geometrical parameters of the VGs. The parameters are taken from Baldacchino et al. [9] and were also used in Seel et al. [10]. "Ctr. CD": Counterrotating Common Down flow positioning of the VGs in the pair.

Number of VGs	10
Shape	Delta
Config.	Ctr. CD
h_{VG}	0.077c
Inclination angle	16.4°
VG Length	3.5 h_{VG}
Intraspacing	3.9 h_{VG}
Interspacing	10 h_{VG}

2. Numerical Methods

In this work, an extruded DU97-W-300 airfoil with a blunt trailing edge of 10% of the chord length (Barone et al. [6]) with and without vortex generators is investigated for an angle of attack of $\alpha = 12.8°$ at a chord-based Reynolds-number of $Re_c = 3.2$ million. Three different cases with VGs at different chordwise positions (20%, 45% and 70%) are considered. The VG geometry parameters used in this work are shown in Table 1. They were already investigated numerically for the DU97-W-300 without flatback [9, 10] and the results showed good agreement with experimental data. The flow field is computed by means of Delayed Detached Eddy Simulation (DDES) and the acoustic noise is extracted with a Ffowcs-Williams and Hawkings (FWH) code. In this chapter both numerical methods are presented and the numerical setup is introduced.

Table 2: Overview of the considered cases

Case	Relative chordwise VG position	Computed time (convection lengths)	Number of cells of setup [Mio.]
noVG	-	0.18496s (10.53)	57.2
VG20	0.20	0.18496s (10.53)	73.4
VG45	0.45	0.18496s (10.53)	74.3
VG70	0.70	0.18496s (10.53)	75.9

2.1 Flow Computations

The computations were performed as fully turbulent DDES including the Bernoulli-shielding (in the following called "BDES") by Weihing et al. [11] and the shear-layer adapted length-scale (short SLA) proposed by Shur et al. [12] with the CFD solver FLOWer. The code was initially

developed at the German Aerospace Center DLR [13] and enhanced at the Institute of Aerodynamics and Gas Dynamics (IAG) of the University of Stuttgart. The two-equation linear eddy viscosity model from Menter, known as Menter-SST turbulence model [14], is applied for closure in the Reynolds-averaged Navier-Stokes (RANS) domain and a Smagorinsky subgrid scale model in the domain of the Large Eddy Simulation (LES). In order to reduce the numerical dissipation, a 5th order WENO (Weighted Essentially Non-Oscillatory) flux discretisation scheme was used in the refinement mesh (Figure 1 in blue).
Details about all considered cases are presented in

Table 2. Before the computation with DDES, pre-curser computations with steady RANS simulations followed by unsteady RANS (URANS) simulations were undertaken. For the URANS computations, four full convections of the flow over the airfoil at a physical timestep corresponding to $t_{phy} = 1/50\, t_{conv}$ were undertaken. t_{conv} corresponds to the time required for the flow to convect over the entire airfoil (i.e. convective length). Finally, the DDES, which is used for all evaluations in this work, was computed at $t_{phy} = 1/140 t_{conv}$ for 1600 physical timesteps. This corresponds to 10.53 full convections of the flow over the airfoil.

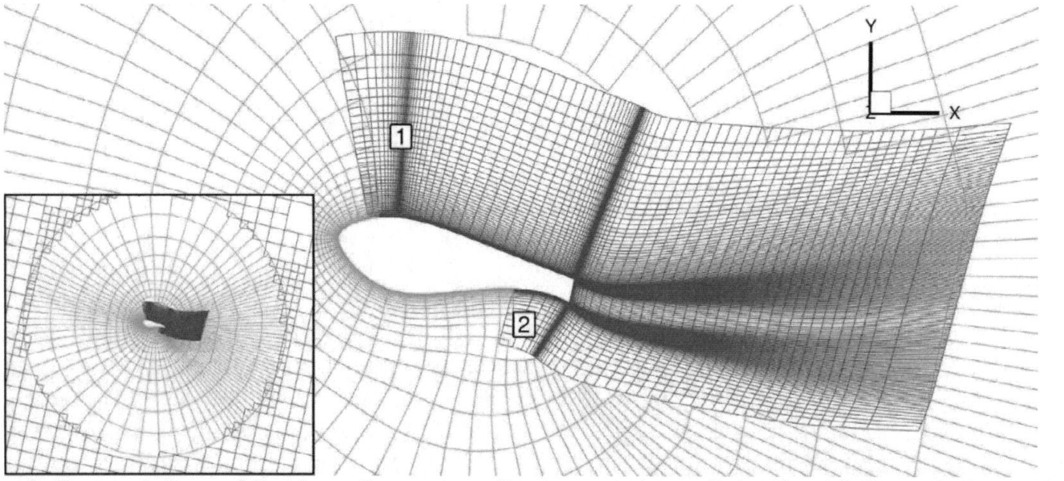

Figure 1: Presentation of the baseline setup: Background mesh (red), airfoil mesh (green) and refinement mesh (blue). Framed 1: Refinement in the vicinity of the VGs. Framed 2: Pressure side is partially included in the refinement mesh. For better visibility, the number of cells is reduced by a factor of four.

2.2 Numerical Setup

The computation of VGs with RANS methods requires specific mesh refinement to resolve the velocity gradients of the streamwise VG vortices. Seel et al. [15] showed that the VG area and the entire propagation area up to the trailing edge of the airfoil requires high mesh resolution in spanwise and streamwise direction. In addition, the VG vortices locally increase the height of the BL and therefore the extrusion in wall-normal direction has to be adapted. Regarding flatback airfoils computed with DDES, the mesh has to be refined in the wake of the airfoil in order to reach cell sizes able to resolve the relevant turbulent structures in the LES area. A rough estimate to fulfil this requirement, proposed by Spalat et al. [16], is that the physical CFL number should not exceed 1. At this point it should be mentioned that the physical CFL number is defined as $CFL_{phy} = u \cdot \Delta t/\Delta x$ with u corresponding to one component of the local flow velocity vector, Δt to the physical timestep and Δx to the cell length in the direction of the flow velocity component. Hence, the CFL number can be different for each direction in space. To ensure a conservative approach, the local CFL number in this work is always defined as

$$CFL_{phy} = \max\left(u \cdot \frac{\Delta t}{\Delta x}, v \cdot \frac{\Delta t}{\Delta y}; w \cdot \frac{\Delta t}{\Delta z}\right) \quad (2)$$

where u, v and w correspond to the local velocities of a cell with the dimension $(\Delta x, \Delta y, \Delta z)$.
When combining VGs and airfoils with flatback, the mesh requirements of both of them have to be met. In order to increase the resolution of the mesh in the areas of interest, an overlapping-mesh technique (chimera) was used.
As presented in Figure 1, the baseline setup (without VGs) is composed of 3 meshes: The background mesh (red, 1.3 Mio. cells), the airfoil mesh (green, 3.7 Mio. cells) and the refinement mesh (blue, 52.4 Mio. cells) which starts upstream of the most forward chordwise position of the VGs (framed 1 for the case with VGs at 20% chord) and extends until 1.5 chord lengths behind the BTE. By doing so, the von Kármán VS in the wake of the airfoil is resolved with high accuracy. As the VS is fed with kinetic energy by the flow from each side of the airfoil alternately, the accuracy of the computation also depends on the resolution of the cells in the rear part of the airfoil pressure side, thus this area (framed 2) was also included in the refinement mesh. The refinement mesh was adapted for each VG case in order to relocate the refinement for the VGs to the corresponding chordwise position. In Figure 1, the refinement mesh is adapted for the case with VGs at 20% of the chord length. For each of the five VG pairs used in the VG cases of this work, an additional mesh (not shown, 3.2 Mio. cells) is embedded in the refinement mesh. For the refinement mesh and the background mesh, hanging nodes are used to further increase the number of cells in the areas of interest. The BL of the airfoil is resolved in the wall normal direction with $y^+ < 0.5$ in order to capture the steep near-wall gradients, which appear due to the momentum redistribution process through the streamwise VG vortices across the BL. The growth rate in the BL is 1.07 and the BTE has 184 cells over its thickness and 640 cells in spanwise direction for an extrusion length of $z_{CFD} = 0.3846c$.

2.3 Acoustic Computations

The acoustic results in this work are computed using the inhouse acoustic code ACCO [17], which is based on the acoustic analogy of Lighthill and the Ffowcs-Williams and Hawkings equation as

$$\begin{aligned} \frac{\partial^2 \bar{\rho}'}{\partial t^2} - c^2 \bar{\nabla}^2 \rho' = & \frac{\partial^2}{\partial x_i \partial x_j}\left[T_{ij} H(f)\right] \\ & - \frac{d}{dx_i}[(p' n_i + \rho u_i (u_n - v_n))\delta(f)] \\ & + \frac{d}{dt}[(\rho_0 v_n + \rho(u_n - v_n))\delta(f)] \end{aligned} \quad (3)$$

where the left-hand side corresponds to the wave operator which describes the propagation of sound in unconfined space and on the right-hand side the source terms generating the sound waves transported by the wave operator. The first term on the right-hand side stands for the quadrupoles in the physical fluid, the second term for the dipoles on the surface and the last term for the monopoles on the surface [3, 17]. f represents the emitting surface which corresponds to the surface of the airfoil in this work. In consequence, the quadrupoles, which are located in the field, are neglected. This is justified by the fact that they become efficient emitters only for higher Mach numbers [18].
The code uses the emitting surface extracted out of the CFD data for each physical time step and computes the sound pressure in time for a predefined observer as an output. Bases on the physical time step used in this work ($\Delta t = 0.0001156s$), a Nyquist-frequency of $4325\ \mathrm{Hz}$ is obtained. As this frequency results from a signal containing only two data points per period, which is not accurate enough, the highest frequency regarded in this work is $1000\ \mathrm{Hz}$ and corresponds to approximatively 8 datapoints for one period.

The observer position was set in accordance with the experimental data [6] at $x = 0.25c$, $y = 3.326c$. The airfoil span in the experimental data is $z_{exp} = 1.969c$. In regard to the computational costs, the spanwise extent of the CFD was reduced to $z_{CFD} = 0.3846c$ (equivalent to 10 VG interspacings) which still exceeds the length of 0.25c, proposed by Bangga et al. [7] for accurate predictions on the same airfoil. In terms of spanwise position, the observer was placed at $z = 0.192c$. In order to obtain comparable results in sound pressure level (SPL), a rescaling was employed as follows:

$$SPL_{scaled} = SPL_{CFD} + 20\log_{10}\left(\frac{z_{exp}}{z_{CFD}}\right) \tag{4}$$

3. Aerodynamic and aeroacoustic validation

In this section, the clean case without VGs ($\alpha = 12.8°$) is first analysed regarding the quality of the numerical grid by means of its physical CFL number. The case is then compared to the experimental data from Barone et al. [6] in terms of lift and drag coefficients and the narrow band SPL spectrum.

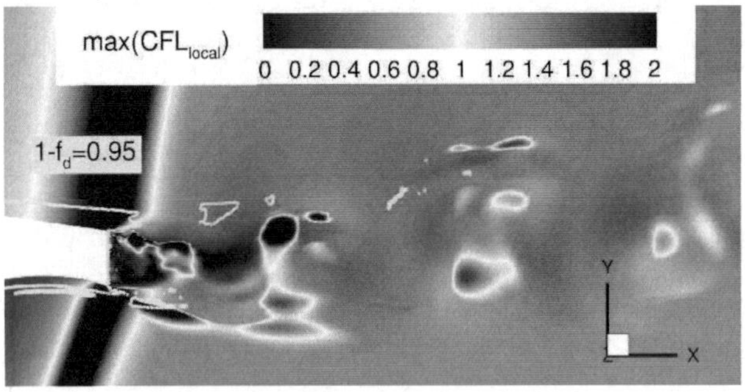

Figure 2: Snapshot of a z-slice of the maximum local CFL number in the wake of the noVG case. The yellow line corresponds to the limit of the DDES Bernoulli shielding.

In Figure 2, the instantaneous distribution of the physical CFL number of each cell as defined in equation (2) and the positions of the edge of the BL shielding are shown. The Bernoulli shielding function works appropriately: The fully attached BL on the suction side and pressure side is included in the RANS mode ($1 - f_d < 0.95$). As soon as the free shear layer starts at the edge of the BTE, the shielding switches to the LES mode. Inside the von-Kármán vortices, the CFL number exceeds a value of two due to the high vertical velocity v combined with the small y-extent of the cells behind the BL (see also Figure 1). Nevertheless, at the blunt trailing edge and in the major part of the wake, the CFL number remains below 1 and thus the combination of the mesh and the physical timestep are suitable for DDES. The large areas of high CFL number starting from the trailing edge in wall normal direction (on suction side and pressure side) are not relevant as the inner flow region is computed with RANS and the outer flow region does not contain turbulence for the regarded cases.

Table 3: Lift and drag coefficients at $\alpha = 12.8°$ for the experiments without VGs [6] and the CFD computation with and without VGs.

Cases	Lift coefficient c_L [-]	$\sigma(c_L)$ [-]	Drag coefficient c_D [-]	$\sigma(c_D)$ [-]
Exp. (trip)	1.83	-	0.0502	-
Exp. (no trip)	1.89	-	0.0578	-
noVG	1.917	0.010	0.0553	0.003
VG20	1.90	0.003	0.0549	0.002
VG45	1.90	0.005	0.0545	0.001
VG70	1.915	0.010	0.0591	0.002

Aerodynamic validation

In Table 3, the time-averaged lift and drag coefficients and the corresponding standard deviations σ are shown. For the noVG case, the computation was fully turbulent. For the experimental data, both the free laminar-turbulent transition case (no trip) and the forced transition case (trip) are listed. In terms of the lift and drag coefficient, the CFD slightly overestimates the experimental tripped results and is rather close to the free transitional results. Nevertheless, the agreement is acceptable for further investigations.

Aeroacoustic validation

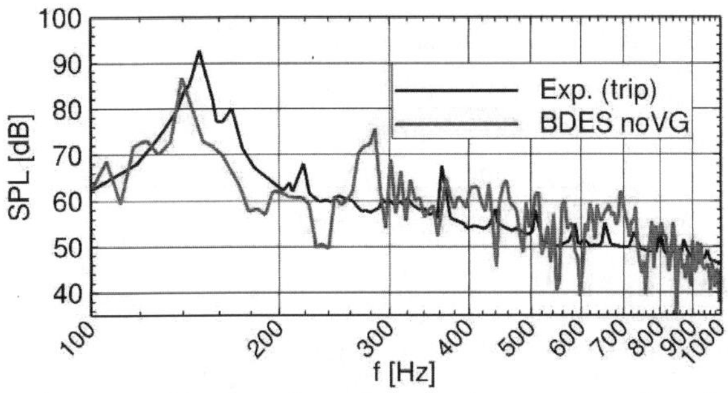

Figure 3: Narrow band sound pressure level at $\alpha = 12.8°$ and $Re_c = 3.2$ Mio. for experimental data [4] (bin resolution 3.125 Hz) and for the BDES (bin resolution 5.58 Hz).

Figure 3 provides the narrow band SPL spectra of the experimental data and the BDES simulation. At a frequency of 148 Hz for the experiment and 140 Hz for the BDES, a tonal peak is visible. The small offset to a lower peak frequency in the simulation is probably related to the higher displacement thickness δ_1 which is in agreement with the slightly higher drag coefficient. Regarding the amplitude of the VS peak, the BDES is lower by 6 dB which could also be related to the differences in the BL state. As there is no additional experimental data available, no further investigations on the discrepancies regarding the peak are made in the present work.

The level of broadband noise for $f > 200$ Hz is in good agreement with the experiment except at around 230 Hz and 280 Hz. The slight broadband offset to higher SPL for the BDES case for $f > 370$ Hz is supposed to result from the postprocessing: The larger width of the frequency bins of the FFT for the BDES case increases the SPL level (bin widths: BDES 5.58 Hz and experiment 3.125 Hz). Due to the high computational costs and the sufficiently long computed time for the further evaluations, it was decided to forego further calculations to achieve matching bin width.

4. Results and Discussion

In this chapter, the results of the numerical study are presented and discussed. In a first stage, the flow topology is analysed qualitatively through snapshots of the vortex structures and the impact of the VGs on the flow field is shown. Then, time-averaged analyses of the lift and drag coefficients, the velocity profiles of the BL and the relevant Reynolds stresses in the wake are shown. As BTEN results from pressure fluctuations next to the BTE, the flow is also characterized in a time dependent manner. Due to the complexity of the flow, including resolved turbulent structures from the LES as well as the streamwise VG vortices and the spanwise vortices from VS, a snapshot POD is used to locate the dominant modes of the flow field and compare them for the different cases. After having shown the aerodynamic impact of VGs, their effects on the aeroacoustics in presented through narrow band SPL spectra of the different cases.

4.1 Flow topology

In order to give a first overview on the flow field, a qualitative insight into the vortex structures for the four cases, through snapshots of λ_2-structures, is given in Figure 4. For the VG cases, the counterrotating VG vortices are clearly visible. All vortex structures are blanked out at a certain spanwise position in order to reveal the von-Kármán VS through the vorticity in z-direction ω_z. Hence, vortices related to the VS from the suction and pressure side are visible. Qualitatively, the VS topology for the noVG case and the VG70 case are similar and correspond to the expected vortex street composed of z-vortices with opposite sign of ω_z (negative for the suction side and positive for the pressure side). For the VG20 case, the VG vortices reduce the strength of the VS from the suction side. Thus, instead of coherent vortex structures with $\omega_z < 0$, only small tattered vortex structures are shed from the suction side and the extent of the vortex street in y-direction is reduced. On the pressure side, the VS is not inhibited in the vicinity of the trailing edge but the shed vortices are less stable in the wake than for the other cases.

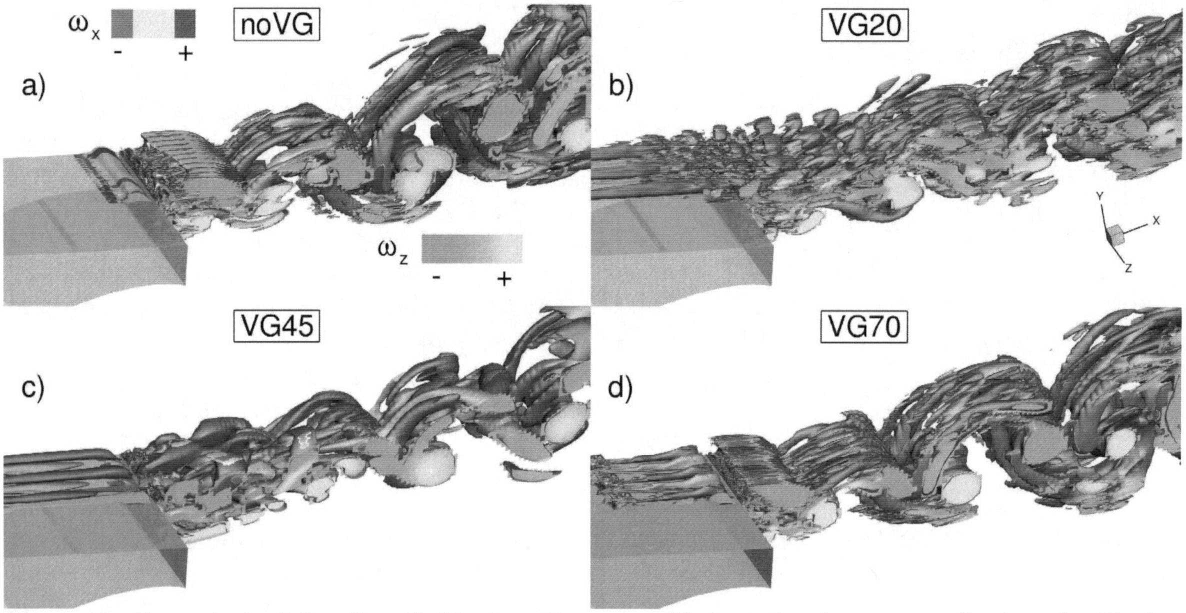

Figure 4: Snapshot of the flow fields for all cases. Vortex structures are displayed with the λ_2-criterion for the half of the span.

4.2 Influence of VGs on Lift and Drag

In Table 3, the time-averaged lift and drag coefficients and their standard deviations σ for all considered cases are shown. VGs are typically used to delay stall onset. In attached flow conditions, as it is the case for the regarded airfoil at $\alpha = 12.8°$, they do not increase lift. Concerning drag, in attached flow conditions, VGs tend to increase the mean spanwise displacement thickness [19] and thus increase pressure drag. For the regarded cases, the time-averaged drag coefficient is not significantly affected by the use of VGs with the exception of the VG70 case. By placing the VGs too far downstream, the shed streamwise VG vortices are not efficient and, as it will be shown for the BL velocities, even decrease the momentum in the BL. This is related to the complex interaction of the VG vortices with the VS, which in the interest of brevity is only briefly discussed in this paper.
Regarding the standard deviations, the VGs slightly reduce the fluctuations for lift and drag coefficients compared to the noVG case, except for the VG70 case. For c_l, the standard deviation increases with the chordwise position of the VGs. This trend can be explained by the reduction of the vertical extent of the VS observed in Figure 4, for the VG20 and VG45 cases in particular. Since fluctuations are the cause of noise emissions, this observation is a first indication of which VG configuration is the most promising for BTEN reduction.

4.3 Averaged boundary layer profiles

In Figure 5, the time and spanwise averaged BL velocity profiles and the corresponding Bernoulli shielding function $1 - f_d$ are shown. For $1 - f_d = 1$ the area is fully shielded and thus, the flow is treated in URANS mode. As soon as the value approaches zero, the flow is treated in LES mode.

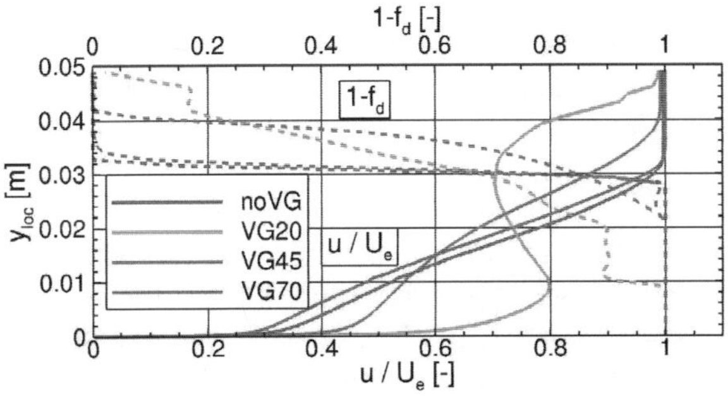

Figure 5: Time and spanwise averaged boundary layer velocity profiles at a chordwise position of x/c=0.9 with BDES

For the VG20 case, the typical S-shape from the VG vortices is visible. This shape results from the momentum transfer from the upper BL towards the lower one. For the VG45 case this shape is much less pronounced. Regarding the VG70 case, the VGs do not create any momentum transfer, and even slightly lower the velocity compared to the noVG case. The reason for the decrease in mixing with increasing chord position of the VGs is twofold: on one hand, low acceleration of the flow in the rear part of the flatback airfoil creates weak main VG vortices and, on the other hand, the distance (convective length) for the VG vortices to generate an efficient mixing shortens.

Regarding the shielding functions, the BL is well-recognized and thus treated in the URANS mode for all cases. For the VG20 case, at the lower local velocity maximum a reduction of $1 - f_d$ to around 0.9 is visible, followed by a further reduction starting from $y_{loc} = 0.02\,m$. Nevertheless, the lower BL ($y_{loc} < 0.01\,m$) is recognized well by the shielding function but the consequences of the shielding behaviour in the upper part of the boundary layer on the turbulent flow requires deeper understanding. Mereu et al. [20] showed promising results on a closed trailing edge DU97-W-300 airfoil with VGs and DDES but the behaviour of the shielding was not addressed in detail.

In the already presented studies, the VG20 showed the highest impact on the flow field. Therefore, and in the interest of brevity, the following studies focus on the VG20 case in comparison to the noVG case.

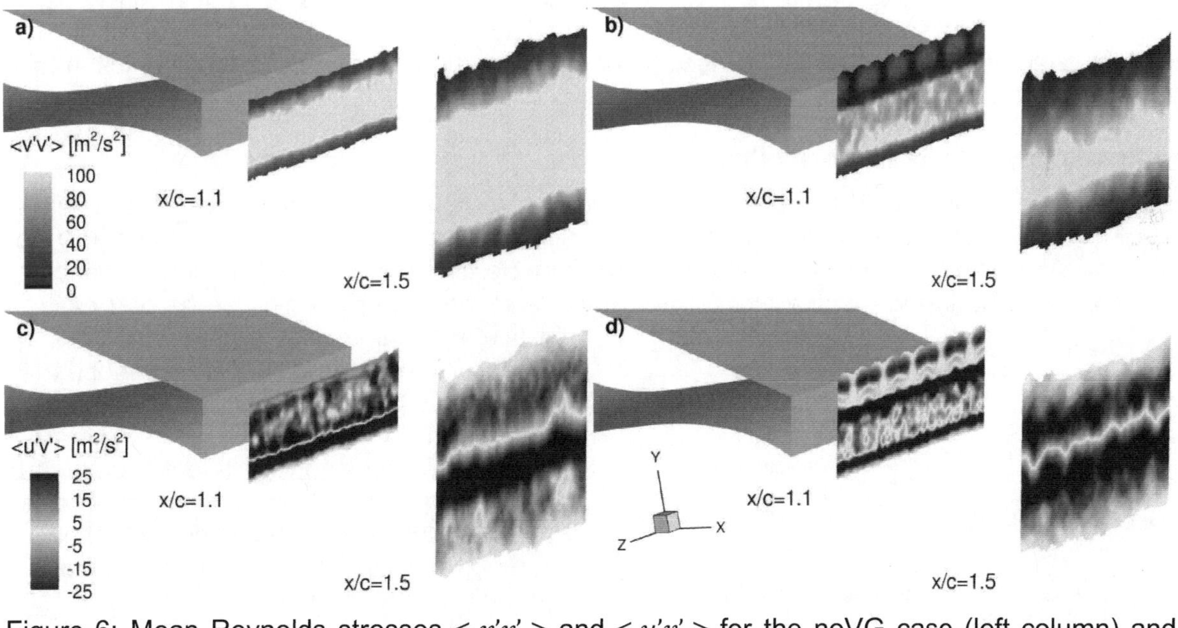

Figure 6: Mean Reynolds stresses $< v'v' >$ and $< u'v' >$ for the noVG case (left column) and VG20 case (right column) at two slices in the wake parallel to the flatback at relative chord positions of 1.1 and 1.5. All values of $< v'v' >$ smaller than one are blanked out for better visibility.

4.4 Mean Reynolds stresses

To quantify the strength of the VS, the vertical part of the turbulent kinetic energy, i.e. the Reynolds stresses parallel to the flat back in vertical direction $< v'v' >$ are displayed in Figure 6 a) and b) for the noVG case and the VG20 case, respectively. On both slices through the wake (x/c=1.1 and 1.5) higher values are reached for the noVG case than for the VG20 case. For the VG20 case, the VG vortices are visible though a local mushroom-shaped increase of $< v'v' >$ for x/c=1.1. Farther downstream, at x/c=1.5, the VG vortices are no longer visible but their induced velocities in y-direction on the surrounding turbulent vortices produce large spanwise fluctuations in $< v'v' >$. Compared to the pressure side, this leads to a less sharp boundary between the VS area and the freestream with low values of $< v'v' >$.

The anisotropic Reynolds stresses $< u'v' >$ are an important indicator for the mixing process in the wake of the blunt trailing edge as they link the flow in streamwise direction and in vertical direction. For the noVG case (Figure 6 c)) at the slice x/c=1.1, $< u'v' >$ shows a very sharp transition from negative values behind the suction side to positive values behind the pressure side. The area of positive $< u'v' >$ values, corresponding to the roll-up process from the pressure side is narrower and the levels are higher than for the suction side. This is due to the smaller wall-normal extent of the BL at high positive angle of attack and it is accentuated by the acceleration from the concave rear section of the pressure side. Vice versa, on the suction side, the adverse pressure gradient reduces the near wall momentum, and consequently, the strength of the roll-up process in the wake.

Regarding the VG20 case, the behaviour is very different and no sharp transition is visible for $< u'v' >$ at x/c=1.1. Instead, the roll-up process from the suction side is higher which is visible through the smaller $< u'v' >$ values behind the suction side. This behaviour results from the momentum transfer of the VG vortices in the BL which is also visible in the slice through the mushroom-like form of low $< u'v' >$. One could intuitively conclude that the VGs increase the VS problem through the momentum transfer. However, as it is visible more downstream in the wake, at x/c=1.5, the transition of change in sign, discussed earlier, is visible but the spanwise flow induced by the VGs leads to larger spanwise variations. These spanwise variations reduce the spanwise coherence of the VS, which is the main driver for the tonal noise emission.

The Reynolds stresses show the impact of the VGs on the mean fluctuations in the wake and give insight in the transfer of turbulent kinetic energy. Nevertheless, as mentioned previously, BTEN has a tonal noise character which results from a strong VS at a constant frequency. Thus, a modal analysis is presented in the next section to give more insight on the dynamics in the frequency domain.

4.5 POD analysis of the velocity fluctuations

In this section, snapshot PODs for the noVG and the VG20 cases are presented at the same slices as for the Reynolds stresses and one additional slice in the direction of the flow (i.e. $z/z_{max} = 0.5$). The VS mechanism at the flat back of the airfoil is unsteady and has a periodic character. The motivation of the POD is to understand the flow dynamics of the two cases. Particularly the most energy containing modes of the vertical velocity component v are important in regard to VS and consequently for BTEN [7].

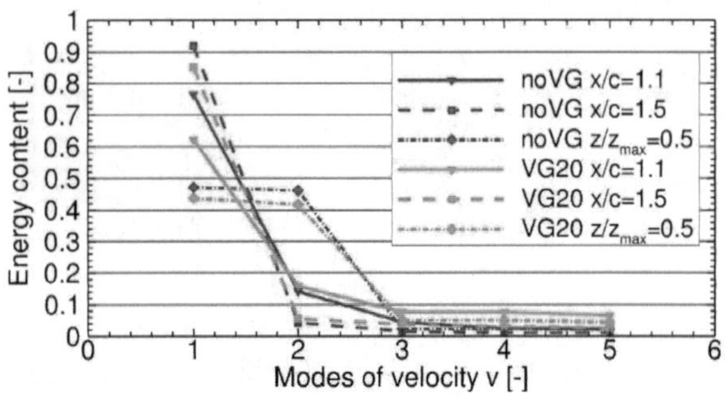

Figure 7: Relative energy content (out of the eigenvalues) of the first five modes for the two cases at different slice positions.

In Figure 7, the energy content (computed with the eigenvalues) of the five first modes of the vertical velocity fluctuations v is shown. As usual in POD, the modes are sorted by ascending number from the largest to the smallest energy content. For the x/c slices, the first mode contains most of the energy for all cases and can therefore be attributed to the VS mechanism. For both x/c-slice positions, the energy content of the first mode is reduced by the VG vortices. For the x/c=1.1 slice, which is closer to the BTE and thus more relevant for this investigation, a reduction of 14% of the first mode is visible for the VG case. The energy is redistributed homogeneously on the higher modes. This is an important observation regarding the acoustics, as a redistribution only in the second mode could enable other noise sources and, in the worst case, a second tonal hump (in addition to the one of the VS). Regarding the z-slice, two modes with a similar energy content are visible. These two modes are also attributed to the VS mechanism. The only difference compared to the x/c-slices is that the change of slice direction reveals the anti-correlation of the vertical velocity related to the alternating upward and downward induced velocities by the VS. The effect of the VGs on the energy content is thus the same as for the x/c-slices.

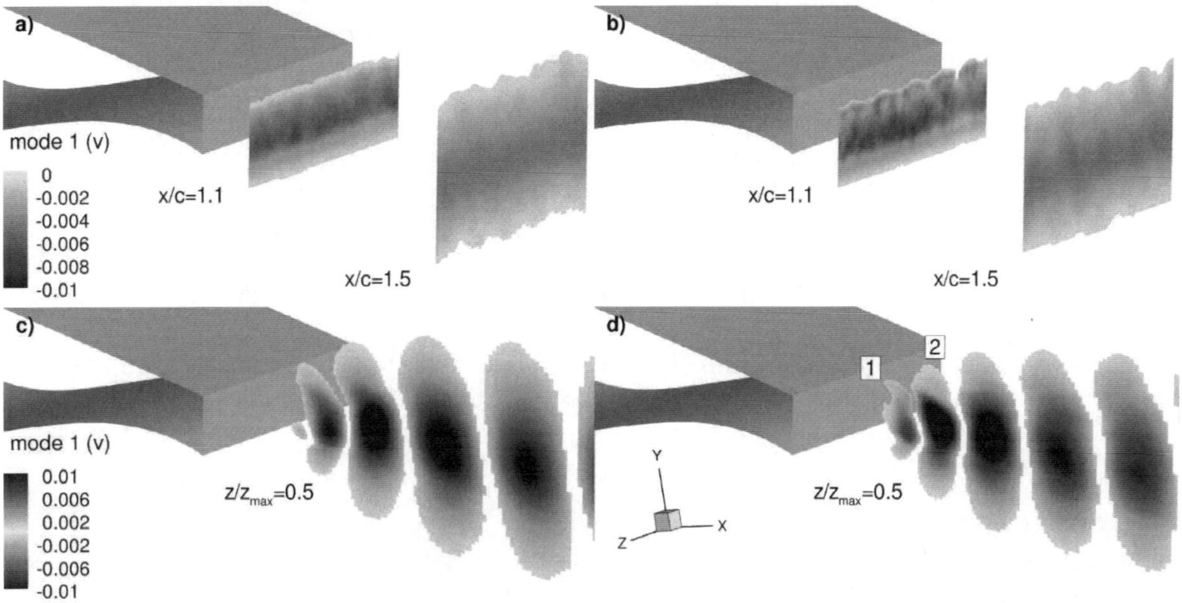

Figure 8: Representation of the first mode of the vertical velocity component v for the noVG case (left column) and VG20 case (right column) at different slice positions (same as in Figure 7).

A closer insight into the distribution of the first v mode in space is given in Figure 8. The first and by far most important mode is correlated in space as the sign is negative for both x/c slices. This was to be expected because VS is a spanwise coherent phenomenon. The small fluctuations in spanwise direction are related to the development of turbulent structures and the VG vortex dynamics (see Figure 4). That same first mode of v for a slice at $z/z_{max} = 0.5$ (Figure 8 c) and d)) is anti-correlated in space and at the same energy level as the second mode (see Figure 7). The anti-correlation is a typical behaviour of convecting structures and can therefore be attributed to the VS. The first spots of high values (framed "1") and low values (framed "2") show the effect of the VG vortices on the VS: The structures are flattened at the top and consequently of smaller extent in vertical direction which corresponds to the observation on the reduced energy content of the first two modes for the VG case in Figure 7. At this point, it should be mentioned that the spanwise behaviour in the VG case is not homogeneous. Hence, the chosen slice at the half of the span is not completely representative for the entire span but the spanwise fluctuations of the vertical velocity component are not very large as seen in Figure 8 b).

4.6 Aeroacoustic noise

In Figure 9, the SPL spectra for all the computed cases and the experiment with and without a splitter plate with a chordwise extent of 0.098c [6] are plotted. Regarding the tonal peak related to the VS at around 150 Hz, a reduction of around 7 dB is obtained for the VG20 and VG45 cases. For the VG70 case, only a minimal reduction of the peak is visible which confirms the very small effect of this configuration of the flow. Nevertheless, an increase in BTEN through the slightly lower BL velocities is not visible. At this point it is clarified that this study only focuses on BTEN and not on trailing edge noise which would be affected by the different BL shapes.

For the frequency band between 150 and 230 Hz, the SPL of the VG20 case is significantly reduced. This is assumed to be related to the interaction of the VG vortices with large turbulent structures reducing their coherence and thus their impact on the wall pressure fluctuations on the airfoil.

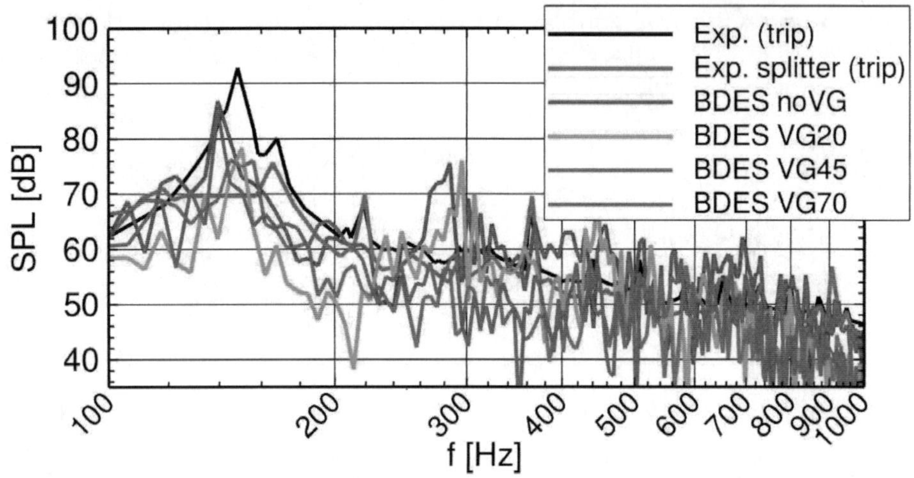

Figure 9: Narrow band sound pressure level at $\alpha = 12.8°$ and $Re_c = 3.2$ Mio. for experimental data with and without splitter plate [6] (bin resolution 3.125 Hz) and for all the computed cases with and without VGs (bin resolution 5.58 Hz).

5. Conclusions

In this work, a flatback airfoil was investigated by means of DDES simulations with shear-layer adaption and the Bernoulli shielding function. The output was processed with a Ffowcs-Williams and Hawkings code in order to investigate blunt trailing edge noise. In the first part, a validation of the numerical results with experiments was performed. Subsequently, the aerodynamic behaviour of four computed cases, one reference case without VGs and three cases with counterrotating delta-shaped VGs at different chord positions on the airfoil, was analysed. The lift and drag coefficients are only merely affected by the VGs. The standard deviation of the lift coefficient decreases with more forward positioning of the VGs. The Reynolds stresses in wall normal direction and the spanwise coherence of the transfer of the stresses between the streamwise and the vertical component $< u'v' >$ (main driver of VS) could be reduced by the VGs placed at 20% relative chord length. A snapshot POD in streamwise and spanwise direction gave insight into the relevant modes of the VS (main driver for the tonal blunt trailing edge noise). For the VS mode a reduction of up to 14% is obtained through the action of the VGs. Finally, the acoustic noise emission shows a reduction of the tonal peak noise at 140 Hz of around 7 dB for the VGs placed at 20% and 45% relative chord length compared to the case without VGs. Furthermore, a remarkable noise reduction in the frequency band from 150 Hz to 230 Hz is obtained for the 20% VG case.
Compared to the noise reduction attained by the splitter plate of around 17 dB, the noise reduction capabilities of the VGs are lower but they represent an additional advantage aside from the main objective of VGs (i.e. to optimise the aerodynamic behaviour of airfoils at high angle of attack by delaying flow separation) for the design of aerodynamically and aeroacoustically optimized flatback airfoils.

Acknowledgments

A part of this work was realized thanks to the project IndiAnaWind (FKZ 0325719F), funded by the German Federal Ministry of Economic Affairs and Climate Action (BMWK). The authors gratefully acknowledge the High-Performance Computing Center Stuttgart (HLRS) for providing computational resources within the project WEALoads.

References

[1] M. Barone und D. Berg, „Aerodynamic and aeroacoustic properties of a flatback airfoil: An update," *47th AIAA Aerospace Sciences Meeting including the New Horizons Forum and Aerospace Exposition,* January 2009.

[2] G. Papadakis und M. Manolesos, „The flow past a flatback airfoil with flow control devices: benchmarking numerical simulations against wind tunnel data," *Wind Energ. Sci.,* pp. 911-927, 2020.

[3] S. Wagner, R. Bareiß und G. Guidati, Wind Turbine Noise, Springer, 1996.

[4] W. Blake, Mechanics of Flow-Induced Sound and Vibration (Volume II), Academic Press, Inc, 1986.

[5] S. Yang und J. Baeder, „Aerodynamic drag and aeroacoustic noise mitigation of flatback airfoil with spanwise wavy trailing edge," *33rd Wind Energy Symposium,* January 2015.

[6] M. Barone, D. Berg, W. Devenport und R. Burdisso, „Aerodynamic and Aeroacoustic Tests of a Flatback Version of the DU97-W-300 Airfoil," Sandia Report, Sandia National Laboratories, 2009.

[7] G. Bangga, F. Seel und T. Lutz, „Aerodynamic and Acoustic Simulations of Thick Flatback Airfoils Employing High Order DES Methods," *Advanced Theory and Simulations,* 2022.

[8] D. Baldacchino, C. Ferreira, D. De Tavernier, W. Timmer und G. van Bussel, „Experimental parameter study for passive vortex generators on a 30% thick airfoil," *Wind Energy,* 2018.

[9] D. Baldacchino, „Experimental benchmark and code validation for airfoils equipped with passive vortex generators," *Journal of Physics: Conference Series,* Bd. 753, 2016.

[10] F. Seel, L. Blazejewski, T. Lutz und E. Krämer, „Numerical Assessment of a BAY-Type Model for different Vortex Generator Shapes Applied on a Wind Turbine Airfoil," *Journal of Physics, Conference Series,* 2022.

[11] P. Weihing, J. Letzgus, T. Lutz und E. Krämer, „Development of Alternative Shielding Functions for Detached-Eddy Simulations," in *Progress in Hybrid RANS-LES Modelling*, Springer International Publishing, 2020, pp. 109-118.

[12] M. Shur, P. Spalart, M. Strelets und A. Travin, „An Enhanced Version of des with Rapid Transition from RANS to les in Separated Flows," *Flow, Turbulence and Combustion,* pp. 709-737, 2015.

[13] N. Kroll, C. Rossow, D. Schwamborn, K. Becker und G. Heller, „MEGAFLOW – a Numerical Flow Simulation Tool for Transport Aircraft Design," *Proceedings of the 23rd International Congress of Aeronautical Sciences, Toronto,* 2002.

[14] F. R. Menter, „Two-equation eddy-viscosity turbulence models for engineering," *AIAA Journal,* Bd. Vol. 32 (8), 1994.

[15] F. Seel, T. Lutz und E. Krämer, „Numerical Study of the Impact of Vortex Generators on Trailing Edge Noise," in *Proceedings from the 9th International Conference on Wind Turbine Noise*, Online, May 2021.

[16] P. R. Spalart, „Young-Person's Guide to Detached-Eddy Simulation Grids," NASA Technical report CR-2001-211032, 2001.

[17] M. Kessler und S. Wagner, „Source-time dominant aeroacoustics," *Computers & Fluids,* Bd. 33, Nr. 5-6, pp. 791-800, 2004.

[18] M. Ghasemian und A. Nejat, „Aerodynamic noise prediction of a Horizontal Axis Wind Turbine using Improved Delayed Detached Eddy Simulation and acoustic analogy," *Energy Conversion and Management,* 2015.

[19] A. Mole, A. Skillen und A. Revell, „The interaction of longitudinal vortex pairs with a turbulent boundary layer," *Journal of Fluid Mechanics,* 2022.

[20] R. Mereu, S. Passoni und F. Inzoli, „Scale-resolving CFD modeling of a thick wind turbine airfoil with application of vortex generators: Validation and sensitivity analyses," *Energy,* 2019.

10th International Conference
on
Wind Turbine Noise
Dublin – 21st to 23rd June 2023

The masking effect of vegetation- and wave noise on wind turbine noise audibility

Lars Sommer Søndergaard[1], Rune Egedal[1], Erik Thysell[1], Franck Bertagnolio[2] and Andreas Fischer[2]
[1] Department of Acoustics, Noise and Vibrations, FORCE Technology, Denmark,
[2] DTU Wind and Energy Systems, Wind Turbine Design Section, Denmark
Corresponding author: lss@forcetechnology.com

Summary

In Denmark and several other countries noise is regulated as absolute levels, hence the audibility of the noise source is not directly handled.
In most countries, including Denmark, wind turbines are often set up in rural areas. Denmark is a flat, windy and in general densely populated country, but the rural areas are less populated and has longer distances between dwellings, and so it is easier to comply with the setback distances and noise demands. Coincidently some of the windiest parts of Denmark are also rural areas.
In the rural areas some of the common sources of noise is vegetation or waves, which masks other environmental noise sources. The effect of masking from vegetation and/or waves has not been studied in much detail in Denmark.
The aim of this project is to gather data from vegetation noise and wave noise and use this to form simple models for both vegetation noise and wave noise. The models are used to estimate the audibility of wind turbines erected in rural areas, considering both temporal effects, spectral effects and effects of wind turbine size, distance and wind shear.
This paper introduces the background and current status of the project, which is based on the Danish noise regulation for wind turbines and a number of investigations/projects.

1. Introduction

Denmark is a windy, densely populated and flat country surrounded by water, where the highest natural points are approximately 170 m above sea level. Figure 1 shows both the population density, the energy production from onshore wind per municipality, onshore and offshore wind turbines operating in 2021 and a wind resource map – all for Denmark. It can be seen that the largest wind resources typically are in the least populated areas of Denmark which also are the areas where most wind turbines are installed. Consequently, some of the areas where wind turbines are installed and have been installed are areas where the background noise level is dominated by not man-made noise sources such as vegetation noise and wave noise + sounds from birds etc. It should be noted that even though these areas are some of the least populated areas in Denmark, dwellings are still spread all over the area. This leads to the fact that distances between dwellings and wind turbines are relatively small.

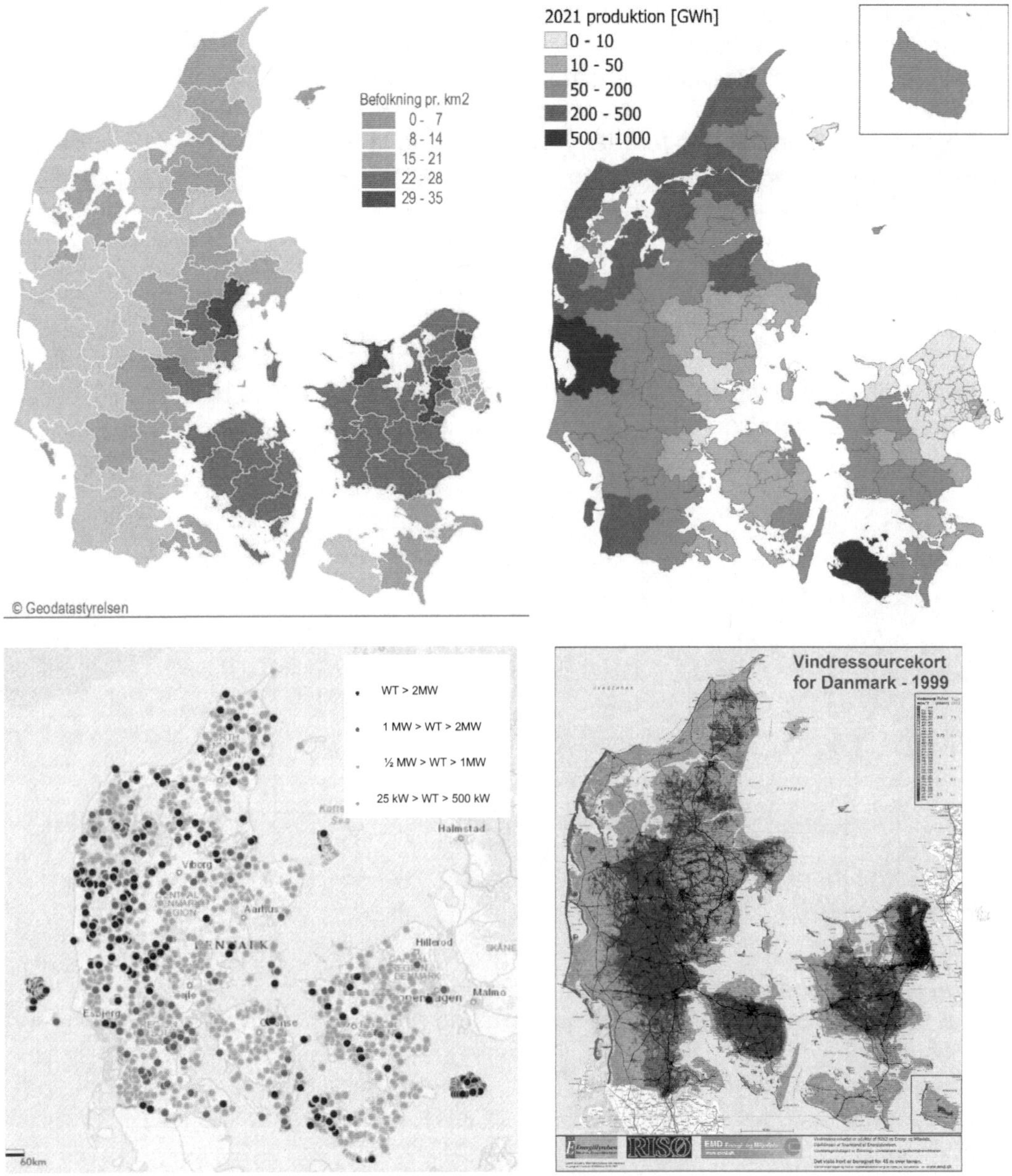

Figure 1: Top left: Population density pr. municipality for rural districts (https://www.dst.dk/da/Statistik/nyheder-analyser-publ/nyt/NytHtml?cid=30696). Top right: Energy production from onshore wind pr. municipality (https://ens.dk/sites/ens.dk/files/Statistik/oversigtskort_over_produktion_pr_kommune_fra_landvind_2021.png). Bottom left: Onshore and offshore wind operating in 2021 (https://www.arcgis.com/home/webmap/viewer.html?webmap=4398142262974c4b9b41ac477d423dcd&extent=7.4005,54.9737,15.514,57.524). Bottom right: Wind resource map for Denmark in 1999 (https://www.emd-international.com/files/windres/images/RES_DK99_25pct.gif).

2. A short introduction to the regulation of noise from wind turbines in Denmark

In Danish, the word "vindmølle" is used both on the modern electrical power producing "wind turbine" and the historical "windmill" even that there now only are a few left of the historical type. Denmark is a windy country with a long history of utilizing the wind. It is one of the first countries in the world to have a specific regulation for noise from wind turbines, the BEK 304 dating 1991 [1]. BEK is a shortening of 'Bekendtgørelse' which translates to 'statutory order'. The following period is characterised as a transition period with much discussion regarding how to handle this "new" noise source. Wind turbine noise can only partially be handled by the noise regulation for environmental noise from industry. Two of the biggest differences are:

- Noise regulation for environmental noise from industry specifies that it should be documented for a wind speed lower than 5 m/s (which only is a bit higher wind speed than the typical cut-in wind speed for many wind turbines at the time)
- The noise changes with wind speed

In the transition period a letter was send out to the county administration on how to handle this in the meantime, dating 1988 [2]. The letter is commonly referred to as 'The county letter'.

Before the first real regulation, BEK 304, and the county letter many investigations and research projects were performed, leading to overall principles of how the wind turbine noise is handled in Denmark. The regulations were finally based on sound power level measurements, which were used as input to a noise calculation of the noise level at receivers position. This is also the general principle for handling environmental noise in general in Denmark.

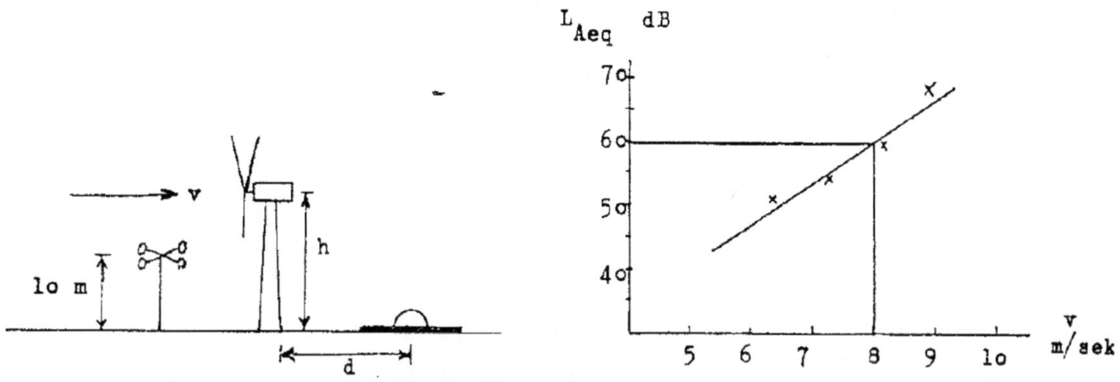

Figure 2: Figure from the county letter [2], where the left side shows the principle of the sound power level measurement, and the right side shows the typical result of such measurement.

At the time the first wind turbine noise regulation were defined, the most typical wind turbine in Denmark was the stall regulated wind turbine. It typically had a noise vs. wind speed pattern as shown in the right side of Figure 2. The noise vs. wind speed relationship can approximately be described as a straight line, where the noise level (L_{Aeq}) at wind speed 8 m/s at 10 m height and the slope was used for noise documentation. The sound power level measurement was a very simplified version of what is now the IEC 61400-11 method [3].

The typical total height of the wind turbines installed at the period when the first noise regulation was defined, was approximately 30-50 m, see Figure 3. Since then, the wind turbines have grown steadily bigger, and now areas for a coming test center are screened to allow for even larger turbines – possibly up to a total height of 450 m. [4]

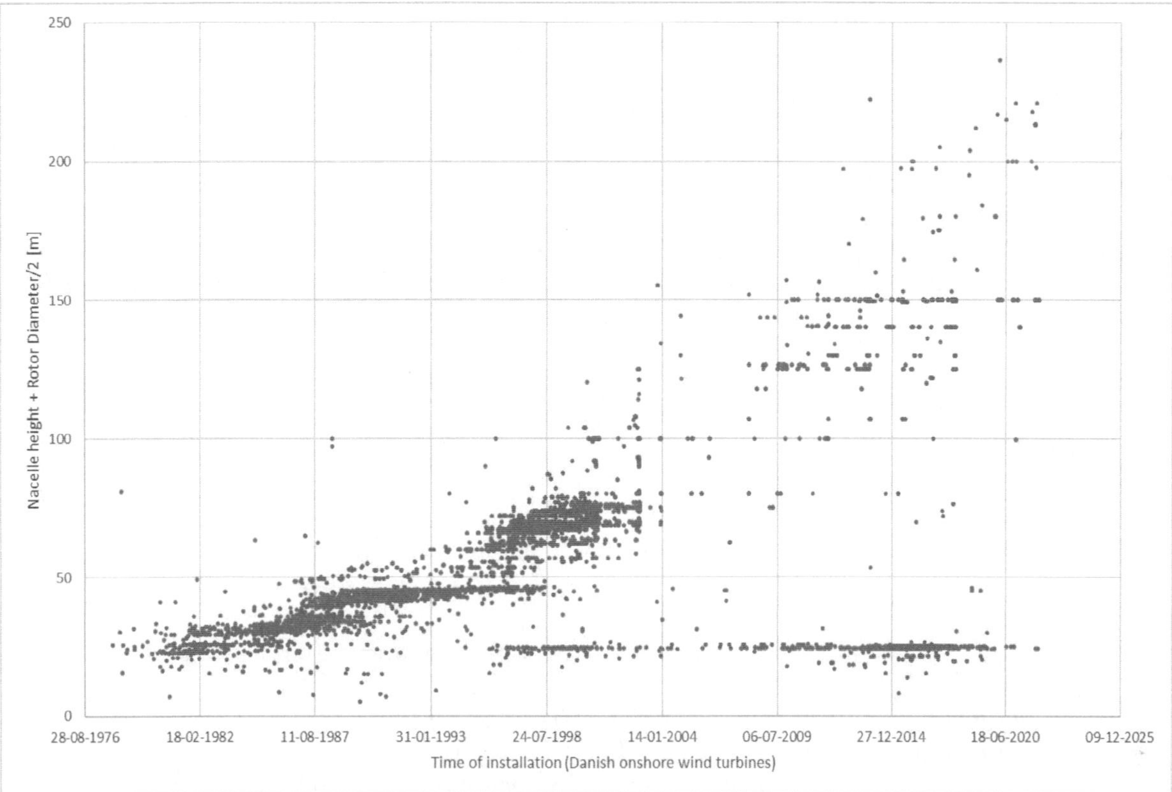

Figure 3: Total height (nacelle + rotor diameter/2) for all land-based wind turbines installed at some point in Denmark. List from https://ens.dk/service/statistik-data-noegletal-og-kort/data-oversigt-over-energisektoren. The list was downloaded on the 13-03-2023. NB: The turbines with a total height larger than 150 m are primarily test turbines primarily placed at the two test centers Høvsøre and Østerild.

The height is of course not the only parameter which has changed over the years. Other relevant changes regarding noise are blade design, drive-train design and naturally wind turbine control. In Denmark, especially in the years up to 2009, there was a large jump in size, but also a change from stall regulated turbines to pitch/RPM controlled turbines. Instead of an approximately linear relationship between noise and wind speed as with the stall regulated wind turbines, typically a linear relationship between noise and wind speed is seen up to the wind speed where rated power is reached – whereas for higher wind speed the noise is approximately constant, see Figure 4.

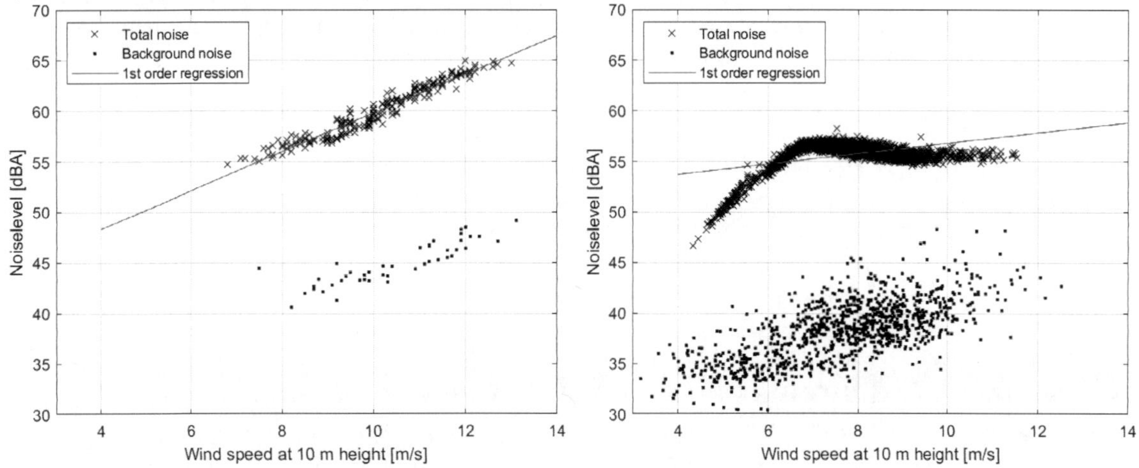

Figure 4: Left: Typical scatterplot (noiselevel versus wind speed) for a stall regulated wind turbine. Right: Typical scatterplot for pitch/RPM controlled wind turbines.[5], [6]

In 2006 a new version of the noise regulation for wind turbines was introduced [7]. Multiple changes were introduced but most importantly in the context of this paper was the change of noise propagation calculation method, the introduction of 6 m/s as a control wind speed and a more detailed sound power level measurement. For the wind turbines installed these years typically the wind speed where the wind turbines reached rated power was between 6 and 8 m/s at 10 m height, and with the noise levels measured at wind speeds of 6 and 8 m/s the noise curve was controlled. Since then, the regulation has been updated several times. In 2012 new noise limits for low frequency noise was introduced [8]. A journal paper [9] was published in 2012 by the Danish EPA describing the Danish noise regulation for wind turbines. In 2019 the noise propagation over water was updated and the tonality method changed [10].

The complexity of the noise regulation has followed the development of the wind turbines. The noise regulation from 1991 had approximately 1800 words. The current noise regulation (from 2019) has approximately 7200 words.

2.1 Current noise regulation

The current regulation for noise from wind turbines in Denmark is from 2019 [10]. In short, it states that the cumulative noise level from all relevant wind turbines may not exceed the following limits:

- At the most noise-exposed point in outdoor living area no more than 15 metres from dwellings in the open countryside:
  (a) **44 dB(A)** at a wind speed of **8 m/s** (10 m height).
  (b) **42 dB(A)** at a wind speed of **6 m/s** (10 m height).
- At the most noise-exposed point in areas with noise-sensitive land use:
  (a) **39 dB(A)** at a wind speed of **8 m/s** (10 m height).
  (b) **37 dB(A)** at a wind speed of **6 m/s** (10 m height).

The total low frequency noise from wind turbines at wind speeds of 6 and 8 m/s (10 m height) may not exceed **20 dB indoors** in neither dwellings in the open countryside nor indoors in areas with noise-sensitive land use. The low frequency noise level is the A-weighted level of the noise in the frequency range defined by the 1/3-octavebands from 10 Hz to 160 Hz, including both.

The noise level at these points is calculated based on sound power level measurements with a prescribed noise propagation method, where there is a method both for noise propagation over land and over water. For the indoor low frequency noise level there are defined sound insulation values both for regular houses and for summer cottages (typically with a lower sound insulation). The regulation also describes how to handle tonality content in the wind turbine noise.

2.2 Setback distance

In Denmark since at least 1999 there has been determined a setback distance as 4 times the total height of the turbine (nacelle height + rotor diameter / 2 for regular wind turbines). The current version [11] describes it as follows (Translation by the author):

- *§ 2. Stk. 2. Permission must not be granted according to the planning act for wind turbines closer to neighboring dwellings than 4 times the total wind turbine height.*

3. Noise from wind turbines and the masking effect of vegetation

Lydteknisk Institut (Now FORCE Technology) performed in 1988 a project for the Ministry of Environment, the Ministry for Energy and Elkraft AmbA (Now Ørsted). The purpose of the project was to examine, to which degree the natural wind noise (vegetation noise) masks the noise from wind turbines. The title of the project report was "Noise from wind turbines and the masking effect of wind noise" [12]. The aim was to investigate to which degree the natural wind noise caused a complete masking of the wind turbine noise, and where it would be fair to relax the noise limits. The project contained a literature study of the theory of masking and its application for wind turbine noise and natural wind noise. In parallel the natural wind noise was measured on different locations. Based on the findings it is generalized that:

- The wind turbine noise is, in a given moment masked by the wind noise, when short time Leq of the turbine noise for all critical bands are at least 4 dB below the sound pressure level of the wind noise.

The generalized wind noise (vegetation noise) is compared to measured wind turbine noise (Nordtank 130 kW, Vestas 75 kW small generator and Vestas V75 kW large generator). It is concluded that the higher the wind speed the better the wind turbine noise is masked, meaning that higher wind speed increases the possibility of the wind turbine noise not being audible. It is also noted that even weak tonal components affect the masking.

Remark that this study is from 1988 and that the measured wind turbines are stall regulated wind turbines, which are characterised by a linear relationship between noise and wind speed, see Figure 4.

4. Wind turbine height and shear

The propagation of sound from different sources like wind turbines is influenced by wind speed and air temperature variations with height above terrain. These variations can be characterized by atmospheric stability. Based on these facts and the findings published in [13], another study for the Danish EPA was conducted "Noise from wind turbines during night". Its purpose was to investigate to which extent the meteorological conditions influence noise from wind turbines, in the surroundings with special regard to the difference between day and night.

A stable atmosphere is characterised by strong shear (a large velocity gradient with respect to height) and low turbulence levels. It occurs typically during night and in clear weather conditions. In a stable atmosphere, the wind speed observed at hub height of a wind turbine, is higher than the one observed in a neutral atmosphere at same wind speed for 10 m height. In an unstable atmosphere, the mixing of the flow between layers of the atmosphere is more present than in stable conditions. As a consequence, the wind shear is usually weak, and the turbulence levels are high. Stable conditions are typically encountered during night-time and especially in clear weather. The wind speed at hub height of a wind turbine equivalently will be lower than observed in a neutral atmosphere with the same wind speed at 10 m height. Since the noise transmission from the wind turbine depends on the wind speeds at the height of the rotor disc, it will therefore be experienced, that the transmitted noise (sound power level) at a constant wind speed at 10 m height will vary depending on the stability of the atmosphere. The biggest difference between wind speed at hub height and at 10 m height will occur during night. Analysis of meteorological data was performed for data from four chosen synoptic Danish weather stations supplemented with extensive meteorological data from Høvsøre (Danish test site for large wind turbines). Two of the stations are located far from the coastline and the rest are located close to the coastline. All locations are spread geographically in Denmark. On the basis of these data, general statistics about wind conditions has been made, and wind speeds for the heights 10 m and 90 m above terrain are compared.
The difference between wind speed at hub height and at 10 m height is seen to be larger at night than during day as described in literature [13]. Generally, wind speed at 10 m height is lower at night than during day, whereas the wind speeds at hub height generally do not differ significantly. In Figure 5 the cumulative frequency of the wind speed for day and night shows that the difference between night and day is small for wind speeds measured in 90 m height but large for wind speeds measured in 10 m height at Høvsøre.

Based on these measurements, the noise emission from wind turbines is not expected to be higher at night than during the day. But if the sound power level of a tall wind turbine is determined on the basis of the measured wind speed at 10 m height, sound power level measurements performed during the night might overestimate the noise emission from the wind turbine significantly compared to the reference conditions from the Danish regulations.

Seasonal variation is seen only to influence the conditions at 10 m height, whereas the influence at wind speed at hub height will be insignificant.
No indication of influential differences is seen between coast near locations compared to inland locations besides the differences to be explained by higher average wind speed.
Lower wind speed at 10 m height is expected to cause a lower background noise level at the residents close to wind turbines. It is possible that the lower masking from background noise level could lead to higher nuisance under these conditions. [14], [15]

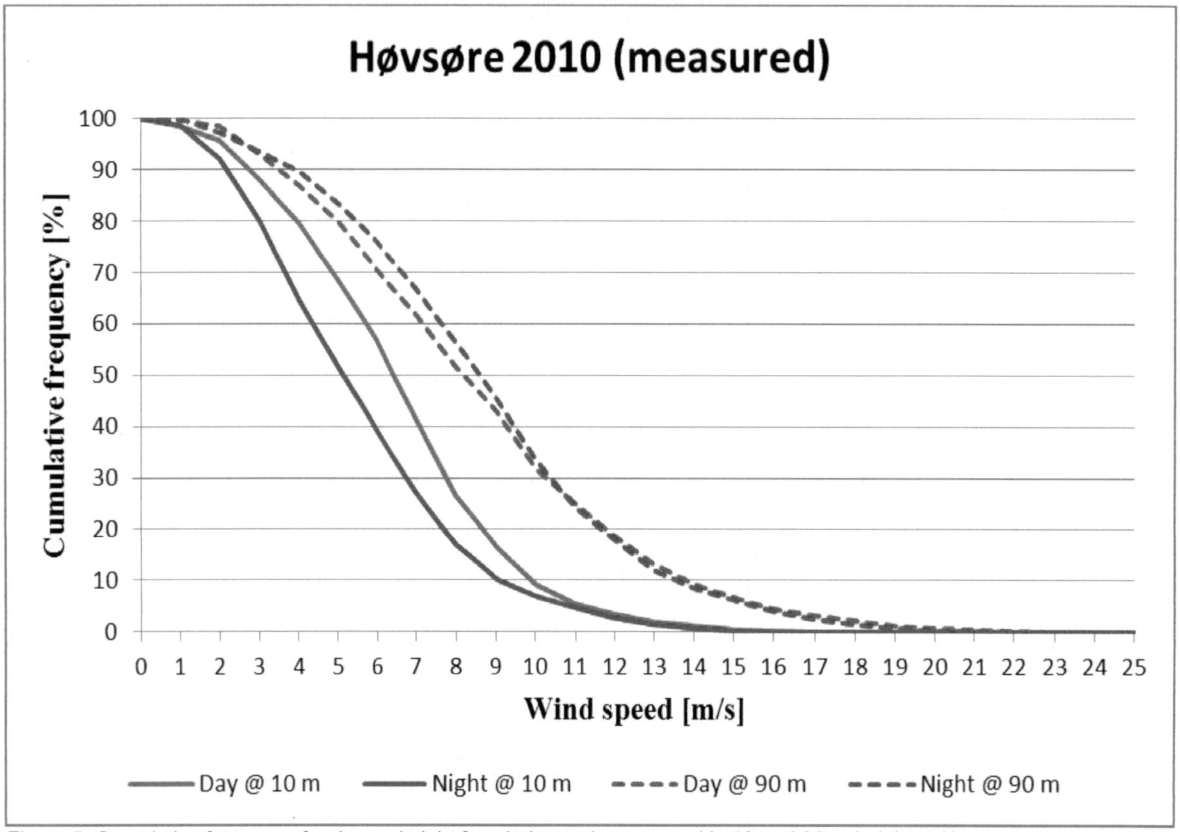

Figure 5: Cumulative frequency for day and night for wind speeds measured in 10 and 90 m height at Høvsøre

5. Wind turbine noise measurements at dwellings versus sound power level measurements and calculated noise levels

As mentioned previously noise from wind turbines in Denmark are controlled based on sound power level measurements. The received noise levels are at receiver position are computed with a predefined propagation method. Hence, the parameter that is in principle controlled is the sound power level from the turbines weighted with a sound propagation function.
There has been uncertainty from residents/neighbours whether the sound propagation methods were correct and questions why the noise is not measured where they live.

Consequently, the Danish EPA initiated three projects to investigate this further and compare results from the default procedure with measurements at residents. The results are reported in three Danish reports and summarized in two conference papers [16]–[20]. In general, it is concluded that it is sometimes possible to measure the wind turbine noise at receivers' position, but that in general that background noise is typically of the same magnitude as the wind turbine noise or even higher. When it was possible to measure the wind turbine noise at the receivers position comparable results to the default procedure (L_{WA} measurement + calculation) were found, but it was more difficult and time consuming than the default procedure.

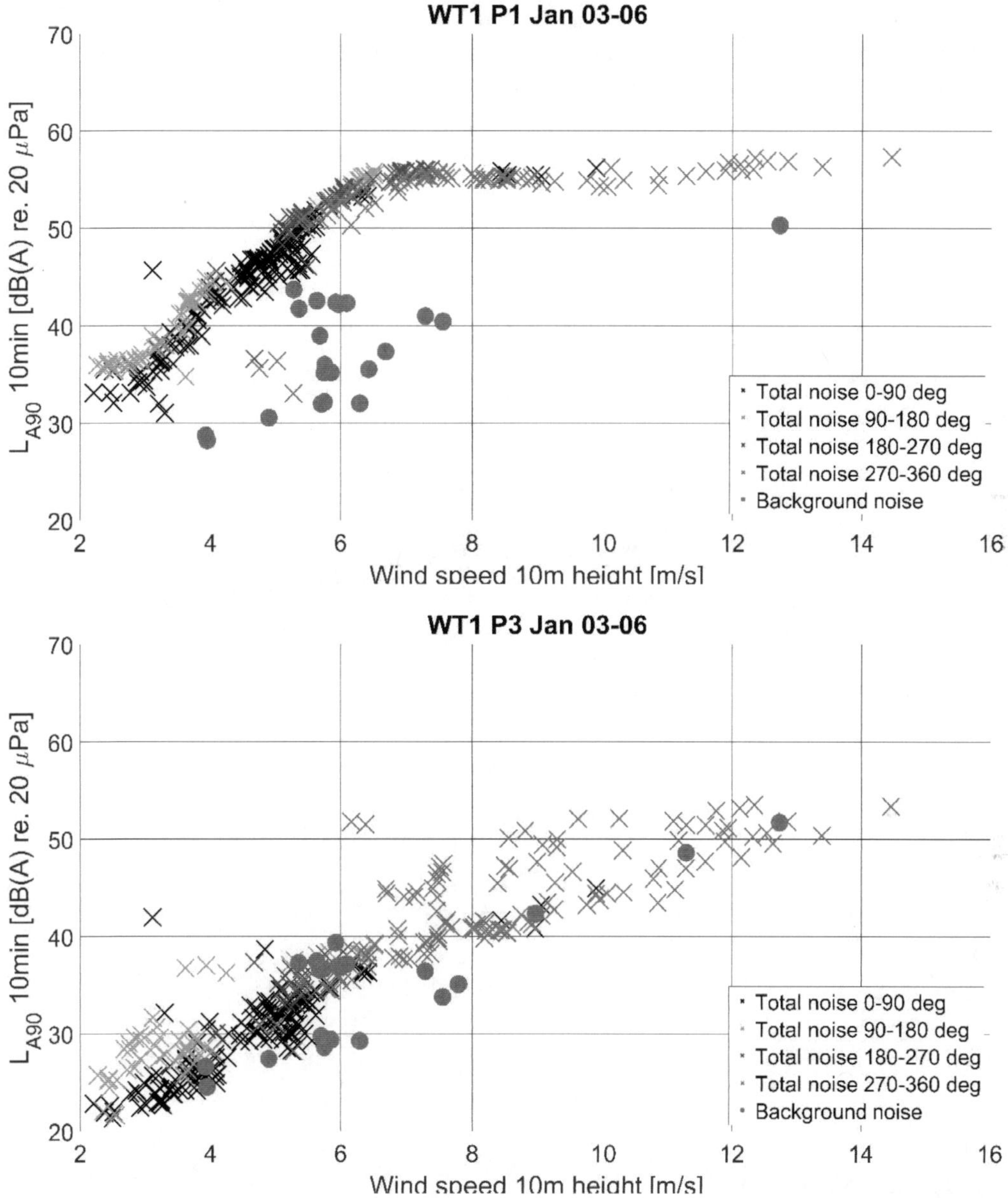

Figure 6: L_{A90} for different wind speeds and direction. Top: Measurement close to the turbine, distance approximately the same as the total height of the turbine. Bottom: Measurement at dwelling, distance approximately 4 times the total height of the turbine.

In Figure 6 a plot from one of the measurements [20] is shown. The top plot shows the results of the measurement close to the wind turbine (approximately a sound power level measurement). There is a clear difference between the background noise and the total noise. For the bottom plot simultaneous data at the nearby dwelling is shown (a distance approximately 4 times the total height of the turbine) where the background noise level is approximately the same as the total noise.

6. Background noise in L_{WA} measurements

When the sound power level of wind turbines is measured the background noise level is also measured. These measurements are performed in a downwind distance from the turbine equal to the total height of the turbine [3], [10]. A Danish study from 2016 [5] compared the sound power level versus wind speed for a large number of wind turbines. In addition, the difference between measured noise levels with the turbines on and off was also shown, Figure 7. Typically, these measurements are performed in daytime for practical reasons. The largest difference in noise level with turbines on and off are in the 6-8 m/s wind speed interval.

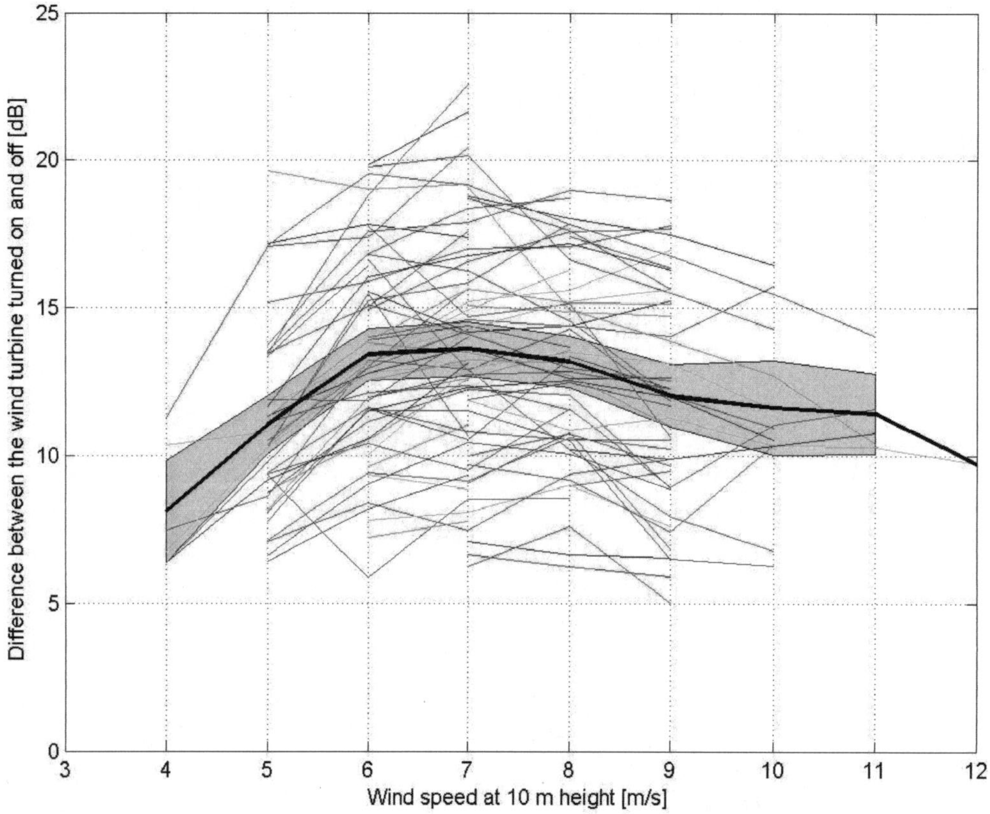

Figure 7: Typical differences between wind turbines on and off versus wind speed in 10 m height for 51 different wind turbines measured over a total of 74 days. The difference for each measurement is shown together with the arithmetic mean + 95 % confidence level.

7. Vegetation noise

Vegetation noise as a function of wind speed is not often documented in Denmark (unless as part of background noise for sound power level measurements for wind turbines). When documented it is typically for short periods (between hours and within a day) where some examples are shown in [21].

As part of the DecoWind project [22] long term monitoring was performed at a relatively quiet site with a single wind turbine. The wind turbine was shut down during several periods of the measurement campaign to determine the background noise. The L_{Aeq} in a frequency range of 50 Hz to 8000 Hz and for 10 seconds periods were computed to characterise the background noise. The L_{Aeq} versus wind speed is shown in Figure 8 for two different microphone positions and filtered according to day and night time. The noise level in general is higher in the day period than in the night period. There is some correlation between wind speed and noise level. However, the scatter of the data is rather large and masks the correlation between wind speed and noise level. This scatter is because some of the measured background noise was caused by human or animal activity at the site. The activity was lower during night than during day. Because of the large amount of data, it is not possible to filter manually for vegetation noise. It is necessary to develop an algorithm that performs this filtering. Filtering might help to reveal the correlation between vegetation noise and wind speed more clearly.

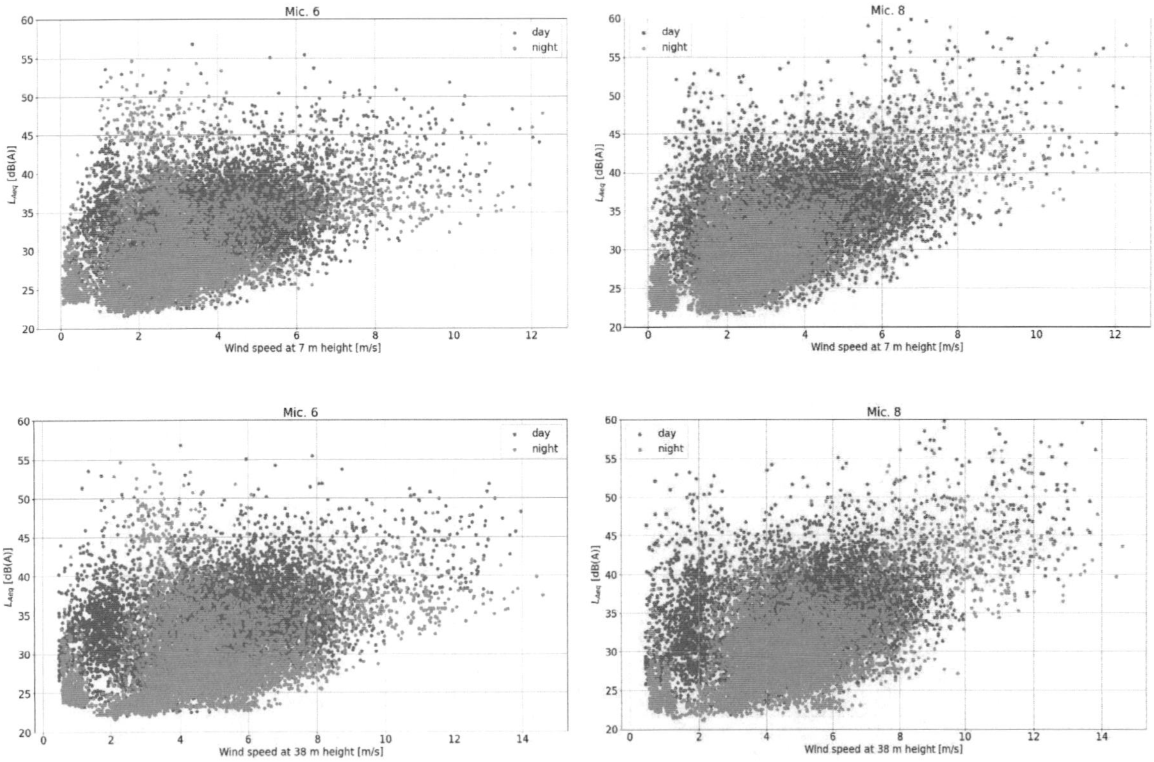

Figure 8: Background noise (primarily vegetation noise) as a function of wind speed measured between March 4th and April 9th of 2020. The data has been filtered for time of day, where day is between 5h to 18h and night between 18h to 5h. The daytime data set contains 7788 samples of background noise, the night time data set contains 7634 samples. The data is plotted both as a function of wind speed determined by a sonic anemometer at 7 m height and a cup anemometer at 38 m height. The instruments were positioned approximately at a distance of 600 meters to microphone 6 and 1000 m to microphone 8.

8. Noise from waves

Most Danish onshore wind turbines are positioned inland, but a few are placed at the beach/harbour. The background noise data obtained from a sound power level measurement on wind turbine located on a beach placed at the North Sea is shown in Figure 9. There seems to be a weak relationship between wind and noise.

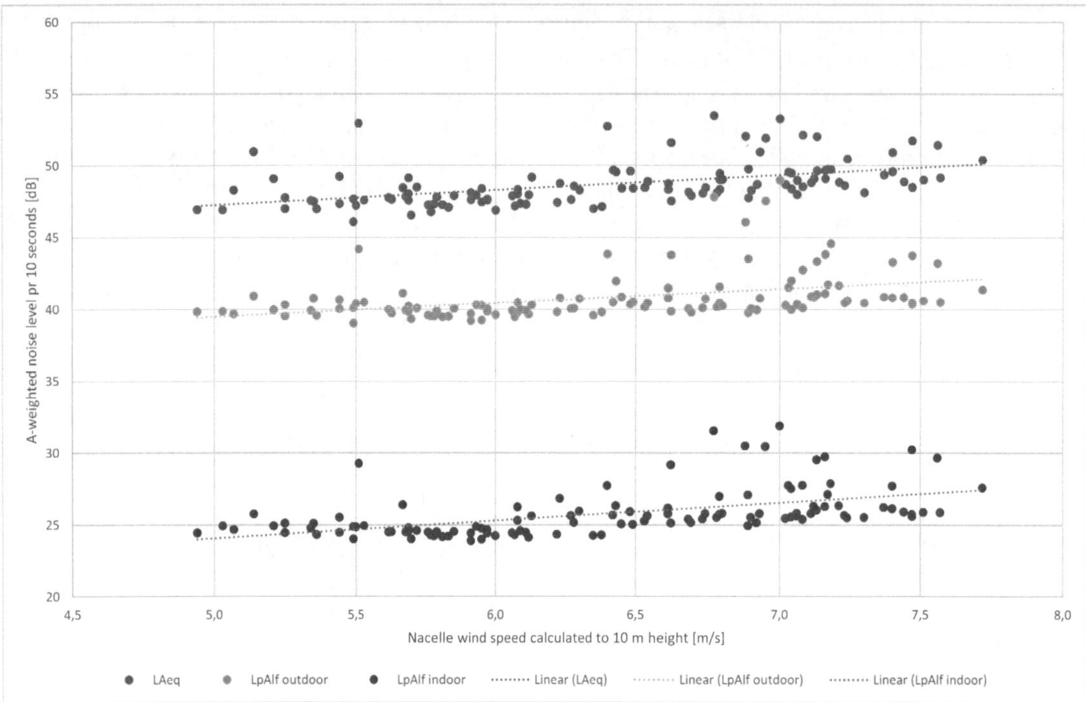

Figure 9: Measurements of background noise over 2 x 10 minutes at a beach around noon at the western part of Denmark performed in relation to sound power level measurements of wind turbines (wind turbines was off during this part of the measurement). The microphone was on a plate on the ground, and there was a small sand dune between the microphone and the water – approximately distance 20-30 m between microphone and water.

Another measurement is part of the DecoWind project [23] where measurements was performed at Dragstrup Vig which is part of Limfjorden. The microphone was positioned just next to the water. The relationship between wind speed and noise level is shown in Figure 10.

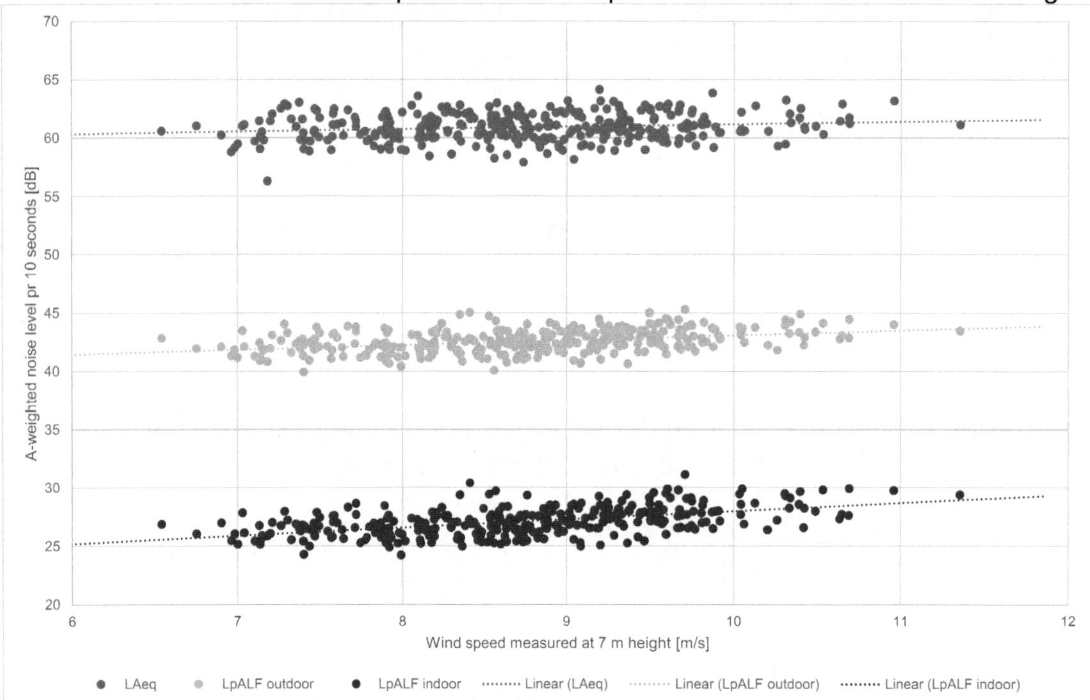

Figure 10: Measurement of background noise for different periods as part of the over water measurements at Dragstrup Vig at the 8 km distance position. Total of approximately 50 minutes during daytime. The microphone was on a tripod with an approximate height of 1.5 m above ground and approximately 3 m from the water.

9. Socio-acoustic study

As part of the DecoWind project [22] a socio-acoustic study was carried out [24]. The annoyance of 68 neighbours to pitch regulated wind turbines over a period of 6 weeks was surveyed by use of an app-based daily questionnaire. Additionally, both the short time noise level and the 24h noise level was modelled. Only neighbours who could hear the turbines were recruited. The wind speed in the period of the study was lower than typical, and as a result only few periods with a maximum level of 44 dB was achieved. The results presented in Figure 11 show that the participants in the study were in general not at all annoyed. It cannot be concluded whether this is also true for higher wind speeds, but at least it can be concluded that wind turbine noise was in general not at all annoying under present conditions during the study. It can also be seen that even estimated noise levels up to 44 dB were found to be not annoying at all – even though few occurrences occurred.

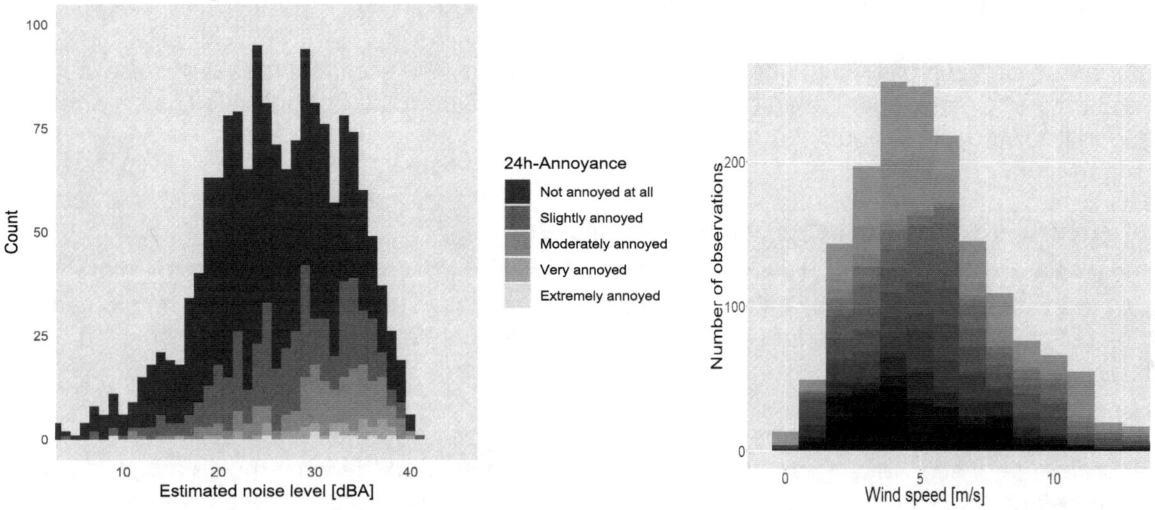

Figure 11: Selected results from the socio-acoustic study in the DecoWind project. Left: Estimated noise level and annoyance. Right: Wind speed

10. Conclusions

To sum the findings from the previous sections, the Danish noise regulations are originally based on assumptions that vegetation noise and background noise in generally to some degree masked the wind turbine noise. An early study of vegetation noise, wind turbines and audibility concluded that the wind turbine noise was masked if the noise level in each critical band was at least 4 dB lower than the masking background noise/vegetation noise. In this document only absolute levels have been shown, and firm conclusion on audibility can therefore not be drawn, but only hinted.

Since then, wind turbines have developed – both in size and in design. One study has investigated differences in wind speed at different height. Based on the findings it can be speculated that there is a risk of the wind turbines being more audible the higher they are.

Other studies have compared wind turbine noise measured at dwellings with calculated noise following the default Danish method for determining the noise level at dwellings, with the general conclusion that often background noise is of the same magnitude as the wind turbine noise.

Measured wave noise shows noise levels higher than the Danish wind turbine noise limits hinting that at least for these locations wind turbines might not be audible. Measured vegetation noise for a single site shows noise levels between 25 – 55 dB with wind speeds of 6-8 m/s hinting that at least some of the time a wind turbine at this position might/will be audible.

A recent socio-acoustic study showed that the participating wind turbine neighbours in general were not at all annoyed by the noise, however the wind speed for the study period was in general lower than the necessary wind speed for the wind turbines to reach rated power.

The possible total height of wind turbines continues to grow, and with the Danish setback distance rule of minimum distance to neighbours of 4 times the total height the possible distance to neighbours continues to grow as well. In Denmark most wind turbines are installed in areas with low population density, and thereby possible sounds which can mask the noise from the turbines are often only natural sounds like vegetation noise and wave noise. Both vegetation noise and wave noise are also wind driven, however with growing height and growing distances there is a risk that correlation between the wind which drives the wind turbines and the wind which excite vegetation and waves grows weaker, and that the masking possibility of vegetation noise and wave noise are diminished.

The purpose of this project is to better describe and quantify the potential masking noise from vegetation and waves, both based on collected data from Denmark, but potentially also with data and knowledge from outside Denmark.

11. Future and on-going work in the project

Focus of the project so far has been on gathering Danish data and knowledge (some of it shown in this paper) and establishing possible collaboration. Further literature will be studied, and by collaboration hopefully data and knowledge from outside Denmark will be gathered as well.

Collaboration with Karl Bolin has been established, who has performed short term measurements of sea wave noise, for the Baltic Sea, and studied the wave noise as a function of wave height [25]. Computation with the empirical model for vegetation noise presented in [26] will be compared to the vegetation noise data gathered in the present study. We further plan to compare our findings with studies of related topics that have been presented in the literature [27]–[30].

In addition to the already gathered Danish data set, measurement systems with microphones close to the shore of a fjord and nearby vegetation have been set up. These systems will have already been collecting data for 3 weeks on the day of the deadline of this paper. It is the intention to measure additional sea noise but at the North Sea, and these data will be compared with the results presented in [25] and [26].

In this paper only the total noise level has been shown, but it is the intention to compare wind turbine noise with the masking noise for critical 1/3 octave bands, and possibly perform listening tests as well.

Acknowledgements
Financial support for this work was partly provided by “Participation to IEA Wind Task 39” (grant 134-21022) under the Energy Technology Development and Demonstration Programme (EUDP).

References

[1] Miljø- og fødevareministeriet, *Bekendtgørelse om støj fra vindmøller nr. 304 af 14/05/1991*. Denmark: Miljø- og Fødevareministeriet, 1991. [Online]. Available: https://www.retsinformation.dk/eli/lta/1991/304

[2] H. Sandbech and P. B. Suhr, *Brev til amtsrådene af 28. april 1988*. 1988.

[3] IEC, "IEC 61400-11 Wind turbines - Part 11: Acoustic noise measurement techniques." 2012.

[4] P. Alexandersen, "Stort behov for et tredje test-center til store vindmøller," *Green Power Denmark*, Aug. 08, 2022. https://greenpowerdenmark.dk/nyheder/stort-behov-tredje-testcenter-til-store-vindmoeller

[5] L. S. Søndergaard and C. Backalarz, "Støj fra store , nyere danske vindmøller som funktion af vindhastigheden," Miljø- og Fødevareministeriet, 2016. Accessed: Apr. 23, 2023. [Online]. Available: https://www2.mst.dk/Udgiv/publikationer/2016/04/978-87-93435-66-7.pdf

[6] L. S. Søndergaard and C. Backalarz, "Noise from wind turbines and health effects - Investigation of wind turbine noise spectra," in *6 th International Conference on Wind Turbine Noise*, Wind Turbine Noise Conference, Glasgow, 2015, 2015.

[7] Miljø- og fødevareministeriet, *Bekendtgørelse om støj fra vindmøller, BEK nr 1518 af 14/12/2006*, vol. 2006, no. 1518. 2016, pp. 1–9.

[8] Miljøministeriet, *Bekendtgørelse om støj fra vindmøller*. 2011.

[9] J. Jakobsen, "Danish Regulation of Low Frequency Noise from Wind Turbines," *Journal of LOW FREQUENCY NOISE, VIBRATION AND ACTIVE CONTROL*, vol. 31, no. 4, pp. 239–246, 2012.

[10] *Bekendtgørelse om støj fra vindmøller*, no. 135. Denmark: Miljø- og Fødevareministeriet, 2019, pp. 1–24.

[11] Erhvervsministeriet, *Bekendtgørelse om planlægning for og tilladelse til opstilling af vindmøller*. 2019.

[12] J. Jakobsen and T. H. Pedersen, "Støj fra vindmøller og vindstøjens maskerende virkning.," Lydteknisk Institut, Apr. 1989.

[13] G. P. Van Den Berg, "Effects of the wind profile at night on wind turbine sound," *J Sound Vib*, vol. 277, no. 4–5, pp. 955–970, Nov. 2004, doi: 10.1016/j.jsv.2003.09.050.

[14] K. D. Madsen and B. Plovsing, "Støj fra vindmøller om natten," 2012. [Online]. Available: https://www2.mst.dk/Udgiv/publikationer/2012/05/978-87-92779-91-5.pdf

[15] L. S. Søndergaard, "Noise from wind turbines under non-standard conditions," in *41st International Congress and Exposition on Noise Control Engineering 2012, INTER-NOISE 2012*, 2012.

[16] B. Søndergaard, J. Heiden, and N. F. Christensen, "Måling af støj hos vindmøllenaboer Mindre vindmøllepark Målinger i dag og nat perioden - Miljøportalen nr. 2048," Nov. 2018.

[17] B. Søndergaard, J. Heiden, and N. F. Christensen, "Måling af støj hos vindmøllenaboer : større vindmøllepark målinger i dagperioden - Miljøprojekt nr. 2096," Miljøstyrelsen, Aug. 2019.

[18] R. Egedal, N. F. Christensen, and B. Søndergaard, "Danish experiences with measuring wind turbine noise at neighbor dwellings," in *8 th International Conference on Wind Turbine Noise Lisbon-12 th to 14 th*, 2019.

[19] C. Søndergaard, Lars Sommer; Egedal, Rune; Hansen, Morten Bording; Pedersen, Torben Holm; Backalarz, "Støjmålinger hos naboer til vindmøller i Nollund," 2017.

[20] R. Egedal, M. B. Hansen, and L. S. Søndergaard, "Wind turbine noise at neighbor dwellings, comparing calculations and measurements," in *INTER-NOISE 2017 - 46th International Congress and Exposition on Noise Control Engineering: Taming Noise and Moving Quiet*, 2017.

[21] L. S. Søndergaard, R. Egedal, and M. B. Hansen, "Variation of wind induced non-turbine related noise due to position, shelter, wind direction and season," in *7th International Conference on Wind Turbine Noise Rotterdam – 2nd to 5th May 2017*, 2017, pp. 1–14.
[22] F. Bertagnolio *et al.*, "DecoWind: Development of low-noise and cost-effective wind farm control technology," in *Inter-noise 2022 Noise Control in a More Sustainable Future*, 2022. Accessed: Apr. 12, 2023. [Online]. Available: https://az659834.vo.msecnd.net/eventsairwesteuprod/production-inconference-public/3889aefd315848d6bf9656a24c0a820d
[23] K. Søndergaard, Lars Sommer; Thysell, Erik; Claumarch, Christian; Vignaroli, Andrea; Thomsen, Carsten; Hansen, "Long distance noise propagation over water for an elevated height-adjustable sound source," *9 th International Conference on Wind Turbine Noise Remote from Europe 18 th to 21st May 2021*, no. May, pp. 1–6, 2021.
[24] L. S. Søndergaard, C. P. Volk, T. R. Hansen, L. Enggaard, T. Sørensen, and A. Pena, "9th International Conference on Wind Turbine Noise," in *Wind farm neighbourship investigated by a daily app questionnaire combining weather, noise, and annoyance*, Online conference: INCE Europe, 2021, pp. 1–15.
[25] K. Bolin and M. Åbom, "Air-borne sound generated by sea waves," *J Acoust Soc Am*, vol. 127, no. 5, pp. 2771–2779, May 2010, doi: 10.1121/1.3327815.
[26] K. Bolin, "Prediction method for wind-induced vegetation noise," *Acta Acustica united with Acustica*, vol. 95, no. 4, pp. 607–619, Jul. 2009, doi: 10.3813/AAA.918189.
[27] S. Cooper, T. Acoustic, and G. Pty, "Hiding Wind Farm Noise in Ambient Measurements - Noise Floor , Wind Direction and Frequency Limitations," in *5th International Meeting on Wind Turbine Noise*, 5th International Meeting on Wind Turbine N, 2013.
[28] M. Terlich, "The Case for Spectral Measurements of Ambient Noise Levels in the Assessment of Wind Farms," *Fourth International Meeting on Wind Turbine Noise*, no. April, pp. 1–6, 2011.
[29] D. Mennitt, K. Sherrill, and K. Fristrup, "A geospatial model of ambient sound pressure levels in the contiguous United States," *J Acoust Soc Am*, vol. 135, no. 5, pp. 2746–2764, May 2014, doi: 10.1121/1.4870481.
[30] H. F. Boersma, "Characterization of the natural ambient sound environment: Measurements in open agricultural grassland," *J Acoust Soc Am*, vol. 101, no. 4, pp. 2104–2110, Apr. 1997, doi: 10.1121/1.418141.

INCE EUROPE

10th International Conference
on
Wind Turbine Noise
Dublin – 21st to 23rd June 2023

Uncertainties on measured and predicted noise levels for wind farms. Do we need it?

Bo Søndergaard, SWECO Denmark A/S bo.sondergaard@sweco.dk

Summary

Assessment of the noise from a wind farm usually gives you a decibel value from which the authorities decide compliance with noise criteria or not. Not many considers the validity of this number or tries to make a qualification through an uncertainty evaluation. Measurement of noise from wind farms can be difficult owing to the dependence of wind speed and direction as well as other meteorological parameters. Most often the noise level is close to the noise limits and the noise limits are close to the background noise in the area. A series of measurement principles are listed in the coming Technical Specification IEC 61400-11-2 which also includes a proposal for an evaluation of the uncertainty. From the discussion behind this document, it is not clear if it is operational in all cases and practice on how to apply these uncertainties is different in different regions. This presentation will focus on the necessity of uncertainties and how to use them if at all.

1. Introduction

Noise from wind turbines has kept consultancies busy for many years by now. Assessment of noise contributions, discussion/interpretations of the results, discussion/interpretation of the regulations, citizens meetings etc. All about whether the noise from a wind farm is too much or not and if there is compliance with regulations or not. The first part is subjective and can not be resolved easily but compliance with regulations should be easy or...

In a Danish perspective you are sometimes surprised with the situation in other countries. The regulation in Denmark is quite simple and does not leave much room for discussion on compliance or not.

From the discussion in the project team PT 61400-11-2; writing a Technical Specification document on “Measurement of wind turbine noise characteristics in receptor position”, it became clear that estimating the uncertainty on the result of a noise measurement can be quite difficult and sometimes not really wanted.

2. Uncertainty

One of the first books I got at the university was "Elementary Uncertainty Calculation" and I guess that everybody working in engineering have had something similar. We learned that the accuracy of a measurement is limited by the measurement principles. This also applies to noise measurements. When you have a measured noise level you should also be able to determine the uncertainty following the noise level.
If you have to make decisions based on the measurement you should know the size of the uncertainty.

In IEC 61400-11 "Acoustic noise measurement techniques": 2018 there is a comprehensive method for the assessment of the averages sound power spectra in individual wind bins and the accompanying uncertainty. The method works well, but this is also close to being an ideal situation, usually with good signal to noise ratio, short distances and well defined measurement conditions.

A simplified version of this method is introduced in TS IEC 61400-11-2. However, noise measurements at receptor positions is more complex with the total sound, the specific sound and the residual sound being very close. In such cases the calculation of the uncertainty from the distribution of the data is more like an estimate.

3. Compliance or no compliance

How can knowledge about the uncertainty be used when evaluating compliance.

If the signal to noise ratio is good, it is possible to estimate the uncertainty from the data based uncertainties and the contributions from measurement equipment and similar.

In many cases the compliance evaluation is evaluated based on the reported noise level alone with reference to a guideline or similar and the data filtering or binning used. This can be entirely acceptable if the applied methodology is well known, well documented and the expected uncertainty is low (from experience?).

3.1 Legal principle

A general legal principle on deciding compliance or not is described in Figure 1.

There 6 possible outcomes.

1) If the measurement result + expanded uncertainty is less than the noise limit there is compliance
2) If the measurement result ÷ expanded uncertainty is higher than the noise limit there is noncompliance
3) If the measurement result is less than the noise limit and the measurement result + expanded uncertainty is larger than the noise limit. Compliance/noncompliance?
4) If the measurement result is larger than the noise limit and the measurement result ÷ expanded uncertainty is less than the noise limit. Compliance/noncompliance?

The outcomes 3 and 4 are sometimes subject to discussion on whether there are compliance or noncompliance. 3) is most often considered compliance while 4) is a little more tricky. However, If the situation is describing the result of a measurement of speeding, the most common procedure is to subtract the uncertainty from the measurement results before comparing with the speed limit. Hence 4) will describe a situation where it is not possible to determine noncompliance (with X% confidence).

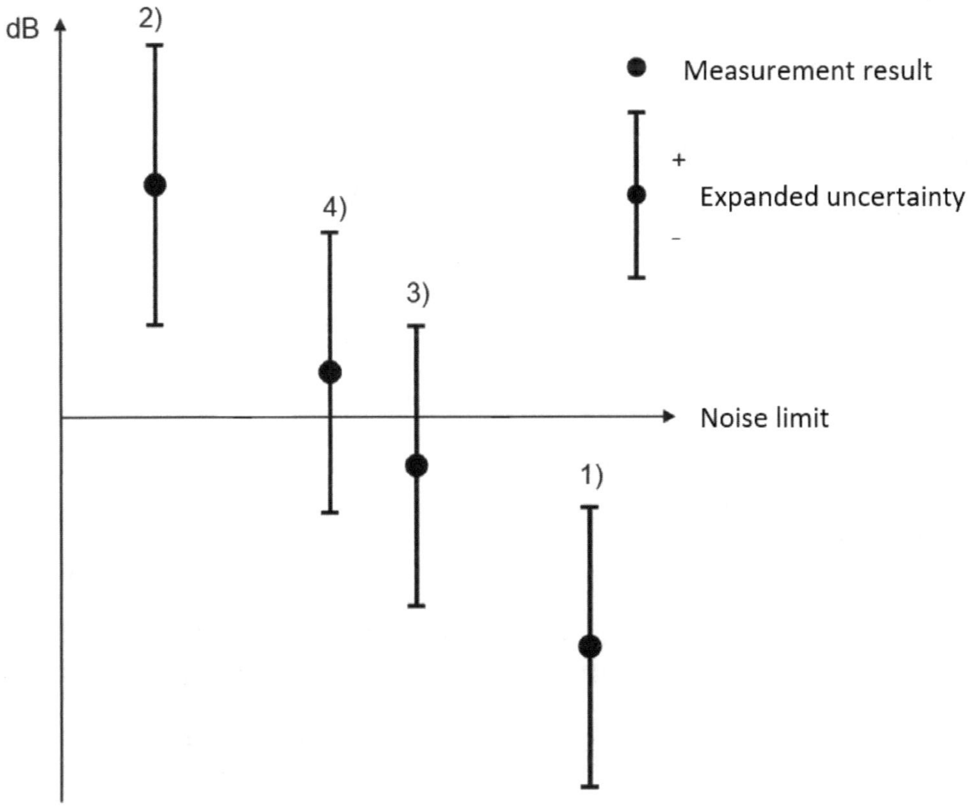

Figure 1 Illustration of measurement results and uncertainty in relation to noise limit.

This principle can only be used if there is an uncertainty connected to the measurement result. The uncertainty can be based directly on the measurement results in which case it will vary depending on the measurement or it can be a fixed uncertainty which is connected to the measurement method. E.g. in Denmark the uncertainty is a fixed value of 2 dB. But if the principle is reversed 2)-4) all lead to noncompliance.

It is important that the uncertainty is small since large uncertainties can lead to lack of confidence in the principle. The uncertainty can usually be reduced by suitable binning and filtering of the data according to operation of the wind turbine and meteorological parameters.

3.2 No uncertainty

If the guideline for the measurements does not include an assessment method for the uncertainty or an estimate of the uncertainty from the selected measurement method only the measurement result is available for comparison with the noise limit. In this case the decision of the authorities on compliance is based on the presentation of the data and the conditions for the measurements but in general this is a question of trust in the company/persons behind the measurements.

4. Other characteristics of wind turbine noise.

In TS 61400-11-2 recommendations for determining the level of the dynamical characteristics like amplitude modulation, tonality and impulsivity presented. The recommendations give detailed guidance on how to determine the levels and statistics for these parameters in bins or in general. This information is not converted into uncertainties making it difficult to evaluate if the levels should lead to noncompliance or not.However

5. Conclusions

The uncertainty on a measured noise level is as important as the noise level itself. If there is no uncertainty reported or related to the measurement guideline all decisions on compliance are reduced to trust.

References and interesting literature related to noise measurements and data reduction

- ISO 1996-1:2016 Acoustics — Description, measurement and assessment of environmental noise — Part 1: Basic quantities and assessment procedures
- Hansen K.L et al Acoust Aust DOI 10.1007/s40857-017-0114-7 "Wind Farm Noise Uncertainty: Prediction, Measurement and Compliance Assessment"
- IEC 61400-11:2012+AMD1:2018+Cor1:2019 Wind turbines — Acoustic noise measurement techniques, edition 3.1
- TS IEC 61400-11-2 Wind energy generation systems – Part 11-2: Measurement of wind turbine noise characteristics in receptor position. To be published.
- Asthiani P. 5th International Conference on Wind Turbine Noise Denver 28-30 August 2013 "Generating a better picture of noise immissions in post construction monitoring using statistical analysis"
- Propagation of wind turbine noise: measurements and model evaluation. Nyborg C.M. et al. To be published. This is part of the Decowind project lead by DTU Wind Energy

10th International Conference

on

Wind Turbine Noise

Dublin – 21st to 23rd June 2023

High resolution analysis of measurements, and comparison of models for long distance noise propagation over water for an elevated height-adjustable sound source

Erik Thysell erth@forcetechnology.com, Rune Egedal rue@forcetechnology.com, Lars Sommer Søndergaard lss@forcetechnology.com, Carsten Thomsen ctho@forcetechnology.com

Acoustics, Noise and Vibrations, FORCE Technology, Denmark.

Thomas Sørensen ts@emd.dk

EMD International A/S

Franck Bertagnolio frba@dtu.dk, Andreas Fischer asfi@dtu.dk

DTU Wind and Energy Systems, Wind Turbine Design Section, Denmark.

Summary

This paper compares the results of a measurement campaign for downwind noise propagation over water for elevated sound sources with different relevant propagation models. The analysis of the measurements utilizes a technique that increases the signal-to-noise ratio when analysing tone sweeps, that were propagated from the sound source, compared to the common 1/3 octave band analysis previously used. Comparison of these results are made to a number of models and methods, both non-country specific models like the Nord2000 model, WindSTAR-Pro and ISO 9613-2, but also country specific models like the models used in Denmark and Sweden. Noise propagation over water for elevated noise sources is specifically relevant for offshore wind turbines, near shore wind turbines or wind turbines on land close to large water bodies. The measurement setup uses a height-adjustable sound source (81 m, 50 m and 30 m above ground) and microphones positioned downwind (at shore and ~100 m inland) from the sound source (~3 km, ~5 km and ~7 km distance over water). The meteorological conditions (wind speed, wind direction, atmospheric stability, temperature, humidity, etc.) were monitored continuously at both ends of the setup, utilizing both a tall met mast, a wind profiler and sonic anemometers at multiple heights.

1. Introduction

In the DECOWIND [1] project measurements of long-range noise propagation over water were carried out, with a loudspeaker as sound source, both octave band filtered noise as well as tone sweeps were used as noise signals. In [1] the analysis was limited to using normal 1/3 octave filters (IEC 61260) in the frequency range 50-630 Hz.

This paper documents analysis of the tone sweeps using high resolution Fast Fourier Transform (FFT) which results in an improved signal-to-noise ratio, compared to 1/3 octave band filters.

Unless otherwise stated, the unit for values is dB (Z-weighted).

2. Measurements

The measurements were carried out at night in the time span between 2020-09-30 20:00 – 2020-10-01 00:30. The wind was measured using multiple instruments see [1] at heights between 5 - 290 m. The wind speed varied between 2.5 – 5.5 m/s at 5 m above terrain and 7.8 – 10.3 m/s at 290 m. The wind direction varied between 105 – 130 °. Compared to the angle from the immission points to the loudspeaker that was about 91° at 3 km, about 100° at 5 km and about 112 ° at 7 km, all immission points were downwind from the sound source.

The temperature gradient was measured between 44 and 118 m and varied between -0.26 and +0.02 K/100m during the whole measurement period.

Measurements were carried out for three separate heights of the sound source: 30 m, 50 m and 80 m above terrain respectively. The sound signal was recorded at three distances from the source: about 3 km, 5 km and 7 km. At each distance, two microphones were positioned: one on the seashore and one at 40-90 m from the shore. The microphones positioned on the seashore is hereafter denoted "Water" and the microphones positioned at a distance from the shore is denoted "Land". All microphones were positioned 1.5 m above terrain. Information about terrain elevation and distance to shore for each position is given in Table 1.

Distance	Position	Terrain elevation above sea level	Distance to shore
3 km	Water	0.1 m	3 m
3 km	Land	0.4 m	91 m
5 km	Water	0.7 m	5 m
5 km	Land	6.8 m	42 m
7 km	Water	0.9 m	4 m
7 km	Land	6.3 m	70 m

Table 1. Terrain elevation and distance to shore for each measurement position.

The sound signal was also recorded about 2 m from the loudspeaker, in order to have a reference signal.

The sweeps consisted of a linear frequency sweep (constant df/dt) with a sinusoidal signal between 40-1100 Hz with a duration of 0.5 s. Sweeps were repeated 18 times for a total of 9 s, then a pause for 3 s. The 9 + 3 s = 12 s was then repeated 31 times for a total of about 6 minutes. For each height, four 6 minutes of sweeps were carried out with about 10 minutes between them, so that for each height the measurements are spread out over about 50 minutes in total. This was done to enable an averaging over the most turbulent changes of the atmosphere [2].

2.1 Analysis of tone sweeps

A repeated swept sine signal was used to achieve a signal to noise ratio about 10-15 dB better than using band limited noise, due to its higher crest factor. Hence the loudspeaker system could be driven at a higher peak level without clipping.

By using a swept signal repeated at a 2 Hz rate, a frequency spectrum with discrete lines spaced at 2 Hz intervals is obtained. This gives an additional signal to noise rate enhancement because the non-sinusoidal random noise between the lines will be reduced by 3 dB for each halving on the frequency resolution. Thus, for example increasing the frequency resolution from 0.2 Hz to 0.1 Hz reduces the random noise by 3 dB and going from 1 Hz to 0.1 Hz reduces it by 10 dB without changing the levels of the peaks of the 2 Hz spectrum. A visual inspection of the peak to valley ratio of 10 dB indicates that an adequate signal to noise ratio is obtained. For example, examining Figure 1 at a 3 km distance, a signal to noise ratio of about 15 dB is obtained. In the actual analysis for this paper, a minimum signal-to-noise ratio of 6 dB was chosen as the threshold for including data.

It is important to note that with a 0.1 Hz frequency resolution (line spacing in the FFT spectrum), the FFT block size is 10 seconds, thus including all the 18 repeated sweeps in each measurement. This large block size results in a line spectrum which does not show any of the sweep characteristics we as humans perceive when listening to the signal. This is because the human auditory perception window is about approximately 30 ms, contrasted to the FFT window of 10 seconds. However, the frequency distribution and levels of the spectrum are correct.

Using a deterministic signal has the additional advantage that only the non-stationary characteristics of the atmosphere will affect the measurement, whereas if a band-limited third octave signal is used, its random characteristics are mingled with those of the atmosphere, and long averaging time would be necessary to separate the two. However, this assumes the atmosphere is stable during this period.

For each height and 6-minute sweep period an FFT analysis of tone each of the blocks of 18 sweeps was carried out with using a frequency resolution of 0.1 Hz giving a block size of 10 seconds. The sampling frequency was 51.2 kHz giving a total FFT block size of 512,000 samples. A flat-top window was use in the FFT analysis to reduce the picket-fence effect.

The FFT gives a spectrum with peaks for every 2 Hz (corresponding to the 2 Hz repetition rate of the sweeps). The spectral peaks at the loudspeaker minus the peaks at receiving position give the propagation loss as a function of frequency. These propagation losses are binned in each 1/3 octave band between 50-1000 Hz and averaged, to compare to more common analysis and prediction methods. This is illustrated for 100 Hz and 1000 Hz respectively in the figure below.

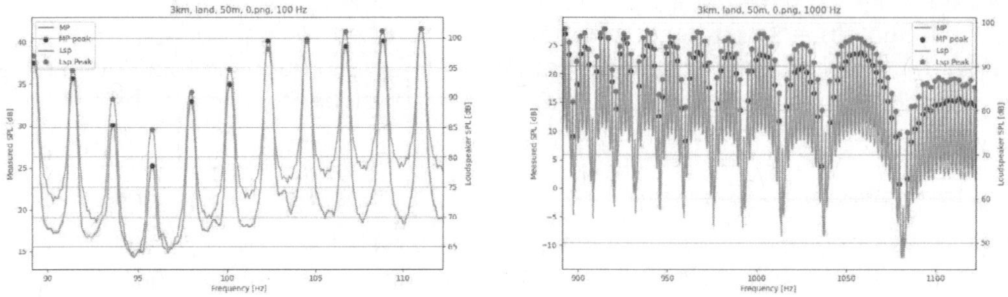

Figure 1. Illustration of FFT analysis for 100 Hz (left) and 1000 Hz (right) 1/3 octave bands. Green curve and points indicate immission point. Red curve and point indicate level measured at loudspeaker. MP = measurement at immission point (not loudspeaker). Lsp = measurement at loudspeaker. Peak (dots) indicate the local peak in for the corresponding curve, that was used to calculate the propagation loss.

For some frequencies and positions the signal to noise ratio was considered too low. This was the case for the 50 Hz 1/3 octave band for all immission points as well as 500 – 1000Hz for the 7km shore immission when the loudspeaker was at 30 and 50 m height.

3. Calculations

3.1 Calculation methods

Calculation of the sound propagation was carried out using the following methods:

- Danish method [3] (both overall as well as low frequency method, denoted DK and DK LF respectively)
- Finnish method [4] (low frequency method, denoted FI)
- Swedish sea method [5] (Denoted SWE Sea)
- ISO 9613-2 G = 0 and ISO 9613-2 A_{gr} = -3 [6] (denoted ISO G = 0 and ISO Agr = -3
- Nord2000 [7], [8]
- DTU Wind STAR-Pro [9]

The Danish method, known as "BEK 135" [3] calculates total level L_{pA} and low frequency L_{pALF} (10-160 Hz) based on 1/1- and 1/3 octave band values respectively. For each frequency, the attenuation is calculated using a logarithmic function of sound distance. A correction is applied for onshore and offshore terrain respectively (difference of 1.5 dB) and a frequency specific air absorption is applied. Specifically for offshore use, a component is added to account for multiple reflections. Multiple reflections will only apply at distances typically >2 km. Beyond the threshold distance multiple reflections will build up as a function of distance. The magnitude of multiple reflections is a function of frequency, elevation of noise source and wind direction. The BEK 135 assumes wind direction is coming from closest offshore turbine to receptor. The low frequency attenuation includes a noise insulation component for indoor sound. This has been removed in these calculations.

The Finnish wind turbine noise guideline from 2014 [4] describes a calculation method for low frequency sound specifically for offshore wind turbines. The method is, as the Danish method, a logarithmic function with distance, but with a different rate. Instead of a fixed correction for terrain, a frequency specific value is used and there is no contribution from multiple reflections.

The Swedish method described in “Mätning och beräkning av ljud från vindkraft” from 2013 [5] is a broadband calculation. It is similar to the Finnish method described above but replaces the frequency specific terrain correction value with a fixed value.

The ISO 9613-2 model [6] is the standard implementation of this model, including frequency specific ground attenuation and atmospheric absorption. In order to make is responsive to wind direction, a correction formulated by the German guideline TA-Lärm [10] has been applied. Two versions of the ISO 9613-2 model are applied: one using a ground factor of 0 and a second replacing the ground attenuation with a fixed value of -3 dB as recommended by the German “Interimsverfahren” guideline [11]. It must be noted that this is a guideline for onshore turbines but aims at negating ground attenuation.

Nord2000 described in [7] and [8] and ISO calculates in 1/3 octave bands and takes both sound speed vertical profile and geometry into account, including terrain as well as source and receiver height into account.

The WindSTAR-Pro model solves the Generalized Terrain Parabolic Equation (GTPE) and is implemented as described in [9]. In the present study, the two-dimensional, wide-angle, Crank-Nicholson, parabolic equation is used with a starter function for modelling a point source. An effective speed of sound is used to account for temperature and wind velocity gradients in the atmosphere. The ground impedance is calculated using the classical Delany-Bazley model, which uses the ground flow resistivity as an input. Further implementation details are given in [12] and [13].

3.2 1/3 octave band comparisons

To lower the displayed dynamic range and ease the effort when comparing 1/3 octave band results, measured and calculated levels are shown relative to free field and without air absorption, using the same distances and air absorption (from [3]) for both measurement and all methods. The resulting values are called “Terrain effect”.

The measurements are compared to the calculations in the following figures; Figure 2 – Figure 7. It can be seen that, none of the prediction methods are very good at predicting the actual measured 1/3 octave band level.

Below are the 3 km results shown in Figure 2 (Water) and Figure 3 (Land). Both Water position but even more so Land position shows a pronounced dip in the 1/3 octave band spectrum. For the Water position this varies between about 315-400 Hz for 80 m source height to about 500-630 Hz for the 30 m source height. This is not accurately reflected in the results from any of the calculation methods.

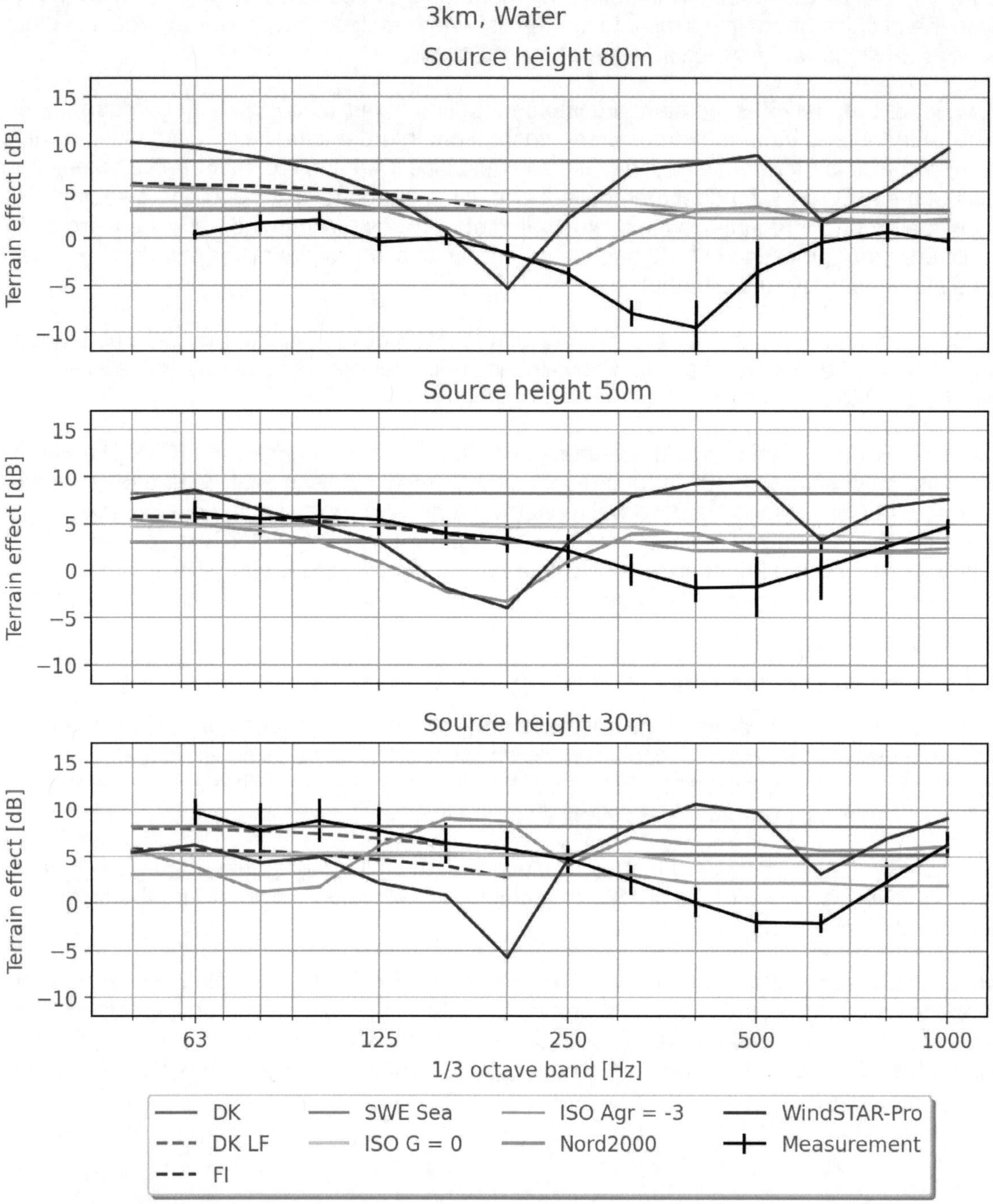

Figure 2. Plots showing measurement and calculation results for the different methods/models with the loudspeaker at 80m, 50m and 30m respectively at the 3km Water position. Error bars on measured curves shows standard deviation for the four sweeps.

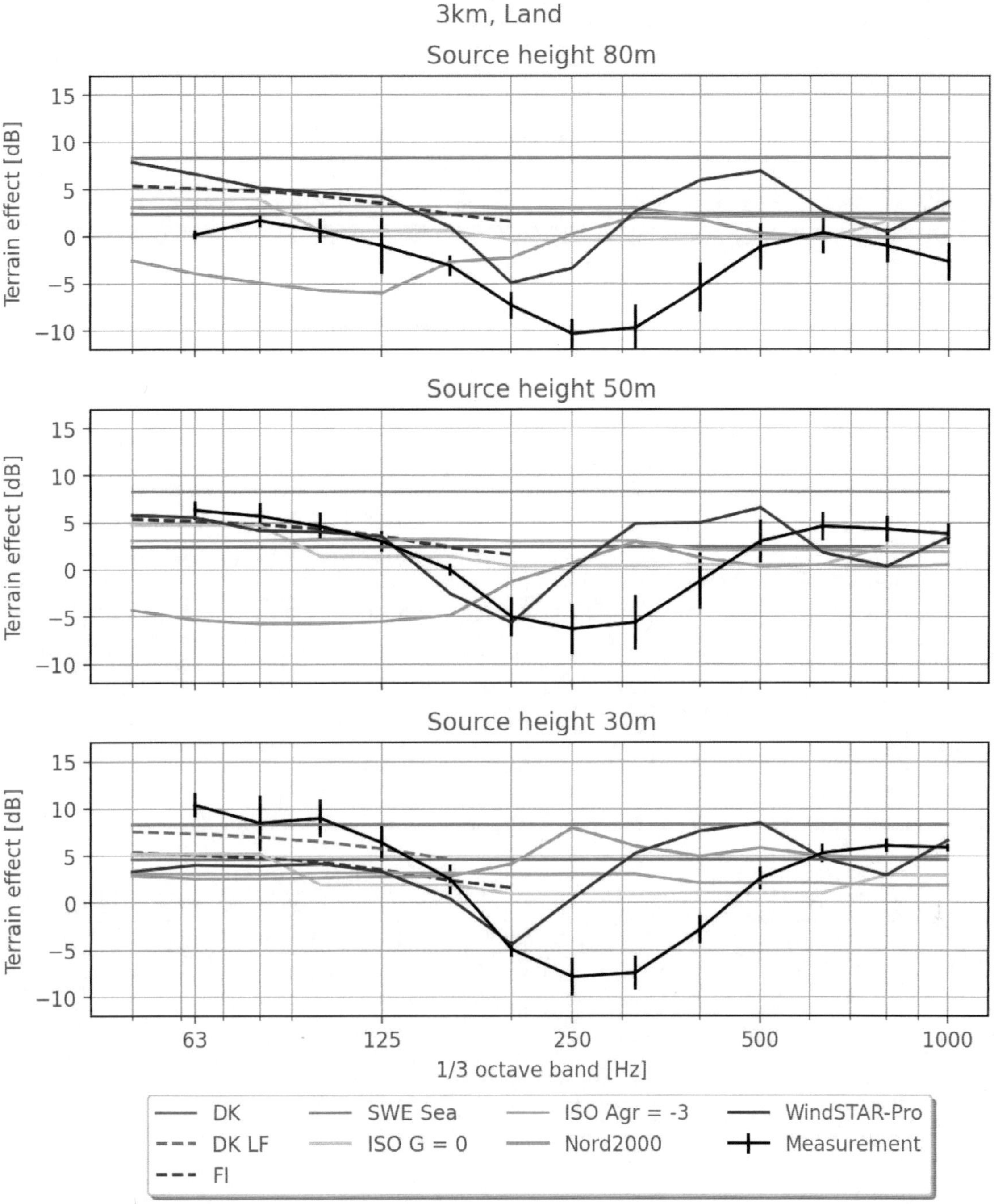

Figure 3. Plots showing measurement and calculation results for the different methods/models with the loudspeaker at 80m, 50m and 30m respectively at the 3km Land position. Error bars on measured curves shows standard deviation for the four sweeps.

Below are the 5 km results shown in Figure 4 (Water) and Figure 5 (Land). Similarly, to 3 km, there are some dips in the 1/3 octave band spectrum for both Water and Land.

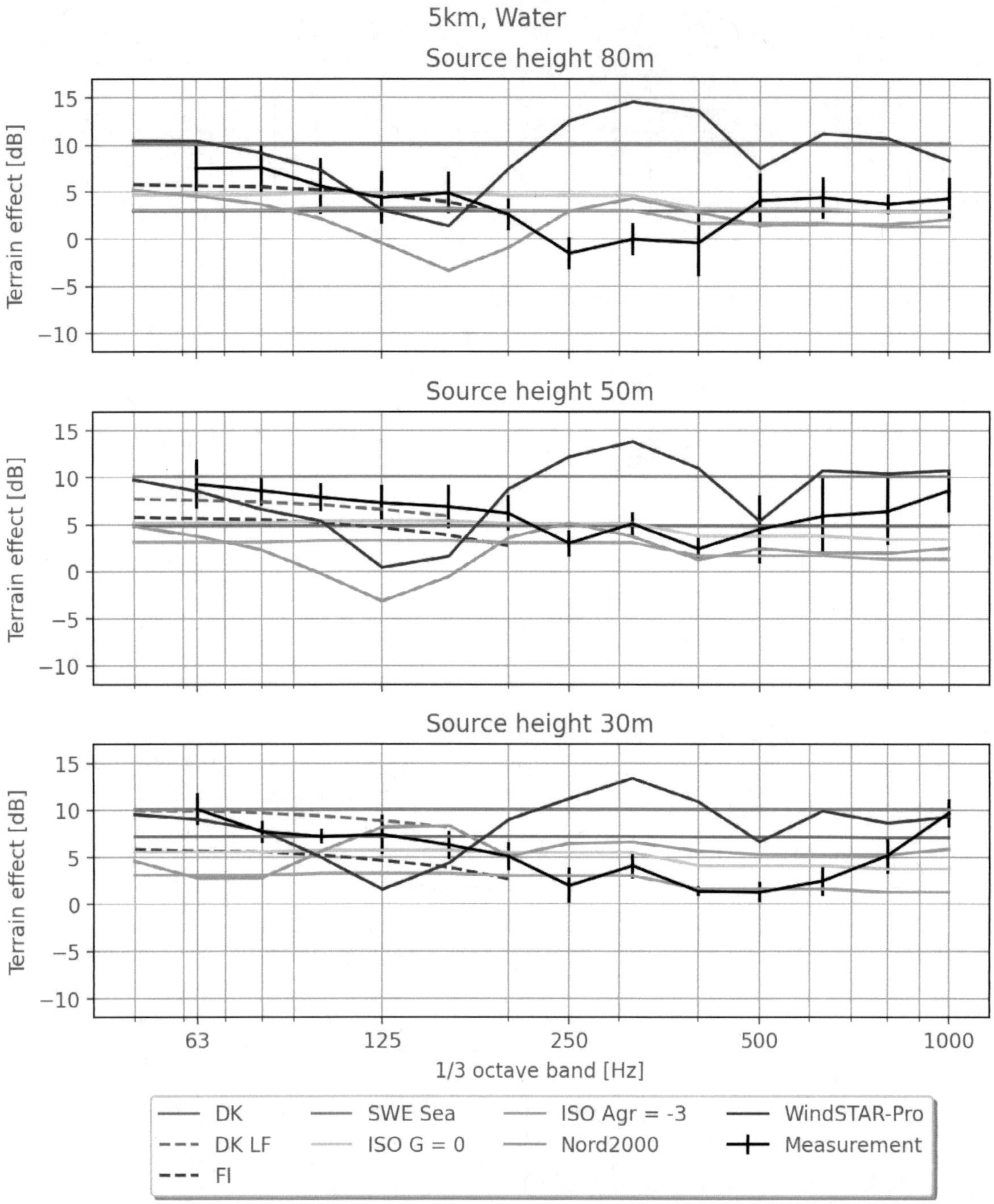

Figure 4. Plots showing measurement and calculation results for the different methods/models with the loudspeaker at 80m, 50m and 30m respectively at the 5km Water position. Error bars on measured curves shows standard deviation for the four sweeps.

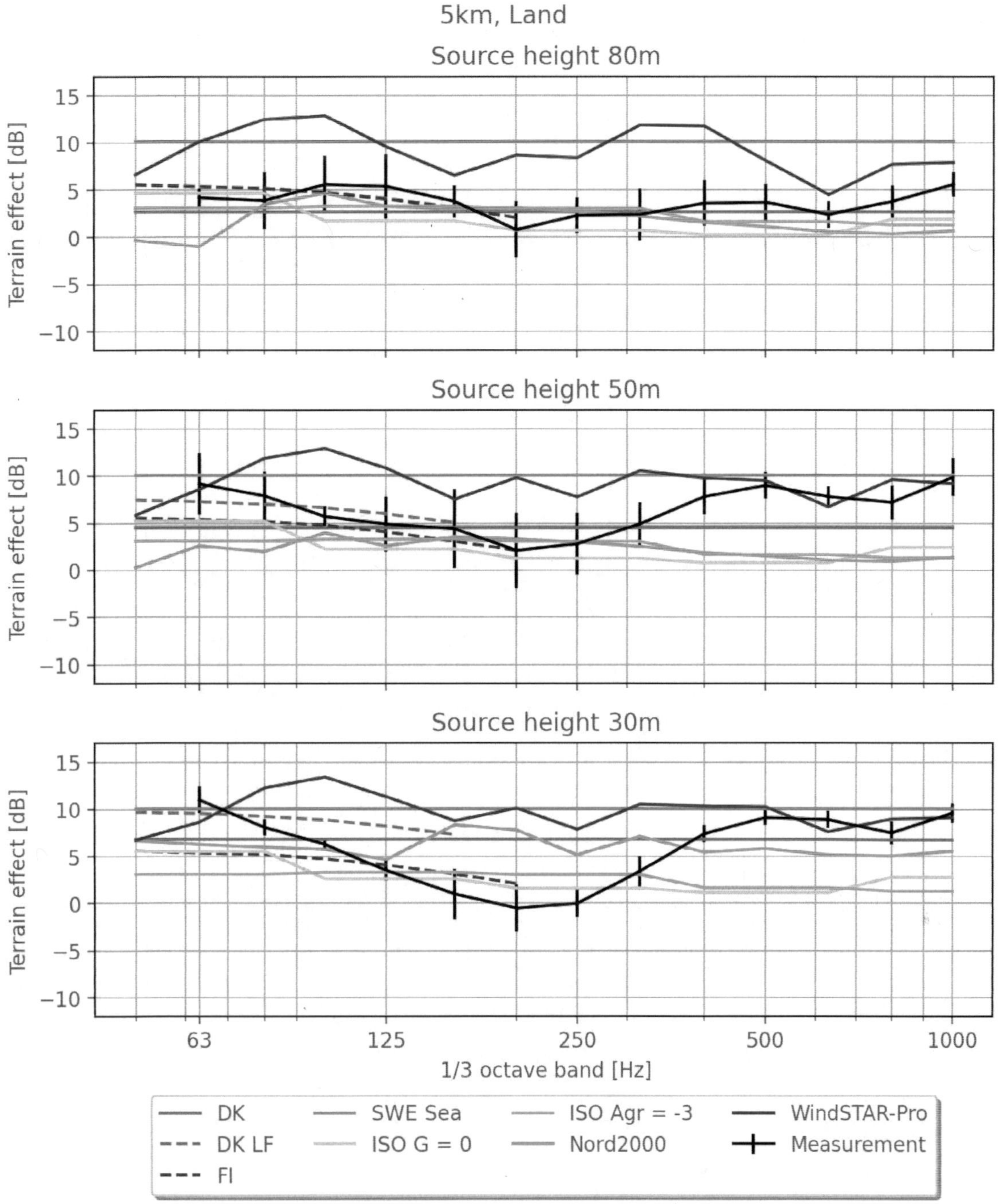

Figure 5. Plots showing measurement and calculation results for the different methods/models with the loudspeaker at 80m, 50m and 30m respectively at the 5km Land position. Error bars on measured curves shows standard deviation for the four sweeps.

Below are the 7 km results shown in Figure 6 (Water) and Figure 7 (Land). As for the other distances the Land position has a dip in the 1/3 octave band spectrum, here at 200 Hz. The dips for the Water position are not as pronounced.

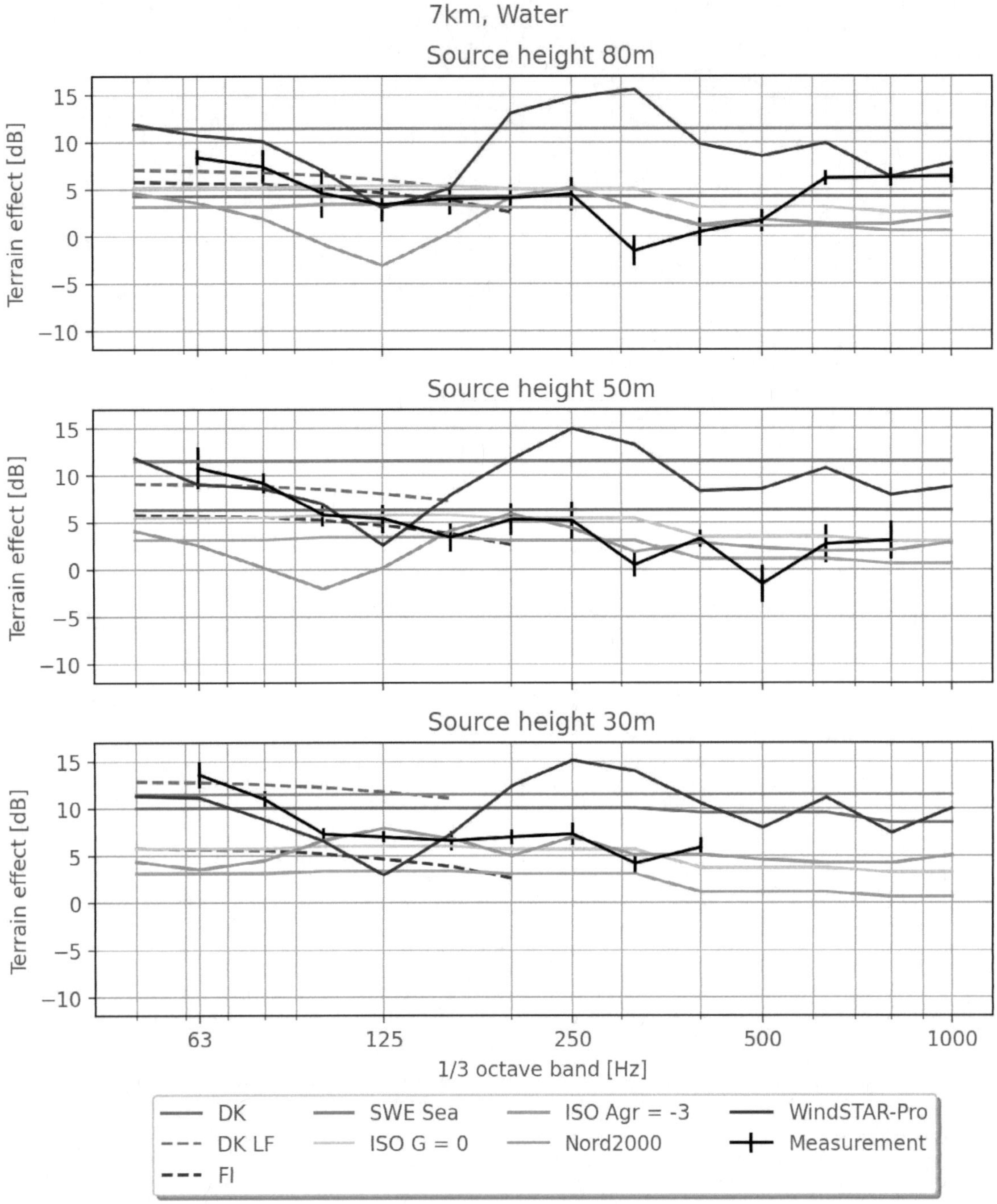

Figure 6. Plots showing measurement and calculation results for the different methods/models with the loudspeaker at 80m, 50m and 30m respectively at the 7km Water position. Error bars on measured curves shows standard deviation for the four sweeps.

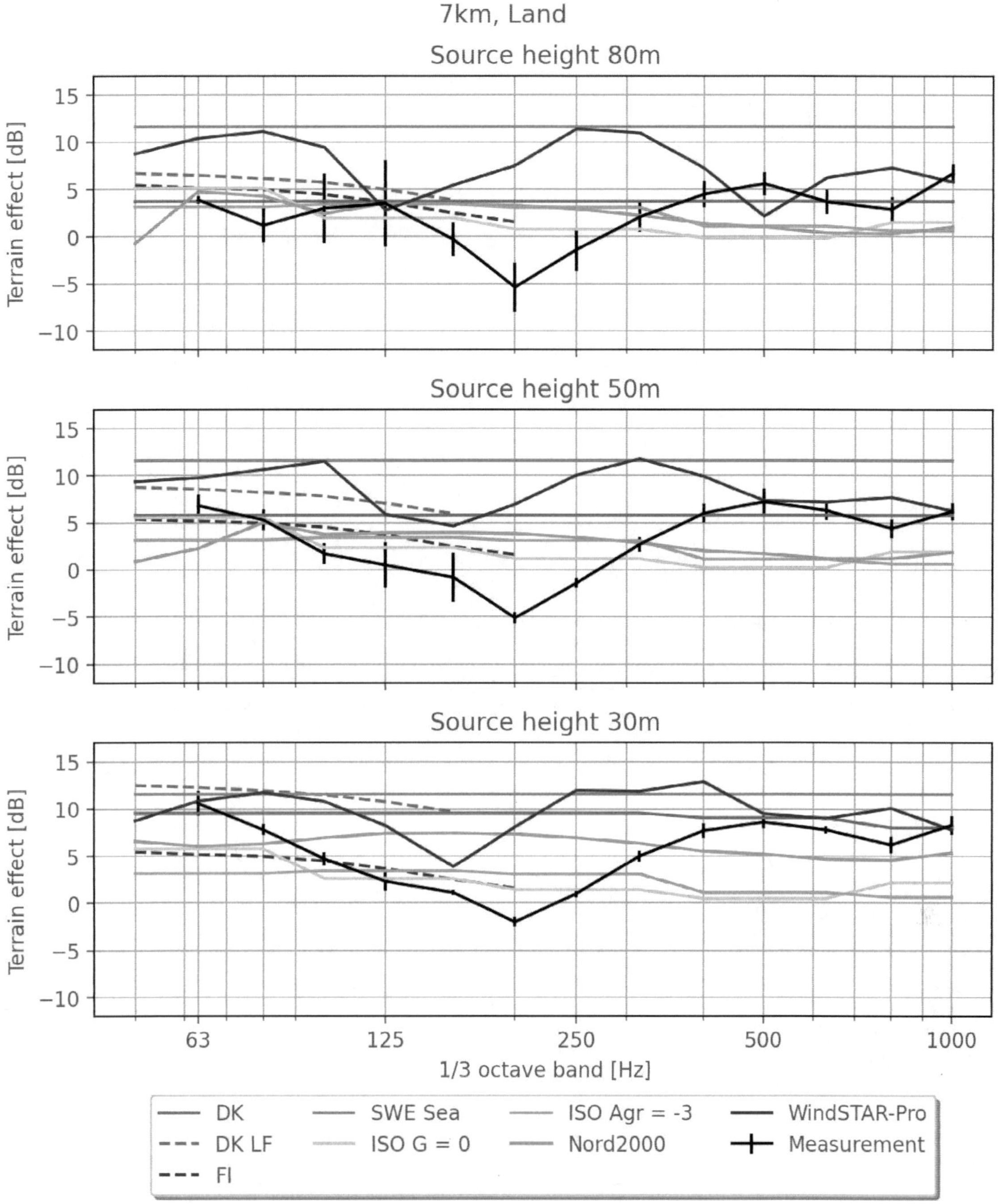

Figure 7. Plots showing measurement and calculation results for the different methods/models with the loudspeaker at 80m, 50m and 30m respectively at the 7km Land position. Error bars on measured curves shows standard deviation for the four sweeps.

The differences in the plots are condensed to averages of differences of measured values compared to each method and 1/3 octave band. Averages are calculated over all measurement points and source heights. These difference averages are shown in in 0 (full frequency range methods) and 0 (low frequency methods).

1/3 octave band	DK	SWE Sea	ISO TA Lärm G = 0	ISO TA Lärm Agr = 3	Nord2000	DTU Wind STAR
63	2.7	-2.3	2.6	4.6	5.0	-1.1
80	1.5	-3.5	1.4	3.4	4.0	-2.1
100	0.4	-4.6	1.7	2.0	3.2	-2.4
125	-0.7	-5.7	0.6	0.9	2.2	-0.4
160	-2.1	-7.1	-0.8	-0.5	0.1	-0.7
200	-4.4	-9.4	-2.4	-2.5	-2.5	-4.0
250	-4.8	-9.8	-2.8	-2.9	-3.6	-7.9
315	-4.7	-9.7	-2.7	-2.8	-3.6	-10.2
400	-3.3	-8.3	-0.4	0.0	-1.6	-7.9
500	-1.7	-6.9	1.0	1.3	0.2	-4.9
630	-0.7	-6.0	2.0	2.3	1.6	-2.6
800	-0.4	-5.7	1.5	2.9	1.9	-2.7
1000	1.3	-4.0	3.1	4.5	3.0	-1.8

Table 2. *Averages of differences for each full frequency range method and 1/3 octave band compared to measured values. Colours indicate high (red) and low (green). Values are in dB.*

1/3 octave band	DK LF	FI
63	0.0	2.3
80	-1.0	1.2
100	-1.7	0.5
125	-2.2	0.0
160	-2.7	-0.4
200		-1.6

Table 3. *Averages of differences for each low frequency range method and 1/3 octave band compared to measured values. Colours indicate high (red) and low (green). Values are in dB.*

These results show that the SWE Sea, and to almost the same extent DTU Wind STAR, methods generally overestimate the terrain effect, whereas the other methods generally underestimate low (below 100 Hz) and higher frequencies (500 – 1000 Hz) with a span in the frequency region between 160-400 Hz where they generally overestimate the terrain effect.

Standard deviations of differences for each method compared to measured values for each 1/3 octave band is shown in 0 (full frequency range methods) 0 (low frequency methods).

1/3 octave band	DK	SWE Sea	ISO TA Lärm G = 0	ISO TA Lärm Agr = 3	Nord2000	DTU Wind STAR
63	2.4	3.3	3.2	3.6	3.8	4.1
80	2.2	2.7	2.4	2.8	3.9	4.1
100	2.7	2.9	2.2	2.4	3.9	4.4
125	2.9	2.9	2.2	2.6	4.3	4.8
160	3.1	3.1	1.9	2.9	4.5	4.2
200	4.4	4.6	2.9	4.7	4.9	6.7
250	4.0	4.2	3.6	4.7	4.4	3.7
315	3.9	4.1	4.6	4.8	4.4	3.3
400	3.8	3.8	5.3	5.0	4.9	4.2
500	3.7	3.5	5.2	4.3	4.5	4.6
630	2.9	2.6	4.1	3.4	3.6	3.1
800	2.0	2.1	2.3	2.6	2.5	2.8
1000	2.5	3.0	3.3	3.7	3.0	3.0

Table 4. Standard deviations of differences for each full frequency range method and 1/3 octave band compared to measured values. Colours indicate high (red) and low (green). Values are in dB.

1/3 octave band	DK LF	FI
63	2.4	3.6
80	2.1	2.7
100	2.6	2.3
125	2.9	2.4
160	2.9	2.5
200		4.2

Table 5. Standard deviations of differences for each low frequency range method and 1/3 octave band compared to measured values. Colours indicate high (red) and low (green). Values are in dB.

The standard deviations of the differences indicates that the largest variations of differences are generally highest between 200-500 Hz, where also the largest overestimations of terrain effect could be seen.

3.3 A-weighted levels

The total A-weighted levels (63 - 1000 Hz and 63 - 200 Hz for low frequency methods) were calculated for both measurements and calculations.

Comparing total A-weighted level differences when subtracting the method values from the measured values (positive values indicate measured levels are higher, whereas negative values indicate predicted values are higher) are shown in 0 (full range frequency methods) and 0 (low frequency methods). The average difference, average of absolute differences and standard deviation of differences are also shown.

	Method						
Position	**DK**	**SWE Sea**	**ISO G=0**	**ISO Agr=-3**	**Nord2000**	**DTU Wind STAR**	**Average**
3km, Water, 30m	-2.2	-1.8	-1.6	0.6	-3.4	-4.6	**-2.2**
3km, Land, 30m	-2.1	-2.5	1.0	0.1	-2.9	-3.0	**-1.6**
5km, Water, 30m	-3.1	-4.3	-0.9	1.6	-2.0	-5.9	**-2.4**
5km, Land, 30m	-1.3	-2.9	3.7	3.0	-0.7	-4.3	**-0.4**
7km, Water, 30m	-3.1	-5.6	1.3	3.9	0.9	-5.8	**-1.4**
7km, Land, 30m	-4.9	-7.0	2.7	1.8	-1.8	-5.8	**-2.5**
3km, Water, 50m	-1.6	-3.4	-2.7	-1.0	-0.9	-5.6	**-2.5**
3km, Land, 50m	-1.1	-3.7	0.3	-1.2	0.7	-2.3	**-1.2**
5km, Water, 50m	0.2	-3.3	0.4	2.5	2.4	-5.4	**-0.5**
5km, Land, 50m	1.3	-2.5	4.4	3.4	3.6	-3.6	**1.1**
7km, Water, 50m	-2.5	-7.5	-1.2	1.2	0.8	-7.3	**-2.7**
7km, Land, 50m	-3.3	-8.9	1.0	-0.2	-0.3	-6.4	**-3.0**
3km, Water, 80m	-5.5	-7.3	-5.7	-4.9	-4.4	-9.0	**-6.1**
3km, Land, 80m	-5.6	-8.2	-3.4	-5.7	-3.6	-7.0	**-5.6**
5km, Water, 80m	-0.6	-6.0	-1.7	-0.1	0.3	-8.9	**-2.8**
5km, Land, 80m	0.1	-5.6	1.8	0.3	0.8	-6.8	**-1.6**
7km, Water, 80m	-1.0	-8.0	-1.4	0.6	0.9	-8.6	**-2.9**
7km, Land, 80m	-2.1	-9.7	0.6	-1.0	-0.6	-6.6	**-3.2**
Average	**-2.1**	**-5.5**	**-0.1**	**0.3**	**-0.6**	**-5.9**	
Average absolut	**2.3**	**5.5**	**2.0**	**1.8**	**1.7**	**5.9**	
Std. Ave	**1.9**	**2.5**	**2.5**	**2.5**	**2.1**	**1.9**	

Table 6. Measured – method, total A-weighted levels (63-1000 Hz). Values are in dB.

	Method	
Position	**DK LF**	**Finnish LF**
3km, Water, 30m	0.8	3.1
3km, Land, 30m	0.7	1.5
5km, Water, 30m	-1.4	2.8
5km, Land, 30m	-3.3	0.1
7km, Water, 30m	-3.3	4.1
7km, Land, 30m	-6.2	0.2
3km, Water, 50m	0.5	0.6
3km, Land, 50m	-0.5	-1.6
5km, Water, 50m	1.2	3.4
5km, Land, 50m	-0.1	1.4

7km, Water, 50m	-1.8	2.0
7km, Land, 50m	-5.0	-2.4
3km, Water, 80m	-4.0	-4.1
3km, Land, 80m	-4.4	-5.3
5km, Water, 80m	1.0	0.8
5km, Land, 80m	0.9	0.5
7km, Water, 80m	-0.9	0.9
7km, Land, 80m	-2.5	-1.9
Average	**-1.6**	**0.3**
Average absolut	**2.1**	**2.0**
Std. Ave	**2.3**	**2.6**

Table 7. Measured – method, total low frequency A-weighted levels (63-200 Hz). Values are in dB.

As was indicated by Figure 2 - Figure 7 the WindSTAR-Pro and SWE sea methods overestimate the terrain effect, and hence also the noise level, at all measurement positions, especially for the 7 km positions and high loudspeaker positions. It also has the largest average difference to the measured values.

All other methods, other than the WindSTAR-Pro and SWE sea method, over- and underestimates the terrain effect.

The 5 km Land position had generally the largest difference (not absolute) between measured and calculated levels for each loudspeaker height. At this position, the terrain was quite steeply sloping upwards the first 10 m from the seashore and then relatively flat from 10 m to the measurement position.

All methods overestimated the terrain effect at 3 km Water and Land positions with the loudspeaker at 80 m.

The ISO method with G = 0 gave the lowest average difference and the Nord2000 method gave the lowest average absolute difference.

4. Conclusions

For this set of measurements some general conclusions are drawn:

- There is a relatively large variance between measured and calculated values.
- The WindSTAR-Pro and the Swedish sea method over-estimates the terrain effect, especially for source heights above 30 m and 7 km distance.
- None of the methods could accurately capture the “phase-dip” that can be seen most clearly Figure 3 (3km Land position) between 200-400 Hz. This position had a fairly long distance of land (abt. 90m from the seashore).
- Steep terrain can result in underestimation of the terrain effect.
- Between the two settings for the ISO method, A_{gr} = -3 gave the lowest values for Water-positions, whereas the ISO method with G = 0 gave the lowest values for Land-positions.
- The Nord2000 was the method that gave the least absolute differences from the measured levels.

There are a number of uncertainties that plays a role, especially for the more advanced methods, Nord2000 and WindSTAR-Pro. These include ground impedance as well as the vertical profiles of wind speed and temperature. All of these contributes to the interference patterns (“phase-dips”) and can have a large effect, at least on the 1/3 octave band differences seen in the comparisons. There is a need for further analysis and investigation into these issues.

5. Acknowledgements

The original measurements were performed as part of the DECOWIND project (Development of low-noise and cost-efficient wind farm technology), where the financial support for measurements was partly provided by File number: 8055-00041B under Innovation Fund Denmark. Supplementary financial support was provided by Danish Agency for Institutions and Educational Grants.

RWE Offshore Wind GmbH has provided financial support to FORCE Technology for the analysis and preparation of this paper and presentation.

6. References

[1] L. S. Søndergaard, E. Thysell, C. Claumarch, A. Vignaroli, C. Thomsen, and K. Hansen, "Long distance noise propagation over water for an elevated height-adjustable sound source," *9th International Conference on Wind Turbine Noise Remote from Europe 18th to 21st May 2021*, no. May, pp. 432–450, 2021.

[2] E. L. Petersen, "On the kinetic energy spectrum of atmospheric motions in the planetary boundary layer," 1975.

[3] *Bekendtgørelse om støj fra vindmøller*, no. 135. Denmark: Miljø- og Fødevareministeriet, 2019, pp. 1–24.

[4] *RAKENNETTU YMPÄRISTÖ YMPÄRISTÖHALLINNON OHJEITA 2, Modellering av buller från vindkraftverk*. Helsinki: Ympäristöministeriö, 2014.

[5] Naturvårdsverket, "Mätning och beräkning av ljud från vindkraft," Stockholm, Jun. 2013.

[6] ISO, "ISO 9613-2:1996, Acoustics — Attenuation of sound during propagation outdoors — Part 2: General method of calculation," 1996.

[7] B. Plovsing, "Proposal for Nordtest Method: Nord2000 – Prediction of Outdoor Sound Propagation,", DELTA Report AV1106/07, 2014.

[8] B. Plovsing and E. Thysell, "Nord2000 – Prediction of Outdoor Sound Propagation. Amendments to Report AV1106/07 revised 2014," FORCE Technology Report TC-101327, 2019.

[9] E. Barlas, W. J. Zhu, W. Z. Shen, K. O. Dag, and P. Moriarty, "Consistent modelling of wind turbine noise propagation from source to receiver," *J Acoust Soc Am*, vol. 142, no. 5, pp. 3297–3310, Nov. 2017, doi: 10.1121/1.5012747.

[10] *Hinweise zum Schallimmissionsschutz bei Windkraftanlagen (WKA)*. 2016.

[11] "Dokumentation zur Schallausbreitung Interimsverfahren zur Prognose der Geräuschimmissionen von Windkraftanlagen," 2015.

[12] M. West, K. Gilbert, and R. A. Sack, "A Tutorial on the Parabolic Equation (PE) Model Used for Long Range Sound Propagation in the Atmosphere," 1992.

[13] E. M. Salomons, *Computational Atmospheric Acoustics*. Dordrecht: Springer Netherlands, 2001. doi: 10.1007/978-94-010-0660-6.

10th International Conference
on
Wind Turbine Noise
Dublin – 21st to 23rd June 2023

Evolution of sound production of onshore wind turbines

Frits van den Berg, Mundonovo sound research: fvdberg@mundonovo.nl
Erik Koppen, Arcadis: erik.koppen@arcadis.com
Casper van Wezel, Arcadis: casper.vanwezel@arcadis.com
Steven Velthuijsen, Bosch & van Rijn: steven@boschenvanrijn.nl

Summary

The sound power Lw of wind turbines (WTs) over time has developed in relation to their electric power P as Lw $\propto$ logP. The spectral distribution of the sound did not change significantly over time. This applies to the average performance of wind turbines, but there are differences between individual WT types. Because of the rapid and substantial growth of onshore wind energy, a greater number of people will be living close to wind farms. This sustains the need for WT sound power reduction. The use of sound reduction measures such as serrations, reduced tip speed and low noise modes, may counteract the development of higher sound power from ever bigger WTs. To investigate this, the sound production of WT types over the last decade is analysed in relation to their size and electric power and the application of sound reduction measures. The analysis includes the broad band A-weighted and the low frequency sound power levels.

1. Introduction

Since the 1980's onshore wind turbines have developed tenfold in size: from about 20 m diameter to present diameters of around 200 m. Electric power capacity in that period has grown ten times faster: from about 50 to 5000 kW. A similar trend was observed in sound power level. In public debates on wind energy plans it is often mentioned that higher wind turbines must be louder and should therefore be placed at a larger distance from residences. In contrast, developers on average do not observe a clear increase in sound power level and thus see no need to place wind turbines at larger distances. Earlier studies showed that the sound power $L_{WA,max}$ of wind turbines from less than 100 kW up to 3 MW increased with size. Relations found between $L_{WA,max}$ and P_{max} were close to $L_{WA,max} \propto 9*\log(P_{max})$.[1,2,3] Individual turbine types could deviate from this relation up to about 7 dB. There is reason to think that larger and more powerful wind turbines may deviate from this trend. One reason is that there is a continuous effort to reduce aerodynamic noise, e.g. by applying and improving serrations to the trailing edge of the blades. Another reason is that the operational design may change as the blades of a higher turbine are at greater heights where wind speed on average is higher and has less diurnal variation. This study aims to show whether changes in wind turbine technology have affected the relation between size and sound power. The focus here is on the development of sound power levels and not on sound character which includes spectral content, tonal sound and amplitude modulation.

2. Sound sources

The main audible components of wind turbine sound are trailing edge sound and in-flow turbulent sound. Descriptions are available in many papers and reports (e.g. Oerlemans, 2011 and van den Berg, 2006)[2,4] from which a summary is given here.

2.1 Dominant sources

Trailing edge sound is produced by the turbulent boundary layer of air at the downstream ('trailing') edge of a rotor blade where the flowing air leaves the blade's surface. This sound is relatively high pitched and is the dominant audible sound from modern turbines at close range. When the angle of attack increases from its optimal value when maximum power output is reached, the turbulent boundary layer on the suction (low pressure) side grows in thickness. For high angles of attack this eventually leads to 'stall': a dramatic increase of boundary layer thickness, drag and sound level and decrease of lift and power performance of a blade. Trailing edge sound level is proportional to $50 \cdot \log V_{in}$, where V_{in} is the velocity of air impinging on the blade.

Apart from this turbulence on the rear surface of the blade, there is also turbulence present in the atmosphere and the interaction of this atmospheric turbulence hitting the blade surface produces in-flow turbulent sound. It is also known as leading edge sound as most of this sound is produced at the front edge of a blade. It is relatively low-pitched and because high frequencies are more strongly attenuated by the atmosphere, in-flow turbulent sound becomes more dominant at larger distances (where at the same time the overall sound level decreases). In-flow turbulence sound level is proportional to $60 \cdot \log V_{in}$, but it also depends on the turbulence intensity of the atmosphere.

Thus, the sound spectrum of a modern wind turbine can be divided in (overlapping) regions corresponding to the two mechanisms mentioned:

- High frequency: trailing edge (TE) sound is noise with a maximum level at 500–1000 Hz for the central octave band, decreasing with 11 dB for neighbouring octave bands and more for further octave bands.
- Low frequency: in-flow turbulent sound is broad band noise with a maximum level of approximately 10 Hz and a slope of 3–6 dB per octave.

In measurements a third sound component may be detected when there are sudden changes in local wind speed, due to a substantial change in wind speed with height, the wake of another wind turbine or the presence of the turbine tower. This thickness sound is visible as regularly occurring peaks at infrasound (<20 Hz) frequencies. It is not relevant for residents as it is far below the perception threshold at residential distances.

Sound mitigation of wind turbines is realized by two different measures. One is to reduce the blade speed and thus lower sound production. This is used in 'low noise' modes, often to meet noise limits in specific conditions. Another measure is the application of noise reducing extensions on or adaptations of the blades. In practice trailing edge serrations (TES) are attached to the blades. TES thus reduces trailing edge sound but has no effect on leading edge sound.

2.2 Relation with size

According to the Betz law a wind turbine extracts an optimum amount of energy from wind at a tip speed ratio $\lambda_{optimum}$ that is equal to $4\pi/N$ with N the number of blades and λ the ratio of blade tip speed V_{tip} and undisturbed wind speed V_{in}. For a three-bladed turbine $\lambda_{optimum} = 4\pi/3 \approx 4.2$, so $V_{tip} \approx 4.2{*}V_{in}$ (in practice it is 25-30% higher: $\lambda_{optimum}$ = 5.2-5.5)[5]. For variable speed wind turbines V_{tip} increases with V_{in} until maximum electric power is reached and then is constant for higher V_{in} until the turbine is stopped to prevent storm damage. Thus, the maximum blade tip speed theoretically does not depend on the size of the turbine or rotor, but it does depend on the wind climate. In practice, the tip speed at maximum electrical power does depend on size: the maximum blade tip speed averaged over a large number of wind turbines and diameters ranging from 40 to 126 m, on average increases from 62 to 86 m/s or with 0,28 m/s per m diameter.[6]

3. Data selection

The main body of data was provided from the WindPRO database by EMD International A/S. This database contains over 1000 wind turbine types. Selected for this study were 1) all onshore, 50 Hz wind turbine types of 2 MW and more, and 2) all wind turbine types placed in the Netherlands. For the second selection a list was provided by Bosch & van Rijn. The two selections overlap for the ≥ 2 MW types. Some types in the WindPRO database have no noise data and were therefore excluded. The resulting wind turbines add up to a total of 245. A complete dataset was not available for all these turbines. Some of the missing data could be supplemented based on manufacturers documentation. Additionally, a set of 8 complementing wind turbine types up to 6.3 MW was added based on available manufacturers documentation. Wind turbine types over 6.3 MW were not added to the WindPRO data set, since due to the very limited number of turbines in this class data may be directly traceable to a specific turbine type or manufacturer; in such cases it was not allowed to present data derived from manufacturers documentation due to nondisclosure agreements. Wind turbine types over 6.3 MW from the WindPRO database were included in the analysis since this data is already public. The only exceptions are the analyses regarding the overall sound power level compared to the low frequency sound power level and regarding the effect of serrations. For these analyses all turbine types over 6.3 MW were excluded due to the very limited data set of turbine types with low frequency noise data and with and without serrations. Due to the very limited number of such large turbine types for which a comparison is possible, it is not clear how representative the comparison would be. Also, a number of these turbine types have not been built yet, which means the noise data is in some cases based on an estimate and not on actual measurements; thus, the sound data (and especially the sound spectrum) may be less reliable.

Of the resulting 253 turbine types, 49 types include noise data both with and without Trailing Edge Serrations (TES). In total 71 types with TES are included in the database. In this paper, a turbine type with or without serrations is only for noise considered as two separate types. This has led to a database consisting of 302 wind turbine types regarding sound power level and 253 wind turbine types with regard to other turbine characteristics.

3.1 Data description

The dataset provides data on a number of wind turbine parameters, including: rated power, rotor diameter, rotational speed (rpm) at rated power and apparent sound power level (L_{WA}) at varying wind speeds. All sound data in the dataset and the present paper are in A-weighted decibel and expressed as dB(A). Differences in level are given in dB. From the available parameters, blade tip velocity and swept rotor area were calculated. To be consistent, the maximum sound power level ($L_{WA,max}$) produced by each turbine was used for analysis. Not all parameters are specified for all turbines. Therefore, each figure in the text below includes a specification on the number of available data points, i.e. turbine types. Furthermore, a total of 171 turbine types have a sound spectrum included in 1/1-octave bands ranging from 63 to 8000 Hz. For these types the maximum low frequency sound power level ($L_{WA,LF}$) was calculated as the sum of the 63 and 125 Hz bands. The range in size of the wind turbines and in their parameters is large. For example, the smallest turbine has a rated power of 80 kW, while this is 7580 kW for the largest.

4. Development of rotor size

4.1 Influence of size

Size can be expressed as diameter or swept area of the rotor, as height of the turbine (including or excluding blade length) or as electric power capacity. Rotor size is related to electric power as a larger rotor catches more wind to be converted to electric energy. Figure 1 shows the relation between the area that is swept by the rotor blades (short: the 'rotor area', A_{rotor}) and the electric power capacity (P_{max}). The correlation coefficient between A_{rotor} and P_{max} is high ($r^2 = 0.96$), which means that the electric power capacity is to a large degree related to the rotor area. The black line is the least squares approximation to the data points and equals $P_{max} = 0.071*A_{rotor}$; it was assumed that this relation includes the origin (point 0,0). Thus, on average every square meter rotor area yields 71 W.

Figure 1. Maximum electric power in kW plotted versus swept area of rotor in m² for 253 turbine types and best linear approximation (black line)

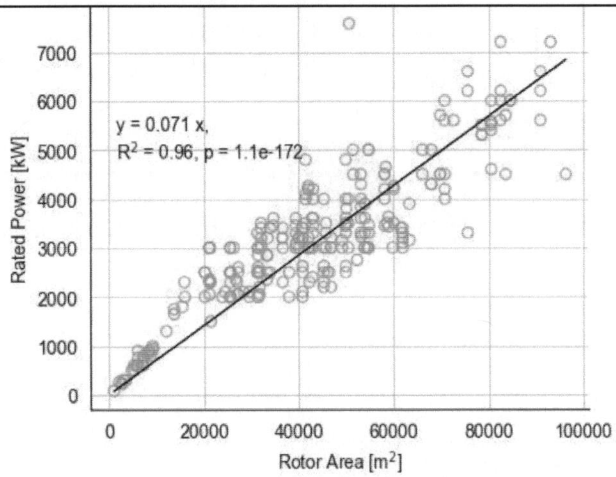

Earlier studies showed that the acoustic or sound power level of a wind turbine is proportional to the logarithm of the electric power: $L_{WA} \propto \log P$. Thus, because of the linear relation between P_{max} and A_{rotor}, the sound power level is expected to be proportional to the rotor area:

$$L_{WA} \propto \log(A_{rotor}).$$

Figure 2 (left) shows the relation between the maximum sound power level $L_{WA,max}$ and the rotor area. To see if the actual data indeed approach a logarithmic fit, a local best fit is applied (LOWESS or Locally Weighted Scatterplot Smoothing) to obtain a smooth line without any assumption about the form of the best fit. This local fit is comparable to a moving average: it is based on a division of all data points in subsequent bands and a low-order polynomial fit is calculated for each band. The fit shows that there is a transition from a steeper gradient below about 6000 m² (≈85 m diameter) to a less steep gradient above about 13000 m² (≈130 m). A comparison with the data points for larger turbine types (≥ 3000 kW: figure 2 right) shows that the transition occurs with these larger types. Although over time the sound power level on average increased from 96 to 107 dB(A), the average increase for the larger types (≥ 3 MW) is much less: the average increase over the entire rotor area range of the larger types is about 1.5 dB. Individual differences between the larger types are larger (7 dB) and not related to size. This is reflected in the low correlation coefficient ($r^2 = 0.04$), indicating that other factors than geometric size are more important.

Figure 2. Maximum sound power in dB(A) plotted versus swept area of rotor in m² for 302 turbine types (left) and 199 types ≥ 3000 kW (right); best logarithmic approximations (black lines); best local fit (left: blue line). Mind the differences in vertical axes.

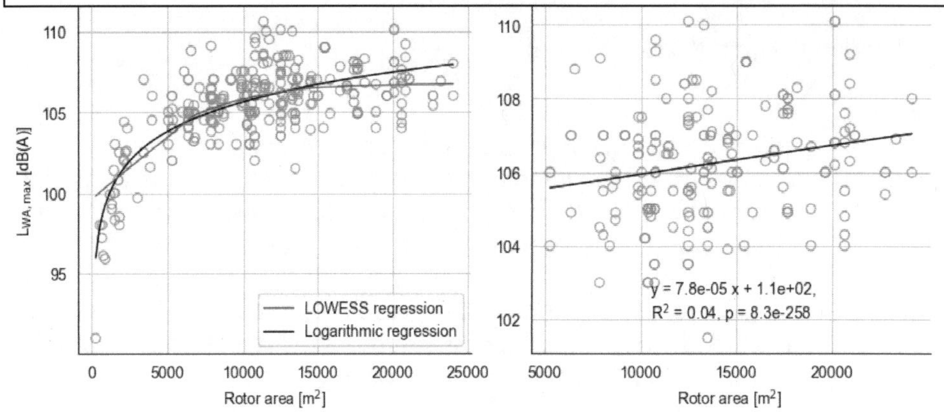

As blade velocity is the most important factor in sound production, a possible reason for the low correlation between the size of the larger turbine types and their sound power level is that blade velocity does not increase with turbine size. This implies that the rotational speed decreases with size. Figure 3 shows that this is true for most turbine types, even for the earlier smaller turbine types: there is a clear tendency that larger wind turbines rotate at lower rotational speeds.

Figure 3. Rotational speed (in rotations per minute or RPM) of 221 turbine types plotted versus rotor diameter and best exponential approximation without outlier (dot at 120 rpm/18 m) (black line)

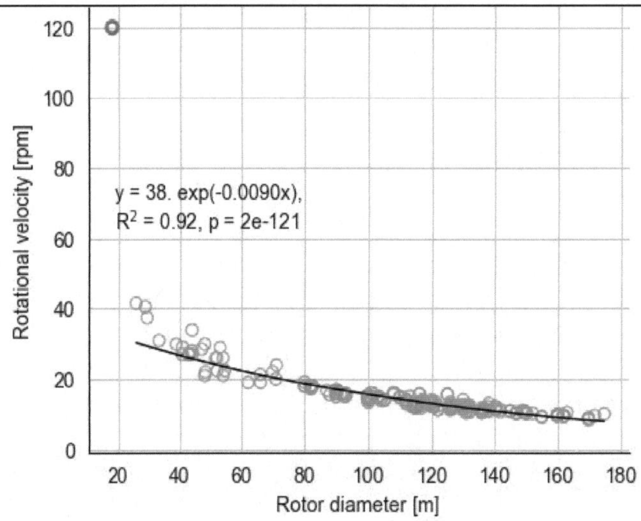

A consequence of a lower rotational speed is a relatively lower blade tip speed. Most of the sound of a wind turbine is produced near the blade tips, and figure 4 shows that for smaller turbines blade tip speed on average clearly increased. However, for turbines over 100 m diameter tip speed increases more slowly with rotor diameter. The local fit in figure 4 (left) shows a transition from a steeper to a less steep gradient. For the larger turbine types (≥ 3000 kW; figure 4 right) the range of diameters (100-175 m) increases with 75%, but the average tip speed increases with 3.8 m/s or 5%. Theoretically, this would lead to a 1.04 dB increase in sound power level.

Figure 4. Blade tip speed in m/s plotted versus rotor diameter in m for 221 types (left) and for 108 types ≥ 3000 kW (right); best local fit (left: blue line) and best linear approximation (right: black line). Mind the differences in horizontal axes.

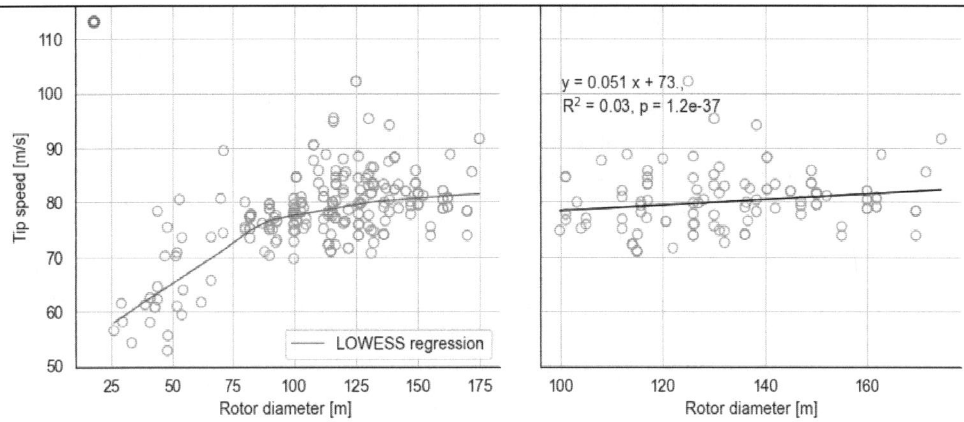

5. Development of sound power level

The relation between sound power level and maximum electric power is plotted in figure 5. Over the entire range the best logarithmic approximation to the data points is:

$L_{WA,max} = 6.8*\log(P_{max}) + 82$ dB(A) (all turbine types)

where P_{max} is in kW.

The local fit shows to what degree the actual data agree with a logarithmic fit. It shows that between 2 and 3 MW a transition is apparent from a steeper to a less steep slope of the fit. Including only turbine types below 2 MW, the best logarithmic fit is:

$L_{WA,max} = 8.9*\log(P_{max}) + 75$ dB(A) (turbine types < 2 MW)

For types of 2 MW and above the local fit again closely fits a logarithmic function:

$L_{WA,max} = 3.0*\log(P_{max}) + 95$ dB(A) (turbine types ≥ 2 MW)

Figure 5. Maximum sound power level in dB(A) plotted versus electric power capacity in kW for all 302 turbine types, best logarithmic approximation (black line) and local fit to data (blue line)

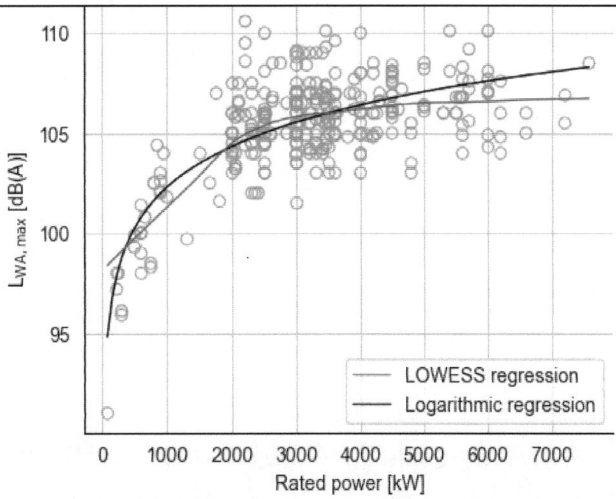

The correlation between sound power level and rated power is strong ($r^2 = 0.72$) for the turbine types <2 MW, but very weak ($r^2 = 0.06$) for the larger types. This means that for the larger types the maximum electric power is not at all a good predictor of the maximum sound power level. Over the entire electric power range from 2 to 7.6 MW the average increase in sound power is only 1.7 dB, whereas the differences between turbines are up to 7 dB. The average increase over the electric power range from 2 to 7.6 MW is considerably less than the 6.4 dB as found by Møller and Pedersen (2011) based on 44 wind turbines ranging from 450 kW to 3.6 MW. Møller and Pedersen found the sound power level increased as 11.0*log of electric power in kW.

Møller and Pedersen also suggested that with the increase of wind turbine size the contribution of the low frequency part would increase relative to the contribution of the higher frequencies. This can be analyzed here by comparing the contribution of the two low frequency 1/1-octave bands (63 and 125 Hz) that are available in the present data. The sum of these frequency bands is calculated for the same conditions as the maximum sound power level.
The result is shown in figure 6 for the entire range of turbine types up to 6.3 MW and for the larger types (2 up to 6.3 MW). The logarithmic fits for smaller types (< 2 MW) show that the increase of the sound power level with rated power was somewhat larger for the LF level when compared to the broad band level: the difference over a tenfold increase in rated power amounts to 1 dB. For the larger turbine types the difference is less (and in fact reversed), but irrelevant due to the very weak correlations.

Figure 6. Maximum sound power level (blue dots) and maximum low frequency sound power level in dB(A) (orange triangles) versus electric power capacity in kW for all 297 turbine types (left) and 267 types ≥ 2 MW (right), and best logarithmic approximations (black lines) based on 30 (L_{WA}) and 18 ($L_{WA,LF}$) turbine types at left and similar 267/145 at right.

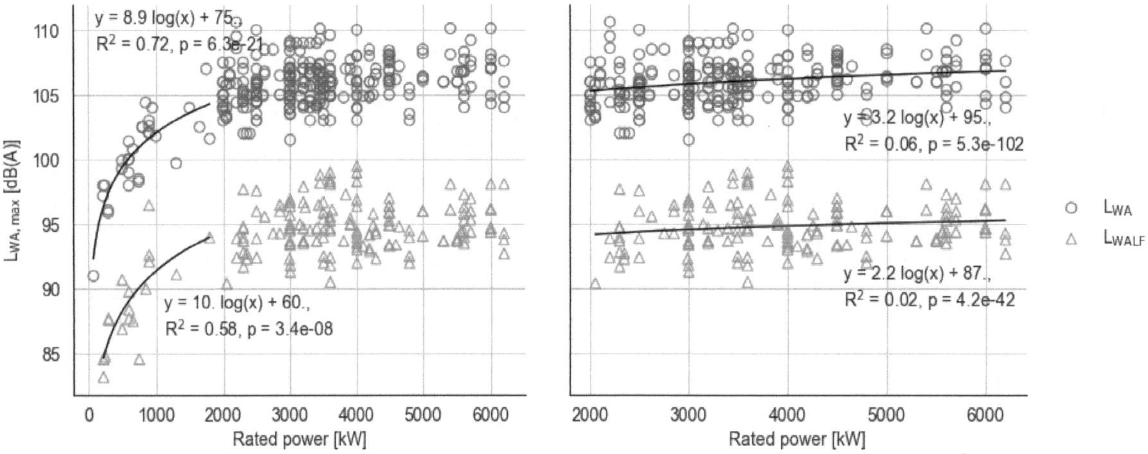

6. Effect of serrations on sound power level

In the present database there are 71 turbine types with trailing edge serrations (TES) and 225 without serrations (non-TES). In figure 7 these selections are plotted separately and this shows that serrations do have an effect on sound power level. When compared to all types (figure 5, TES +non-TES), the non-TES types on average have a higher, the TES types a lower maximum sound power level. Figure 7 also shows that TES only have been applied to the larger turbines (≥ 2 MW).

In figure 8 only turbine types ≥ 2 MW are plotted and the effect of serrations on the broad band sound level $L_{WA,max}$ as well as on the low frequency sound level $L_{WA,LF}$ is shown. Comparing the broad band sound power levels ($L_{WA,max}$), the difference between TES and non-TES on average is 1.75 dB. Differences between the slopes of the logarithmic fits (i.e. (the coefficient of the log function) are statistically not significant (p=0.49). The difference between the average $L_{WA,max}$ and $L_{WA,LF}$ (i.e. between the logarithmic fits) is 10.9-12.0 for the non-TES types and 11.0-11.6 for the TES types.

Figure 7. Maximum sound power level in dB(A) of turbine types with serrations (orange triangles) and without (blue dots) versus rated power in kW, and best logarithmic approximations (black lines) based on 226 (without) and 71 (with serrations) turbine types.

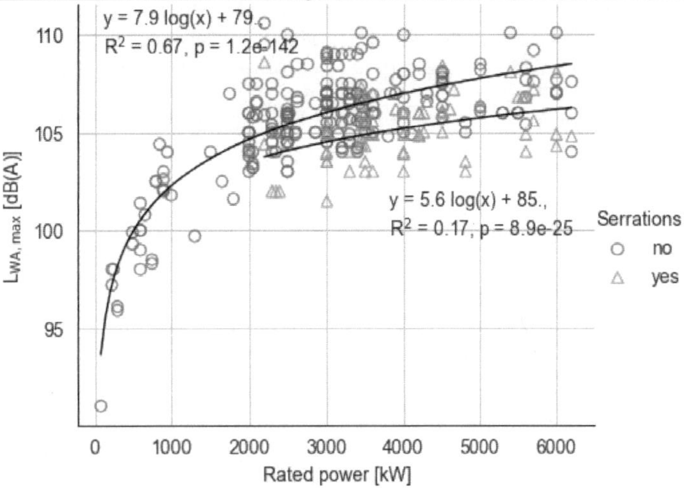

Figure 8. Maximum sound power level (blue dots) and maximum low frequency sound power level (orange triangles) in dB(A) versus rated power in kW for all turbines without TES (left) and with TES (right) and best logarithmic approximations (black lines) based on 196 (L_{WA}, no TES), 93 (L_{WALF}, no TES), 71 (L_{WA}, TES) and 52 (L_{WALF}, TES) wind turbine types.

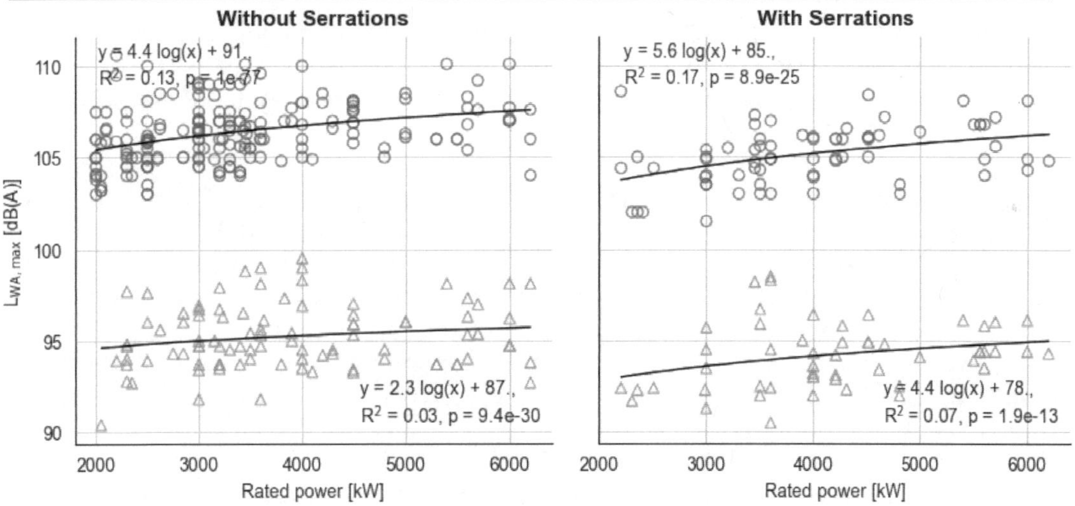

7. Effect of serrations at neighbours

The analysis above concerns the effect of trailing edge serrations on the sound power level of wind turbines at the source, without any influence of the environment. At residential locations the sound level will be lower because of geometrical spreading, but also the spectral content will change as higher frequencies will be attenuated more than lower frequencies. Serrations reduce only trailing edge sound, not the more low-frequency leading edge sound. Thus, the effect of serrations on residential sound level is expected to be less than the effect on sound power. To assess such propagation effects, the immission sound level within two kilometres of a typical large wind turbine (5 MW class) with and without serrations has been calculated. For this turbine type trailing edge serrations reduce sound power level with 2.0 dB. The ISO 9613 calculation model is used with a ground factor of 0.5 in a downwind situation. Figure 9 shows the results for a receiver height of 1.5 m. The effect of the serrations is presented in the figure at the right: close to the wind turbine the effect is about 2 dB and this effect is reduced to 1.5 dB at 2 km distance. At 4.0 m receiver height the difference is slightly (0.0-0.1 dB) less.

Figure 9. Left: immission sound level in dB(A) at distances up to 2 km from a class 5 MW wind turbine with (orange line) and without trailing edge serrations (blue line). Right: difference in immission levels due to presence of serrations.

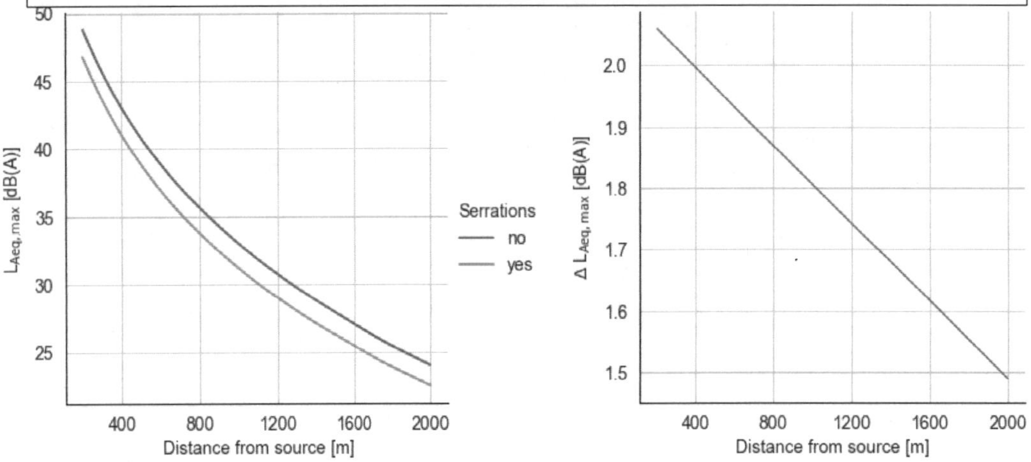

8. Discussion and conclusion

The results show that for all wind turbine types included sound power level increases with rated electric power with a slope of 6.8: $L_{WA,max} = 6.8*\log(P_{max}) + 82$ dB(A) (P_{max} in kW). However, this increase has changed over time: there is a transition zone from a higher slope (8.9) for turbine types < 2 MW to a lower slope (3.0) for types ≥ 2 MW.

For 78 wind turbine types from 80 to 3000 kW and for sound power levels at 7 and 8 m/s (i.e. near rating power) Van den Berg et al. (2008)[7] found a slope of 9.9 and 10.0, respectively. Møller and Pedersen (2011) found a slope of 11.0 (according to their figure 13) based on 44 wind turbines ranging from 450 kW to 3.6 MW. Søndergaard (2015) found a slope of 8.9 based on wind turbines (no number mentioned) ranging from about 300 kW to 3 MW. For the development of wind turbines from 2 to 6 MW this transition means that where an increase of sound power level of 4.2 – 5.2 dB was expected (based on a slope of 8.9 – 11.0), the actual average increase is 1.4 dB (based on a slope of 3.0).

One reason for the later development appears to be the use of trailing edge serrations (TES) on turbine types of 2 MW and above. On average they lead to a sound reduction of 1.75 dB. The increased use of TES leads to a lower increase with rated power. However, also the larger turbine types that do not use TES have a lower slope than found before: 4.4 instead of 8.9-11.0. The same is true for the larger turbine types with TES. A possible reason for this appears to be a slower increase of blade tip speed with size. For the larger turbine types (≥ 3 MW) the range of diameters (100-175 m) increases with 75%, but the average tip speed increases with 3.8 m/s or 5%. For a 5% increase in tip speed the expected increase in sound level is $50*\log(1.05) = 1.04$ dB.

Møller and Pedersen (2011) analyzed data from 44 wind turbines, of which 11 at a rated power > 2 MW, and found that the increase of the low frequency part of the sound power level $L_{WA,LF}$ was significantly higher than the broad band level L_{WA}. The average difference between $L_{WA,LF}$

and L_{WA} was 11.6 dB for the smallest turbines and 9.7 dB for the large turbines (taken from their figure 1). The differences in level in the Møller and Pedersen study and the present study cannot be compared directly as the spectral LF range of Møller and Pedersen (10-160 Hz 1/3-octave bands) differs from the present study (63 and 125 Hz 1/1-octave bands, which is the same as 1/3-octave band 50-160 Hz) in including the lowest 1/3-octave bands (10-40 Hz). However, at these low frequencies the A-weighting is quite large and their contribution to the broad band level very limited. Søndergaard (2015) repeated the analysis of Møller and Pedersen with more larger turbines and found similar differences between $L_{WA,LF}$ and L_{WA} (12.0 dB for small, 10.3 dB for large turbines), but this difference was not significant. The present study finds a difference of about 11.0-11.3 dB and no significant difference over the range of rated powers.

Trailing edge serrations reduce the relatively high frequency trailing edge sound, not the relatively low frequency leading edge sound. When the sound propagates to neighbouring locations, atmospheric and ground absorption reduce higher frequencies more effectively than lower frequencies. Calculations show that the effect of serrations becomes less at larger distances, but even at 2 km distance the effect still is fairly high. Trailing edge serrations on a typical class 5 MW wind turbine reduces sound power level with 2.0 dB and the effect at 2 km still is 1.5 dB.

Acknowledgments

The authors thank Jaap Boon and Madelon Ekelschot-Smink from Arcadis for their support, suggestions and comments with respect to the analysis.

References

[1] Søndergaard, B (2015) *Low frequency noise from wind turbines: do the Danish regulations have any impact? An analysis of noise measurements*. Aeroacoustics 14(5 & 6), 909-915

[2] Van den Berg, F (2006) *The sound of high winds: the effect of atmospheric stability on wind turbine sound and microphone noise*. Thesis University of Groningen

[3] Møller, H, Pedersen, CS (2011) *Low-frequency noise from large wind turbines*. J. Acoust. Soc. Am. 129 (6), 3727-3744

[4] Oerlemans, S (2011) *Wind turbine noise: primary noise sources.* Report no.NLR-TP-2011-066 National Aerospace Laboratory NLR

[5] Ragheb, M, Ragheb, AM (2011) *Wind turbines theory – The Betz equation and optimal rotor tip speed ratio*. In: Carriveau, R (ed.) *Fundamental and advanced topics in wind power.* InTech

[6] EWEA (2008?) *Tip Speed Trends*. wind-energy-the-facts.org/tip-speed-trends-7.html, consulted 6-3-2023

[7] van den Berg, F., Pedersen, E., Bouma, J., Bakker, R. (2008) *WINDFARMperception: Visual and acoustic impact of wind turbine farms on residents*. Final report FP6-2005-Science-and-Society-20, EU Specific Support Action Project no. 044628

10th International Conference
on
Wind Turbine Noise
Dublin – 21st to 23rd June 2023

Wind energy and health: the Dutch case

Ric van Poll, National Institute for Public Health and the Environment, the Netherlands (RIVM): ric.van.poll@rivm.nl.
Anne Bolders, RIVM: anne.bolders@rivm.nl.
Marije Reedijk, RIVM: marije.reedijk@rivm.nl.

Summary

Tulips, wooden shoes, cheese and windmills form the classic image of the Netherlands. But Dutch policy is also focused on developing more sustainable forms of energy. To meet the challenge of producing more sustainable energy, wind turbines have become an increasingly common feature of the Dutch landscape in recent years.
As a result of the national energy strategy, the expansion of wind energy (number, output) on land will increase in the coming years. This raises concerns and questions about the safety and health of local residents and the need for accessible information on the subject. For these reasons, the Wind Energy and Health Competence Centre (2021) has been established at the RIVM.
The aim of our center is to build, secure and share knowledge about wind turbines and health (e.g. noise, shadow, external safety risks, visual aspects, erosion). This includes relevant factors that influence this relationship (such as non-acoustic and non-visual aspects, e.g. level of participation, noise sensitivity, attitudes towards wind turbines). In the competence centre, we also work with the municipal health services (GGDs) in the Netherlands, who receive many questions from citizens and local governments on the topic of wind turbines and health.
Examples of our knowledge building and dissemination activities include quarterly alerts of new scientific and grey literature (published on the RIVM website), structured literature reviews, conference visits and conducting research (perception research, exposure-response research).
The presentation will cover the background, structure and activities of the Wind Energy and Health Competence Centre. Furthermore, the content of some of our current projects (e.g. literature review, research studies) will be explained in more detail.

1. Introduction

In this paper, we briefly discuss some aspects of the energy transition in the Netherlands as it relates to onshore wind turbines. We discuss the policy of the national government, the strategy used (RES) and the role of the Wind Energy and Health Competence Centre (WE&H-CC). We discuss some of the stakeholders and the research that has been carried out at RIVM in recent years.

2. Dutch Case

2.1 Dutch Policy[1]

The Climate Agreement is the Dutch elaboration of the international climate agreements of Paris (2015). The government, provinces, municipalities, businesses and civil society organisations have signed the agreement. The agreement aims to drastically reduce the amount of greenhouse gases within 10 years. Almost half of the 1990 CO_2 emissions by 2030. Nearly zero CO_2 emissions by 2050. In the climate agreement it was agreed that at least 35 terawatt hours of sustainable electricity must be produced on land (wind and sun) by 2030. Wind energy is an important form of renewable energy to meet this target. In addition, the Energy Agreement states that the provinces will each provide a share to realise a total of 6,000 MW of wind energy (in 2022 this target has been achieved). These agreements on onshore wind energy are necessary to achieve the Dutch targets for growth in sustainable energy and reduction of CO_2 emissions. The Dutch government considers wind energy to be a reliable and efficient source of sustainable energy that is already widely available. The technology for onshore wind is sufficiently developed for wind energy to be used on a large scale. It is also one of the cheapest ways of producing sustainable energy. In 2022 there were about 2700 wind turbines on land

There is a downside of these developments. People who live near wind turbines, or near possible future sites for wind turbines, are concerned about their health. Their concerns include the effects of blade noise or shadow flicker, amplitude modulation, property values and aesthetic degradation of the surrounding area.

2.2 Strategy: RES[2]

There are many measures in the climate agreement. Factories will make products in a cleaner way. Agriculture and horticulture sectors looking at ways to reduce CO_2 emissions. We will drive more electric cars, save as much energy as possible and generate more energy sustainably. By 2030, this will mainly be done with proven technologies: solar panels and wind turbines. Most wind power is generated at sea. That is not enough to meet our target. That is why we agreed in the climate agreement that we will also produce solar and wind energy on land.

How much and where the Dutch government will produce solar and wind energy on a large scale is set out in the Regional Energy Strategy, or RES for short. 30 energy regions in the Netherlands have drawn up an RES 1.0. The energy regions worked together with governments, residents, companies, grid operators, energy cooperatives and civil society organisations. A RES describes where solar or wind projects can or cannot be built and what the impact on the energy infrastructure will be. It also identifies the best sustainable heat sources to heat neighbourhoods and buildings. Residents, businesses and social groups are also involved in these decisions. The decisions are also coordinated with other regions. The RES focuses on 2030 and, if possible, 2050.

2.3 Wind Energy and Health Competence Centre (WE&H-CC)[3]

Until the end of 2020, RIVM organised the Expertise Network for Onshore Wind Development. This was available for the whole range of wind turbine topics. The funding for this expertise network has ended.

---

[1] https://www.rijksoverheid.nl/onderwerpen/duurzame-energie/windenergie-op-land

[2] https://regionale-energiestrategie.nl/ga+de+res+gespreksassistent/ga+wat+is+de+res/default.aspx Consulted: April 9th 2023.

[3] In Dutch EP-WE&G: ExpertisePunt WindEnergie en Gezondheid

As the expansion of onshore wind energy will continue to increase in the coming years as a result of regional energy strategies and may lead to concerns and questions from local residents it was decided to set up the Expertise Centre for Wind Energy and Health.
The ministries asked the RIVM, in cooperation with the GGDs (municipal health services), to keep the knowledge base up to date, so that the RIVM has the most up-to-date knowledge on wind turbines and health (e.g. noise, shadow casting, etc.), including relevant influencing factors (e.g. level of involvement).
RIVM uses this knowledge to communicate with GGDs and the national government. In turn, the GGDs use this knowledge in their communication with civil servants and administrators in the local authorities and with citizens. And also: to be able to interpret recently published literature or statements in the media, both against the background of existing scientific literature.
The members of the WE&H-CC meet on a monthly basis to review recent developments and determine actions (reaction, pro-action), and a list of relevant (scientific) literature is produced every three months for the benefit of interested parties. The quarterly overviews are available to everyone via the RIVM website.

2.4 Stakeholders

In addition to the government (national, provincial, municipal) and the RIVM, there are many other stakeholders involved in the issue of wind turbines. All from their own point of view. Some of them are mentioned here, to give some background to the participants in the wind turbine debate.

Public health care
Municipal health services (GGDs[4]) are committed to the health of everyone living in the Netherlands. For those who can choose a healthy lifestyle, but especially for those who need extra help to lead a healthier life. Public health care revolves around the question of how we can protect the entire Dutch population from risks over which people have little or no control. There are 25 GGDs in the Netherlands that work for and are paid by the surrounding municipalities. The GGDs participate in the WE&H CC (see 2.3). They inform municipalities and provinces about the possible health impacts of wind turbines

Residents/professionals
The Dutch Wind Farm Residents' Association (NLVOW[5]) defends the interests of (future) residents of wind farms, wind turbines or solar parks. Their mission is to give this group as full a voice as possible when they are confronted with plans for a park or turbines. They do this by sharing information and legal knowledge, and through consultations, conferences and presentations. They act as a federation for action groups and individuals who are faced with a wind farm, wind turbine or solar farm in their area and want to fight back.
WindWiki[6] is a group of concerned doctors who have been studying the safety of wind turbines for people and the environment. Some of them have been confronted with this issue by their patients, others by their own living environment. Together they aim to provide scientific information on the health and environmental effects of wind turbines. They believe that there is a lack of knowledge among scientists, politicians, doctors and the general public about the effects of living near wind turbines.

Trade association
The Dutch Wind Energy Association (NWEA[7]) is the trade association for the wind energy sector. NWEA promotes the development of wind energy with a view to a sustainable Dutch

---

[4] https://www.ggd.nl/
[5] https://www.nlvow.nl/
[6] https://www.windwiki.nl/
[7] https://www.nwea.nl/

energy supply. For and with its members, NWEA works on a strong wind sector and the new policies needed for this. On behalf of the sector, NWEA is a discussion partner for ministries and other organizations on all sorts of topics related to wind energy. NWEA maintains contacts with national and regional governments and politicians, with policy makers, with scientists and knowledge institutes, with companies and with nature and other (social) organizations in the Netherlands. NWEA members thus have their own voice in the national energy transition and climate discussions.

3. So far

In the following sections, we present some of the work that has been done at RIVM in recent years.

3.1 Survey (Bolders et al. 2022)

As part of a larger study into perception of possible risks in the environment RIVM conducted a nationwide survey of 3500 people living within 5 km of wind turbines, of whom 662 (19%) participatedhe survey included, amongst others, questions about annoyance and sleep disturbance by wind turbine noise (on a 0-10 scale),. Results showed that within 5 km of a wind turbine, 3.2% of participants were very disturbed by wind turbine noise and 2.4% very sleep disturbed. For the subset of participants living within 2.5 km of a wind turbine, a larger percentage (5.2%) were highly annoyed and 3.8% were highly sleep disturbed. Earlier data from the Netherlands, collected in 2007 among people living within 2.5 km of wind turbines, showed that 2.9% were very annoyed by wind turbine noise when indoors and 4.1% when outdoors. Direct comparison of results between studies is problematic because of differences between the studies. These differences included variations in the way the annoyance questions were phrased (indoors/outdoors vs. at home), the use of different response scales for these questions (4-point vs. 11-point scale), and differences in the sampling method used (stratified vs. non-stratified sampling methods). In addition, the cut-off point for high annoyance (score ≥8, equivalent to ≥72.72 on a 100-point scale) was slightly lower than the cut-off point for the 2007 data (score ≥4, equivalent to ≥75 on a 100-point scale). With these caveats in mind, the results seem to suggest similar levels of annoyance or a slight increase in annoyance over time, which may be worth exploring further.
Looking at the difference between two distance categories (2.5-5 and < 2.5 km from the nearest wind turbine), an increase in the percentage of high annoyance and high sleep disturbance was observed with increasing distance. High annoyance and high sleep disturbance responses further away from a wind turbine (2.5-5 km) were low, 1.6% and 1.4% respectively, compared to the responses closer to a wind turbine (< 2.5 km) described above. A similar pattern was found for the percentages of annoyance and sleep disturbance using a lower cut-off point (score ≥5). The results are consistent with those of a recent Finnish study, which found an inverse relationship between distance to the nearest wind turbine and both annoyance and sleep disturbance. If distance is taken as a proxy for sound exposure, the results are also consistent with the evidence for an association between annoyance and exposure to wind turbine sound reported in recent reviews on the subject (see below).

3.2 Literature review (van Kamp & van den Berg, 2020, 2021)

RIVM and Mundonovo sound research collected scientific literature on the effects of wind turbines on annoyance, sleep disturbance, cardiovascular disease and metabolism. They also looked at what is known about annoyance caused by visual aspects of wind turbines and other non-acoustic factors, such as the local decision-making process.
The literature review clearly showed that annoyance is a result of noise: the louder the noise (in dB) from wind turbines, the greater the annoyance response. The literature did not show that so-called "low-frequency sound" (low-pitched sound) causes additional annoyance compared to "normal" sound. For other health effects, the results of scientific research were mixed: these

effects are not a clear consequence of sound levels, but in some cases are related to the annoyance people experience. These results confirm previous conclusions from a similar study conducted three years before this one. The literature also showed that residents experience less annoyance when they are involved in the siting process. The ability to participate in the siting process, and in weighing up the costs and benefits, reduces the level of annoyance experienced by residents. It is therefore important to take the concerns of local residents seriously and to involve them in the planning and siting of wind turbines.
We are now in the process of finalising the results of the next update (2020-2022).

3.3 Exploration (Zock et al. 2022)

Previous research has already revealed gaps in our knowledge on health effects. Annoyance is a proven effect, but there is insufficient scientific evidence with regard to other effects such as disturbed sleep. Nor is it clear how the specific noise generated by wind turbines can be studied among local residents. And finally, we do not know at what levels of noise and at what distances in the Netherlands adverse effects on health might arise.
Commissioned by the national government RIVM explored health research options around wind turbines. RIVM has explored the possibilities in terms of researching the health effects of wind turbines in the Netherlands. To this end, input was sought from organisations involved in this topic, such as local residents' organisations, the Municipal Public Health Authorities (GGD) and provincial executives. RIVM started by charting the most important questions: can wind turbines have adverse effects on local residents' health, and if so, how does this happen and how many people are affected? For each question RIVM specified the types of research that will be able to provide an answer, as well as the pros and cons thereof.
One of the suggested research options is to monitor the health of a large group of people over several years. Another is a retrospective study looking at whether there is an increased incidence of certain health problems in the vicinity of wind turbines. Yet another option is to get a panel of people who live near wind turbines to regularly self-report any health problems they have. Lastly, correlations between wind turbine exposure and current health issues could be investigated.
Based on the exploration the Ministry has decided what research is to be conducted. It already commissioned a retrospective study. In the near future it will commission research on perception of wind turbines and exposure response relation (noise annoyance and sleep disturbance).

3.4 Perception of the Home Environment (van Poll & Simon, 2022)

The Ministry of Infrastructure and Water Management (I&W) wants to know how residents perceive their home environment. It has been conducting surveys in this area since 1977. Since 2003, these surveys have been carried out by the RIVM. The RIVM bases its findings on the results of a questionnaire (OBW[8]) that measures, among other things, how residents perceive noise, vibrations, odours and safety in their living environment. In 2021, almost 1,990 Dutch residents aged 16 and over took part in the study. The survey was conducted by RIVM and Statistics Netherlands (CBS). Noise annoyance (0.4% very annoyed) and sleep disturbance (0.2% very sleep disturbed) due to wind turbine noise are relatively low at national level compared to, for example, annoyance (9.4% very annoyed) and sleep disturbance (5.9% very sleep disturbed) due to traffic noise. Issues that receive a lot of media attention in the Netherlands, such as wind turbines, appear to have relatively low and sometimes even decreasing (severe) annoyance scores at the national level. Nevertheless, this issue can cause a lot of social unrest at regional or local level. The monitoring report provides a national picture ofannoyance and sleep disturbance, as experienced by a representative sample of residents in the Netherlands. The environmental problems surveyed (noise, odours, vibrations) and 'quality of the living environment' are mainly problems that occur and are experienced at local level.

[8] Study of Perception of the Home Environment (OBW = Onderzoek Beleving Woonomgeving).

therefore differences between local situations are to be expected. Issues that score highly nationally may be less important locally where other issues, such as wind turbine noise, dominate perceptions of the local home environment.

3.5 Emissions of chemical substances from onshore wind turbines (Hof et al., 2022).

RIVM conducted an exploratory study on the release of chemicals and microplastics from onshore wind turbines. Te research was commissioned by the ministry of Economic Affairs and Climate change. Various materials and chemicals are used in the production of wind turbines. For example, the protective coating used on the tower and the blades contains various chemicals. These substances, which can be harmful to the environment, and the microplastics contained in the coating used on the blades, can be worn away by the elements.
As a result, these chemicals are likely to end up in soil and water. However, we still know very little about which substances end up there and how much of each is released. We need this information to determine whether substances from wind turbines are harmful to human health and the environment. It is also important to refine estimates of how much microplastic is released.
Coatings may contain chemical substances that are classified as hazardous to human health and/or the environment under European legislation. These substances can be washed into the environment by rainwater. Laboratory studies of this 'leachate' from coatings used on water and road structures have shown that the substances can indeed be harmful to organisms and that the damage caused can vary greatly depending on the type of coating. It is still unclear whether the same is true for run-off from wind turbine coatings.
Each wind turbine is estimated to release between 3 grams and 14 kilograms of microplastics per year. How much is released depends, among other things, on whether techniques have been used to reduce wear and tear. It is not currently known to what extent the industry in the Netherlands uses such techniques on wind turbine blades. Various materials and chemicals are used in the production of wind turbines. For example, the protective coating used on the mast and blades contains various chemicals. These substances, which can be harmful to the environment, and the microplastics contained in the coating used on the blades, can be worn away by the elements. As a result, these chemicals are likely to end up in soil and water. However, we still know very little about which substances end up there and how much of each is released. We need this information to determine whether substances from wind turbines are harmful to human health and the environment. It is also important to refine estimates of how much microplastic is released.

4. Conclusion

The energy transition contributes to a more sustainable use of resources. On the other hand, there are also drawbacks. There are concerns about the siting of wind turbines, noise from wind turbines and hazardous materials that can be eroded by wind turbines. The Wind Energy and Health Competence Centre aims to build and support a knowledge base so that there is a common, factual knowledge base on which stakeholders can base their discussions.

References

Bolders et al. (2022). Annoyance and sleep disturbance responses in people living in the vicinity of wind turbines in the Netherlands.
https://az659834.vo.msecnd.net/eventsairwesteuprod/production-inconference-public/1c79b99e049f4c4db71c7ee1c9abd0fd.

Hof, M., et al. (2022). Eerste inzicht in emissies van chemische stoffen bij windturbines op land. Resultaten quickscan. RIVM-report 2022-0235.

van Kamp, I. and G. van den Berg (2020). "Health effects related to wind turbine sound: an update." RIVM-report 2020-0150.

van Kamp, I. and F. van den Berg (2021). "Health effects related to wind turbine sound: An update." International Journal of Environmental Research and Public Health 18(17): 9133.

van Poll, R. and S. Simon (2022). Onderzoek Beleving Woonomgeving (OBW). Hinder en slaapverstoring, de 2021-cijfers. RIVM-report 2022-0083.

Zock, J. P., et al. (2022). Verkenning van opties voor gezondheidsonderzoek rond windturbines. RIVM-report 2022-0041.

INCEEUROPE

10th International Conference
on
Wind Turbine Noise
Dublin – 21st to 23rd June 2023

Aero-acoustic prediction tool for complex wind turbine blade configurations

Guannan Wang, LM Wind Power guannan.wang@lmwindpower.com
Jeremy Hurault, LM Wind Power Jeremy.hurault@lmwindpower.com
Andreas Herrig, LM Wind Power andreas.herrig@lmwindpower.com
Ramakrishnan R., LM Wind Power ramakrishnan.r@lmwindpower.com
Marc Canal, LM Wind Power marc.canal@lmwindpower.com
Carlo-Enrico Carcangiu, LM Wind Power carlo-enrico.carcangiu@lmwindpower.com

Summary

A semi-empirical aeroacoustic prediction package was developed in LM Wind Power to compute the wind turbine blade noise emission with various aerodynamic add-ons, including serrations, Leading Edge Protection (LEP) and Lightening Protection System (LPS). Excellent agreement between the prediction and field measurement was achieved on both the apparent sound power level (Lwa) and the spectral shape. The serration and LEP noise prediction models are the foci of this paper and will be discussed further in detail.

1. Introduction

Acoustic emissions from wind farms have a negative impact on social acceptance of wind energy and can be a barrier for its further deployment. Many countries and communities around the globe have or are drafting noise regulations to minimize their residents' exposure to the noise. To comply with these local noise regulations, wind farm operators often curtail the rotor rotational speed (RPM) of their fleet, resulting in a loss of AEP (Annual Energy Production) and an increase of LCOE (Levelized Cost Of Energy). And wind turbine manufacturers must make trade-offs between performance and noise in the design process to ensure competitiveness of their onshore products. Considering the dominant noise source on a modern wind turbine is the blade-generated aerodynamic noise, an accurate, fast, and robust blade noise prediction tool is a critical part of the wind turbine blade design process.

Many studies (Boorsma and Schepers, 2011, Bertagnolio et. al., 2017, Bortolotti et. al., 2020, Wagner et. al. 1996) have been done in the past to model the wind turbine blade aerodynamic noise emission. The typical prediction strategy is to first obtain the blade aerodynamic operation conditions such as the Angle of Attack (AoA) from a turbine aerodynamic/aeroelastic solver. The second step is to break down the blade into a series of 2D airfoil sections and compute their self-noise individually including all relevant noise mechanisms: trailing edge noise, inflow turbulence noise, blunt trailing edge vortex shedding noise and separation noise (Brooks et. al. 1989). The overall blade noise can then be readily obtained through a spatial integration along span coupled with a tip vortex noise model. In general, additional corrections are also applied in the calculation to account for blade rotation effects such as Doppler shift and amplitude amplification (Oerlemans and Schepers, 2009). The final sound pressure level at the far field receiver location can be computed using a sound propagation model (ISO 9613-1, 1993).

At LM, a semi-empirical rotor noise prediction tool based on a similar scheme has been developed. The tool uses the turbine operation condition from industry-standard aeroelastic codes such as HAWC2 or SIMPACK, which simulates acoustic-relevant blade structural deformations, and then computes the boundary layer characteristics using an in-house airfoil section panel code. The individual noise sources are then calculated based on these inputs. In particular, the trailing edge noise is predicted by an internally developed model based on the TNO-Blake (Parchen, 1998) model. The tool has been extensively calibrated and validated using data from aeroacoustic wind tunnel measurements, high fidelity Computational AeroAcoustics (CAA) simulations and field measurements.

The challenge, however, is that modern wind turbine blades are equipped with various add-ons, such as trailing edge serrations and LEPs, many of which have a direct impact on the noise emission of the blade. Hence, it is imperative to model the noise signature of these devices quickly and reliably in the early blade design stage, to enable a holistic blade optimization. For this reason, a semi-empirical serration delta spectral correction model and LEP noise model are developed and incorporated into the production workflow.

2. Add-on Noise Modelling and Validation

Blade add-ons located in the outboard part of the blade generally have an impact on blade noise. The trailing edge serrations are designed to mitigate blade noise, which constitute a critical part of the overall rotor noise performance calculation. On the other hand, the LEP film, which adds an additional thin layer of material over the leading edge of the blade and creates small backward facing steps at the edges of the film, alters the characteristics of the boundary layer transition and growth over the blade, hence influencing the trailing edge noise indirectly.

2.1 Serration Spectral Correction Model

The trailing edge noise is the dominant noise source on the wind turbine blade (Wagner et. al. 96). As an effective and robust noise reduction device, sawtooth shaped trailing edge serrations have been developed and widely applied in the wind industry for more than a decade (Oerlemans et. al., 2009). Numerous theoretical (Lyu et. al., 2016), numerical (Romani et. al. 2021), and experimental (Oerlemans et. al., 2009) studies have been carried out in the past to investigate the optimal geometry, performance, and noise reduction mechanism of the serrations.

In LM, a large wind tunnel, CAA, and field acoustic measurement database have been established over the years with various serration geometries, which provides the foundation for the development of a semi-empirical serration spectral correction model.

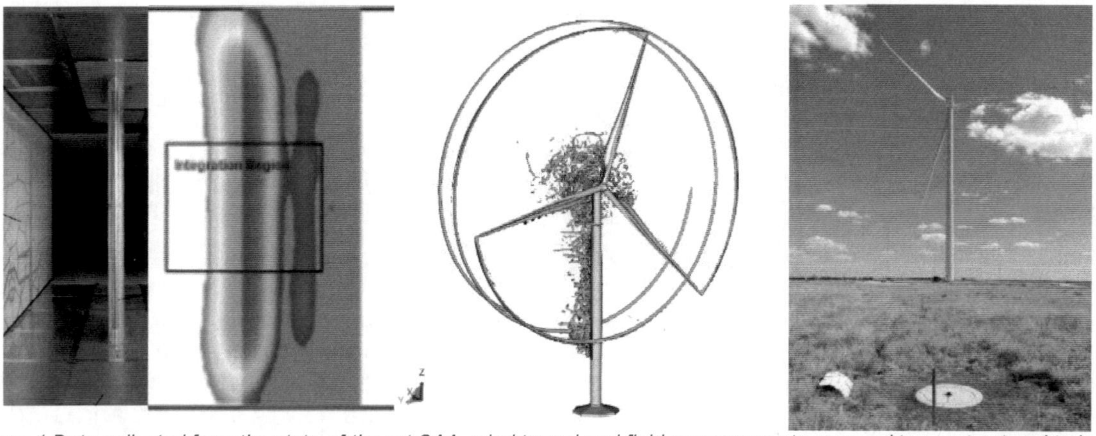

Figure 1 Data collected from the state-of-the-art CAA, wind tunnel and field measurements are used to construct and train the prediction model

The correction model is built upon our latest understanding of the physical mechanism of noise reductions by serrations. The hypothesis is that the serration offers noise mitigation through its ability to reduce scattering efficiency of the hydrodynamic turbulent boundary layer pressure fluctuation into the propagating acoustic pressure fluctuation. This reduction is in principle broadband and covers a wide range of frequencies (Howe, 1991). However, the presence of the serration also affects the aerodynamics both locally around the trailing edge and globally on the airfoil. Locally, when the serration teeth are misaligned with the trajectory of the original airfoil wake, flow will be pushed through the gaps between the teeth. When vortical structures in this crossflow interact with the serration edges, broadband high frequency self-noise is generated. Globally, the serration acts like a traditional flap, which changes the effective camber and chord of the airfoil, and in turn modifies the pressure distribution and boundary layer development around the airfoil. These aerodynamic effects play an important role in modelling serration performance and are one of the main reasons that the semi-empirical approach is preferred to the analytical scattering calculation which generally ignore the aerodynamic impact.

In addition, the serration has been shown to be an effective device to suppress the blunt trailing edge vortex shedding, likely due to the breakdown of the spanwise coherence of the original trailing edge. The summary of this overview is highlighted in Figure 2.

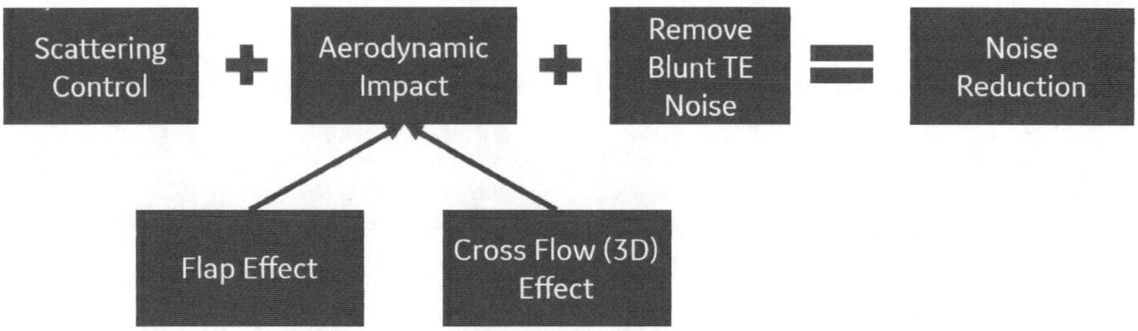

Figure 2 Schematic of the serration noise reduction mechanism

Based on these understandings, a semi-empirical spectral level correction model is constructed to capture the most relevant physical mechanism and calibrated using CAA simulations. It takes into consideration the effects of the serration length and flap angle under various flow conditions, and therefore can be used as both a powerful analysis/prediction tool and a serration layout design tool. Finally, the model is validated by comparing its predictions against both wind tunnel and field measurements. An example of each is discussed in detail in this paper.

The wind tunnel measurements were carried out in the Virginia Tech Stability Wind Tunnel with a large aperture phased microphone array (Devenport et. al., 2013). An LM proprietary airfoil is tested with/without standard LM serrations under clean and tripped conditions over a wide range of angles of attack. The Reynolds number based on chord is 2.4 million. The overall SPL (OASPL) is computed by integrating the third octave band spectra between 400 Hz and 3150 Hz, which is the valid frequency range of the beamforming measurement. The center of the microphone array is located approximately 90 degrees above the trailing edge on the suction side of the airfoil model. Both natural transition (clean) and forced transition (tripped) cases are measured and assessed in this validation study, but only the clean case is discussed in the paper.

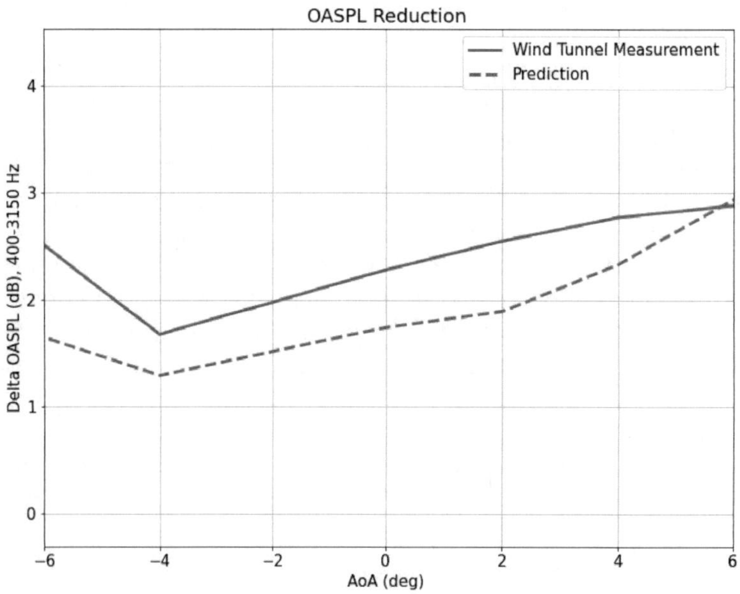

Figure 3 Serration noise reduction as a function of AoA

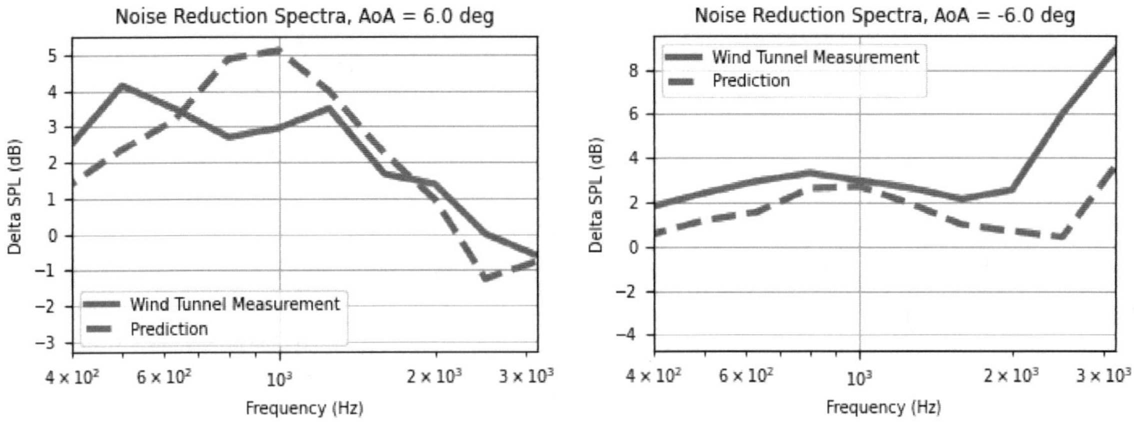

Figure 4 Serration noise reduction spectra at AoA = 6 and -6 degrees

The serration OASPL reductions from -6 to 6 degrees angles of attack are shown in Figure 3 for both the wind tunnel result and the prediction. The noise reduction is computed by subtracting the serration noise from the "naked" airfoil noise. The comparison clearly shows that the amplitude and the sensitivity of the serration noise delta to the angle of attack have been very well predicted compared to the measured delta. The difference between the prediction and the measurement is smaller than 1 dB across all angles of attack, which is within the uncertainty of the measurement.

In addition, a one-third octave band delta spectra comparison is shown in Figure 4 at two extreme angles of attack, i.e., -6 and 6 degrees. The general delta spectral shape is well predicted by the model at both angles of attack. The peak noise reduction frequency and the cross-over frequency where the serration benefit vanishes are both captured. At -6 degree angle of attack, there is a blunt vortex shedding noise at high frequency on the baseline airfoil which is eliminated by the serration. That's actually the primary reason for the large noise reduction at high frequency.

With the high prediction accuracy on the 2D airfoil level, the model is then implemented on the standard LM blade noise prediction code and compared against various field measurements. In general, the field noise measurement and data analysis follow the IEC 61400-11 standard (IEC, 2012). When available, machinery sound power level is added to the rotor apparently sound power (Lwa) noise prediction to be as consistent as possible.

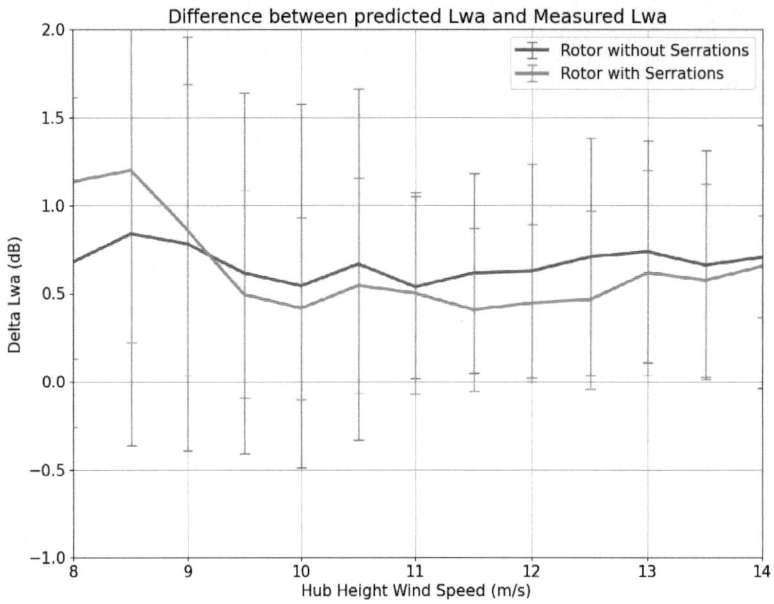

Figure 5 Averaged difference between the absolute predicted Lwa and measurements across 6 turbines

Figure 5 shows a summary of the prediction accuracy of the tool with and without serration correction model. The absolute difference between the predicted and measured apparent sound power level is generally smaller than 1 dB across 6 different onshore wind turbines both with and without serrations at critical wind speeds.

As an example, the validation against a research prototype turbine set up within the TIADE project (3.8MW, 130m rotor) with and without serrations is demonstrated here. Figure 5 shows the comparison between predicted and measured Lwa noise curves as a function of hub height wind speed. The turbine operation and aerodynamic properties are obtained from an aeroelastic solver as inputs to the noise simulation. Excellent prediction accuracy is achieved at all wind speeds for cases with and without serrations. The noise reduction comparison highlights again that the result from the serration noise prediction model matches the test value within the measurement uncertainty.

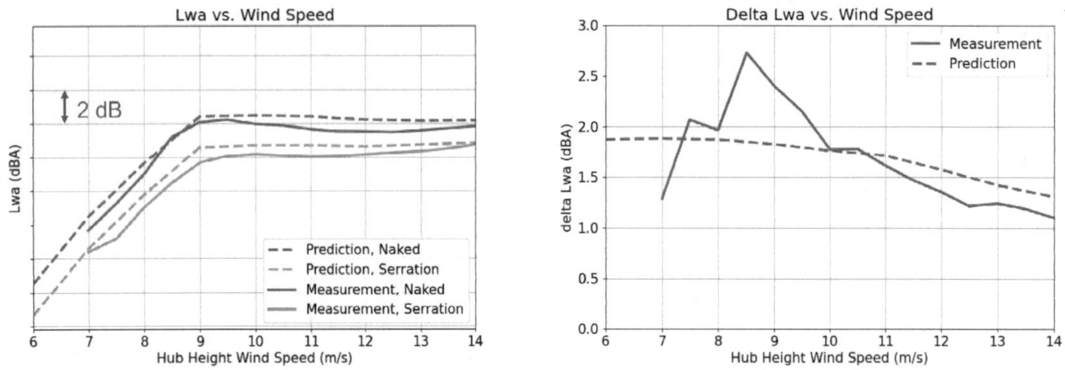

Figure 6 Predicted Lwa versus measured Lwa on the TIADE turbine with/without serrations and the serration noise reduction comparison

An A-weighted third octave band spectra comparison at 10 m/s is shown in Figure 7. An excellent match between the prediction and measurement is achieved. The delta spectra comparison in Figure 8 however shows a slight frequency shift of the predicted serration benefit comparing to the measured delta spectral curve.

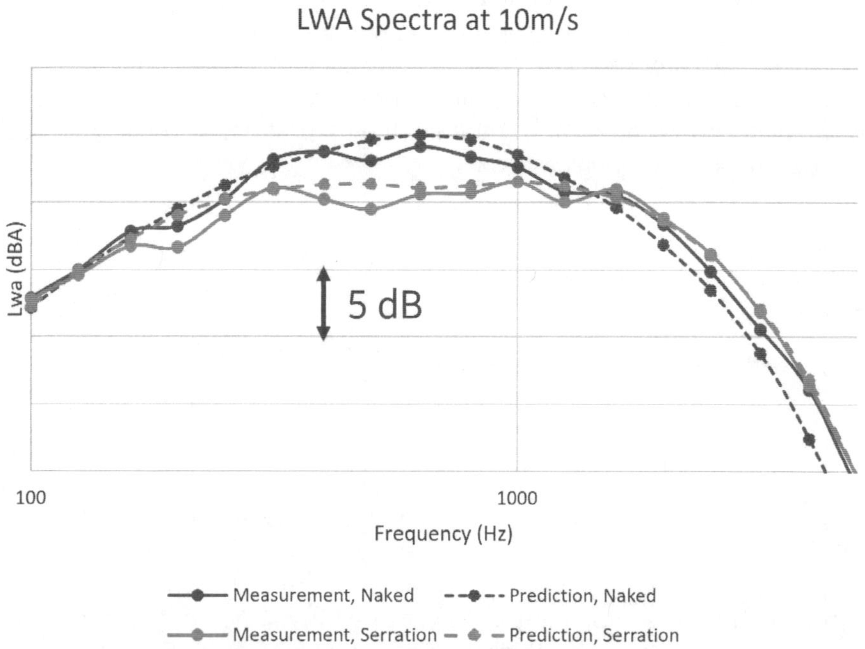

Figure 7 Spectral comparison of prediction versus measurement at 10 m/s

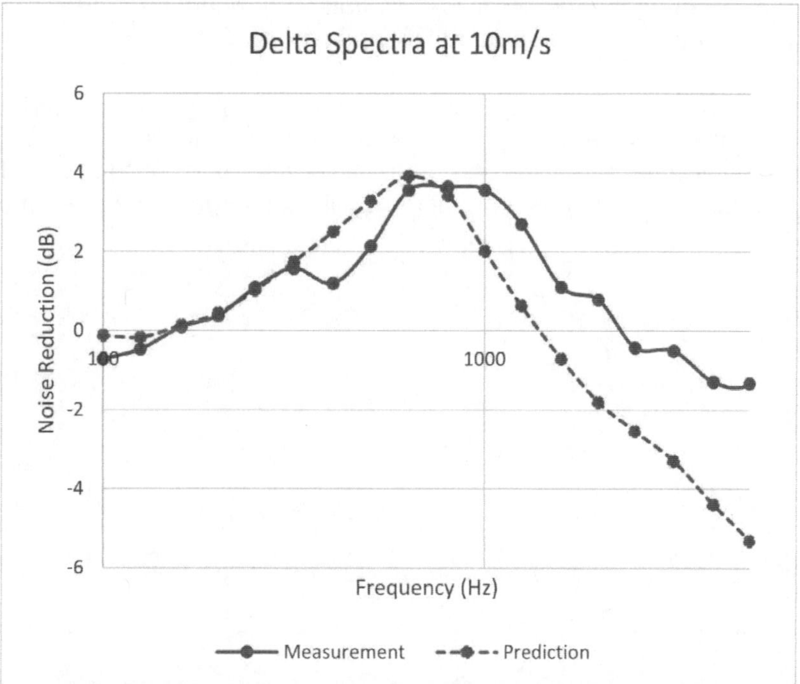

Figure 8 Noise reduction spectral comparison of prediction versus measurement at 10 m/s

2.2 LEP Correction Model

The tape-type LEP is a popular solution for erosion protection in the wind industry. A layer of thin polymeric film wraps around the leading edge along the outboard part of the blade to protect it from erosion. However, the step at the end of film typically triggers a laminar to turbulent bypass transition, if the boundary layer has not naturally transitioned into turbulence at the location of the step. Moreover, when the step thickness is large compared to the local boundary layer thickness, it was observed in wind tunnel and CAA studies that the downstream turbulent boundary layer grows more rapidly than in the natural transition case, and that leads to an increase in the trailing edge noise emission.

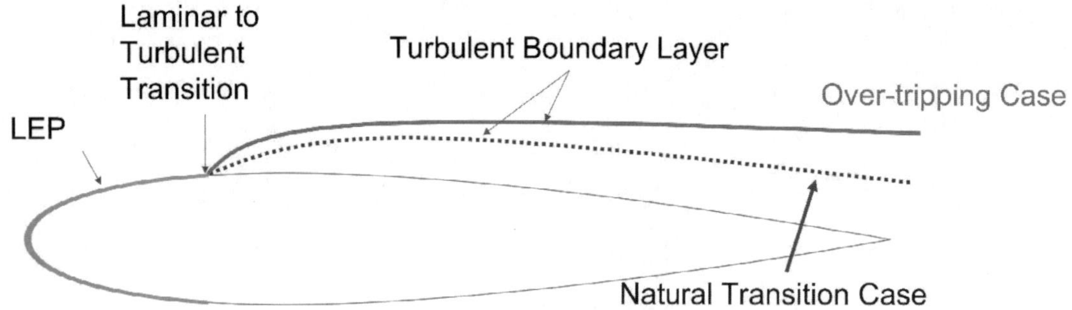

Figure 9 LEP effect on the boundary layer growth

An empirical correction model is implemented to capture the change of the transition location as well as the increased boundary layer thickness at the trailing edge. Boundary layer thickness and displacement thickness computed from the panel code are corrected based on the nondimensionalized step height and chordwise location. The model parameters are tuned based on the CAA simulations and wind tunnel measurement.

Validation against field data has been carried out using data from the TIADE turbine. Multiple IEC noise measurements have been performed before and after application of LEP on all three blades. LEP coverage starts at the outer part of the blade and extends to the tip. The relative chordwise coverage varies along the span, but generally has larger coverage on the suction side than the pressure side.

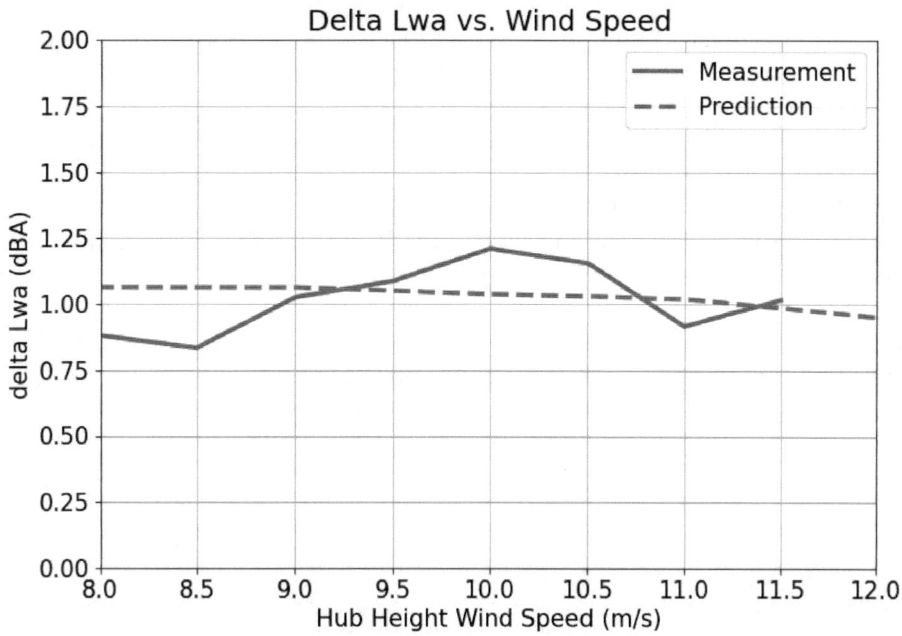

Figure 10 Comparison of the predicted noise increase against measurements due to LEP

Figure 10 shows a comparison between the predicted noise increase due to the LEP to the measurement. In the measurement, a noise increase of 1 dBA is observed across a wide range of wind speeds, and this value is well predicted by the empirical model. Moreover, the delta spectra comparison at 10 m/s wind speed shown in Figure 11 demonstrates that the delta spectral shape is also well predicted. This indicates that this simple model is capable of capturing the key part of the noise increase mechanism due to the application of the LEP.

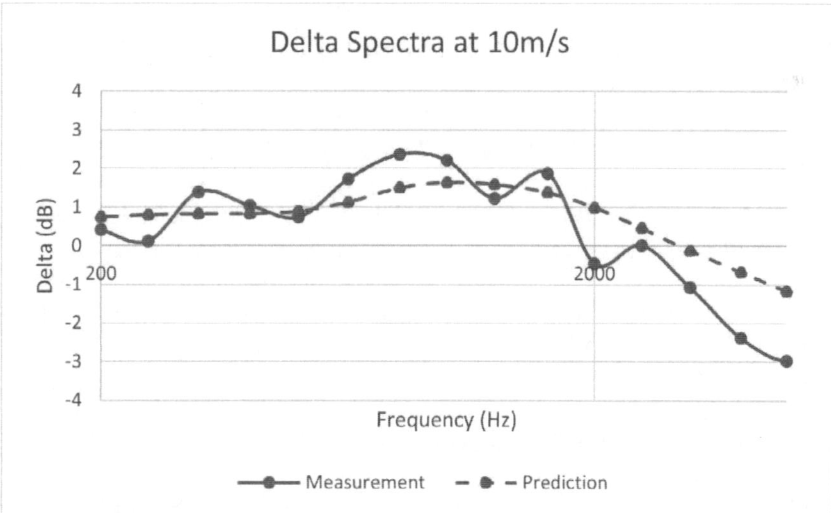

Figure 11 LEP noise increase spectra at 10 m/s wind speed, prediction vs. measurement

3. Conclusions

Semi-empirical serration and LEP noise models are developed, tuned, and validated against measurements in LM. Excellent agreement is achieved on a sound power level and spectral basis for both models. In addition, an LPS noise model is available but omitted from the present paper for conciseness. These advancements enable a holistic blade design approach taking the effect of serrations and LEP into consideration in the early blade design phase, and therefore potentially lead to blades with better overall performances and generally improve the design fidelity. Moving forward, more models for various types of blade add-ons, e.g. vortex generators, are being developed to open up even larger design space for blade optimization.

Acknowledgement

Authors would like to thank Dr. Drobietz and many other GE Onshore wind acoustic team colleagues for their valuable inputs; and Dr. Borgoltz, Dr. Intaratep and their colleagues at the Virginia Tech Stability Wind Tunnel for their efficient collaboration in the wind tunnel measurements over the years. In addition, the collection of field validation data alongside the TIADE project would not have been possible without the financial support from the Netherlands Enterprise Agency (RVO) for establishing this research prototype.

References

Bertagnolio F, Madsen HAa, and Fischer A (2017) *A Combined Aeroelastic-Aeroacoustic Model for Wind Turbine Noise: Verification and Analysis of Field Measurements* 20(8), 1331-1348 doi: 10.1002/we.2096

Boorsma K. and Schepers J.G. (2011) *Enhanced wind turbine noise prediction tool SILANT* 4th International Meeting on Wind Turbine Noise

Bortolotti P, Branlard E, Platt A, Moriarty PJ, Sucameli C, and Bottasso CL (2020) *Aeroacoustics Noise Model of OpenFAST* NREL/TP-5000-75731

Brooks TF, Pope DS, and Marcolini MA (1989) Airfoil self-noise and prediction NASA Reference Publication 1218

Howe MS (1991) *Aerodynamic noise of a serrated trailing edge*, Journal of Fluids and Structures, Vol. 5, Issue 1.

IEC (2012) *Wind turbines – Part11: Acoustic noise measurement techniques* IEC 61400-11

International Organization for Standardization (ISO) (1993) ISO 9613-1; *Acoustics of sound during propagation outdoors*

Lyu B, Azarpeyvand M and Sinayoko S (2016) *Prediction of noise from serrated trailing edges.* Journal of Fluid Mechanics, 793, 556-588. doi:10.1017/jfm.2016.132

Moriarty PJ, Guidati G, and Migliore P (2005) *Prediction of Turbulent Inflow and Trailing-Edge Noise for Wind Turbines* 11th AIAA/CEAS Aeroacoustics Conference

Oerlemans S and Schepers JG (2009) *Prediction of wind turbine noise and validation against experiment* NLR-TP-2009-402

Oerlemans S, Fisher M, Maeder T and Koegler K. (2009) *Reduction of wind turbine noise using optimized airfoils and trailing-edge serrations* AIAA Journal, Vol. 47, No. 6

R. Parchen (1998) *Progress report DRAW: A Prediction Scheme for Trailing-Edge Noise Based on Detailed Boundary-Layer Characteristics.* TNO Rept. HAG-RPT-980023

Romani G, Casalino D, and van der Velden W.C.P. (2021) *Numerical Analysis of Airfoil Trailing-Edge Noise for Straight and Serrated Edges at Incidence* AIAA Journal 2021 59:7, 2558-2577

Wagner, S., Bareiß, R. and Guidati, G. (1996), Wind Turbine Noise, Springer-Verlag

Devenport, WJ, Burdisso RA, Borgoltz A, Ravetta PA, Barone MF, Brown KA, Morton MA (2013) *The Kevlar-walled anechoic wind tunnel,* Journal of Sound and Vibration, Volume 332, Issue 17, 2013

10th International Conference
on
Wind Turbine Noise
Dublin – 21st to 23rd June 2023

Computational study on the aeroacoustics of the X-Rotor – a hybrid vertical - horizontal axis wind turbine

Yan Wu, Flow Physics and Technology Department, Faculty of Aerospace Engineering, Delft University of Technology, Delft, The Netherlands: Y.Wu-10@tudelft.nl
Francesco Avallone, Politecnico di Torino, Dipartimento di Ingegneria Meccanica e Aerospaziale, Torino, Italy: francesco.avallone@polito.it
Daniele Ragni, Flow Physics and Technology Department, Faculty of Aerospace Engineering, Delft University of Technology, Delft, The Netherlands: D.Ragni@tudelft.nl
Damiano Casalino, Flow Physics and Technology Department, Faculty of Aerospace Engineering, Delft University of Technology, Delft, The Netherlands: D.Casalino@tudelft.nl

Abstract

This study presents a computational aeroacoustic analysis of the X-Rotor Offshore Wind Turbine. The X-Rotor is a new offshore wind turbine concept consisting of an x-shaped vertical axis wind turbine (VAWT) combined with two horizontal axis wind turbines (HAWT). In order to evaluate performance and noise emitted from the X-Rotor baseline concept, a full-scale high-fidelity computational fluid dynamics (CFD) simulation was carried out with the Lattice-Boltzmann very large eddy simulation (LB-VLES) solver 3DS PowerFLOW. The results from the numerical simulation were used to predict the far-field sound pressure using the Ffowcs-Williams & Hawkings (FW-H) acoustic analogy implemented in the Opty∂B® software. In the present study, the X-Rotor was operated at the designed maximum power extraction condition. The rotational tip Mach number of the VAWT and HAWT are 0.19 and 0.59, respectively. The corresponding Reynolds number (Re) based on the blade sectional chord and the local flow velocity of the VAWT and HAWT are over 10 and 1 million, respectively. The results show that the VAWT is the main noise source of the X-Rotor causing high amplitude noise in the low-frequency region. Additionally, the overall sound pressure level (OSPL) of the VAWT and HAWT at a virtual microphone located at an upwind location is higher than the one computed when the virtual microphone is located at a downwind location.

1. Introduction

The X-Rotor project is a Horizon 2020 funded project which aims to develop a new concept of offshore wind turbine with improved performance and self-starting capabilities. This concept was originally developed by (Leithead et al., 2019) with a VAWT combined with two HAWTs. A 3D sketch is shown below in Figure 1. In the original design, the primary rotor was designed for the wind power collection, while the secondary rotors (the smaller HAWTs installed at the tips) had the purpose of providing the power for take-off. The whole system is kept in equilibrium by balancing the primary rotor's torque with the loading of the secondary rotors'.

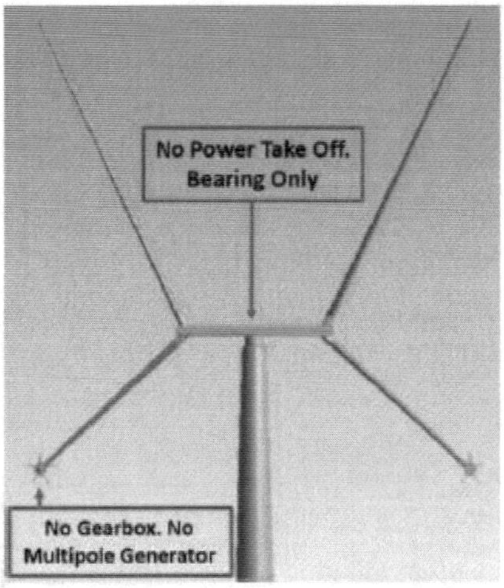

Figure 1. Baseline concept of X-Rotor, provided by. (Leithead et al., 2019)

The current design offers a significant advantage by reducing energy consumption, operational costs, and maintenance expenses. This is achieved by eliminating the need for components such as gearbox or mulita-pole generator (Leithead et al., 2019), which not only reduces energy requirements but also results in a reduction of mechanical noise. However, it is essential to consider aerodynamic noise sources, particularly the noise produced by the interaction of airflow with the blades. Whereby the present work only focuses on the evaluation of the aerodynamic noise.

Since both primary and secondary rotors of the baseline concept are designed to operate at high Reynolds numbers (Re > 1×10^7 for the primary rotor, Re > 1×10^6 for the secondary rotor), the turbulent boundary layer – trailing edge noise (TBL-TE) is one of the most dominant noise sources with broadband noise spectral feature according to (Brooks et al., 1989). For the VAWT, previous studies (Boorsma & Schepers, 2011; Pearson, 2014) showed that separation stall noise (SS) and turbulent inflow noise (TI) are the major noise sources. They largely vary with the tip speed ratio (TSR). As a matter of fact, the angle of attack changes as a function of azimuthal angle and the blades will experience, in a range of azimuthal angle, separation stall at low TSR. However, the SS can also potentially be a noise source for the secondary rotors. The angle of attack, unlike regular HAWT, changes with the azimuthal positions of the primary rotor's lower blade. At higher TSR, the TI becomes more relevant, as both the primary rotor and secondary rotor blades are subjected to the turbulent wake of the blades and tower.

The work described in this report provides an investigation of the aerodynamic and aeroacoustics performance of the X-Rotor baseline concept at full scale. This is achieved by using the Lattice-Boltzmann Method / Very Large Eddy Simulation (LBM/VLES) approach to compute the flow field and FW-H acoustic analogy (Ffowcs Williams & Hawkings, 1969) for the prediction of far-field acoustic pressure. The paper is organized as follows, Section 2 describes the methodology and the simulation setup. The CFD results are presented in Section 3 and the acoustic results are presented in Section 4. The last section is the conclusion of the work.

2. Methodology

2.1 Flow solver

The commercial software 3DS SIMULIA PowerFLOW 6-2021-R2 is employed. The software solves the discrete, explicit, transient, compressible LB equations by statistically tracking the streaming and collisions of fluid particles for a finite number of directions. The macroscopic flow properties can be derived from the discrete form of the mesoscopic kinetic equation. A more detailed description of this method is presented in the works of (Succi, 2001) and (Shan et al., 2006), or in a review by (Chen & Doolen, 1998). The implementation for the present case consists of 19 discrete velocities in 3 dimensions (D3Q19), involving a third-order truncation of the Chapman-Enskog expansion which has been shown to provide sufficient lattice symmetry to recover the NS equation for a low Mach number isothermal flow (Chen et al., 1992).

A Very Large Eddy Simulation (VLES) model is implemented to account for the effect of the sub-grid unresolved scales of turbulence. Following (Yakhot & Orszag, 1986), a two equation $k-\epsilon$ re-normalization group is used to compute a turbulent relaxation time that is added to the viscous relaxation time. In order to reduce the computational cost, a pressure-gradient-extended wall-model is used to approximate the no-slip boundary condition on solid walls (Teixeira, 1998; Wilcox, 2006). The model is based on the extension of the generalized law-of-the-wall model (Launder & Spalding, 1983) to take into account the effect of the pressure gradient.

2.2 Acoustic solver

The acoustic pressure at the far field is computed using FW-H acoustic analogy, the formulation 1A developed by (Farassat & Succi, 1980) is employed as a solution of FW-H equation and solved in a forward-time (Casalino, 2003) by integrating the unsteady pressure on the solid surfaces of the wind turbine blades. The approach is carried out using a software Opty∂B®-PFNOISESCAN (a toolkit embedded in the automatic eVTOL aeroacoustic PowerFLOW workflow developed by Casalino) for predicting the secondary rotors' noise.

3. Computational setup

3.1 Geometry and computational setup

There are 3 local reference frame (LRF) coordinate systems set for the simulation of the X-Rotor baseline, as is shown in Figure 2. The primary rotor with a radius of $R_t^{(p)} = 75$ m, rotates around the $\mathrm{x}_p -$ axis at a fixed speed $\Omega_p = 0.838$ rad/s, resulting in a tip Mach number of 0.19. The secondary rotors are located at the lower blade tips, and have a radius of $R_t^{(s)} = 4.69$ m. They rotate around the x_{s_1} and $\mathrm{x}_{s_2} -$ axis, respectively, at a constant speed of $\Omega_s = 43.27$ rad/s, resulting in a tip Mach number of 0.59. The incoming freestream velocity is set at $V_\infty = 12.5$ m/s towards the positive z – direction in the global coordinate system.

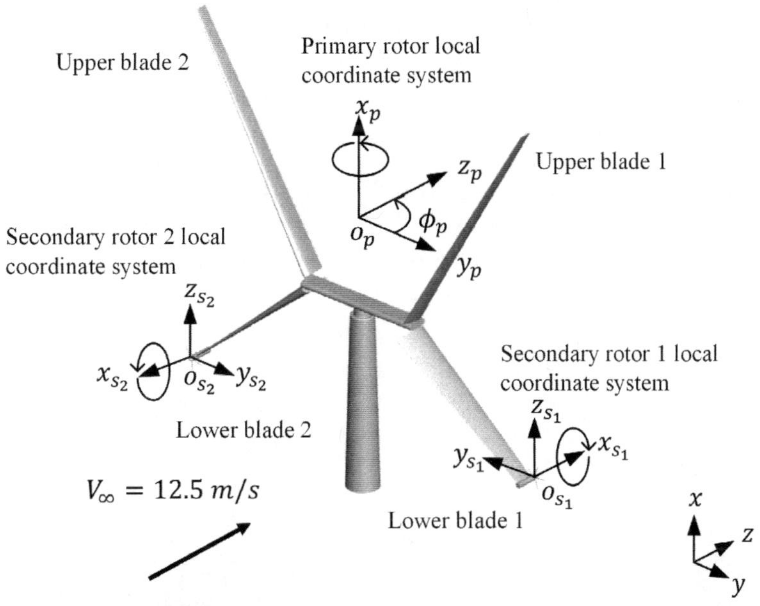

Figure 2. The X-Rotor baseline concept geometry and the local reference frame coordinate systems

The blade profiles used in this simulation were the same as in the previous study (Leithead et al., 2019). The upper and lower blades of the primary rotor are 100 m and 65 m long, respectively, with coning angles of 30° and 50°. The airfoil profiles are the NACA 0025 and NACA 0008, respectively. The chord is linearly reduced from the blade roots to the tips for both the upper and lower blades. They are reduced from 10 m to 5 m for the upper blades and 14 m to 7 m for the lower blades. No twist and pitch angles are applied along the radial direction of the primary rotor blades. For the secondary rotor, the airfoil profile FFA_W3_241 is used for the entire blade, the twist and the chord length distributions are shown in Figure 3.

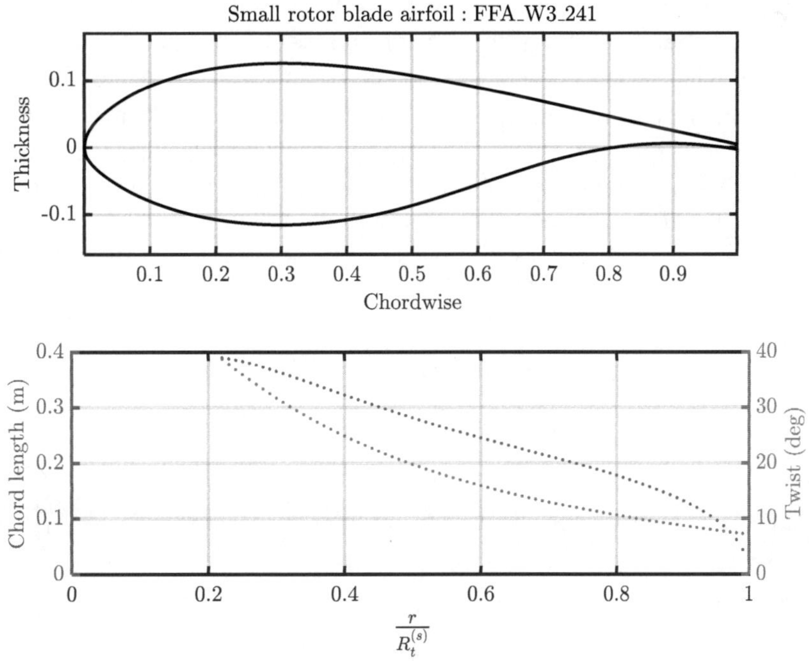

Figure 3. Secondary rotor blade profile.

3.2 Simulation domain setup

The computational domain is discretized in different variable resolution (VR) regions. There are 14 VR regions set for this simulation, ranging from the coarsest resolution VR0 to the finest resolution VR14, with the grid resolution varying by a factor of 2 between adjacent VR levels, see Figure 4. Here the smallest voxel size is defined as δx in VR14, and the largest voxel size is δx^{13} in VR0. As shown in the figure below, due to the proximity of the places of interest to the rotor blades, VR regions 14 to 12 with the smallest voxel size are located near the secondary rotor blade, while VR regions 11 to 10 are placed closer to the primary rotor blades. It has been guaranteed that there are at least 6 voxels between 2 adjacent resolution regions. The grid resolution is defined as the number of voxels per characteristic length, where the characteristic length is considered as the chord length at $0.75R_t^{(s)}$ of the secondary rotor.

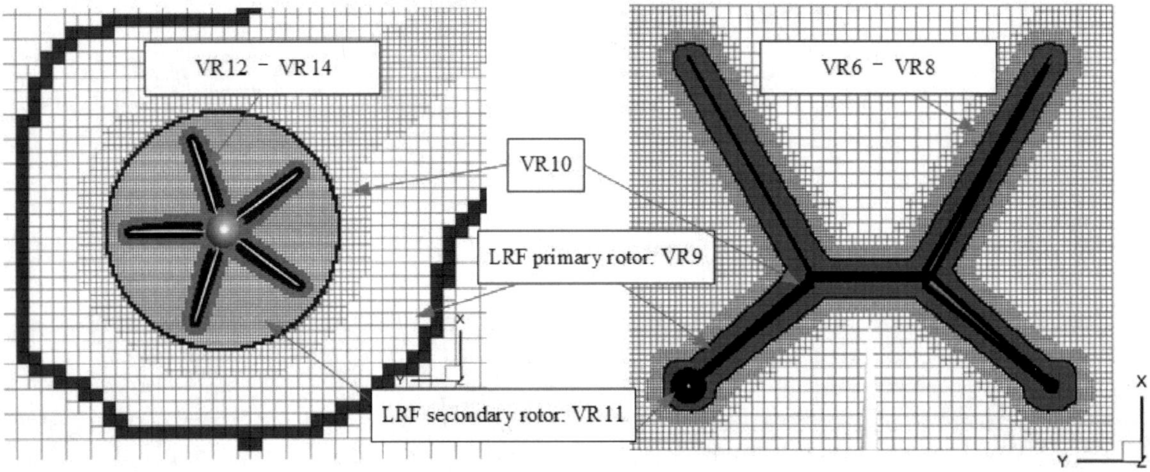

Figure 4. Change in grid resolution with VR region levels

The simulation domain is a rectangular box, as shown in Figure 5. The origin of the primary rotor coordinate system O_p is placed at $18.5R_t^{(p)}$ downstream of the inlet and $37R_t^{(p)}$ away from other edges. In order to minimize the impact of vortex shed from the bottom side of the tower on the flow field around wind turbine, the height of the tower is set to $11.437R_t^{(p)}$. In addition to this, two spherical-shaped acoustic sponge regions, with exponentially increasing the value of damper parameters, are set at a distance $4.7R_t^{(p)}$ and $12.45R_t^{(p)}$ from O_p. At the inlet, the freestream velocity V_∞ , the turbulent intensity $\mathrm{I} = 10.2\%$ and turbulent length scale $L = 145$ m are set as boundary condition. The turbulent intensity and turbulent length scale are evaluated using the method described by Zhu et al. (Zhu et al., 2005). At the outlet, ambient pressure $p_\infty = 1.00125 \times 10^5$ Pa is set for a pressure boundary condition. For other walls at the edge, the slip wall boundary conditions are applied.

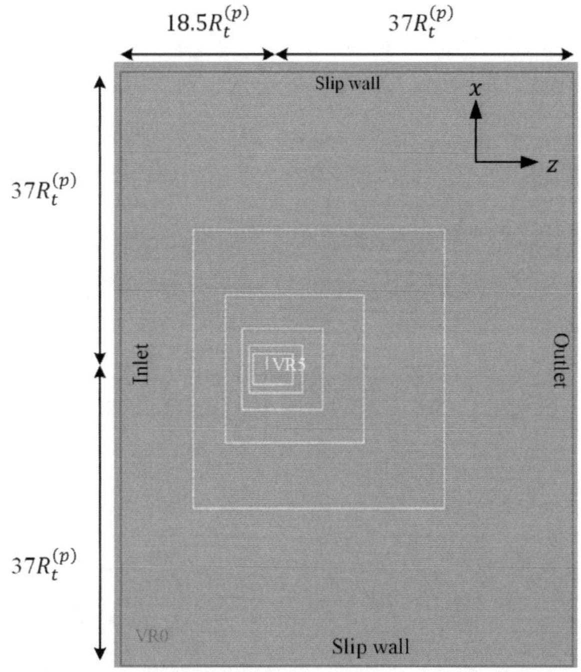

Figure 5. Simulation domain and distribution of variable resolution regions

The current design exhibits multiple scales, with a chord length ratio of 39 between the primary and secondary rotor blades. Additionally, the primary and secondary rotor blades are designed for high Reynolds number conditions, which result in a turbulent flow over the blade surfaces. Therefore, resolving small-scale eddies in the turbulent flow through simulations can be extremely costly. To balance accuracy with computational efficiency for this preliminary simulation, the finest voxel size was set to 9.52 mm, which corresponds to 25 voxels per average chord of the secondary rotor blade. And the physical time step, corresponding to a Courant-Friedrichs-Lewy (CFL) number of 1 in the finest mesh resolution is 1.439×10^{-5} second. This resolution prioritizes resolving the large eddies in the flow, which predominantly determine the overall performance of the current design. Furthermore, the resolution is expected to sufficiently resolve the tip vortex and blade wake due to their significant sizes. The total number of voxels and surface elements for this simulation is 73.45 million and 7.7 million respectively. The flow-simulation time is 90.20 seconds for this case, which is equivalent to 12 primary rotor revolutions requiring 4.558×10^3 CPU hours per revolution using a Linux Xeon® Gold 6130 2.1GHz Platform. The unsteady forces on the solid surface of both primary and secondary rotor blades were measured at every 1-degree revolution of their respective rotors.

3.3 Noise measurement setup

As shown in Figure 6, the far-field acoustic pressure is measured in both polar ($x - z$ plane) and azimuthal direction ($y - z$ plane), at 8 virtual microphones in each plane. The measurements are performed at two distances with the microphones located 200 m (2.5 $R_t^{(p)}$) from the rotor center O_p.

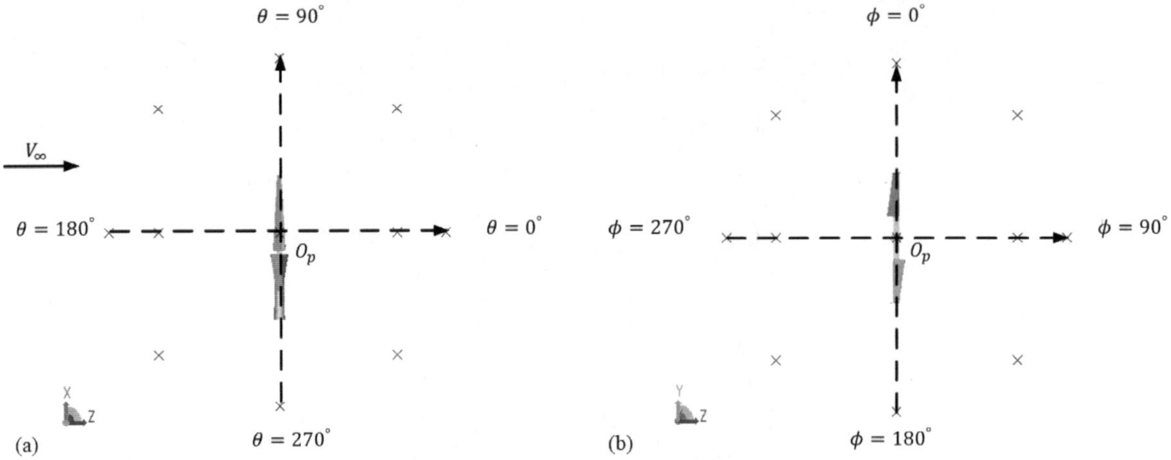

Figure 6. 15 microphones are placed at (a) polar angles of X-Rotor in $x-z$ plane, and (b) azimuthal angles of X-Rotor in $y-z$ plane respectively with a distance of 200 m (2.5 $R_t^{(p)}$) from O_p

The unsteady pressure on a solid surface is sampled for 22.5 seconds, corresponding to 3 revolutions of the primary rotor after the flow convergence. Details on the sampling frequency are reported in Table 1.

Table 1. Sampling frequency for different resolution regions

Surface	VR regions	Sampling frequency, Hz
Primary rotor	VR10	4.342×10^3
Secondary rotor	VR12	1.7370×10^4

4. Results and discussion

4.1 Aerodynamic performance and flow field description

The time history data of the force components in three axial directions have been analysed to check for time convergence of the simulation, as shown in Figure 7. Here F_x, F_y and F_z are the forces in x_p, y_p and $z_p$directions of the global coordinate system. As presented in sub-figure (a) and (b), the force generated from the primary rotor and the secondary rotor 1 is periodic in the rotating azimuthal angle ϕ_p , and the values become stable after the 4th rotation of the primary rotor. The axial force F_x is observed to be much smaller in magnitude than the other components for both primary and the secondary rotor.

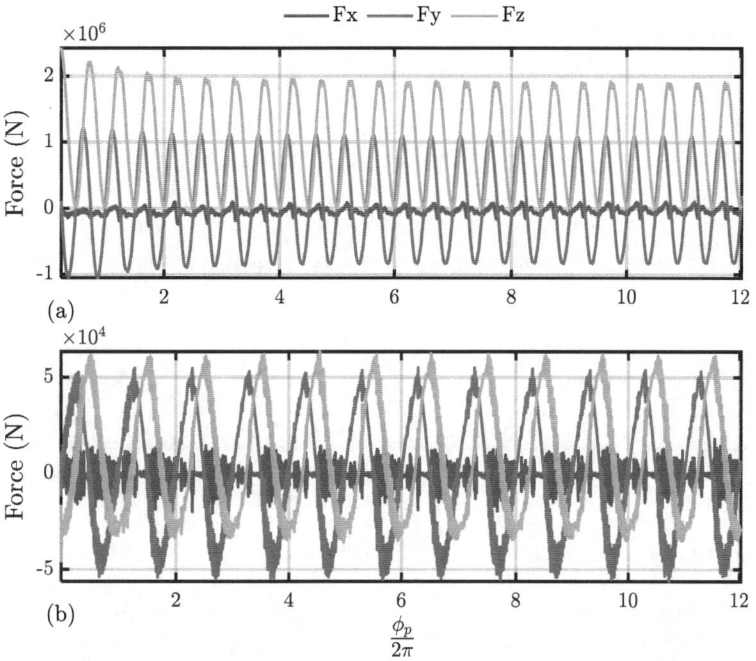

Figure 7. Time history data, for a) total forces generated from the upper and lower blades with the generators included, b) forces generated from secondary rotor 1. The forces are all based on the global coordinate system, and the time scale for (a) and (b) is normalized to the revolutions numbers of the primary rotor.

In order to evaluate the rotor performance conveniently, the following non-dimensional parameters are introduced.

The tangential force coefficient C_{T} is defined as in eq. (1)

$$C_{\mathrm{T}} = \frac{\mathrm{F_T}}{0.5\rho_0 A_{rot} V_\infty^2}, \quad (1)$$

The pressure coefficient, C_p, is expressed as in eq. (2)

$$C_p = \frac{p - p_\infty}{0.5\rho_0 V_\infty^2}, \quad (2)$$

where $\mathrm{F_T}$ is the tangential force around the primary rotor axial axis x_P, ρ_0 is the air density and A_{rot} is the swept area of the rotor, and p is the static pressure.

Figure 8 presents the phase-averaged C_{T} of different components of the X-Rotor over 6 rotations. The results show that the upper blade 1 and lower blade 1 have the maximum and the minimum C_{T} values, respectively, at azimuthal angles around $\phi_p = 280^\circ$ and $\phi_p = 190^\circ$. Additionally, a slight fluctuation in C_{T} can be observed when the lower blade1 passes through the downwind location at approximately $\phi_p = 85^\circ$, reaching its local minimum value. Compared to the primary rotor blades, the C_{T} variation of the secondary rotor 1 is relatively small with a difference of 0.1 between the maximum (occurring at $\phi_p = 180^\circ$) and minimum (occurring at $\phi_p = 0^\circ$) value. However, fluctuations can be found between $\phi_p = 180^\circ$ and $\phi_p = 360^\circ$, as the secondary rotor 1 traverses the upwind locations of X-Rotor. Conversely, only two fluctuations occur at around $\phi_p = 45^\circ$ and $\phi_p = 105^\circ$ when the secondary rotor 1 is at the downwind locations.

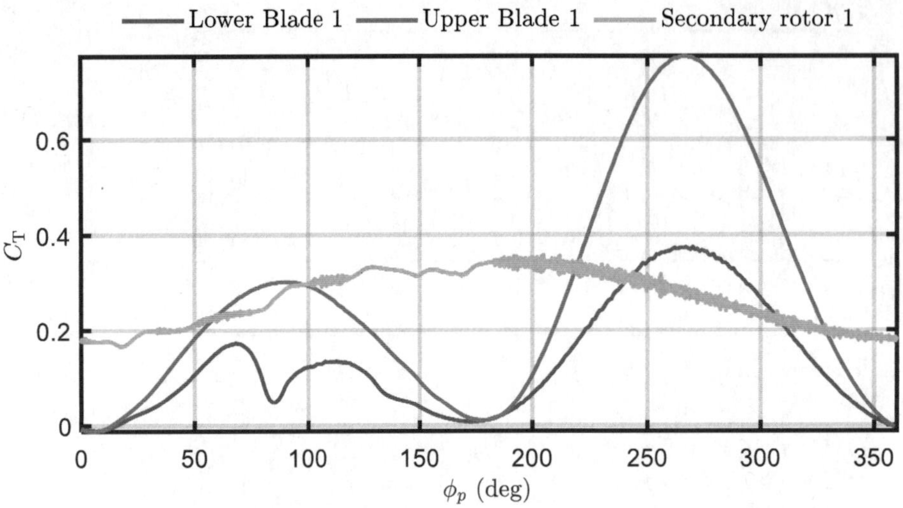

Figure 8.Plot of the primary rotor phase averaged tangential force coefficient C_T against the azimuthal position ϕ_P.

Figure 9 shows an instantaneous flow field using the iso-surface of the constant lambda-2 criterion to display turbulence structures. The colour contours represent the non-dimensional velocity magnitude V_{mag}/V_∞. The results indicate that most turbulence structures are generated near the tip and root of the upper and lower blades in the near-wake region. The upper blade tips generate large, coherent vortex structures, while the other components generate small-scale vortices. It is evident that the coherent helical-vortex structures are emerging from the blade tips of the secondary rotors, which interact with the lower blades and further convect downstream together with the vortices shed from the generators, finally forming a large-scale vortex ring at the far-wake region. Furthermore, it is also clearly seen that the helical structures generated from the secondary rotor at $\phi_p = 0^\circ$ consist of more small turbulence structures than those visible at $\phi_p = 180^\circ$, since the freestream velocity and the incoming velocity caused by the rotation of the primary rotor are in the same direction for the former while in the opposite direction for the latter.

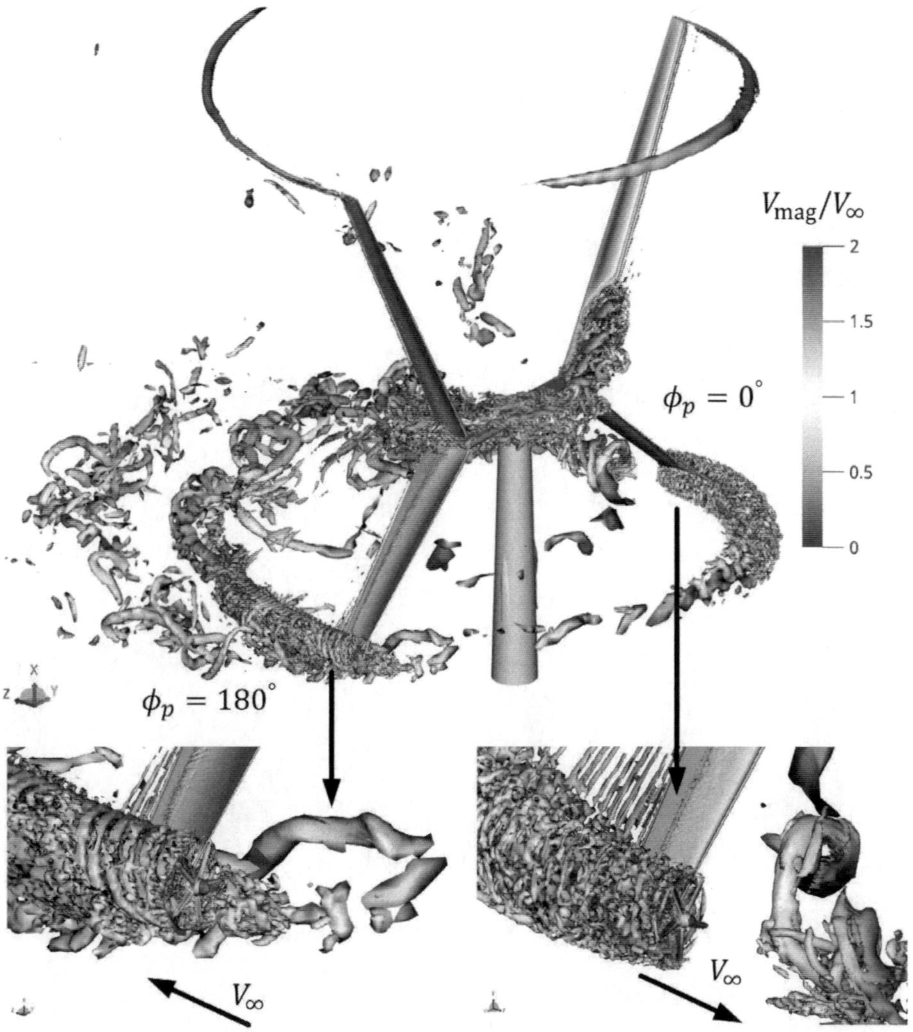

Figure 9. Instantaneous turbulence structure visualization using iso-surface constant lambda-2 criterion ($\lambda_2 = -2.5\ s^{-2}$), the colour contoured with the non-dimensional velocity magnitude V_{mag}/V_∞.

As depicted in the figure above, the presence of wind turbine configurations results in blade-wake interaction (BWI) and blade-vortex interaction (BVI) for each blade of the X-Rotor, particularly for the lower blades and secondary rotors. The instantaneous flow fields for the wind turbine at different azimuthal angles are demonstrated in Figure 10, where the contours of the non-dimensional velocity magnitude V_{mag}/V_∞ are generated at the plane located at the same height of the secondary rotor rotational center. Sub-figure (a) shows that the wake shed from the secondary rotor and generator at azimuthal angle $\phi_p = 0^\circ$ interacts with secondary rotor at $\phi_p = 180^\circ$. In sub-figure (b), the secondary rotor begins to interact with its own wake at $\phi_p = 45^\circ$, as well as with the high-velocity region of the wake generated from the tower. In sub-figure (c), the vortex ring produced by the secondary rotor and the generator in the upwind region travels along the freestream direction, interacting with the tower and causing the shedding of a Kármán vortex street from the tower. This vortex street subsequently impinges on the secondary rotor at $\phi_p = 90^\circ$. In sub-figure (d), the secondary rotor is once again exposed to the high-velocity region of the tower's wake, thereby providing the explanation for the force fluctuations observed in Figure (8) at the corresponding locations.

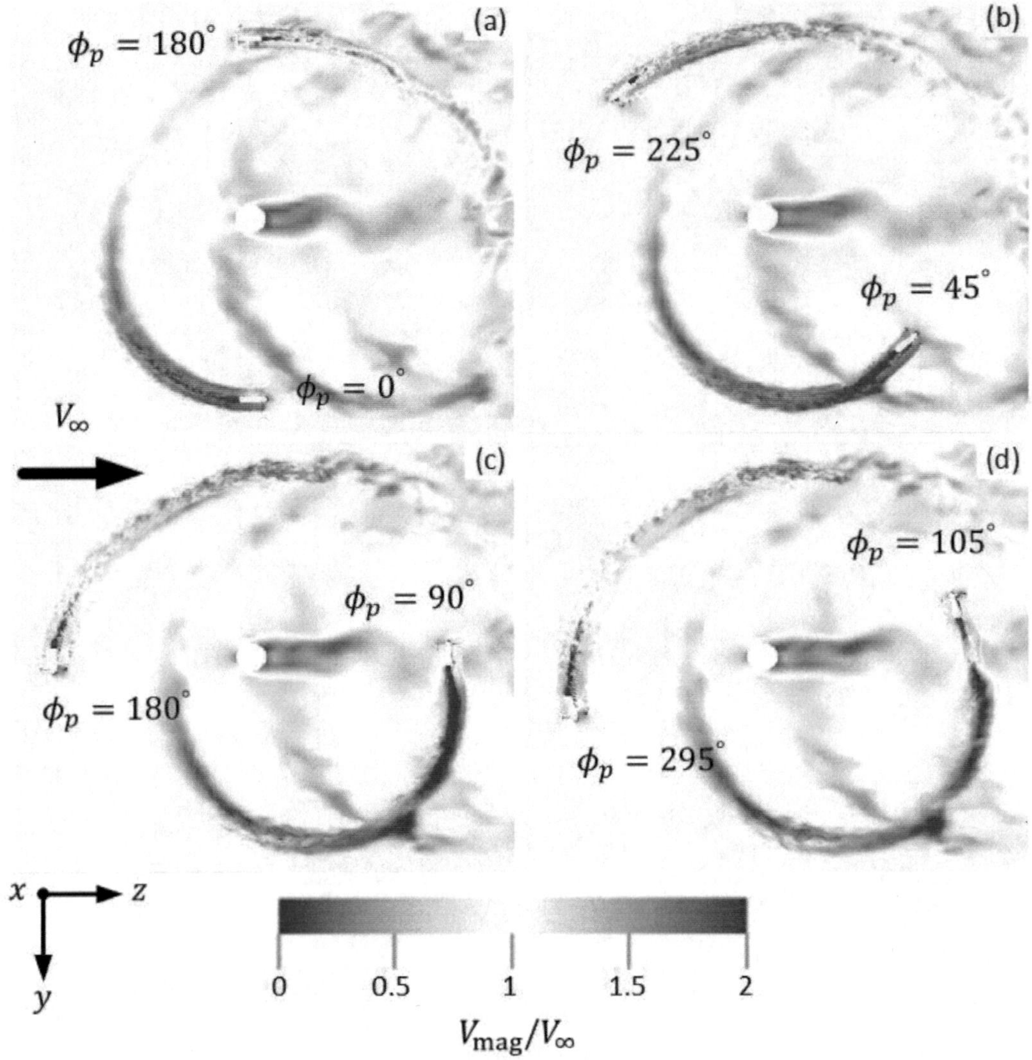

Figure 10. Instantaneous flow field visualized through the non-dimensional velocity magnitude V_{mag}/V_∞, all the sub-figures are at the plane of the center of the secondary rotor.

Figure 11 illustrates the instantaneous flow field around four cross-sections of the lower blade, at azimuthal angles of $\phi_p = 85^\circ$ and 280°, using a non-dimensional chord-wise velocity field V_c/V_∞ and pressure coefficient C_p. This visualization provides insights into the physical behaviours that result in a local minimum and global maximum value of C_T at the corresponding angles shown in Figure 8.

As shown in sub-figure (a), the lower blade at a downwind location $\phi_p = 85^\circ$ experiences a stall near the root region at $36.7\%R_t^{(p)}$. This is due to a fully separated flow at approximately 40% of the blade chord on the blade outside surface, which is characterized by a negative chord-wise flow velocity. Consequently, a small negative pressure region emerges near the blade leading edge. Between 56.7% to $76.7\%R_t^{(p)}$, the separation point moves towards the blade trailing edge, leading to a growing negative C_p region on the blade outside surface. Due to the induction by the secondary rotor, the flow around becomes turbulent and reversed at $96.7\%R_t^{(p)}$, resulting in negative C_p regions on both outside and inside blade surfaces.

At an upwind location of $\phi_p = 280^{\circ}$, Sub-figure (b) reveals a large high-speed velocity region at the leading edge of the blade, spanning from 36.7% to 76.7%$R_t^{(p)}$, with the onset of flow separation occurring close to the trailing edge as negative chord-wise flow appears. Meanwhile, a pressure difference is revealed between the outside and inside surfaces of the blade, and the difference increases towards to the blade tip region. This large pressure difference can be attributed to the blade blocking effect, as the wind is being forced to slow down and change direction abruptly due to the presence of the blade in its path. This effect is also evident in the velocity contours, which show a lower velocity over the outside surface and higher velocity over the inner surface of the blade. These results suggest the presence of strong aerodynamic loads on the blade at this location.

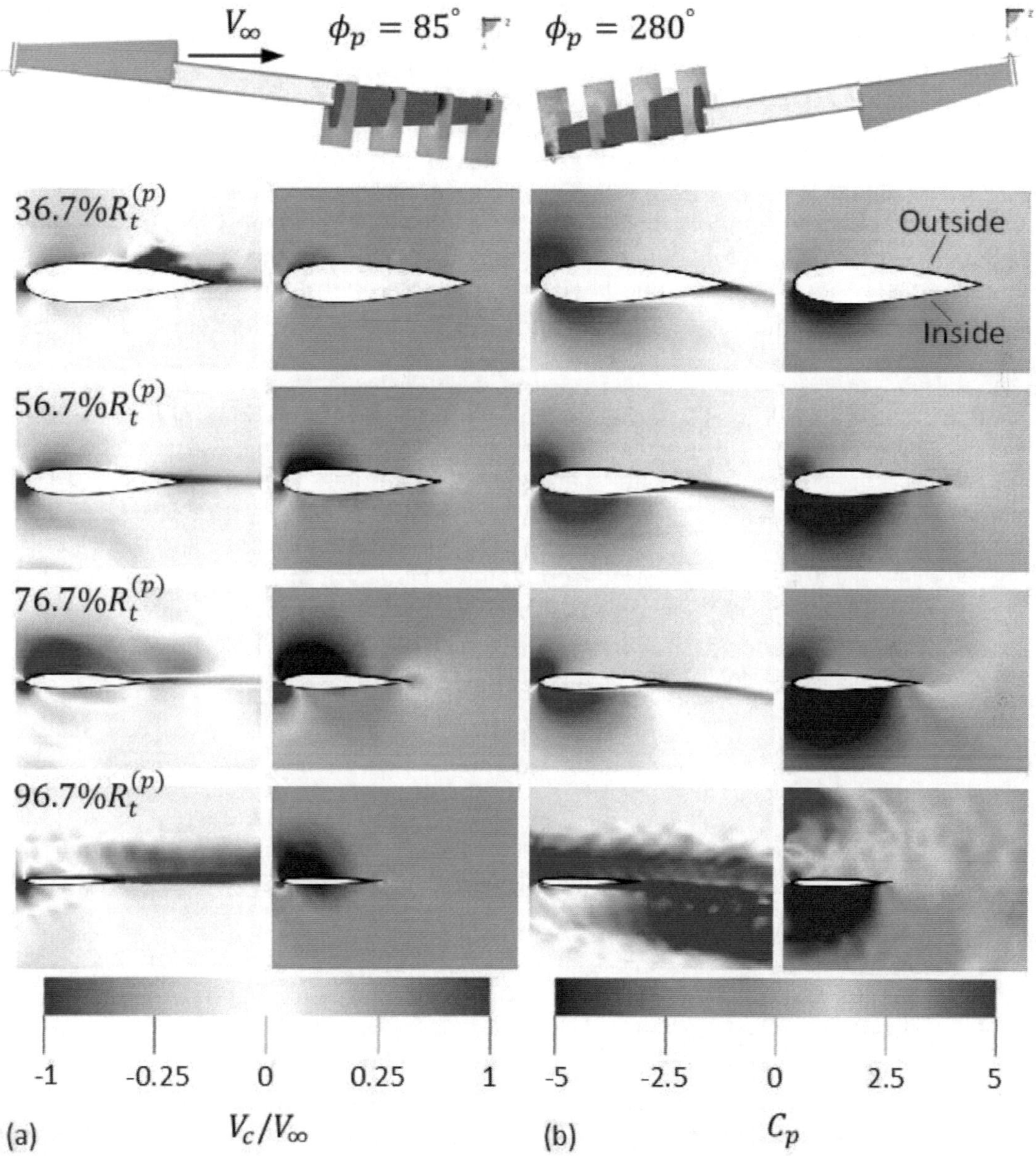

Figure 11. Sectional instantaneous flow field visualization using non-dimensional chord-wise velocity field V_c/V_∞ and the pressure coefficient C_p of the lower blade at different azimuthal angles for (a) $\phi_p = 85^{\circ}$, and (b) $\phi_p = 280^{\circ}$.

4.2 Noise results

The noise results were obtained through the method described in the previous section, utilizing the FW-H acoustic analogy. In total, 22.5 seconds of data were sampled, corresponding to 3 revolutions of the primary rotor. Figure 12 shows the overall sound pressure level (OSPL) in a cross-plane, expressed in dB with reference pressure $p_{\text{ref}} = 2 \times 10^{-5}$ Pa, and integrated from 20 Hz to 1000 Hz (up to the 29th blade passing frequency (BPF) of the secondary rotor). The results show that the noise emitted by the primary rotor blades contributes the most to the total noise. Despite the higher rotational speed of the secondary rotors, the loading makes the primary rotor to be more important than the secondary ones. The directivity of the OSPL shows little dependence on the observer locations, with noise from the secondary rotors observed at the upwind location at $\theta_p = 180^{\circ}$ being 5.7 dB higher than at the downwind location $\theta_p = 0^{\circ}$, while the noise observed from the primary rotor blades at these locations shows only 1.5 dB difference.

Figure 13 illustrates the sound pressure level (SPL) spectra at different polar angles, the acoustic date was processed using Welch's power spectral density (PSD) estimate with a Hanning window of 50% overlap and 1 Hz bandwidth. As expected, the spectra produced by the primary rotor blades and the secondary rotors have broadband characteristics. The results show that the primary rotor blades noise dominate in the low frequency region at all observer angles, maintaining a constant level of about 90 dB, and then rapidly declining. In comparison, the secondary rotor's noise contributes significantly above 400 Hz. The spectra also clearly show the directivity of the secondary rotor noise, as the increase in the level of the broadband noise at the upwind location $\theta_p = 180^{\circ}$ is higher than that at other locations. At $\theta_p = 90^{\circ}$, which corresponds to the observer location directly above the X-Rotor, the first BPF tonal noise from the secondary rotor is presented at the level of 72 dB.

Figure 14 compares the SPL produced by the primary rotor and the secondary rotor blades with 20 Hz bandwidth at various polar angles. The results indicate that the lower blade is the most dominant source of noise among all blades at all observer angles. Furthermore, the SPL of the upper blade reveals a considerable discrepancy in the frequency range of 63 Hz to 200 Hz, decreasing from approximately 80 dB to 50 dB, and then gradually declining with increasing frequency. Additionally, the SPL generated from the five blades of the secondary rotor 1 show a similar trend at $\theta_p = 0^{\circ}$ to $\theta_p = 90^{\circ}$, however, a noticeable difference at low frequency region can be observed for the microphone at the most upwind location, $\theta_p = 180^{\circ}$.

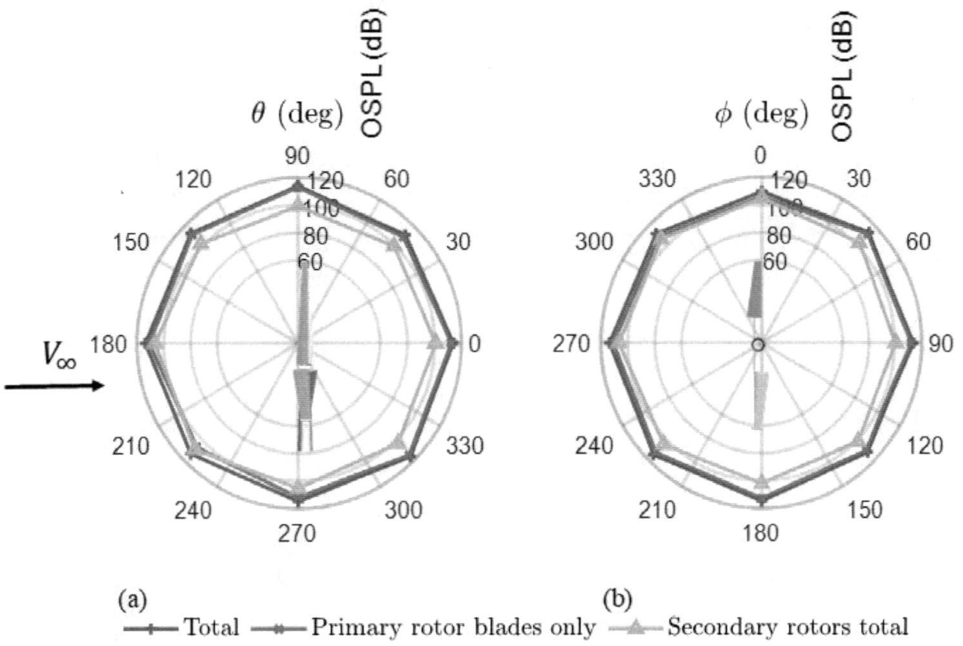

Figure 12. Directivity pattern of the OSPL, for (a) polar angle of the X-Rotor in $x-z$ plane, and (b) azimuthal angles of X-Rotor in $y-z$ plane.

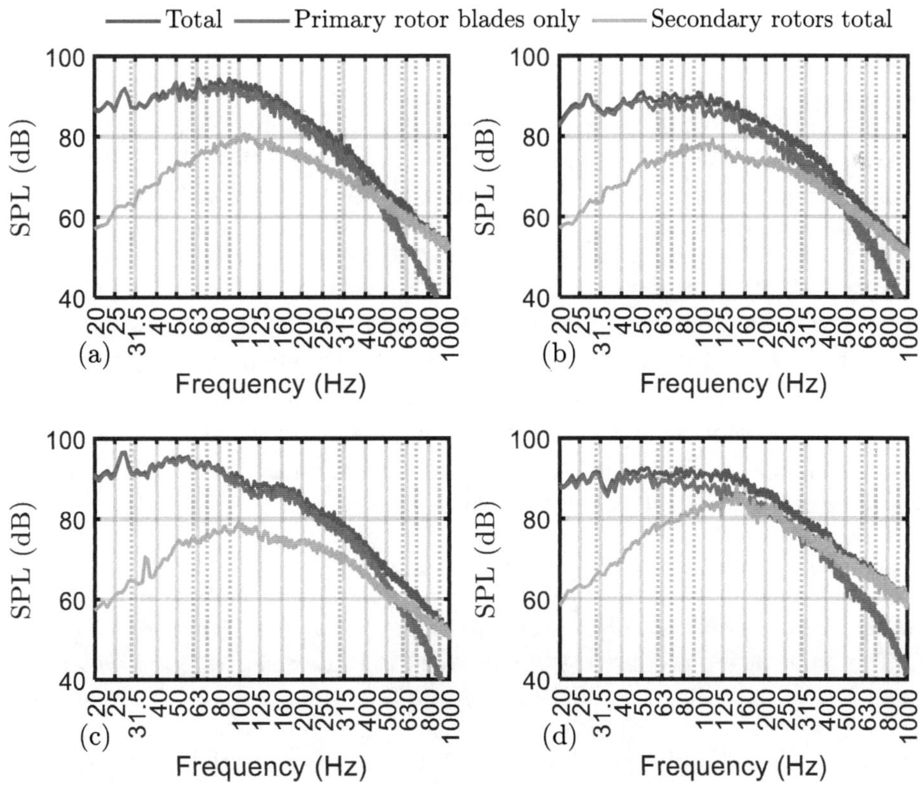

Figure 13. Comparison of sound pressure level spectra (1 Hz bandwidth) produced by different components at different polar angles. For (a) $\theta_p = 0^{\circ}$, (b) $\theta_p = 45^{\circ}$, (c) $\theta_p = 90^{\circ}$,(d) $\theta_p = 180^{\circ}$.

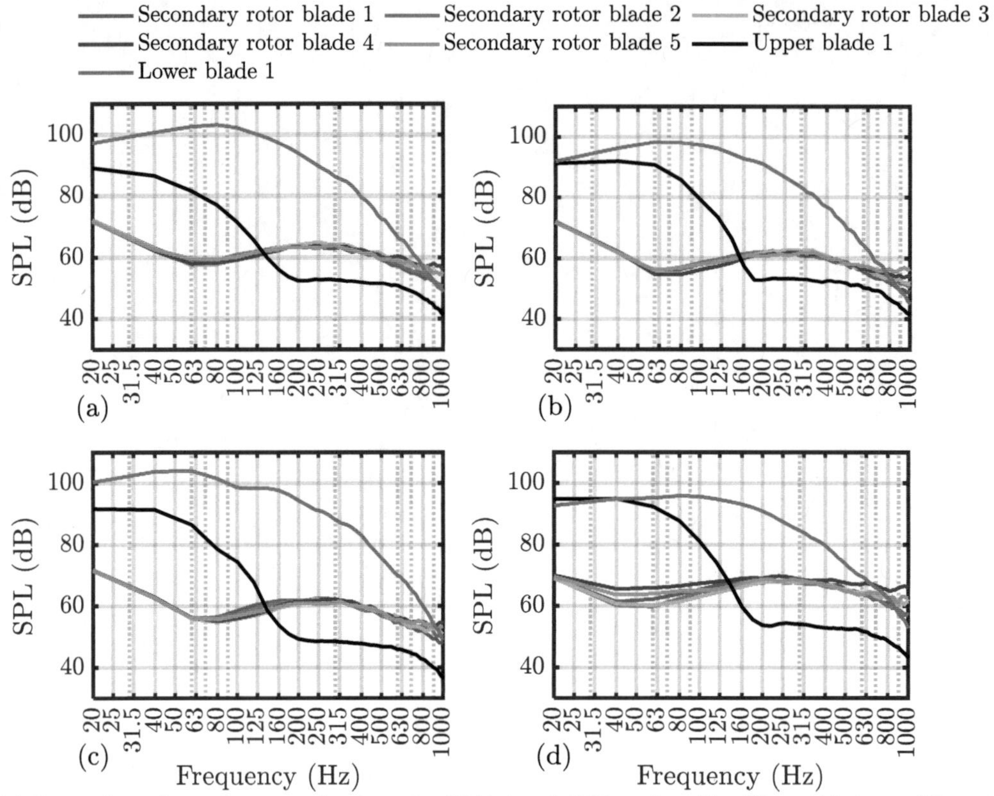

Figure 14. Comparison of sound pressure level spectra (20 Hz bandwidth) produced by different blades at different polar angles. For (a) $\theta_p = 0^\circ$, (b) $\theta_p = 45^\circ$, (c) $\theta_p = 90^\circ$,(d) $\theta_p = 180^\circ$.

5. Conclusions

This paper presents a numerical study of a novel X-shaped offshore wind turbine concept, which incorporates a VAWT and two HAWTs designed to operate at a high Reynolds number condition. The approach uses a high-fidelity CFD simulation based on the LBM/VLES method, and the far-field acoustics are computed using the FW-H acoustic analogy. The results allow to evaluate the aerodynamic and aeroacoustic performance of the current design.

A detailed analysis of the instantaneous flow field contours at various azimuthal angles demonstrates that force fluctuations of the secondary rotor at downwind locations around $\phi_p = 45^\circ$ and $105^\circ$are likely caused by the interactions with its own wake and wake from the tower. Additionally, the secondary rotor at the $\phi_p = 180^\circ$ is subjected to the wake generated from the secondary rotor and generator at $\phi_p = 0^\circ$. Additionally, the occurrence of separation stall near the root of the lower blade at downwind location is clearly demonstrated, this separation stall leads to a local minimum value C_T for the lower blade.

The OSPL results demonstrate that the noise generated by both the primary rotor and the secondary rotors exhibits little dependence on directivity pattern across various observer locations, with the noise level at the most upwind location being 5.7 dB higher for the secondary rotors and 1.5 dB higher for the primary rotor compared to the most downwind location. Notably, the lower blades of VAWTs produce a substantial amount of noise that dominates the noise level at all observer locations, while the upper blades also significantly contribute to the noise in the lower frequency range. In contrast, the secondary rotors primarily contribute to the noise level for frequencies above 400 Hz and at most observer angles.

Acknowledgement
The authors wish to acknowledge the European Union's Horizon 2020 Programme for Research and Innovation for funding the X-Rotor project under the grant Grant Agreement No. 101007135.

References

Boorsma, K., & Schepers, J. (2011). Enhanced wind turbine noise prediction tool SILANT. 4th International Meeting on Wind Turbine Noise,

Brooks, T. F., Pope, D. S., & Marcolini, M. A. (1989). *Airfoil self-noise and prediction*.

Casalino, D. (2003). An advanced time approach for acoustic analogy predictions. *Journal of Sound and Vibration*, *261*(4), 583-612.

Chen, H., Chen, S., & Matthaeus, W. H. (1992). Recovery of the Navier-Stokes equations using a lattice-gas Boltzmann method. *Physical review A*, *45*(8), R5339.

Chen, S., & Doolen, G. D. (1998). Lattice Boltzmann method for fluid flows. *Annual review of fluid mechanics*, *30*(1), 329-364.

Farassat, F., & Succi, G. P. (1980). A review of propeller discrete frequency noise prediction technology with emphasis on two current methods for time domain calculations. *Journal of Sound and Vibration*, *71*(3), 399-419.

Ffowcs Williams, J. E., & Hawkings, D. L. (1969). Sound generation by turbulence and surfaces in arbitrary motion. *Philosophical Transactions of the Royal Society of London. Series A, Mathematical and Physical Sciences*, *264*(1151), 321-342.

Launder, B. E., & Spalding, D. B. (1983). The numerical computation of turbulent flows. In *Numerical prediction of flow, heat transfer, turbulence and combustion* (pp. 96-116). Elsevier.

Leithead, W., Camciuc, A., Amiri, A. K., & Carroll, J. (2019). The X-rotor offshore wind turbine concept. Journal of Physics: Conference Series,

Pearson, C. (2014). *Vertical axis wind turbine acoustics* University of Cambridge].

Shan, X., Yuan, X.-F., & Chen, H. (2006). Kinetic theory representation of hydrodynamics: a way beyond the Navier–Stokes equation. *Journal of Fluid Mechanics*, *550*, 413-441.

Succi, S. (2001). *The lattice Boltzmann equation: for fluid dynamics and beyond*. Oxford university press.

Teixeira, C. M. (1998). Incorporating turbulence models into the lattice-Boltzmann method. *International Journal of Modern Physics C*, *9*(08), 1159-1175.

Wilcox, D. C. (2006). Turbulence modeling for CFD. La Canada, CA: DCW Industries. *Inc, November*.

Yakhot, V., & Orszag, S. A. (1986). Renormalization group analysis of turbulence. I. Basic theory. *Journal of scientific computing*, *1*(1), 3-51.

Zhu, W. J., Heilskov, N., Shen, W. Z., & Sørensen, J. N. (2005). Modeling of aerodynamically generated noise from wind turbines.

AUTHOR INDEX